Life Science

FIFTH EDITION

bju press
Greenville, South Carolina

Life Science
Fifth Edition

Coordinating Writer
Rachel Santopietro, MEd

Writers
Christopher D. Coyle
Elwood Groves, MA

Contributing Writers
Kevin Anderson, PhD
Jeremy Byrd, MD
Robert Carter, PhD
Marc Chetta, MD
Georgia Purdom, PhD
John Sanford, PhD

Biblical Worldview
Bryan Smith, PhD
Tyler Trometer, MDiv

Academic Oversight
Jeff Heath, EdD

Project Editor
Rick Vasso, MDiv

Cover, Design, and Interior Concept Design
Sarah Lompe
Points & Picas

Page Layout
Carrie Walker

Illustration
John Cunningham
Sarah Lompe
Ben Schipper
Dana Thompson

Permissions
Sharon Belknap
Tatiana Bento
Sylvia Gass
Carrie Hanna
Lily Kielmeyer
Hannah Labadorf
Rita Mitchell
Ashleigh Schieber

Project Coordinators
Tony Every
Don Simmons

Photo credits appear on pages 472–78.

The front cover photo shows a husky (*Canis lupus familiaris*). The back cover photo shows an Atlantic Portuguese man o' war (*Physalia physalis*). The spine photo shows a horned melon (*Cucumis metuliferus*).

ISBN 978-1-62856-401-3

15 14 13 12 11 10 9 8 7 6 5 4 3

Contents

LAB EXERCISES

CASE STUDIES

ETHICS

WORLDVIEW SLEUTHING

CAREERS

LET'S TALK.

Welcome to *Life Science* 5th Edition! I'm sure you have some questions about living things. Maybe things like …

Where did everything come from?
How do living things stay alive?
How do living things compare?
How do we classify living things?
How do living things interact with their environment?
What is right and wrong when it comes to using living things?

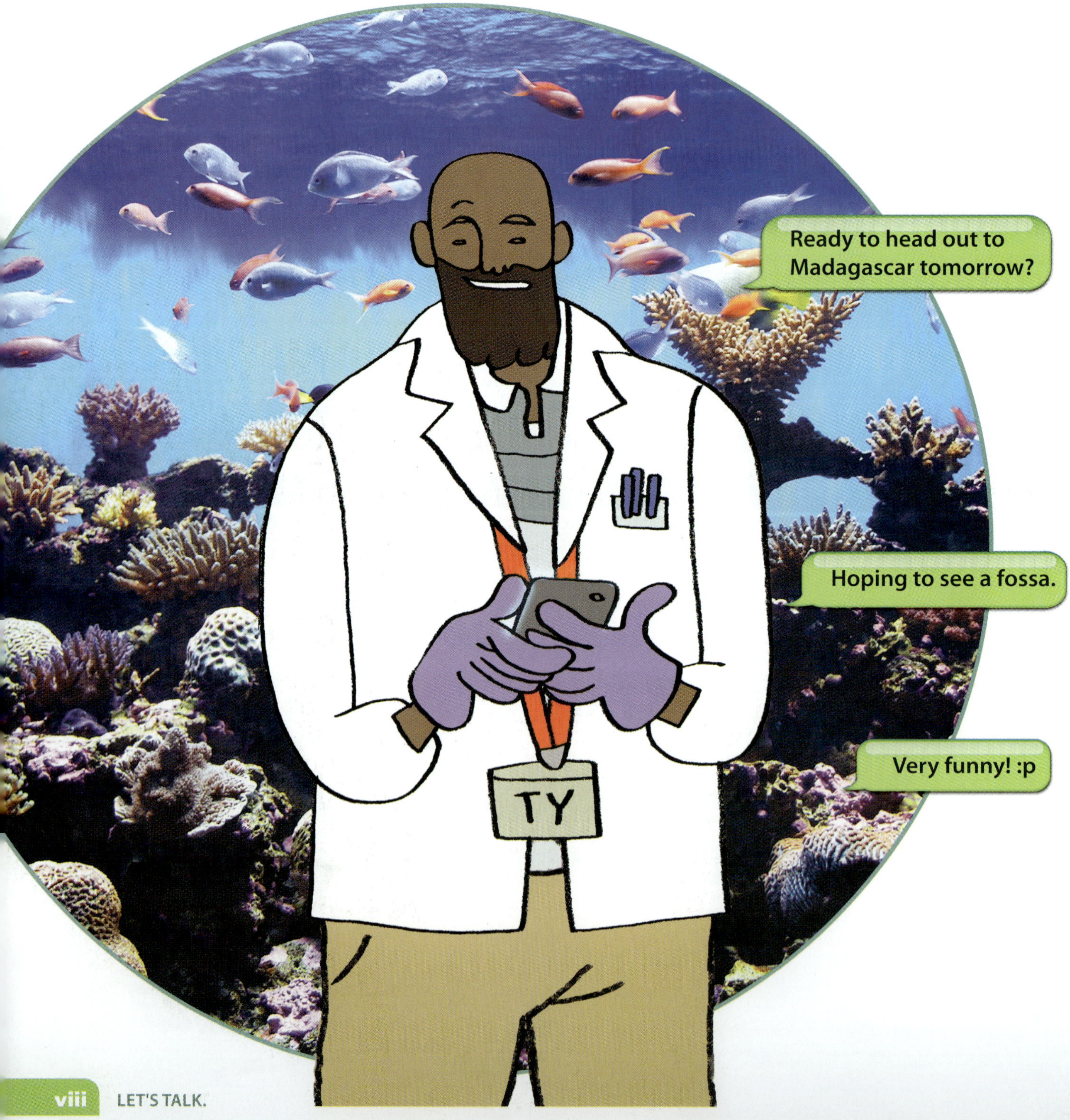

Don't worry! You're not the only one with questions. Two biologists will be joining you in your quest to make sense of the living world. They're both nice guys and friends, though they have different views of life. Norm, the biologist on the right, is an evolutionist. His buddy Ty, on the left, believes the Bible and is a creationist. You'll run into them from time to time in this textbook. Sometimes they agree, and sometimes they can't. Think about what they say. You'll see that people's beliefs affect how they see the world and answer the questions they have about life, like the questions you're thinking about. Use the information that you have in this textbook to responsibly answer your questions, help others, and glorify God.

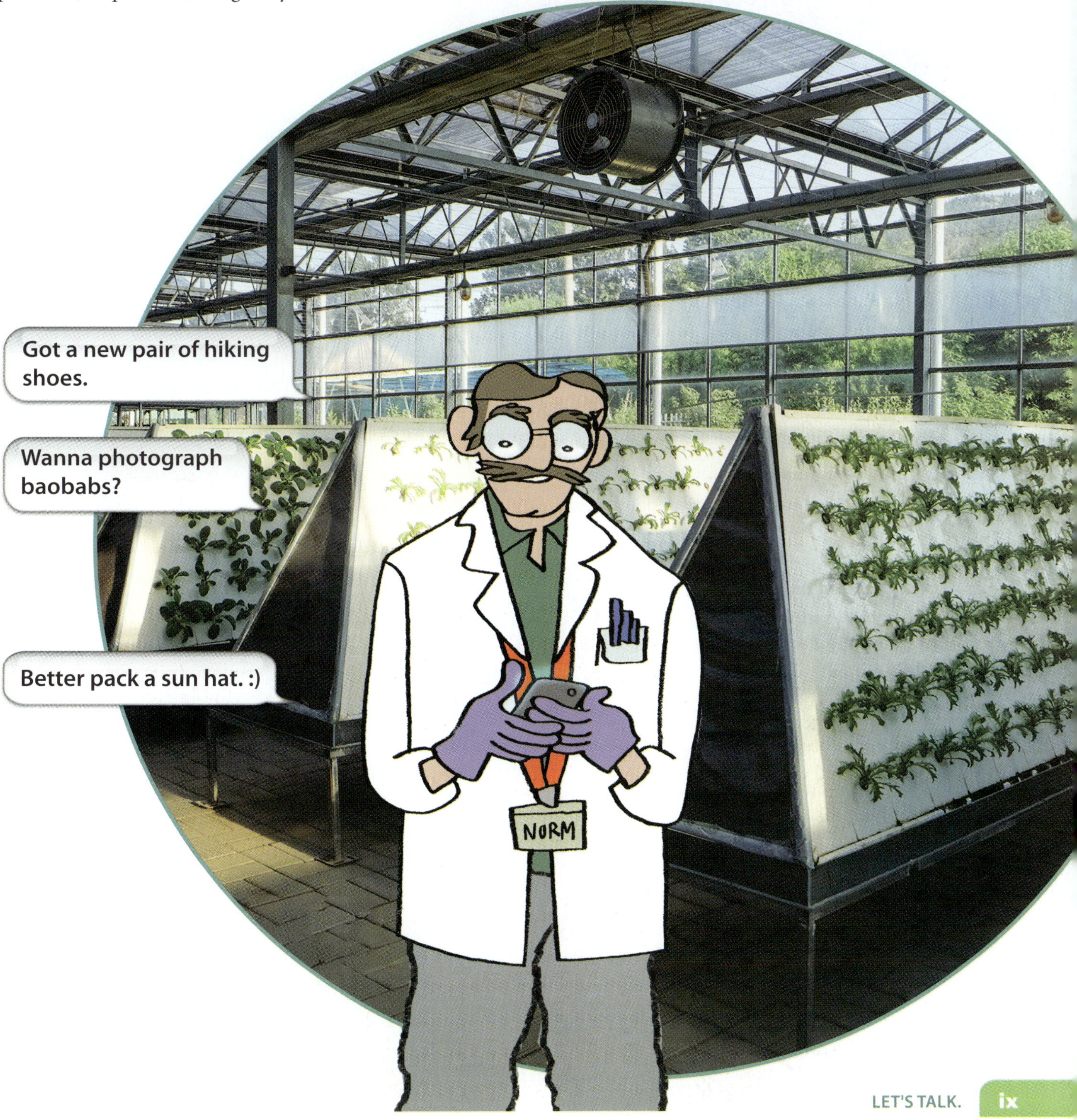

TAKE A PEEK INSIDE!

We've designed this textbook with you in mind. We hope it will help you appreciate the wonders of God's creation even more. Flip through the following pages to see the features that we've included to help you succeed in life science. In the back of the book you'll find appendixes, a glossary, and an index.

1 GOD'S LIVING WORLD

ISLAND OF LIFE ❶

Madagascar is an island of life. This island off the eastern coast of Africa is about the size of California. About 80% of the plants and animals that live here appear nowhere else on Earth! Baobab trees, lemurs, tenrecs, and rosy periwinkles are some of the beautiful organisms that call Madagascar home.

But despite this lush biodiversity, its people are some of the poorest in the world. In 2004, a team of adventurers and agri-culturists hiked the length of Madag miles! Their goal? To educate the M their land. Malagasies need to succ land of life so that people for generati here. But why is this important? This question.

1A Seeing Life Through the Bible's Story 3
1B Understanding Life 8
1C Learning About Life 14

❶ **Chapter Opener**—highlights issues and developments in life science that need to be examined from a biblical worldview

8A
CLASSIFYING PLANTS

❷ *How are plants different from other living things?*

8A Questions
- What is a plant?
- How are plants different from other living things?
- How are plants classified?
- How are vascular and nonvascular plants different?

❸

Corn is just one of the many thousands of species of living things that we call *plants*. As you'll soon see in this and the following chapter, they serve the creation in a variety of vital ways. God's original good design for plants included that they would provide food for both humans and animals (Gen. 1:29–30). In addition to food, plants and plant products have been used for fuel, shelter, and clothing. Plants provide oxygen and recycle carbon dioxide. Many plants, such as beautiful flowers or majestic trees, are also pleasing to the eye and inspirational to the human spirit. But plants and the ways that we use them have also been affected by the Fall. Just as corn can be used to make a tasty tortilla chip or illegal moonshine, many plants can be used for either good or evil.

8-1
Like other instances in which humans have tried to classify living things, some plants bend the rules. For instance, these plants at left have lost the ability to photosynthesize and live as parasites on fungi.

156 CHAPTER 8

❷ **Essential Question**—the big question that you will learn about in a section

❸ **Key Questions**—the smaller questions that you can ask along the way through a section to help you answer the essential question

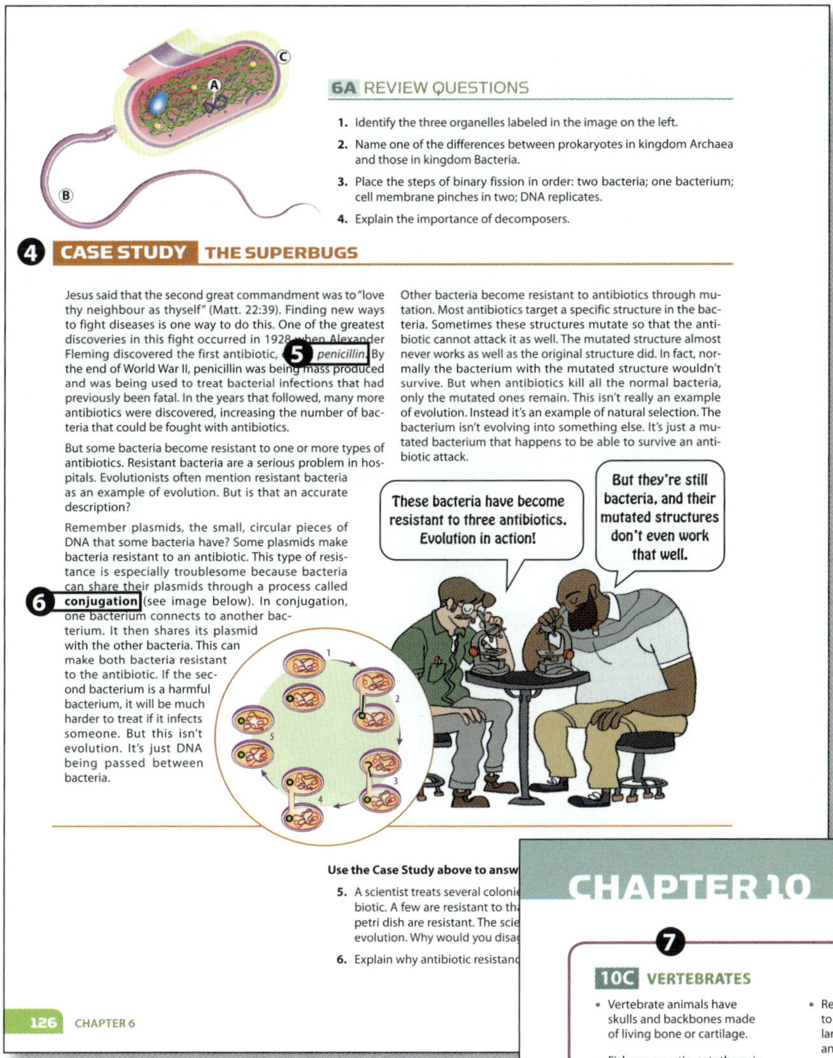

6A REVIEW QUESTIONS

1. Identify the three organelles labeled in the image on the left.
2. Name one of the differences between prokaryotes in kingdom Archaea and those in kingdom Bacteria.
3. Place the steps of binary fission in order: two bacteria; one bacterium; cell membrane pinches in two; DNA replicates.
4. Explain the importance of decomposers.

CASE STUDY THE SUPERBUGS

Jesus said that the second great commandment was to "love thy neighbour as thyself" (Matt. 22:39). Finding new ways to fight diseases is one way to do this. One of the greatest discoveries in this fight occurred in 1928 when Alexander Fleming discovered the first antibiotic, *penicillin*. By the end of World War II, penicillin was being mass produced and was being used to treat bacterial infections that had previously been fatal. In the years that followed, many more antibiotics were discovered, increasing the number of bacteria that could be fought with antibiotics.

But some bacteria become resistant to one or more types of antibiotics. Resistant bacteria are a serious problem in hospitals. Evolutionists often mention resistant bacteria as an example of evolution. But is that an accurate description?

Remember plasmids, the small, circular pieces of DNA that some bacteria have? Some plasmids make bacteria resistant to an antibiotic. This type of resistance is especially troublesome because bacteria can share their plasmids through a process called **conjugation** (see image below). In conjugation, one bacterium connects to another bacterium. It then shares its plasmid with the other bacteria. This can make both bacteria resistant to the antibiotic. If the second bacterium is a harmful bacterium, it will be much harder to treat if it infects someone. But this isn't evolution. It's just DNA being passed between bacteria.

Other bacteria become resistant to antibiotics through mutation. Most antibiotics target a specific structure in the bacteria. Sometimes these structures mutate so that the antibiotic cannot attack it as well. The mutated structure almost never works as well as the original structure did. In fact, normally the bacterium with the mutated structure wouldn't survive. But when antibiotics kill all the normal bacteria, only the mutated ones remain. This isn't really an example of evolution. Instead it's an example of natural selection. The bacterium isn't evolving into something else. It's just a mutated bacterium that happens to be able to survive an antibiotic attack.

Use the Case Study above to answ...

5. A scientist treats several coloni... biotic. A few are resistant to th... petri dish are resistant. The scie... evolution. Why would you disa...
6. Explain why antibiotic resistanc...

126 CHAPTER 6

❹ Case Studies—opportunities to investigate specific areas in life science to apply what you have learned in a chapter

❺ Italic Terms—terms that are important in scientific fields

❻ Bold-Faced Terms—glossary terms that you need to know

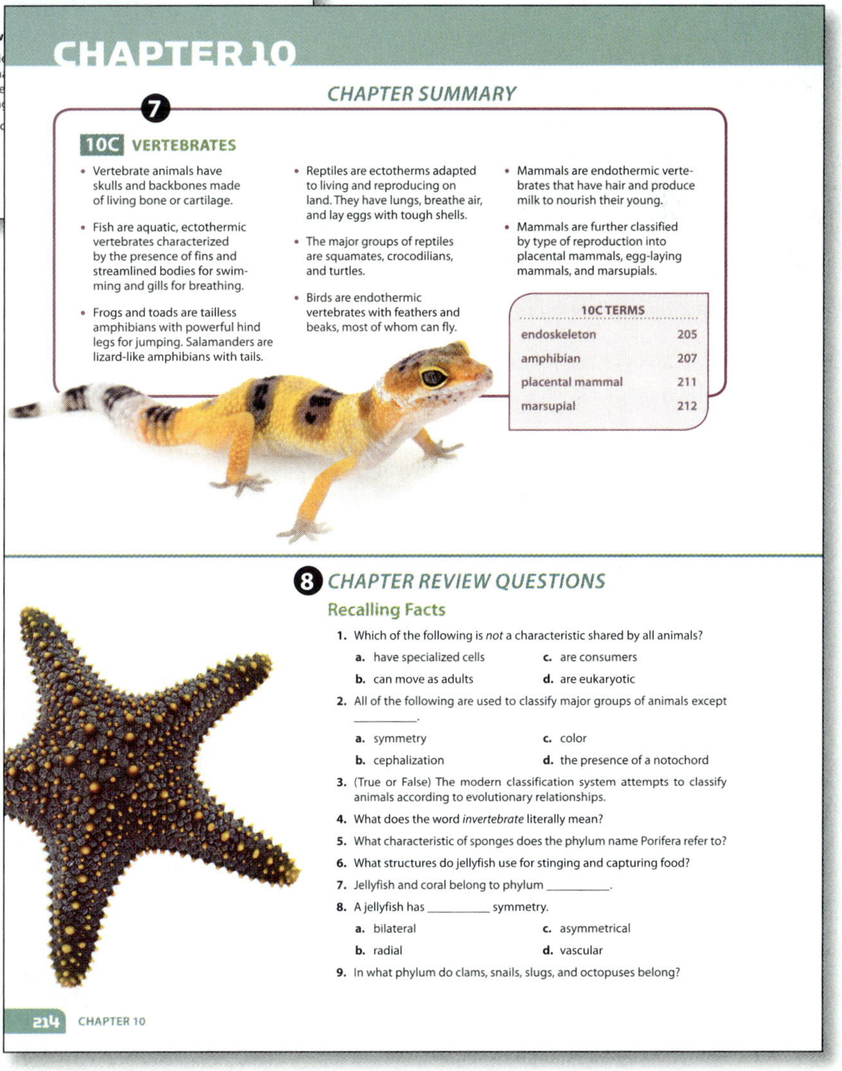

CHAPTER 10

CHAPTER SUMMARY

10C VERTEBRATES

- Vertebrate animals have skulls and backbones made of living bone or cartilage.
- Fish are aquatic, ectothermic vertebrates characterized by the presence of fins and streamlined bodies for swimming and gills for breathing.
- Frogs and toads are tailless amphibians with powerful hind legs for jumping. Salamanders are lizard-like amphibians with tails.
- Reptiles are ectotherms adapted to living and reproducing on land. They have lungs, breathe air, and lay eggs with tough shells.
- The major groups of reptiles are squamates, crocodilians, and turtles.
- Birds are endothermic vertebrates with feathers and beaks, most of whom can fly.
- Mammals are endothermic vertebrates that have hair and produce milk to nourish their young.
- Mammals are further classified by type of reproduction into placental mammals, egg-laying mammals, and marsupials.

10C TERMS	
endoskeleton	205
amphibian	207
placental mammal	211
marsupial	212

CHAPTER REVIEW QUESTIONS

Recalling Facts

1. Which of the following is *not* a characteristic shared by all animals?
   a. have specialized cells c. are consumers
   b. can move as adults d. are eukaryotic
2. All of the following are used to classify major groups of animals except ________.
   a. symmetry c. color
   b. cephalization d. the presence of a notochord
3. (True or False) The modern classification system attempts to classify animals according to evolutionary relationships.
4. What does the word *invertebrate* literally mean?
5. What characteristic of sponges does the phylum name Porifera refer to?
6. What structures do jellyfish use for stinging and capturing food?
7. Jellyfish and coral belong to phylum ________.
8. A jellyfish has ________ symmetry.
   a. bilateral c. asymmetrical
   b. radial d. vascular
9. In what phylum do clams, snails, slugs, and octopuses belong?

214 CHAPTER 10

❼ Chapter Summary—handy statements of the big ideas of the chapter, including terms

❽ Review Questions—practice in applying what you've learned in a section or a chapter (*Note*: Questions with a blue box require some extra thinking.)

4C REVIEW QUESTIONS

1. Define *genetic drift*.
2. (True or False) Genetic drift is measured by the ratios of organisms with a specific phenotype.
3. Define *natural selection*.
4. Compare genetic drift and natural selection.
5. (True or False) Natural selection can remove but not add genetic information.
6. A population of deer mice living in the dark soil of Nebraska move onto the sand dunes deposited by glaciers. While most of the mice are dark, some are light and blend in well with the light-colored sand. Predict which color of mouse will be more common in the sand dunes after fifty years. Explain your answer.
7. How do creationists and evolutionists differ in their views of natural selection?

4-7 Which mouse will become more common in an area with light-colored sand?

9 Ethics Boxes—opportunities to apply a biblical worldview to ethical issues in life science

9 ETHICS Genetic Engineering

The gene gun mentioned in the Chapter opener is one tool used in genetic engineering. Genetic engineering is the process of deliberately manipulating the genes in an organism in ways other than natural processes like reproduction. It involves changing an organism's DNA by inserting new genetic material. Genetic engineering is a blossoming field of biology with real potential to solve significant real-world problems. It has been extensively applied to enhancing food production by modifying the genes of plants.

Scientists are convinced that genetically engineered food crops (called *GMOs*—genetically modified organisms) are safe to eat after extensive testing. However, some people have become very concerned about GMOs. They fear that these food crops may cause long-term health problems. But GMOs have been used in food for over a decade; so far, no evidence of ill effect has been discovered.

As the human population grows, we must think about the amount of food needed to keep everyone healthy. Genetic engineering may be the answer to this potential problem.

Questions 8–13.

erm "genetic engineering." Write a brief ned.

portant ethical issue regarding genetic

say on this issue?

eptable options on this issue?

nces of the acceptable options?

u take with regard to this issue?

10 LAB EXERCISE

LONG-LASTING FLOWERS

What makes cut flowers last longer?

Roses are beloved by many people, and florists sell millions of them every year, especially around Valentine's Day. But cut flowers rarely last much longer than a week. Many popular ideas suggest adding substances to the water in the vase to make the flowers stay fresh longer. But do any of these actually work?

Equipment E

roses of one color with fertilizer packet (12)

vases, 1 L (4)

aspirin tablet, 325 mg

lemon-lime soda, 60 mL

water

PROCEDURE

1. Pour 700 mL of water into each 1 L vase.
2. Pour the fertilizer packet from the bouquet into Vase 2.
3. Drop an aspirin tablet into Vase 3.
4. Pour 60 mL of lemon-lime soda into Vase 4.
5. Place three roses in each of the four vases.

1. Which vase do you think will do the best? The answer to that question is your hypothesis.
2. After one day, do any of the vases of flowers look any worse than the others? If so, which one(s)?
3. After five days, have any of the roses started to fade? If so, which one(s)?
4. After ten days, do any of the vases of roses look better than the others? If so, which one(s)?
5. Rank the vases of roses from those that remained fresh looking the longest to those that faded most quickly.
6. Compare your observations to your hypothesis. Try to explain the differences, if any.
7. What else could you test to lengthen the life of cut flowers?
8. Why is this scientific work valuable from a Christian worldview?

10 Lab Exercises—short hands-on activities to get you thinking and acting like a scientist

11 FROM SLAVE TO SCIENTIST: SERVING AS A HORTICULTURALIST

Cotton was king in Alabama at the beginning of the twentieth century. In fact, it had been king for decades. But all the years of growing nothing but cotton had removed most of the nutrients from the soil. Then around 1915 boll weevils appeared in the state and destroyed much of the cotton crop. Farmers and sharecroppers, already struggling due to the poor soil conditions, were facing bankruptcy.

At the same time, George Washington Carver worked as a teacher and horticulturalist at Tuskegee Institute (now Tuskegee University). He realized that planting peanuts instead of cotton would help farmers. Peanuts return nitrogen to the soil, improving farmland. They also are not affected by boll weevils. To convince people to begin planting peanuts, Carver began publishing information on how to grow and use peanuts. He also traveled throughout Alabama to teach farmers the importance of growing more than one crop. As a result, the state's economy improved greatly.

Some horticulturalists, like George Washington Carver, find new and better ways to grow crops. Others experiment with new ways to control crop pests, while some help create better landscaping techniques in urban areas. Still others care for lawns and fields for private landowners or the government.

GEORGE WASHINGTON CARVER'S TESTIMONY

"I was just a mere boy when converted, hardly ten years old. There isn't much of a story to it. God just came into my heart one afternoon while I was alone in the loft of our big barn while I was shelling corn to carry to the mill to be ground into meal.

A dear little white boy, one of our neighbors, about my age came by one Saturday morning, and in talking and playing he told me he was going to Sunday school tomorrow morning. I was eager to know what a Sunday school was. He said they sang hymns and prayed. I asked him what prayer was and what they said. I do not remember what he said; only remember that as soon as he left I climbed up into the loft, knelt down by the barrel of corn, and prayed as best I could. I do not remember what I said. I only recall that I felt so good that I prayed several times before I quit.

My brother and myself were the only colored children in that neighborhood and, of course, we could not go to church or Sunday school, or school of any kind.

That was my simple conversion, and I have tried to keep the faith."

—G. W. Carver; Letter to Isabelle Coleman; July 24, 1931

11 **Career Boxes**—feature careers in life science (could be yours!) that can be pursued to wisely use God's world and help people

12 WORLDVIEW SLEUTHING: MEDICINAL PLANTS

Did you know that a dog could be poisoned by a common ornamental flower? The foxglove flower is beautiful but deadly. But it can also be used to save lives. The foxglove's poison, digoxin, is used to make a drug that treats heart issues. Many drugs are poisons when taken in large doses. And many poisonous plants can be used to make medicines when processed properly.

INTRODUCTION

You are a science writer for a website. Your next assignment is to prepare a short presentation on poisonous plants that are used as medicines.

TASK

Research a plant that produces a poison that can also be used as a medicine. Examples include the opium poppy, deadly nightshade, and rosy periwinkle. Create a PowerPoint® or Prezi™ presentation that is based on your results. Include in your presentation where the plant grows, when its medical use was discovered, and what diseases the medicine treats.

PROCEDURE

1. Do a web search using the keywords "medicines from poisonous plants." Find a plant that you would like to research.
2. Do keyword searches using the name of the plant and the name of the medicine that is made from it.
3. Note how the worldviews of the authors of the websites affect their view of using plants for medicine.
4. Plan your presentation and collect any photos or videos you need. Be careful to give proper credit to your sources.
5. Create your presentation and show it to another person.
6. Deliver your presentation to your class or family.

CONCLUSION

Many poisonous plants have been used throughout the centuries for medicines. Pharmacists and scientists suspect that many more exciting medicines are waiting to be discovered.

8A REVIEW QUESTIONS

1. Which of the following is *not* true of most plants?
   a. They have tissues.
   b. They are eukaryotic.
   c. They have plastids.
   d. They are consumers.
2. Recall some of the characteristics of fungi from Chapter 7 (pp. 142–49). Describe at least two major differences between fungi and plants.
3. You also saw in Chapter 7 that some protists, such as euglenas, can produce their own food through photosynthesis. Why are they not considered plants?
4. What two characteristics are used to classify plants into their three primary groupings?
5. Classify each of the plants in the images on the right as either a nonvascular plant, seedless vascular plant, or seed plant.
6. What are the two major groups of seed plants, and what is the difference between them?

a.

b.

c.

THE PLANT KINGDOM 165

12 **Worldview Sleuthing Boxes**—inquiry-based investigations that help you think through controversial areas of life science through the lens of Scripture

unit 1 THE PATTERN OF LIFE

FROM A MICROBIOLOGIST TO YOU

I received my scientific training in the area of bacterial genetics. With that background, I got involved in the search for the tissue and proteins that we find in dinosaur fossils. Imagine that—stretchy tissue still inside dinosaur bones! Yet evolutionists say that these fossils should be at least 65 million years old. How could this tissue have survived so long? I'm doing research to show just how rapidly tissue will degrade. This means that tissue in dinosaur bones is clear evidence that dinosaurs lived only a few thousand years ago.

We need more Christians to become scientists and use their training to show the next generation the truth about Creation. Maybe you will be one of them.

Dr. Kevin Anderson

1 GOD'S LIVING WORLD

ISLAND OF LIFE

Madagascar is an island of life. This island off the eastern coast of Africa is about the size of California. About 80% of the plants and animals that live here appear nowhere else on Earth! Baobab trees, lemurs, tenrecs, and rosy periwinkles are some of the beautiful organisms that call Madagascar home.

But despite this lush biodiversity, its people are some of the poorest in the world. In 2004, a team of adventurers and agriculturists hiked the length of Madagascar, a distance of 2234 miles! Their goal? To educate the Malagasies to wisely farm their land. Malagasies need to successfully conserve this island of life so that people for generations to come can flourish here. But why is this important? This chapter will answer that question.

1A

SEEING LIFE THROUGH THE BIBLE'S STORY

Why should a Christian study life science?

1A Questions

- Why do people view the world so differently?
- How different are people's views of life science?
- How does the Bible affect how Christians view life science?

1.1 YOUR VIEW OF THE WORLD

Whether you are a scientist trekking across Madagascar or a middle school student studying life science, you have a view of the world. It begins with the big story that's playing in your head, the story that tells you where this world came from and where it's headed. That narrative or story affects how you answer the key questions that you'll face in life. "What is most important to me?" "What is true?" "Who can I trust to tell me what is true?" The way that you answer those questions will determine how you think and how you live. So that big story is very important, and we have a special name for it—**worldview**, the overarching narrative that a person uses to see and interpret the world. It shapes your beliefs and values, and it will guide you as you make choices in life.

Life science is the study of life. It includes what you know about life and the processes that you use to learn about living things. Another name for life science is *biology*. But life science isn't just about facts; it also involves making sense of these facts. This involves your worldview.

So how different are people's worldviews when it comes to life science? Well, we all assume that the world is real and that we can make sense of it. If we didn't believe these two things, then we couldn't even do life science! But that is where the similarities end.

Some scientists believe that living things came from completely natural processes. This is called *naturalism*. Naturalists might believe in God, or they might be atheists. But the result is the same: in their worldview, God is not life's Designer. To them, life is not the product of a Divine Engineer; instead, it is simply the product of evolution. They view science as the authority on truth.

1-1
Tenrecs are one of the unusual creatures that live in Madagascar.

1-2
A person's worldview affects how he sees life.

Here's an example. After Charles Darwin wrote *On the Origin of Species*, an orchid grower sent Darwin a sample of a star-shaped orchid from Madagascar. This orchid was unusual because it had a spur 20–25 cm long in which it stored nectar. What kind of insect could possibly pollinate a flower like that?! Darwin assumed that some insect that had yet to be discovered had evolved along with the orchid to pollinate it. In 1992, 130 years after Darwin had made the prediction, scientists made a discovery. They observed a Morgan's sphinx moth with a sucking mouthpart 20 cm long feeding from this orchid. Is this a proof of evolution?

1.2 SCIENCE AND THE BIBLE'S BIG STORY

A person with a biblical worldview trusts in the Bible as God's Word. God reveals truth to us in the Bible, and its story can transform the way you think about life science. Darwin's moth and orchid aren't products of random chance—God designed them!

Creation

The Bible tells us that God has been personally involved with creation from Day 1. A straightforward reading of Genesis 1 shows us that God spoke everything into existence from nothing in six days. God made a good creation because He is a good God. Because He is good, He provides for and cares for His creation. His creation glorifies Him by displaying His goodness and the perfections of His character.

One of the ways that God provides for His creation is through man, His highest creation. He made only man in His image, separating him from the rest of the living things. Our nature is a dim reflection of what God is like. The image of God in us produces our personalities and emotions, our abilities to plan and reason, our sense of eternity, and our desire to create, love, and be loved. These things show how we are similar to God.

After God created Adam and Eve, he gave them a job to do—He commanded them to fill the earth and have dominion over it (Gen. 1:26–30). This command is often called the **Creation Mandate**.

Fall

But man hasn't obeyed God's commands. When Adam and Eve disobeyed God and ate from the tree of the knowledge of good and evil, all of creation was cursed. Death is the direct result of sin. But that doesn't mean that the Creation Mandate has gone away.

Day 5: Animals of the sky and waters

Day 7: God rested.

Day 4: Sun, moon, and stars

The work that God expected from the beginning (Gen. 2:15) became more difficult because of the Curse (Gen. 3:17–19). It's harder now to wisely care for and use God's world, but we must do it for our good and His glory.

Day 6: Land animals and people

Redemption

Though we live in a world that is a broken version of God's very good creation, we have hope. God has not abandoned His creation. He provided an escape from sin through His Son Jesus, the seed of the woman (Gen. 3:15). Jesus came to Earth to restore creation through His sacrifice for sin. By being a sacrifice for sin on the cross, Jesus provided a redemption that will one day reverse all the effects of the Fall. The creation will be restored. We help prepare for this restoration in small but significant ways by living out the Creation Mandate through good works and sharing the gospel.

What Creation, the Fall, and Redemption Mean to Life Science

Imagine two scientists excavating fossils in the hills of China. There is a dinosaur graveyard there. It contains hundreds of bones of dinosaurs that used to be alive but are now buried in rock layers. One of the scientists working at the site is a naturalist and the other is a creationist. Both scientists have the same burning question—how did these dinosaurs die?

The naturalistic scientist will fit this evidence into what he already believes. He will say that these dinosaurs are the product of millions of years of evolution. He expects to see large differences in the organisms from one rock layer to the next. If he doesn't, he assumes that the layers were laid down quickly in a violent local storm.

1-3

A person's worldview affects how he sees the world.

The creationist scientist fits this evidence into what he already believes too. When he sees the dinosaur graveyard, he links this place with other places around the world containing fossil graveyards. These aren't just the effects of many local storms. He is thinking about the Bible's account of Noah's Flood. In this global catastrophe God judged mankind for sin. So the Bible's story can change the way that we make sense of life in God's world. Those who believe God's Word are in the best position to understand His world.

But that's not all. Those who believe God's Word also have the best reasons to do life science. Life science is an incredible tool to obey the Creation Mandate. Christians should seek places of influence so that they can help others wisely use God's world. This could include helping Malagasy farmers better use their land or using the unique flora and fauna of Madagascar to produce medications. And as we obey God's command to wisely use His living world, we can also help people who bear His image. Loving other people is another one of God's most important commands (Mark 12:31).

1.3 SCIENCE, ETHICS, AND THE BIBLE

The work of life science involves making decisions where the right thing to do can be hard to figure out. The study of **ethics** is all about using moral principles to help people apply life science in good ways. Ethics tries to answer the question, "Is this right or wrong?" It tries to define our moral duty to living things, especially people. People try to make moral rules on the basis of things that they agree on. They also try to make others happy and keep them from pain because that's what people want too. But this doesn't always work because people have different worldviews. What they want might not be the best. We need a distinctly biblical way to evaluate ethical issues in life science. In order to make truly good and right decisions in God's eyes, we should follow biblical principles and seek biblical outcomes.

Biblical Principles

We need good laws. God created the world, and life works best when people live the way that He wants them to live. He didn't make life a mystery—He shows us in His Word! This is the place to find what God says is good and evil. But there aren't always direct statements in the Bible for every ethical issue we bump into. We should study God's Word to glean general principles. So the first thing that we should ask when we encounter an ethical issue is, "*What does the Bible say about this?*" Three biblical principles lead us to clear thinking: the image of God in people, the Creation Mandate, and considering the whole truth of the Bible.

Biblical Outcomes

We also must have good goals. The Bible tells us the goals that God has for His creation. Targeting right outcomes is a key part of making wise ethical decisions. The next question we should ask when faced with an ethical issue is, "*What are the results?*" Clear goals follow from the biblical outcomes of helping people flourish, helping creation thrive, and giving God glory.

ETHICS Looking to the Bible

BIBLICAL PRINCIPLES

"What does the Bible say?"

❶ ***Image of God.*** A key biblical starting point is to treat people with respect since we all bear God's image. Because of this characteristic, we have a special position and responsibility for how we treat people (Gen. 1:26–28).

❷ ***Creation Mandate.*** We have a duty to wisely care for God's creation. We are not free to use it any way we want to. It doesn't belong to us, and we are accountable to God for how we use and care for His world (Gen. 1:26–28).

❸ ***God's Whole Truth.*** God's image in people and the Creation Mandate touch many ethical issues. But other parts of Scripture also give us helpful insights about what God wants us to do. Part of making wise ethical decisions is developing a broad understanding of God's Word and applying it to life in God's world.

BIBLICAL OUTCOMES

"What are the results?"

❶ ***Human Flourishing.*** As soon as God created mankind, He blessed him (Gen. 1:28). Passages like Psalm 1 and the Beatitudes teach that God wants people to enjoy blessing, or flourishing. He wants people to savor the joy and peace that comes from living life to the fullest. However, flourishing in a sin-cursed world sometimes looks very different from what we think it would (see Matthew 5:1–11). We need to align our ethical decisions with God's will. By looking at the outcomes of an ethical decision in life science, we can maximize human flourishing.

❷ ***A Thriving Creation.*** God's first blessing of mankind included the Creation Mandate. It is our job to make creation thrive (Gen. 1:28; 2:5, 15). We are to wisely use and cultivate creation so that it flourishes.

❸ ***God's Glory.*** The ultimate goal of all that we do should be God's glory (Lev. 10:3; Ps. 115:1; Rom. 11:36). Our decisions should show God that we love and honor Him. This includes what we do at school, at home, and at work. Some decisions may seem to be ethical because they help creation thrive. But if they leave God out—*if they fail to show God the honor He is due*—they are not truly right decisions.

Making ethical decisions in life science isn't easy. Things can get sticky in the real world. We need to make careful choices about the actions we take using our knowledge of life science. You'll get some practice throughout this textbook analyzing sensitive issues in life science. By tackling these issues you'll see that God's Word is the best tool to help us wisely use God's world and love people through life science.

1A REVIEW QUESTIONS

1-4
Aquarius is an underwater laboratory located off the Florida Keys.

1. What is a person's worldview? How does it affect him?
2. How can we know whether a person has a biblical worldview?
3. We've considered how people's views of life are different. How are their views of *death* different?
4. Outline the Bible's story in three words.
5. Read the Creation Mandate (Gen. 1:28). Summarize it in your own words.
6. Which of the following actions of life science can be justified using the Creation Mandate? Give reasons for your choices.
   a. teaching people from Madagascar good farming practices
   b. building underwater laboratories to study marine animals
   c. analyzing the flight of birds to make better aircraft
   d. developing medications to prevent childhood diseases
   e. classifying living things into categories
7. Look up the following verses and tell which of the three biblical principle(s) of ethics they support: the image of God, the Creation Mandate, or God's whole truth.
   a. Genesis 9:1–7
   b. Acts 20:17–18, 27
   c. Matthew 5:43–48
8. How does ethics relate to the Creation Mandate?

1B Questions

- How can I tell whether something is alive?
- How does something stay alive?
- What are the marks of God's design on life?
- How are people different from other living things?

1B UNDERSTANDING LIFE

Why is life so special?

1.4 THE CHARACTERISTICS OF LIFE

What exactly *is* life? We can't exactly define it, but we can describe it. What makes living things different from nonliving things? Biologists call living things **organisms**. Organisms have certain characteristics that make them different from things that aren't alive. What makes a sculpture of a dragonfly different from the real thing? The answer to this question can get you thinking in the right direction about the attributes of life. Most scientists agree that an organism needs to have *all* the characteristics on the facing page to be considered alive.

CHARACTERISTICS OF LIFE

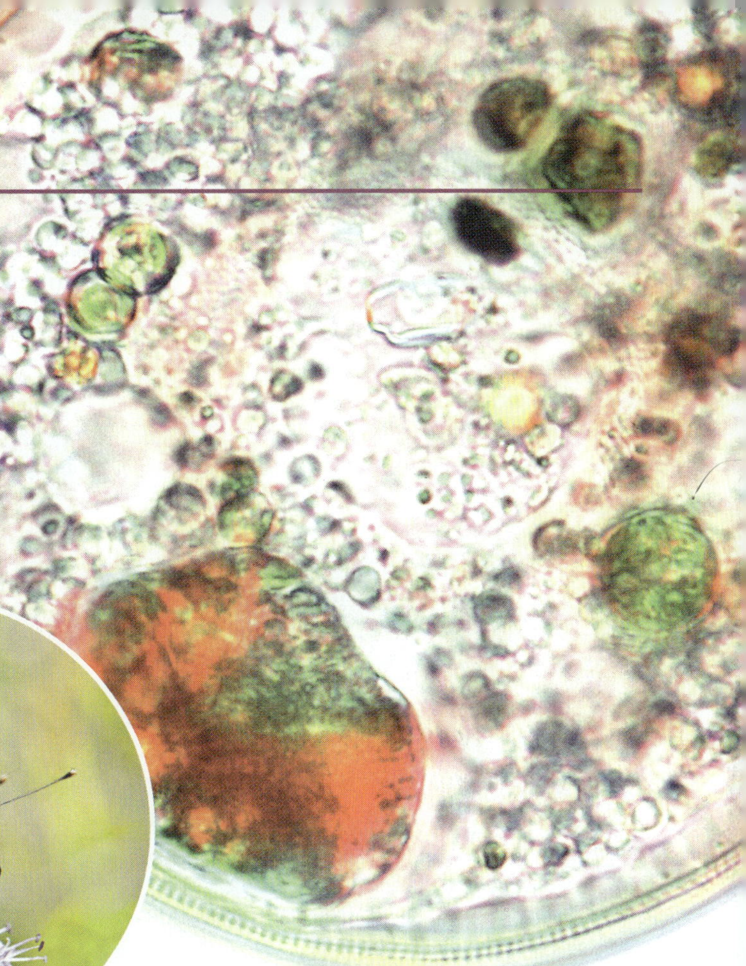

1 ***Organization.*** All living things are made of **cells**. Cells are the smallest working unit of an organism—microscopic machines. They have parts that work together to sustain life. Organisms that are made of only one cell are called **unicellular** organisms. **Multicellular** organisms have many cells. Different cells have different jobs in the organism, such as allowing movement or digesting food. Cells also contain information that spells out a plan for these functions. An organism's structure comes from the parts of the organism that serve a specific function.

2 ***Energy.*** Organisms need energy for movement and for carrying out other life processes. Plants get this energy through photosynthesis; animals and people eat food to gain energy. Because of organization, organisms can take in and use energy from food. Cells have the molecular machinery needed to process and use energy.

3 ***Growth.*** As you can see in the butterfly shown above, organisms grow. They get bigger in size, and as they do they produce more cells and eliminate worn-out or damaged cells. Many organisms don't just grow; they develop. Just like the caterpillar changes into a butterfly, other organisms change as they get older.

4 ***Reproduction.*** Every organism alive today came from another one of the same kind. Life comes only from life. Different organisms reproduce in different ways, passing down information to the next generation.

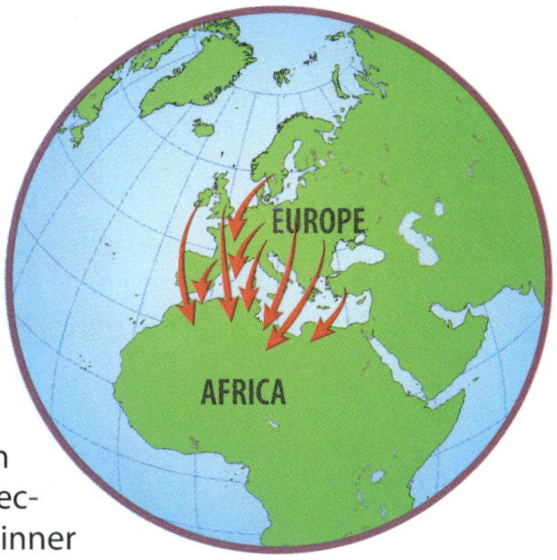

5 ***Response.*** Organisms have the ability to respond to their environment. For example, the painted lady butterfly shown above smells with its feet. When it smells nectar or fruit, it knows it's dinner time! It also migrates 9000 miles in the fall (see map at right). Painted lady butterflies fly 30 miles per hour at an altitude of 3000 feet to escape the approaching winter. Changes called *stimuli* hint that winter is approaching. Stimuli can come from the environment or from within an organism's body.

1.5 THE BALANCE OF LIFE

So now you know how to tell whether something is alive. But how does an organism stay that way? An organism's body and its environment are constantly changing, so the organism's body has to continually adjust. There are life processes that work to keep an internal balance. That's what **homeostasis** is—an organism's ability to keep its internal processes stable. The body needs and looks for information about its internal or external state and for ways to respond to changes.

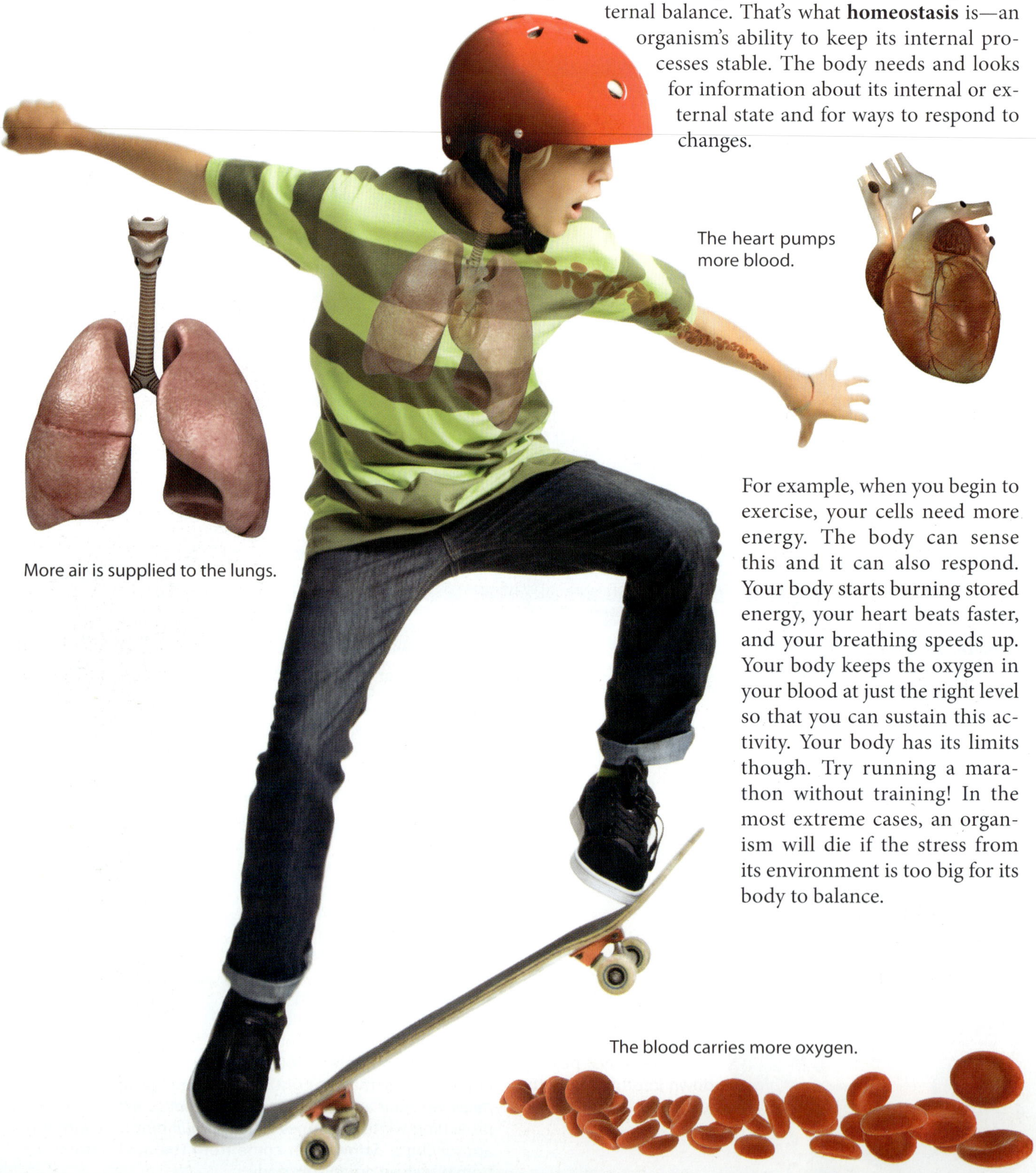

More air is supplied to the lungs.

The heart pumps more blood.

For example, when you begin to exercise, your cells need more energy. The body can sense this and it can also respond. Your body starts burning stored energy, your heart beats faster, and your breathing speeds up. Your body keeps the oxygen in your blood at just the right level so that you can sustain this activity. Your body has its limits though. Try running a marathon without training! In the most extreme cases, an organism will die if the stress from its environment is too big for its body to balance.

The blood carries more oxygen.

1.6 THE VALUE OF LIFE

Your body is made of about 60 trillion cells! God has engineered all these cells to work together to maintain your body's homeostasis. He didn't do this just for you; God provides this for all living things. Life is special because the Bible tells us that God created life. It has been divinely engineered with everything it needs down to the tiny cell. The energy and information required to sustain life show us that it is no coincidence. The complexity required for even a unicellular organism to maintain homeostasis is too great for it to be the product of random chance. God provides for life.

God doesn't just provide for you—He loves you and made you in His image. Your body reflects His divine design, and you reflect His image. We will explore many living things in this textbook. Remember your place in God's creation: you are His special design.

1B REVIEW QUESTIONS

1. A snowflake moves, grows, and has an organized crystal structure. Why isn't it considered alive?
2. Humpback whales migrate to the coast of Madagascar to give birth. A mother won't eat for several months until she and her nursing calf migrate to the Antarctic to feed on schools of fish. Prove that the humpback whale is alive using the characteristics of life.

3. How is the ability of living things to reproduce linked with the Bible's story of Creation (see Gen. 1: 24–30)?
4. For an organism to maintain homeostasis, it needs a way to (1) collect information about the environment and (2) make changes to stay balanced. Identify which of the following machines perform both functions and explain your answer(s).
   - **a.** thermostat
   - **b.** electric mixer
   - **c.** autopilot in a flying airplane
   - **d.** cruise control in a car
   - **e.** hairdryer
5. How do living things show that God has divinely engineered them?
6. How are people different from other living things?

LAB EXERCISE

Equipment

- two hand lenses of different magnifications
- your textbook for viewing
- tape measure or ruler
- black construction paper
- tape

FOCUSING ON LIFE: ENGINEERING A MICROSCOPE

How do microscopes work?

Ever since a Dutchman named Antonie van Leeuwenhoek discovered bacteria through his simple microscope, biologists have relied on this science tool to focus our attention on life. A microscope uses curved pieces of glass (*lenses*) that bend light to make things look bigger than they really are. This is called *magnification*. A fuzzy magnified image doesn't do us very much good though! We need to be able to see the finer parts of the image. This image sharpness is called *resolution*. We need both magnification and resolution to gather data using a microscope. In this hands-on activity, you get to work with various lenses and mirrors to create your own microscope.

1. Look at the diagram on the next page. How many primary sets of lenses does light pass through in this microscope? Give their names.

PROCEDURE

❶ Experiment with both lenses one at a time and from several distances to see how well they magnify the words in your book.

❷ Experiment with both lenses together and from several distances to see how well they magnify the words in your book. Experiment with having a different lens on top.

❸ Now measure the distance between the two lenses using the tape measure.

2. Which lens did you have on top? How far apart were the lenses? Draw a sketch on another piece of paper.

❹ Now build it! Fold your construction paper so that its width is equal to the distance that you measured. Then roll the paper into a tube. Make sure that you roll the tube in the direction perpendicular to the measured side so that your tube will be the desired length. Next, tape the lenses to the tube. Test how well your microscope works. You may need to adjust the length of the tube after testing your microscope.

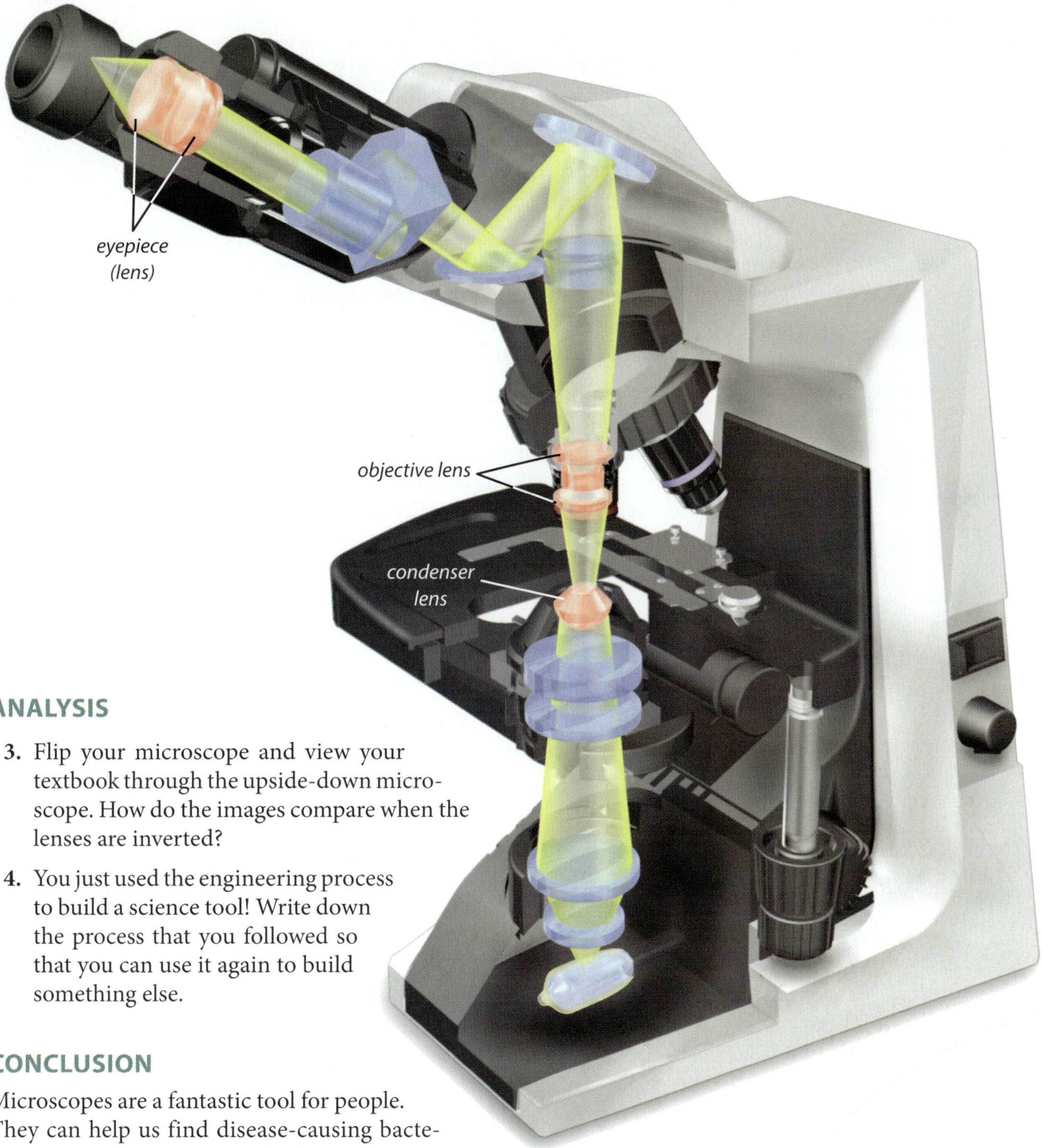

ANALYSIS

3. Flip your microscope and view your textbook through the upside-down microscope. How do the images compare when the lenses are inverted?
4. You just used the engineering process to build a science tool! Write down the process that you followed so that you can use it again to build something else.

CONCLUSION

Microscopes are a fantastic tool for people. They can help us find disease-causing bacteria in people's bodies and do detailed surgeries. They can help us spot defects in manufacturing and in studying fossils. Microscopes are a tool that we can use to wisely manage God's world.

GOING FURTHER

Do an Internet search using the keywords "virtual microscope." Experiment with an interactive virtual microscope so that you can be well prepared to use the real thing. Your teacher may also provide you with a compound microscope for you to view some slides.

1C Questions

- How do people learn about life science?
- How does scientific inquiry work in life science?
- How do biologists classify living things?

1C

LEARNING ABOUT LIFE

How do biologists work in the real world?

1.7 LIFE SCIENCE AS MODELING

Organisms in Madagascar tend to be huge. One of the largest moths in the world, the comet moth (below), lives here. In the past, the 10-foot-tall elephant bird, gorilla-sized lemurs, and the giant tortoise lived here but became extinct in the last few centuries. In fact, today there are no large carnivorous animals in Madagascar.

So how do we know about these animals if they aren't alive anymore? Scientists have studied the bones, lake sediments, and species of Madagascar to gather data about its history. **Data** is information that people collect. The data of life science can be qualitative or quantitative. For example, scientists can describe the color or the texture of the bones they find in the caves of Madagascar. This is **qualitative data**, data communicated using words and descriptions. Scientists can also measure the bones they find to communicate data with numbers. This is **quantitative data**.

Scientists can take the data that they find, the length of the leg bone of an elephant bird for example, and use it to draw conclusions. One of the reasons that we think the elephant bird was 10 feet tall is the sizes of the bones that we have found.

Scientists have collected lots of data about the animals of Madagascar to answer one question: Why did life change so dramatically here? To answer this question, they pull all the data together with a theory. A **theory** is an idea that scientists use to *explain* the data that they have collected. There are several theories that scientists use to explain why animals have changed so much in Madagascar. It could have been a natural disaster, like a flood or fire. It could have been drought or disease. Perhaps the earliest human inhabitants of Madagascar hunted these animals into extinction. Or it could have been a combination of these.

A scientific **law** is an idea that scientists use to *describe* what they observe. There is a lot of data behind the idea in a scientific law. One of the laws that scientists use as they make sense of what has happened to living things over time is the *law of superposition*. The basic idea of this law is that organisms buried deeper in undisturbed rock layers are older than living things buried higher in the ground. Naturalists think that thousands to millions of years passed before the next rock layer was laid down. However, those who believe God's Word recognize that these rock layers were

laid down very quickly during the Flood. Higher layers may indeed be younger than older layers, but only by a matter of hours or days.

You can see that life science relies on our observations of the world. Those observations are shaped by our worldviews. But we are limited in our ability to observe and then relate and explain what we observe. To understand complicated things about the world, we sometimes have to simplify the problem by leaving some information out. This means that an explanation of something usually does not give us the whole picture. A **model** simplifies something in the world to explain, describe, or represent it. A model can be a physical object, a graph, a computer program, or even just a number. Let's look at how life science uses models.

1-5

Animals in Madagascar are still at risk of extinction, partly because of the poverty of the people there who have hunted them. The ploughshare tortoise (below left) and the ring-tailed lemur (below right) are both endangered.

MODELING THE ELEPHANT BIRD

How do we know how big the elephant bird really was?

1 ***Historical Documents.*** Scientists have had to do some detective work to find out how big the elephant bird really was. Since this bird went extinct sometime during the seventeenth and eighteenth centuries, we have had some reports of elephant bird sightings. A French governor wrote about it in 1659. Marco Polo reported stories about it. The only thing they told us is that it was really big! Since nobody actually measured one, scientists needed more information.

2 ***Eggs.*** Scientists have also found elephant bird eggs that are not fully fossilized. People have even found beads that the native peoples made out of egg shells. Some eggs are as long as 33 cm—that would make the volume 160 times greater than that of a chicken egg. Imagine one egg that could make omelets for 53 people!

3 ***Bones.*** Scientists have found a whole skeleton's worth of elephant bird bones! Though each individual was different, these bones give us an idea of what the elephant bird was like. The tallest skeletons are almost 3 m tall, and we can measure the length and thickness of the bones to determine that elephant birds probably weighed between 350 to 500 kg. That's one big bird!

1.8 SCIENTIFIC INQUIRY

The world is full of data. Because of this, we need to have a systematic way to examine the world so that we can learn about it. **Scientific inquiry** involves using a variety of tools and skills to answer a question or solve a problem. This is how science works in the real world. This could involve using technology or building things. It often involves observing and collecting data.

Sometimes people need to create an experiment in order to learn how something works. A **controlled experiment** is a way for a scientist to test the way he thinks something works by examining the effect of one factor at a time. The factors that change are called **variables**. The scientist changes one variable, the *independent variable*. This can generate changes that the scientist did not directly cause in other variables. These are called *dependent variables*. Scientists measure changes in dependent variables, but other things can change that scientists don't measure. Let's look at an example of the process of doing a controlled experiment to answer the scientific question, "How can we use plants to fight cancer?"

USING A CONTROLLED EXPERIMENT TO FIGHT CANCER

1. ***Observe.*** Rosy periwinkle is a flowering plant native to Madagascar. It's toxic for people to eat. But native peoples have used it topically and have even eaten small amounts to treat diabetes. These people observed side effects, however, including nausea, hair loss, and even death. Scientists looking for medicines in plants in the 1950s thought that there must be some chemical in this plant that is good for treating disease. This is an example of a **hypothesis**, an explanation that is based on limited observations. It was time to test this hypothesis.

2. ***Test.*** Drug companies can take twenty years to get a new medicine approved. One of the problems that they face is that they can't experiment with people right away! Chemists working with rosy periwinkle extracted different chemicals from the plant. Then they grew groups of cancer cells. One group of cancer cells would be allowed to grow without any new chemicals applied. This is called the **control group**. The other groups of cancer cells had all of the same conditions as the control group. However, each of these groups of cells were treated with one of the chemicals isolated from rosy periwinkle. These were the **experimental groups**. The researchers measured cell growth for both the control and experimental groups. So cell growth is the dependent variable. The treatment of the cells with chemicals extracted from the rosy periwinkle is the independent variable.

3 *Analyze and Draw Conclusions.* After all this testing, researchers pinpointed two anticancer chemicals in rosy periwinkle. The cancer cells in these experimental groups either didn't grow very much or died. You can see the molecular structure of one of these chemicals at right. Researchers did more tests to confirm their results. This process involved testing the drugs on mice and then testing it on people in the final stages of approval.

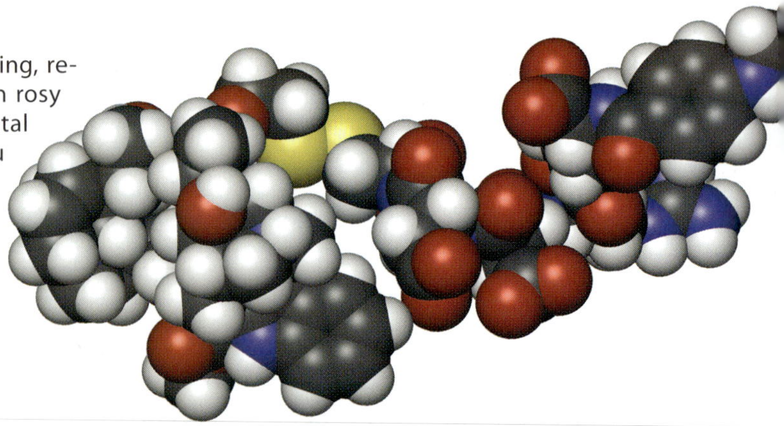

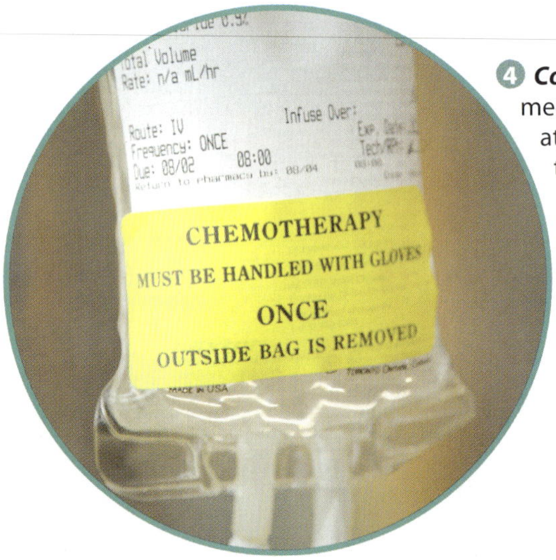

4 *Communicate.* The chemists working on developing anticancer medications from rosy periwinkle presented their findings in 1958 at a conference in New York. The US Food and Drug Administration approved anticancer drugs from the rosy periwinkle in the early 1960s. Doctors have been using these chemicals in chemotherapy to treat blood and bladder cancers ever since. This is an example of how we can use the process of science to help people.

1.9 CLASSIFYING LIFE

Why Classify?

People name pets, cars, roads, and even machines. Why do they do this? They do this to help themselves understand the world. We make the world more useful through naming. After God finished His creation, He commanded every living thing in the world to reproduce after its own kind (Gen. 1:11, 21, 24). God established groups of organisms, or *kinds*, that can reproduce. And one of the first things that Adam did after God gave mankind the Creation Mandate was to name the animals. This job was important because it involved classifying creation in a way that helped mankind manage it.

There are about 8.7 million types of living organisms in the world! Even more than that are preserved in the fossil record. Classification is just a handy way to understand the variety of life in the world. Classification helps us generalize among groups of organisms to see how they are *similar*. It also helps us compare similar organisms to see where they are *different*. If we find a new organism, we can look at other organisms that are similar to it to learn more about it. Going through this process helps us name them.

How Do Scientists Classify?

There are many ways to categorize organisms—classifying is just a tool. We can choose to classify organisms in ways that help us understand them the most. The science of sorting organisms into groups is called **taxonomy**. Organisms with similar characteristics are often classified together. The system that we use today was started by a Swedish botanist named Carolus Linnaeus in the 1700s. This system has eight basic levels: *domain*, *kingdom*, *phylum*, *class*, *order*, *family*, *genus*, and *species*.

Scientists classify organisms into domains and kingdoms on the basis of three traits:

1. the types of cells they have
2. the number of cells they have
3. how they get energy

Let's consider these three traits. Organisms can have one of two types of cells: eukaryotic (yoo KEHR ee AHT ik) and prokaryotic (PROH kehr ee AHT ik). The difference between the two has to do with the cell's **nucleus**, the dense part of the cell that contains the information it needs to carry out cell processes. **Eukaryotic cells** are cells with a nucleus. **Prokaryotic cells** are cells with no nucleus. Since they don't have a nucleus, they are organized differently so that they can still carry out life processes.

As for the other traits, we've looked at unicellular and multicellular organisms. All organisms get energy from their environment. Organisms that make their food using energy from their environment are called **autotrophs**. Organisms that get energy by eating other organisms are called **heterotrophs**.

Table 1-1

Domains of Life

Traits	**Archaea** ❶ prokaryotic cells	**Bacteria** ❶ prokaryotic cells	**Eukarya** ❶ eukaryotic cells
Kingdom	**Archaea** 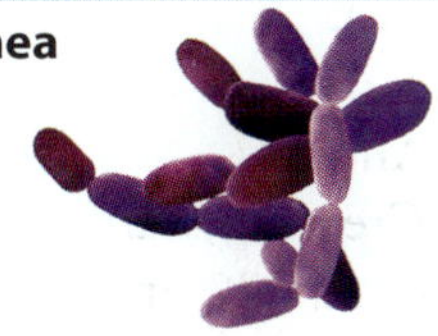	**Bacteria** 	**Chromista** 
Traits	❷ unicellular ❸ autotrophs and heterotrophs	❷ unicellular ❸ autotrophs and heterotrophs	❷ unicellular and multicellular ❸ most are autotrophs
Examples	bacteria living in extreme conditions	*E. coli*, bacteria found in yogurt, *Salmonella*	water molds, diatoms

Traits	**Eukarya** ❶ eukaryotic cells			
Kingdom	**Protozoa** 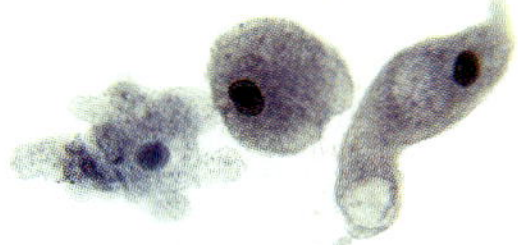	**Fungi** 	**Plantae** 	**Animalia** 
Traits	❷ unicellular and multicellular ❸ most are heterotrophs	❷ unicellular and multicellular ❸ heterotrophs	❷ multicellular ❸ autotrophs	❷ multicellular ❸ heterotrophs
Examples	euglenas, amoebas, *Plasmodium*	mushrooms, yeast	trees, ferns, flowering plants, mosses	sand dollars, sponges, worms, fish, reptiles, birds, mammals

CLASSIFYING THE HERMIT CRAB

The animal at the bottom of this page is a hermit crab. But if you were to do an Internet search on this common name, you would see that lots of crabs have this name. A **scientific name** identifies an organism uniquely. The first part gives the genus, and the second part is the species. This crab's scientific name is *Coenobita perlatus*. Notice that the genus comes first and is capitalized, while both parts are italicized.

DOMAIN: EUKARYA
(4–10 million species)

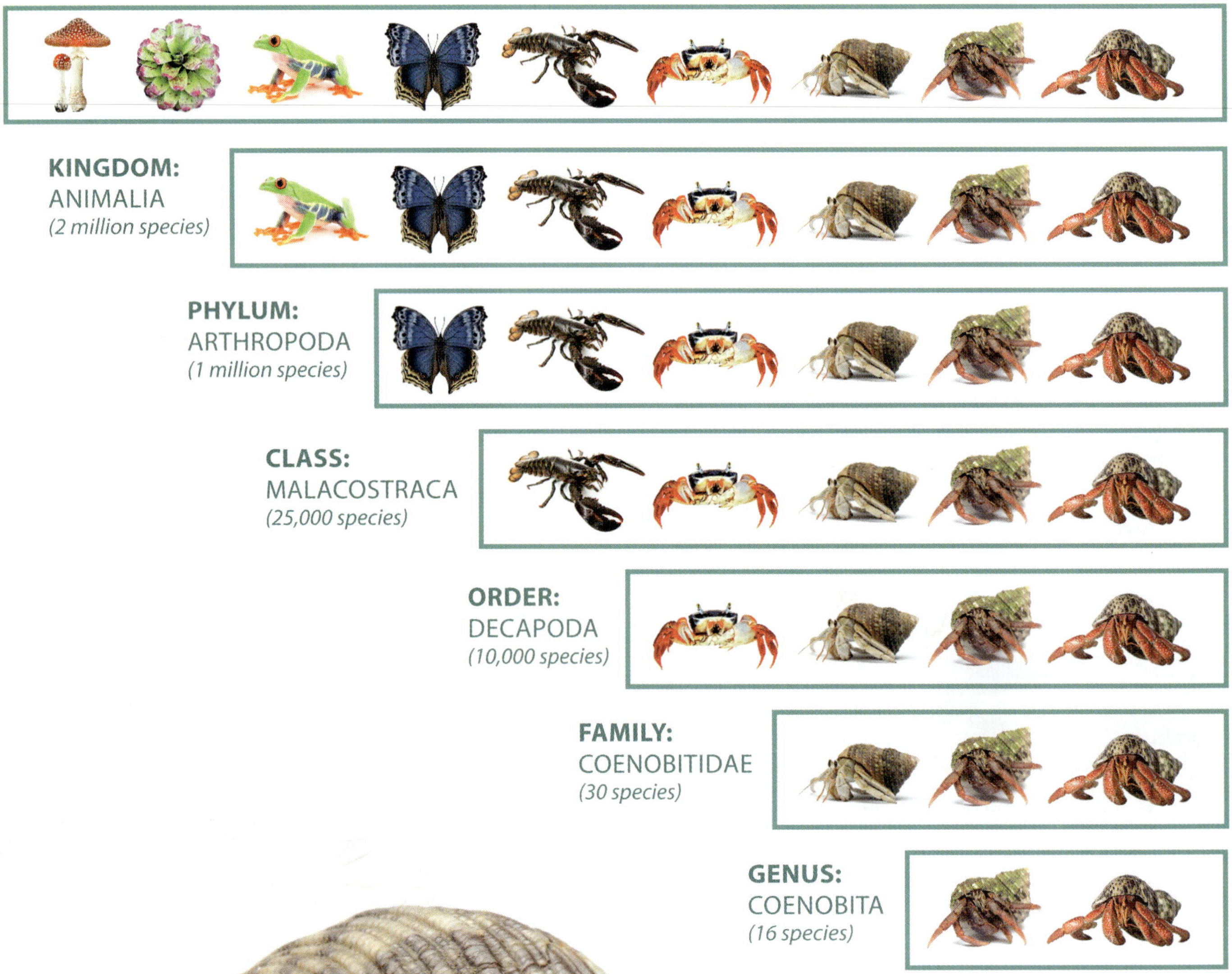

KINGDOM: ANIMALIA
(2 million species)

PHYLUM: ARTHROPODA
(1 million species)

CLASS: MALACOSTRACA
(25,000 species)

ORDER: DECAPODA
(10,000 species)

FAMILY: COENOBITIDAE
(30 species)

GENUS: COENOBITA
(16 species)

SPECIES: PERLATUS

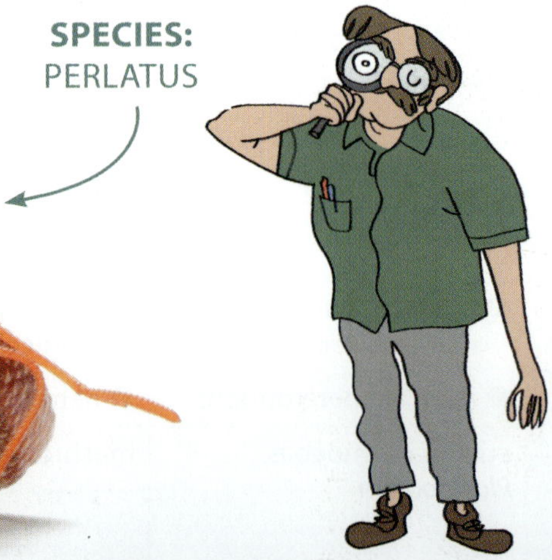

CLASSIFYING THE FOSSA

What is the animal at the bottom of this page—a cat, a weasel, or a mongoose? It's a fossa from Madagascar. It's been hard to classify! Scientists have had to study the genetics of this animal in order to classify it. Its scientific name is *Cryptoprocta ferox*. At a length of 60–90 cm, it is Madagascar's largest carnivore.

DOMAIN: EUKARYA
(4–10 million species)

Classification and Worldview

While taxonomists work to understand the world of living things through classification, some have another goal. Evolutionists want to develop a classification system that shows the evolutionary relationship of organisms. It links them back to a common ancestor and how one kind of organism evolved into another kind. This creates problems for two reasons. First, it violates God's law of nature that organisms reproduce after their kind. Second, scientists are trying to connect organisms that really aren't related. Evolutionists say that these organisms came from a common ancestor. They use genetics to support their revamp of the Linnaean classification system. Trying to use a classification system this way might make it less useful or helpful.

Viewing God's living world in the light of His Word frees us up to understand it better. Though evolutionists are using genetics and classification to support their theory, classification can also support a biblical view of life. The diversity of life that we see around us isn't a mark of evolution. It's the Creator's mark. He has included a lot of information in the cells of each of His creatures. That information continues to reveal itself in new species and varieties, but not new kinds of organisms. The same trait in very different creatures is evidence of a common designer instead of a common ancestor. The Divine Engineer chose to give certain traits, such as the ability to fly, to a variety of organisms so that they could thrive in His world. When we view life this way, we are in the best position to wisely manage God's living world.

1C REVIEW QUESTIONS

1. Which of the following activities of life science involve using a model?
   a. building an interactive computer program to learn about the cell
   b. graphing the populations of migrating monarch butterflies in the United States each year for a decade
   c. dissecting a frog to learn about its organs
   d. creating a replica of a whole dinosaur skeleton from only a few fossils
2. Explain how you would make a model to determine the height of a seventh grader.
3. Is a controlled experiment the same thing as scientific inquiry? Explain.

Use the Using a Controlled Experiment to Fight Cancer box on pages 17–18 to answer Questions 4–6.

4. What was the independent variable?
5. What was the dependent variable?
6. Why can't pharmaceutical companies test drugs on people in the early stages of development? Use biblical principles and biblical outcomes to support your answer.
7. List the levels of the modern classification system in order of increasing number, starting with the level that contains the *fewest* organisms.
8. A polar bear has fur, eats other animals, and provides milk for its young. Do you think that its classification would be more like the hermit crab's or the fossa's? Explain.

CHAPTER SUMMARY

1A SEEING LIFE THROUGH THE BIBLE'S STORY

- A person's worldview is the big story that he uses to make sense of the world, to know what to believe and value, and to make choices.
- People share some views of life, but they can have big differences that are based on their view of the Bible and its story.
- The Bible says that God created the world. It tells us that the world fell because of Adam's sin, but God is working to restore His creation through the person of Jesus Christ.
- The Bible's story puts us in the best position to do life science and gives us the best reasons to do the same.
- The three biblical principles of ethics concern the image of God in mankind, the Creation Mandate, and God's whole truth.
- The three biblical outcomes of ethical decisions include human flourishing, a thriving creation, and bringing God glory.

1A TERMS	
worldview	3
life science	3
Creation Mandate	4
ethics	6

1B TERMS	
organism	8
cell	9
unicellular	9
multicellular	9
homeostasis	10

1B UNDERSTANDING LIFE

- Most scientists agree that an organism needs to have all the characteristics of life for it to be considered alive.
- The characteristics of life are organization, energy, growth, response, and reproduction.
- Homeostasis is how an organism's body constantly adjusts to keep its internal processes stable.
- Life's complex parts show that it has been divinely engineered and is not the product of random chance.
- People are different from the rest of creation because they are made in God's image and have been put in charge of His world.

CHAPTER 1

CHAPTER SUMMARY

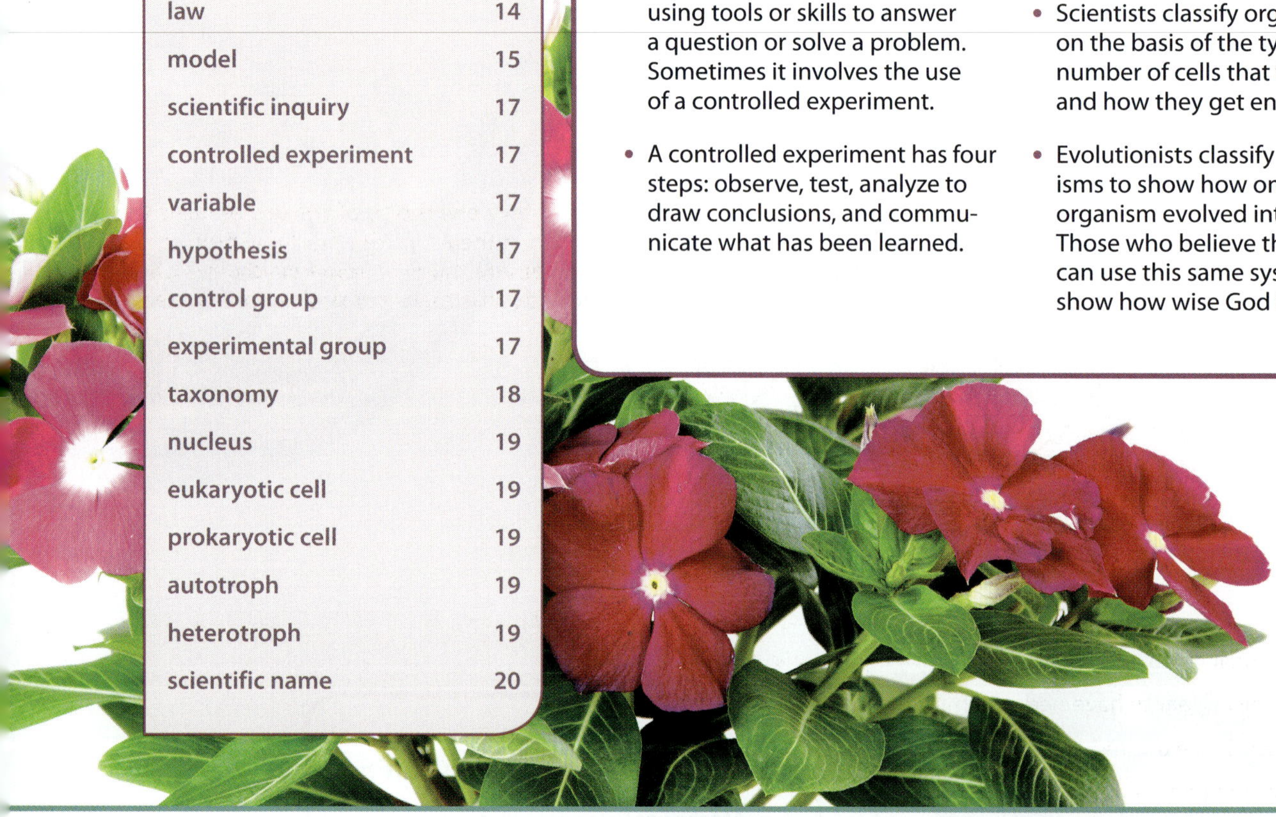

1C TERMS

data	14
qualitative data	14
quantitative data	14
theory	14
law	14
model	15
scientific inquiry	17
controlled experiment	17
variable	17
hypothesis	17
control group	17
experimental group	17
taxonomy	18
nucleus	19
eukaryotic cell	19
prokaryotic cell	19
autotroph	19
heterotroph	19
scientific name	20

1C LEARNING ABOUT LIFE

- We extend our knowledge of life science by collecting data and then using it to make theories, laws, and models.
- Scientific inquiry involves using tools or skills to answer a question or solve a problem. Sometimes it involves the use of a controlled experiment.
- A controlled experiment has four steps: observe, test, analyze to draw conclusions, and communicate what has been learned.
- Scientists classify living things into a classification system with eight basic levels: domain, kingdom, phylum, class, order, family, genus, and species.
- Scientists classify organisms on the basis of the type and number of cells that they have and how they get energy.
- Evolutionists classify organisms to show how one kind of organism evolved into another. Those who believe the Bible can use this same system to show how wise God is.

CHAPTER REVIEW QUESTIONS

Recalling Facts

1. Define *worldview* in your own words.
2. Name the three major themes of the Bible's story.
3. List the three biblical principles of ethics.
4. List the three biblical outcomes of ethics.
5. List the characteristics of life.
6. Give your own definition for *homeostasis*.
7. How are people different from the rest of creation?
8. What is every law, theory, and model based on?
9. What is a model?
10. What are the steps of a controlled experiment?
11. What are the eight basic levels of the classification system?
12. What three things do scientists consider when classifying an organism?

Understanding Concepts

13. Find another Bible verse that supports the idea of the image of God in mankind. (*Hint*: Check the Ten Commandments [Exod. 20:1–20] or the Sermon on the Mount [Matt. 5–7].)

14. How does your body automatically maintain homeostasis when you get hot or cold?

15. Suppose an object you're observing appears to have all of the characteristics of life except one. Is it considered alive? Explain.

16. Prove that the organism in Figure 1-6 is alive using the characteristics of life.

1-6
The snowy owl is one of the few animals that remain in the Arctic during the winter when it reaches adulthood. Parents build nests on the ground in the tundra since there are no trees.

17. What does it mean to say that life has been divinely engineered?

18. How does a cell appear to have been divinely engineered?

19. How does life science use models?

20. Classification is changing all the time. To stay familiar with how organisms are classified, scientists need to keep up with journal articles that researchers are publishing. The *Catalogue of Life* website tracks these publications and maps out the classification tree. Do an Internet keyword search to find this website. Press the **Access** button, and search for the giraffe. Write all eight levels of classification for this organism.

21. Birds, reptiles, and mammals have backbones and are classified in similar ways. Explain the reasoning behind why evolutionists classify them this way.

Critical Thinking

22. Create a table that compares a naturalistic worldview with a biblical one for the following issues: origin of life, the role of science, and the value of people compared with other living things.

23. Some Christians are suspicious of science because many scientists do not have a biblical worldview. Use the Bible's teachings to show why life science is worth doing.

24. Life comes only from life. Give an example of how some scientists would disagree with this statement. You may need to do some research.

Graphic Organizers

A graphic organizer is a way to organize ideas by arranging them on a page in different ways to help you understand them better. The table that you create in Question 22 is an example of a graphic organizer. To get some tips on how do do this, check Appendix B.

ETHICS Eating Animals

Do you know anyone who is a vegan (VEE gan) (a person who eats plant products only, avoiding anything that comes from animals, including meat, eggs, and dairy products)? People choose such a lifestyle for a variety of reasons. They might like this kind of food or think it is trendy to eat like this. Maybe they have friends who eat this way. Some have health reasons. But some people choose such a diet because they think that it's wrong to use animals at all. Other vegans think that farming and raising animals is harmful to the environment. People who follow a vegan diet can reduce their chances of heart disease, but a vegan diet isn't healthy for children or pregnant women because it lacks certain nutrients. Less than 2% of Americans are vegans, and 70% abandon this diet at some point because it becomes expensive and hard to maintain.

Use the Ethics box above to answer Questions 25–28.

25. What does God's Word say about eating animals? Consider the three biblical principles.

26. What are the results of choosing a vegan diet?

27. From the perspective of the Bible's story, what is the most important thing to consider when determining whether it is right to eat animals?

28. What is your opinion about eating animals and animal products? Describe how you would respond to someone who is a vegan because he thinks that it is wrong or bad for the environment to use animals for food. As you write, incorporate the principles of ethics and biblical teachings.

Use the Case Study below to answer Questions 29–38.

29. How did the researchers working with biobots use technology?

30. Is this an example of a controlled experiment? Justify your answer.

31. What is the scientific problem?

32. What was the scientists' hypothesis?

33. What was the independent variable in this study?

34. What was the dependent variable in this study?

35. What was the control group?

36. What was the experimental group?

37. What did the researchers conclude?

38. How did researchers communicate their results?

CASE STUDY SCIENTIFIC INQUIRY WITH BIOBOTS

Scientists are looking to an unusual resource for search and rescue: insects. They are hoping to find a way to control insects' movements to find people in an emergency. Scientists were wondering whether they could use electrical signals to do this with insects equipped with cameras.

Researchers raised a colony of Madagascar hissing cockroaches, an insect about the length of your finger. They kept all the roaches in the same kind of environment, providing them with food, moisture, and warmth in a way that mimicked their natural habitat. Twelve of the most active roaches were outfitted with electrodes attached to their antennae. Researchers tested different electrical pulses to find the best way to steer the roaches to follow a specific path. They found that a current of 150 microAmps at pulses of 50 Hz worked the best. It was kind of like driving a cockroach!

Because of this research, published in a scientific paper in 2015, scientists are finding possible ways to use a roach's ability to squeeze into tight places to help people in an emergency situation.

actual size

39. Design an experiment to find out which fertilizer causes the tomato plants in your garden to grow the best.

2 CELL STRUCTURE

DIVINELY ENGINEERED

The F-22 Raptor is the US Air Force's most powerful fighter. Equipped with stealth technology and able to travel over twice the speed of sound, the F-22 is a jaw-dropping marvel of design. It has hundreds of thousands of parts, each one performing a special and often critical role. It has lots of safety features to prevent something from going wrong, a good thing since a mechanical failure could destroy both pilot and plane. No one would believe that an F-22 created itself. A machine of this power and complexity had to have been designed by someone.

Cells are even more complex than an F-22. A cell consists of many smaller parts. Each part depends on all the others to work properly. The theory of evolution proposes that the parts of a cell came about slowly. It says that life as we know it evolved by many small steps, each step improving the system. But this theory faces a great challenge when it comes to the structure of a cell. A cell without any one of its parts wouldn't be able to survive. Has the cell evolved, or has it been divinely engineered?

2A

CELLS AND LIFE

How do cells make up all living things?

2A Questions

- What does cell theory tell us about cells?
- Why is the cell theory a model?
- What is the difference between an animal cell and a plant cell?
- How do cells work together to form tissues?

It's pretty easy to see an F-22. It's big! And it has many parts: wings, a tail, and engines, among others. But cells are small. In fact, cells are *tiny*! When you look at a cactus, a guppy, or a human, you can't see all the millions of cells that make up that organism. Imagine the surprise for scientists when they discovered that living things were made of vast numbers of tiny parts. They were probably even more surprised when they looked closer and saw that those extremely small parts were made of even tinier parts! Even to this day, scientists are still learning about how all these parts work together.

2.1 CELL THEORY

Although God created the first cells during Creation Week, humans really didn't know about them until the 1600s. English scientist Robert Hooke studied cork through his microscope and saw structures that he called *cells* because they reminded him of rows of cells in a prison. Although he didn't know it, Hooke was seeing the dead remains of cells, not the actual living ones. As microscopes improved, people could see cells in more detail and learn more about how they work. In time, this knowledge formed what we call the modern **cell theory**.

MODELING THE CELL

In the 1800s, four German scientists made the discoveries that later became the central ideas of the cell theory.

It is important to remember that a model like the cell theory is not the same thing as truth. Scientists cannot say with certainty that all living things are made of cells. In fact, scientists cannot even agree on how to define life. For example, viruses are not made of cells, yet many scientists classify them as living things. But this is a violation of the cell theory. Nevertheless, the cell theory is useful since it explains the vast majority of our observations and can be used to make predictions. Like all models, though, it is subject to being changed or replaced.

2.2 CELL TYPES

Eukaryotes and Prokaryotes

Although cells form all living organisms, not all cells are the same. In fact, cells often differ greatly, depending on what type of creature the cell belongs to. Even so, a cell can be classified as either eukaryotic if it has a nucleus or prokaryotic if it lacks one.

All the organisms that you can plainly see, such as dogs, daisies, and deer ticks, are made of eukaryotic cells. Many microscopic life forms are eukaryotic as well. But all prokaryotic life is microscopic, and most prokaryotes are bacteria.

2-1
All cells are either eukaryotic or prokaryotic, depending on the presence or absence of a nucleus. You'll learn more about these two types of cells in Section 2B.

Unicellular and Multicellular Organisms

All prokaryotes and some eukaryotes are unicellular. A unicellular organism consists of just one cell that can live by itself and perform all the functions needed in order to live, grow, and reproduce. Other eukaryotes are multicellular, meaning that the organism is made of many cells that work together to perform all the tasks needed to survive.

In a multicellular organism, cells form tissues. A **tissue** is a group of similar cells that work together. So the muscle cells in the heart all work together to cause a heartbeat, while cells in the tissues of the skin work together to provide a protective covering for the body. Cells in one tissue perform tasks that are often very different from the roles of other tissues. As a result, cells from different tissues often look and behave very differently. For instance, a muscle cell and a skin cell look very different because they are each part of tissues doing different jobs. Groups of tissues together form an **organ** such as the heart or the skin. You will learn more about different tissues and organs in Chapter 13.

2A REVIEW QUESTIONS

1. List the main ideas of the cell theory.
2. How would you respond to a friend who tells you, "The cell theory is settled science"?
3. You are studying a microscopic, unicellular organism that has a nucleus. What type of cell is this organism?
4. What type of multicellular organism is composed of prokaryotic cells?
5. Arrange the following terms from lowest to highest level of organization: tissue, organ, cell.
6. You are looking through a microscope at two human cells that look very different. Explain why they probably did not come from the same tissue.

2B
CELL STRUCTURES

How do the parts of a cell work together?

2B Questions
- What are the main cellular organelles?
- What does each organelle do?
- How do cells keep some molecules out and let others in?
- How do creationists and evolutionists differ in their views of the cell?

2.3 CELL PARTS

In the previous subsection, you learned that a eukaryotic cell has a nucleus, while a prokaryotic cell does not. There are other differences as well. A eukaryotic cell has many structures that perform special functions in the cell. These structures, which are usually surrounded by a membrane, are called **organelles**. Prokaryotic cells lack organelles, but the two types of cells do share some structures.

PROKARYOTIC CELL STRUCTURES

1 The **cell membrane** surrounds all cells and regulates what comes in and goes out. The membrane is made mostly of long carbon and hydrogen molecules called *lipids*. The membrane also has many large molecules made of carbon, hydrogen, oxygen, and nitrogen—called *proteins*— embedded in it.

2 Some cells have an additional rigid **cell wall** that protects them. Most bacteria, algae, fungi, and plants have cell walls, but animal, human, and protozoan cells do not.

3 Some bacteria surround themselves with a slimy layer called a capsule that protects the bacterium from harmful substances and prevents it from drying out.

4 The inside of the cell is called the cytoplasm. In eukaryotic cells, it includes all the organelles, except the nucleus, as well as jellylike cytosol, which surrounds them.

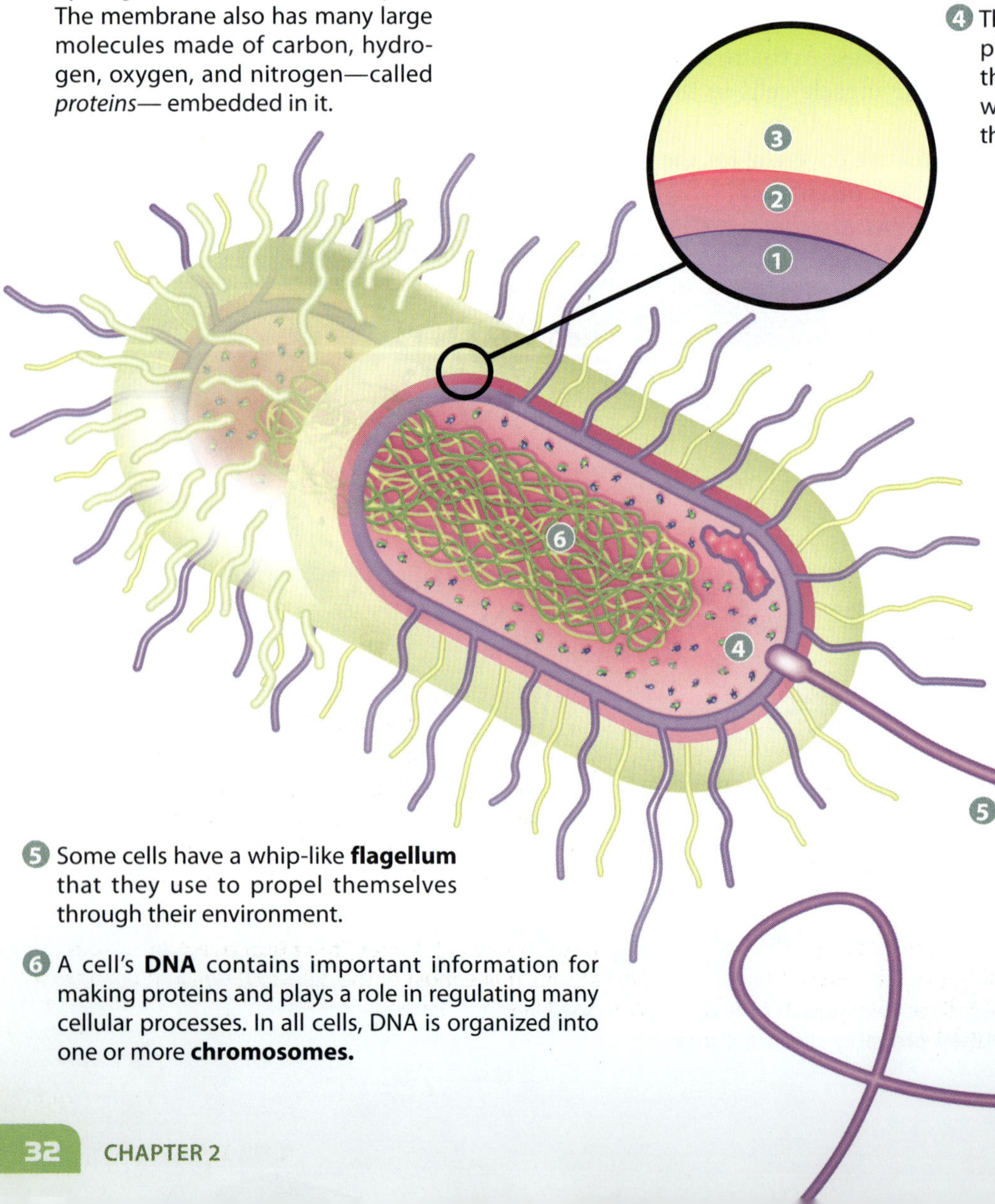

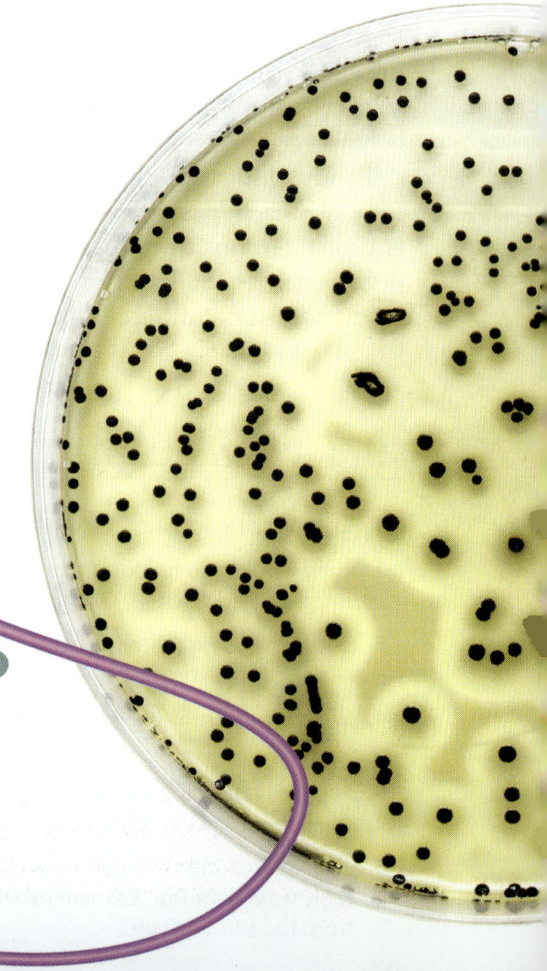

5 Some cells have a whip-like **flagellum** that they use to propel themselves through their environment.

6 A cell's **DNA** contains important information for making proteins and plays a role in regulating many cellular processes. In all cells, DNA is organized into one or more **chromosomes.**

EUKARYOTIC CELL STRUCTURES

7 The nucleus holds the DNA inside a nuclear membrane, which controls what enters and exits the nucleus.

8 The **endoplasmic reticulum**, or ER, is a network of tubes that help produce and transport proteins and other molecules throughout the cell. Rough ER has protein-producing ribosomes attached to its surface. ER without ribosomes, called smooth ER, makes fat molecules for the cell membrane.

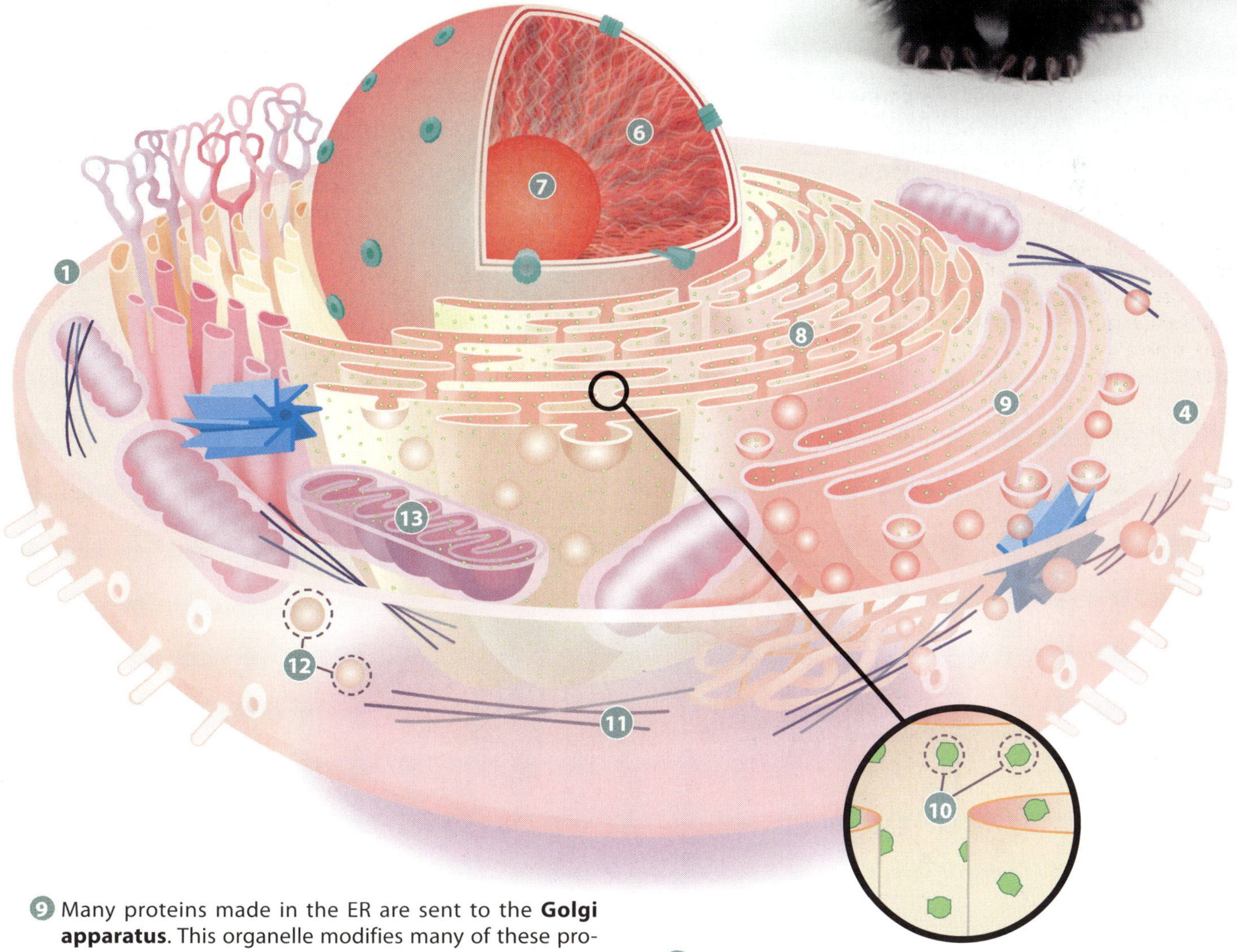

9 Many proteins made in the ER are sent to the **Golgi apparatus**. This organelle modifies many of these proteins and packages them into vesicles, small bubble-like structures that transport the proteins to their final destinations.

10 **Ribosomes** are small structures that assemble proteins. Both eukaryotes and prokaryotes have them.

11 The **cytoskeleton** consists of several types of long, thin fibers anchored to various parts of the cell. It helps the cell maintain and change its shape. It also serves as a highway for moving vesicles and organelles.

12 The lysosomes are small organelles that act as the demolition crews of the cell. They contain enzymes, which are proteins that speed up or slow down chemical reactions. The lysosomes dispose of old cellular structures and destroy foreign substances such as bacteria or viruses.

13 The "powerhouses of the cell," the **mitochondria** (s. mitochondrion), make most of the cell's ATP, which is used to power cellular processes.

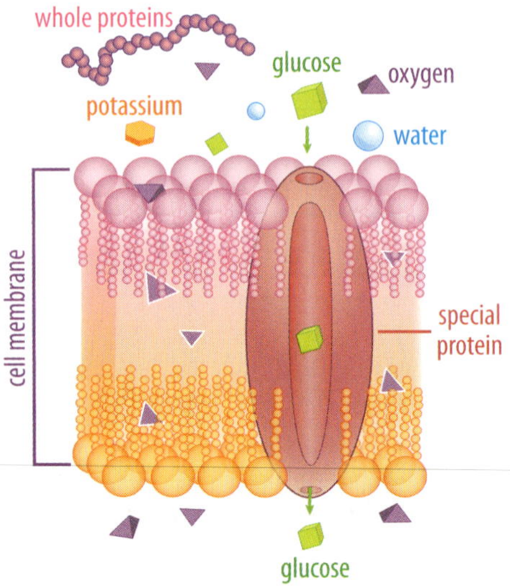

2.4 ENTERING AND EXITING THE CELL

A cell's organelles work properly only when the cytoplasm has the correct substances. A cell must be able to bring in just enough of certain molecules but keep others out. If a cell builds molecules that are needed elsewhere in an organism, those have to be allowed out of the cell too. The task of monitoring passage into and out of the cell is the job of the cell membrane. There are several ways that it performs this task.

1 The cell membrane is **selectively permeable**. It allows some substances to pass through but keeps others out. Oxygen molecules can pass through easily. Glucose molecules can pass through by means of special proteins in the membrane. But whole proteins cannot pass through the membrane at all.

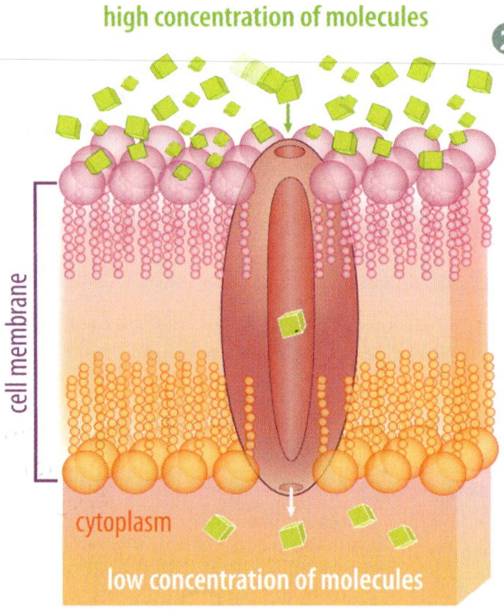

2 Molecules are constantly in motion. When a molecule such as glucose strikes one of the protein channels in the cell membrane that allows the molecule to enter the cell, the molecule moves into the cytoplasm. When a certain type of molecule is more densely packed on one side of the membrane than on the other, a **concentration gradient** is formed.

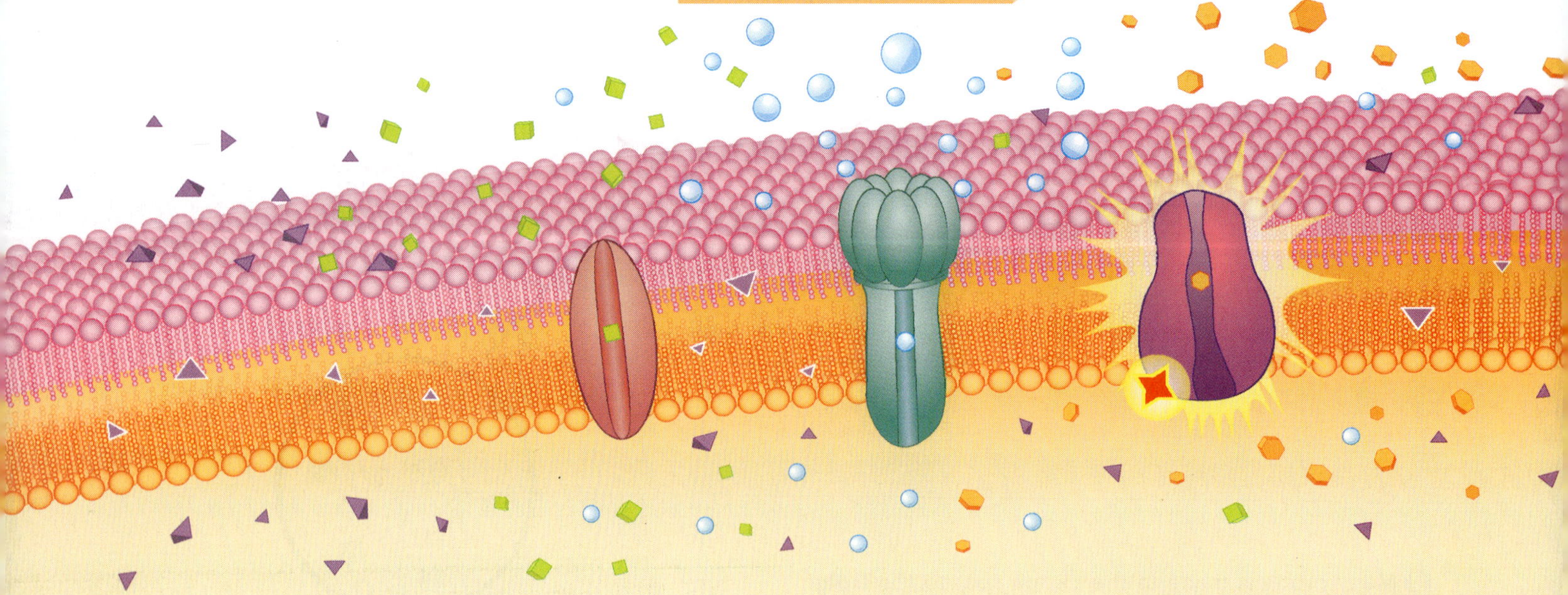

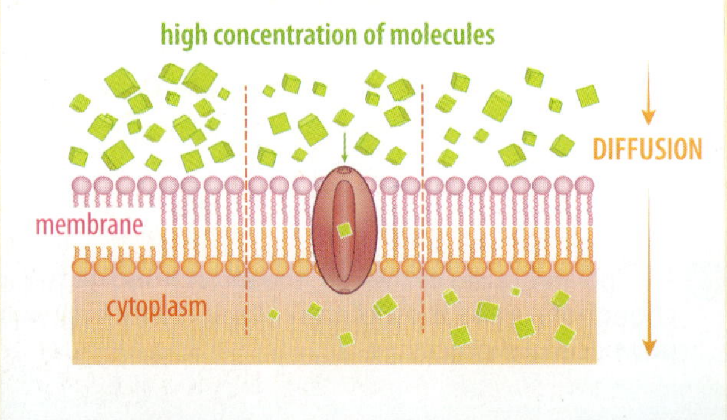

3 When there is a higher concentration on the outside of the cell than on the inside, more molecules on the outside bump into protein channels than molecules on the inside. As a result, more glucose molecules enter the cell than leave it. The concentration of glucose molecules on the outside of the cell decreases, while that of glucose molecules on the inside of the cell increases. This movement of molecules through a membrane is called **diffusion**. The greater the concentration gradient is, the faster diffusion takes place.

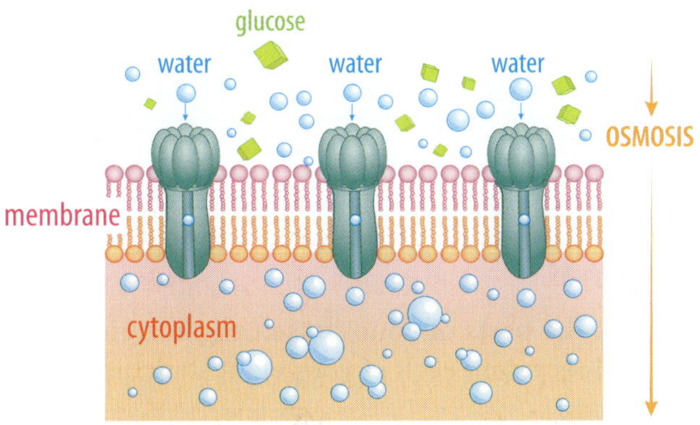

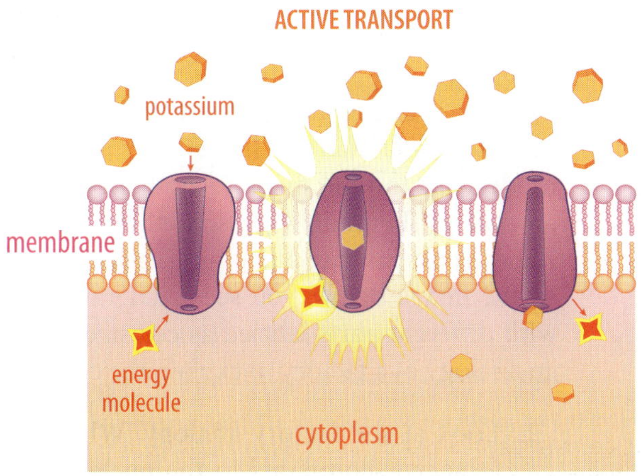

4 Like other small molecules, water molecules can also move across a membrane from an area of high concentration to an area of low concentration. This movement of water across a semipermeable membrane is a special type of diffusion called **osmosis**. Many cellular processes depend on osmosis, which is why all living things need adequate amounts of water in order to survive.

Osmosis and diffusion happen without the cell doing any work. They are both forms of **passive transport.**

5 But osmosis and diffusion can't transport everything a cell needs. For instance, cells need lots of potassium ions, but they tend to flow out of the cell by diffusion because there are normally more of those ions inside than outside. Cell membranes contain protein structures that pump in potassium ions from outside the cell. Since this pump requires energy to work, it is an example of **active transport**.

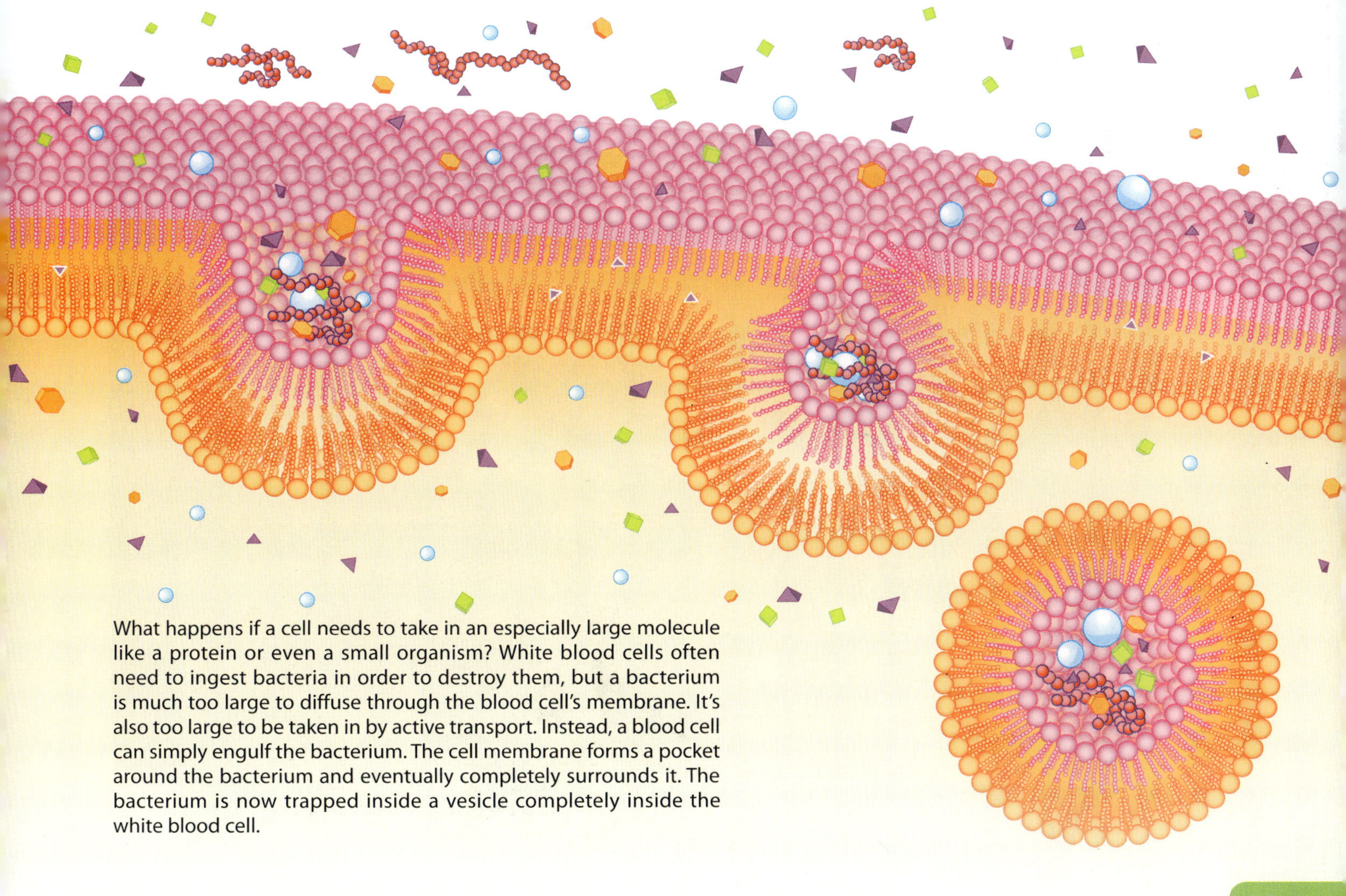

What happens if a cell needs to take in an especially large molecule like a protein or even a small organism? White blood cells often need to ingest bacteria in order to destroy them, but a bacterium is much too large to diffuse through the blood cell's membrane. It's also too large to be taken in by active transport. Instead, a blood cell can simply engulf the bacterium. The cell membrane forms a pocket around the bacterium and eventually completely surrounds it. The bacterium is now trapped inside a vesicle completely inside the white blood cell.

LAB EXERCISE

MODELING LIFE: MAKING AN ANALOGY FOR THE CELL

How do the cell structures compare to a large organization?

The cell is a very complex structure, and it can be hard to understand. That is why scientists make models like the cell theory. Another type of model is an analogy. In an analogy, something hard to understand is compared to something easier to understand. In the image below, you see a diagram of a factory with different parts labeled as cell structures. This is an analogy of a factory to explain the cell.

1. Look at the factory analogy. What part of the factory is compared to the nucleus? What part is compared to the Golgi apparatus?

PROCEDURE

❶ Choose a large organization, other than a factory, to make an analogy of the cell. It can be a house, a city, a castle, or any other organization you want.

❷ Your analogy must include the nucleus, ER, Golgi apparatus, and mitochondria. Choose four other structures to include in your analogy.

2. List the four additional structures that you have chosen.

3. List the parts of your chosen organization that will represent each of your eight structures.

❸ Now go make your analogy! You may draw a picture like the one below, or you can write a report describing how your organization is like a cell. Or you could make a video explaining your analogy. With your teacher's permission, you could also prepare a presentation on your analogy to present to your class.

GOING FURTHER

Do an Internet search using the keywords "cell analogy." See what other analogies people have created to explain the cell.

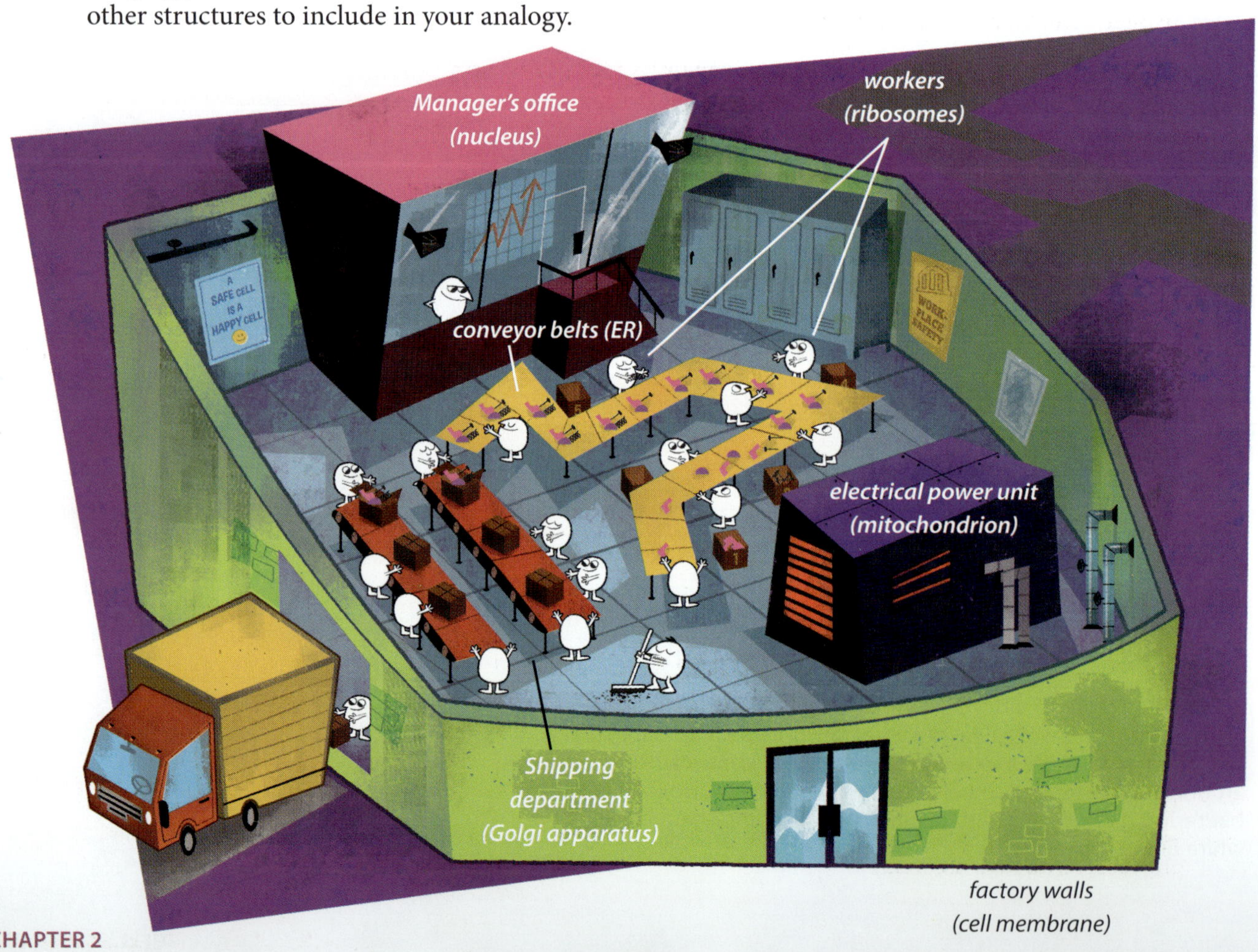

2.5 ORIGINS OF THE CELL

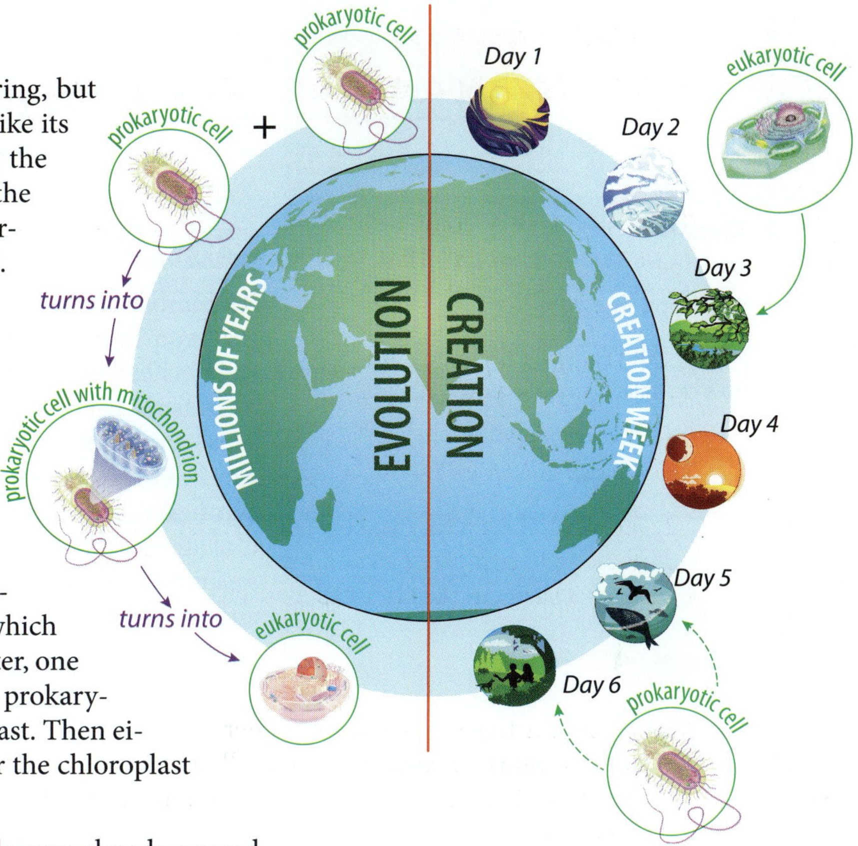

Remember the F-22? It's a marvel of engineering, but the cell is even more amazingly designed. It's like its own little country. The cell membrane keeps the right chemicals in the cytoplasm and keeps the wrong ones out. The DNA provides the information for many important parts of the cell. Ribosomes act like factories, using the information originating from DNA to produce proteins. The ER carries many of those proteins much like a subway system. The Golgi apparatus performs the work of a processing plant or the postal service. And the cytoskeleton forms the train tracks of the cell.

Eukaryotic cells and prokaryotic cells are two distinct groups. One current evolutionary theory is that one prokaryote engulfed another, which eventually became the first mitochondrion. Later, one of the new eukaryotic cells engulfed a second prokaryote, which ultimately became the first chloroplast. Then either the cell membrane, the mitochondrion, or the chloroplast produced the ER and Golgi apparatus.

Of course, all this assumes that prokaryotic cells were already around as a result of life evolving from nonliving materials. Although this has never been observed—and a method of making life this way is not known—the theory of evolution depends on it.

The Bible indicates that God created prokaryotic and eukaryotic cells during Creation Week. The first eukaryotic cells were plant cells created on Day 3. Although the Bible doesn't mention any prokaryotic organisms, they were probably created either on Day 5 or Day 6. Those first eukaryotic and prokaryotic cells would have looked much like the cells we see today.

2B REVIEW QUESTIONS

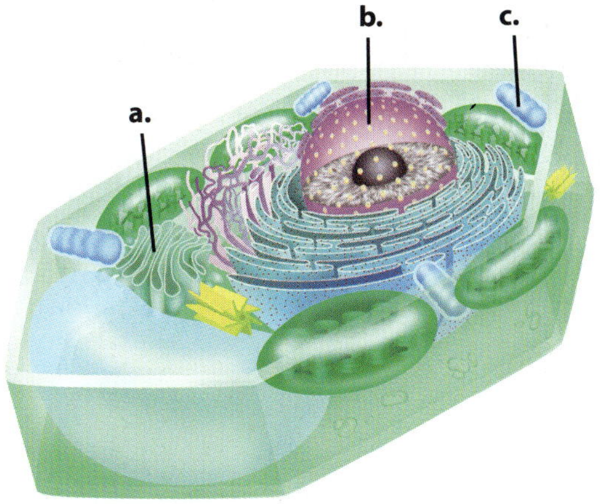

1. Identify each labeled organelle in the drawing to the right.
2. Indicate the function of each structure below.

   a. ribosome **b.** capsule **c.** mitochondrion
3. Cells use sodium ions to signal other cells and to control the amount of water in the cell. Most sodium ions are on the outside of the cell, but some manage to move into the cytoplasm. This is an example of what process?
4. The cell uses energy to move sodium ions from the cytoplasm to the outside of the cell. This is an example of what process?
5. A plant is placed in corn syrup, and after a day the plant has wilted and lost weight. What process accounts for this? Explain your answer.
6. How would a creationist and an evolutionist answer the following question: Did eukaryotic cells or prokaryotic cells come first? What would be the reasons for their answers?

2C

CELL FUNCTION

How does a cell power its processes?

2C Questions

- How does a cell get energy?
- What is photosynthesis?
- What is cellular respiration?
- How do cells turn energy into ATP?

2.6 PHOTOSYNTHESIS

Remember active transport? It's just one of many cellular processes that require energy. In fact, all organisms need energy to survive. Organisms called *consumers* eat other organisms to get energy. Other organisms, called **producers**, harness the sun's energy to make food. Most producers are either plants or algae. Ultimately, all organisms get the energy they need from the sun through the process of **photosynthesis**. Photosynthesis uses energy from the sun to turn carbon dioxide and water into glucose, a type of sugar. This sugar is one of several important *carbohydrates*, molecules made of carbon, oxygen, and hydrogen.

PHOTOSYNTHESIS

HARNESSING THE SUN'S ENERGY

1. Sunlight enters the chloroplasts and is absorbed by the green pigment **chlorophyll**. This is called the *light-dependent phase*.
2. The chlorophyll uses the energy from the sun to split water into protons, electrons, and oxygen.
3. The protons and electrons enter the *light-independent phase*, the second stage of photosynthesis.

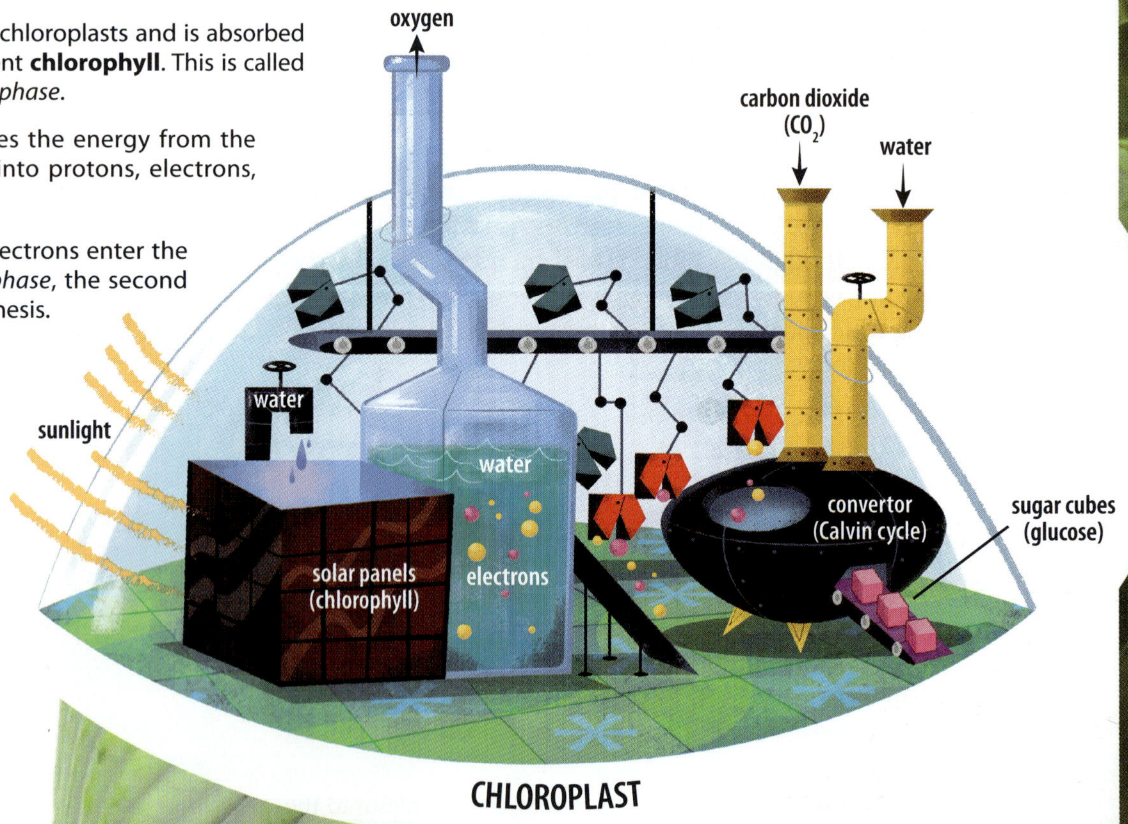

4. During the light-independent phase, the electrons and protons are used to convert carbon dioxide and water into glucose, a sugar that cells use as food.

The entire process of photosynthesis can be represented as shown below.

carbon dioxide + water + light energy = sugar + oxygen

2.7 CELLULAR RESPIRATION

A plant uses the glucose that it makes during photosynthesis for its own energy needs. A lot of energy is stored in the chemical bonds of a glucose molecule, but it's not easy to use. If all the energy in one molecule of glucose were released in a cell at once, it would destroy the cell. So eukaryotic cells don't use glucose directly. Instead, they use the energy within a glucose molecule to make around thirty molecules of a different energy molecule called **ATP**. ATP is the "energy currency" used to power the active transport proteins and many of the other processes in the cell that need chemical energy. When chemical energy is needed for a process, ATP can transfer some of its energy to other molecules such as proteins. In the process, ATP becomes ADP. The cell then uses the energy from glucose molecules to convert ADP back into ATP. The process of breaking down molecules to produce ATP is called **cellular respiration**.

Bacterial Cellular Respiration

Although prokaryotic cells do not have mitochondria, they can still perform cellular respiration. The entire process occurs in the cytoplasm of the bacterium.

CELLULAR RESPIRATION

HARNESSING GLUCOSE'S ENERGY

1. Glucose arrives in the cytoplasm and is broken down into fragments. This process, called *glycolysis*, produces a small amount of ATP. Because glycolysis does not use oxygen, it is an **anaerobic** process.

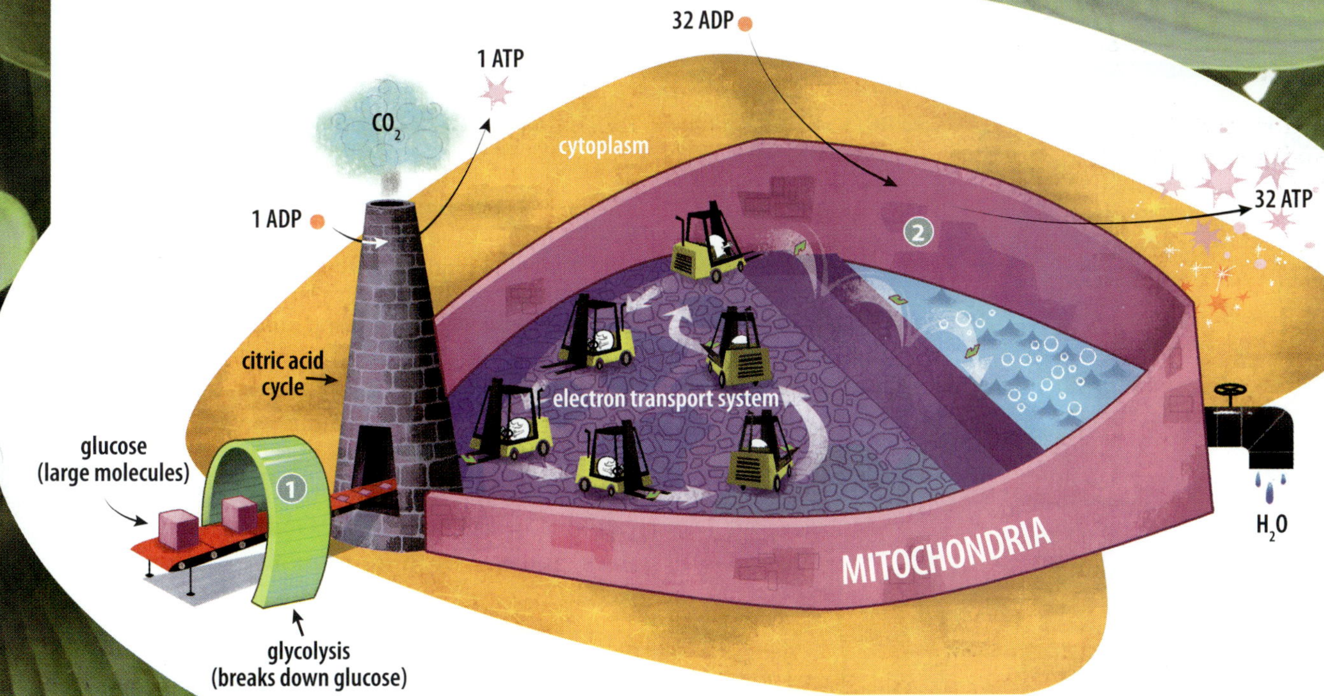

2. The glucose fragments are carried into the mitochondria where they are used along with oxygen to produce a large amount of ATP. Water and carbon dioxide are released as waste products. Because the process uses oxygen, it is called *aerobic respiration*.

As you saw on the previous page, aerobic respiration uses oxygen. When you breathe, oxygen flows into your lungs and is carried by your blood to all the cells in your body to work in cellular respiration. But some cells live in places with very little oxygen. How do they survive? They make ATP using a process called *fermentation*.

FERMENTATION

Fermentation occurs in two forms, depending on the cell. Yeast produce ATP through glycolysis just like other eukaryotes. But then they must have a way to eliminate the glucose fragments that are left over. To do this they transform the glucose fragments into alcohol and carbon dioxide, which are released from the cell. This process is used in making bread. When yeast break down the sugar in the bread dough, the carbon dioxide that they release causes the bread to rise.

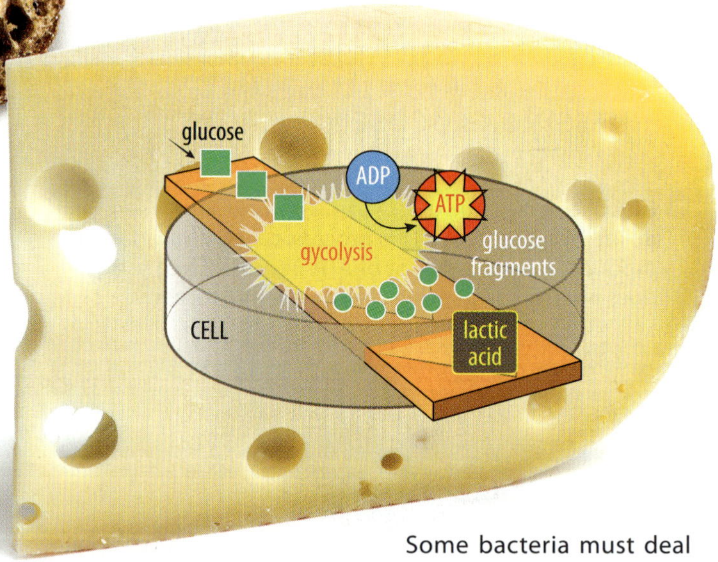

Some bacteria must deal with the same problems as yeast. But instead of turning the glucose fragments into carbon dioxide and alcohol, they produce *lactic acid* from the fragments. Bacteria that produce lactic acid this way are used in part of the cheese-making process.

So plants use carbon dioxide and water to make glucose and oxygen. Glucose and oxygen are used by both plants and animals to make ATP, releasing carbon dioxide and water in the process. This is one example of the cycles that God has established in nature. You will learn more about these cycles in Chapter 19.

As we look around, we see God's amazing provision for His creatures. Plants perform photosynthesis, providing food for themselves and other creatures. Animals are equipped to use the products of photosynthesis to power their cells. The earth shows the truth of Isaiah 45:18: "God himself that formed the earth and made it; he hath established it, he created it not in vain, he formed it to be inhabited."

DRUG DETECTIVES: SERVING AS A PHARMACOLOGIST

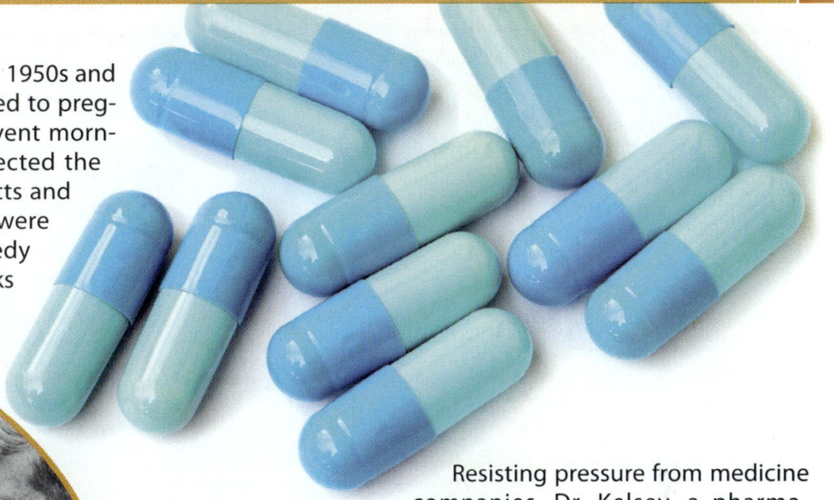

Cellular functions are affected by drugs. In the 1950s and 1960s, a drug called *thalidomide* was marketed to pregnant women to help them sleep and to prevent morning sickness. However, thalidomide also affected the unborn babies, resulting in severe birth defects and many deaths. Approximately 10,000 babies were injured by the drug in Europe. But this tragedy was largely averted in the United States thanks to the efforts of Dr. Frances Oldham Kelsey.

Resisting pressure from medicine companies, Dr. Kelsey, a pharmacologist with the FDA, demanded that experiments be performed to test the safety of thalidomide. These tests showed the dangers of the drug, which was eventually banned in the country.

Some pharmacologists, such as Dr. Kelsey, work in regulatory agencies to make sure that medicines are safe before they can be sold to people. Others develop new medicines to treat diseases, while some test these new medicines to see whether they work and are safe for people. Other pharmacologists study how drugs affect the body.

Dr. Frances Oldham Kelsey was awarded the President's Award for Distinguished Federal Civilian Service from President Kennedy in 1962.

2C REVIEW QUESTIONS

1. Photosynthesis uses sunlight to make glucose from ______.
2. Label each substance as a waste product of photosynthesis, cellular respiration, or both.
   a. oxygen
   b. carbon dioxide
   c. water
3. The energy molecule of the cell is ______.
4. In which part of the cell is most of the ATP produced during cellular respiration?
5. Which phase of cellular respiration also occurs in fermentation?
6. Why would plants be unable to carry on cellular respiration for long if photosynthesis ceased to occur?

Use the Ethics box on pages 42–43 to answer Questions 7–8.

7. What might happen in Step 2 if Step 1 is not answered thoroughly?
8. Why is Step 5 necessary?

Do you remember the discussion of ethics in Chapter 1? If you are a little rusty, you should review pages 6–8 because we are going to study a strategy for evaluating the ethics of issues in life science. Then we will apply this strategy to evaluate stem cell research.

THE STRATEGY

Strategies are helpful because they provide a way to deal with the complexities of ethics and because strategies help people organize their thoughts. This strategy is organized into five steps.

❶ *What information can I get about this issue?*

Often you may need to do some research on a topic before evaluating whether it is right or wrong. For this exercise, you can find the needed information below.

❷ *What does the Bible say about this issue?*

Study the Bible to see what it says about this issue. Sometimes God's Word directly addresses an issue. Many times, you have to study the Scriptures to learn how its principles apply to the issue. Consider the biblical principles of the image of God in man, the Creation Mandate, and God's whole truth.

❸ *What are the acceptable and unacceptable options?*

When trying to solve an ethical dilemma, you must look at all the options. Some of these potential solutions may not be obvious, while some may be talked about by almost anyone discussing the issue. Using the biblical teaching and principles that you've considered in Step ❷, decide which of the options are acceptable from a biblical point of view. Discard those that are biblically unacceptable.

❹ *What may be the consequences of the acceptable options?*

Remember the biblical outcomes of human flourishing, thriving creation, and God's glory? Some of the acceptable options may have consequences that fit with the biblical outcomes. Others may not. Reject the options that have consequences that are inconsistent with human flourishing, a thriving creation, and God's glory.

❺ *What action should I take?*

Suggest a course of action or form an opinion. Be prepared to justify the action that you suggest on the basis of the analysis in the previous steps.

THE ISSUE: STEM CELL RESEARCH

1. What information can I get about this issue?

Multicellular organisms have to produce a lot of different types of cells. To perform this task, multicellular organisms have special cells called **stem cells**. Stem cells don't look like specialized cells such as skin cells or muscle cells. But when the body needs more specialized cells, the stem cells divide and mature into the type of cells needed. For instance, if a damaged muscle needs new muscle cells, stem cells in the muscle divide. Some of the stem cells turn into new muscle cells that begin to repair the injured muscle.

Doctors and scientists are investigating ways of removing stem cells from a patient's or donor's body and using them to create replacement body parts. This is currently done to treat one form of blood cancer. Other treatments are being researched.

But this type of adult stem cell treatments has limitations. A stem cell from a muscle cannot become a skin cell or a nerve cell. Adult stem cells also do not last long in the laboratory.

Some scientists experiment with embryonic stem cells. These stem cells can become almost any other type of cell and can be kept in a lab for years. But there is one major problem. Embryonic stem cells come from human embryos; experimenting with embryonic stem cells *requires the death of the embryo*, meaning that scientists take a human life to do their work.

2. What does the Bible say about this issue?

There are actually two important ethical issues with stem cell research. (1) Developing new treatments for diseases recognizes the innate value of human beings as bearers of God's image. In Mark 3:4, Jesus said that it is always lawful to do good. (2) The Bible also clearly condemns killing, most famously in Exodus 20:13: "Thou shalt not kill." Each person is a human being from conception (Ps. 139:13–16), and killing human embryos violates that principle.

3. What are the acceptable and unacceptable options?

On the basis of the biblical principles and passages, adult stem cell research is an acceptable option. Treatments for diseases such as SCID, spinal cord injuries, and MS might be developed from this research. These treatments would greatly benefit people who suffer from these diseases. On the other hand, biblical principles make embryonic stem cell research unacceptable. Killing an embryo denies that soul the respect due to him as fully human, and, of course, violates the sixth commandment. Developing new treatments for sick people honors God's Word, but killing embryos is forbidden by Scripture.

4. What may be the consequences of the acceptable options?

The consequences of adult stem cell research may be the treatment or even cure of diseases. This would certainly increase human flourishing. Christians have the opportunity to give greater glory to God as new ways to use His glorious creation are explored.

5. What action should I take?

Biblical principles all suggest that adult stem cell research should be pursued. However, embryonic stem cell research violates God's Word because it involves taking a human life. You can educate others on the problems of embryonic stem cell research. Many charities collect money for research for treatments of diseases; it would be wise to carefully research charities before donating to them to ensure that they do not support embryonic stem cell research. Finally, you can contact your elected officials and ask them to pass laws that encourage adult stem cell research and ban embryonic stem cell research.

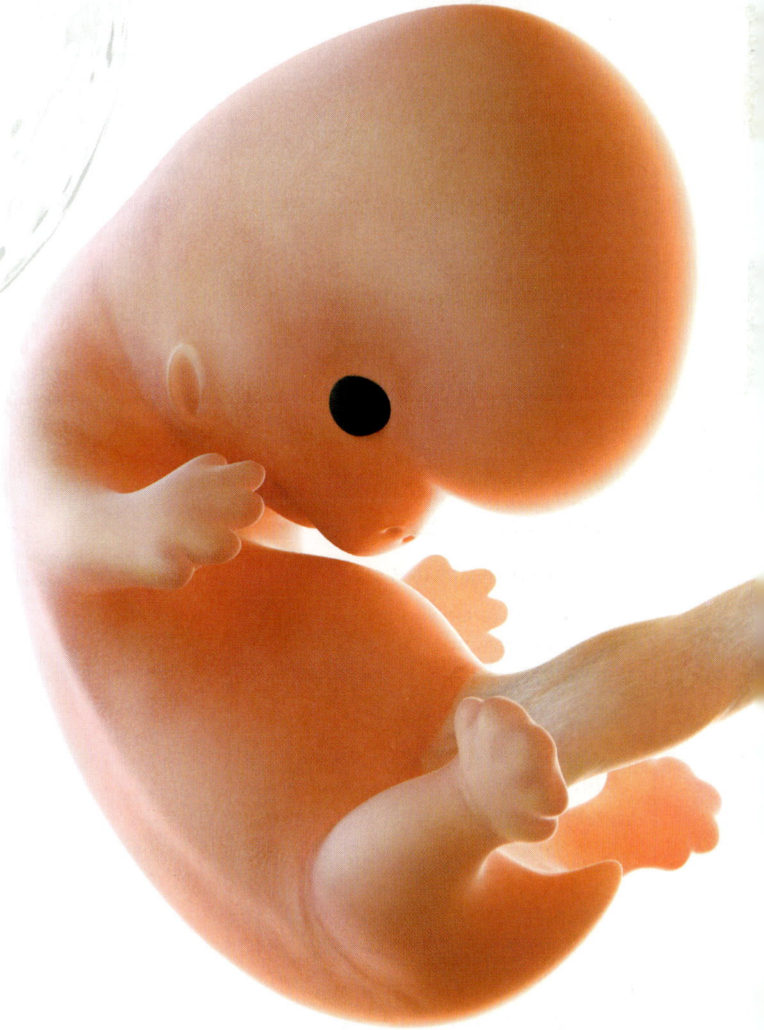

CHAPTER 2

CHAPTER SUMMARY

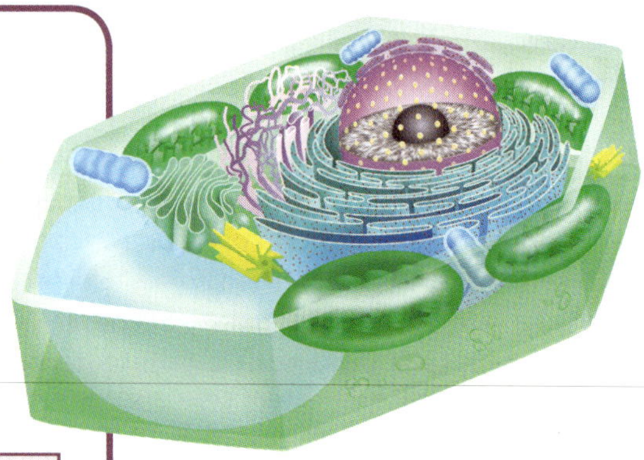

2A CELLS AND LIFE

- The cell theory states that all cells come from other cells, make up all living things, and perform the functions of all living things.
- The cell theory is a model that explains many observations but may change in the event of new discoveries.
- Cells with nuclei are eukaryotic; cells without nuclei are prokaryotic.
- Unicellular organisms have only one cell. Multicellular organisms have many cells organized into tissues and organs.
- The cells of multicellular organisms are grouped into tissues made of similar cells.

2A TERMS

cell theory	29
tissue	31
organ	31

2B CELL STRUCTURES

- Eukaryotic cells have structures called organelles.
- Each organelle performs a specific task within the cell.
- Cell membranes regulate what substances enter and exit a cell through passive and active transport.
- God created both eukaryotic and prokaryotic cells during Creation Week.
- Cells give evidence of being engineered systems.
- Evolutionists have not yet come up with a model that can explain the origin of life from nonliving materials.

2B TERMS

organelle	32
cell membrane	32
cell wall	32
capsule	32
cytoplasm	32
flagellum	32
DNA	32
chromosome	32
endoplasmic reticulum	33
Golgi apparatus	33
ribosome	33
cytoskeleton	33
lysosome	33
enzyme	33
mitochondrion	33
selectively permeable	34
concentration gradient	34
diffusion	34
osmosis	35
passive transport	35
active transport	35

2C CELL FUNCTION

- Photosynthesis uses light energy to turn water and carbon dioxide into glucose and produces oxygen as a waste product.
- Cellular respiration uses glucose and oxygen to convert ADP into ATP, the energy molecule of the cell, with water and carbon dioxide as waste products.
- Some cells perform fermentation, which does not use oxygen, instead of aerobic respiration.

2C TERMS

producer	38
photosynthesis	38
chlorophyll	38
ATP	39
cellular respiration	39
anaerobic	39
stem cell	42

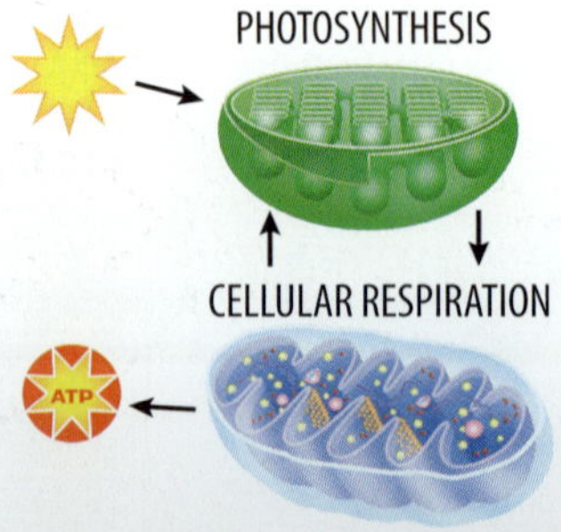

CHAPTER REVIEW QUESTIONS

Recalling Facts

1. Is the organism shown on the right made of prokaryotic or eukaryotic cells? Explain.
2. You want to determine whether a cell under a microscope is eukaryotic or prokaryotic. What structure should you look for?
3. List five organelles found in eukaryotic cells along with their functions.
4. Water enters the cell by a process called _____.
5. According to a common evolutionary model, how did mitochondria come about?
6. Photosynthesis takes energy from _____ and stores it in _____.
7. Indicate whether each substance listed below is needed for photosynthesis or cellular respiration.

   a. carbon dioxide **b.** water **c.** oxygen
8. When ATP binds to an active transport protein, it powers the protein, and the ATP becomes _____.
9. Cellular respiration takes energy from _____ and temporarily stores it in _____.

Understanding Concepts

10. Which statement, if true, would cause the cell theory to be changed?
    - **a.** A bacteria split into three separate cells.
    - **b.** A microscopic organism was discovered that was composed entirely of silicon fibers.
    - **c.** An algae digested a protein inside a special mitochondrion.
11. Would you expect a bone cell to look like a nerve cell? Explain.
12. An organelle has stopped working, and now the cell's ATP supply is decreasing. What organelle most likely has stopped?
13. A cell needs a protein carried in from outside its cell membrane. There are already more protein molecules inside the cell than outside and the protein is too big to diffuse through the cell membrane. How will the protein most likely enter the cell?
14. You are studying a group of organisms that live in a cave where the air is almost entirely carbon dioxide. Would you expect these organisms to perform aerobic respiration or fermentation? Explain.

For Questions 15–20, use the letter for each element below to fill in the graphic organizer.

a. sunlight **c.** oxygen **e.** ATP

b. glucose **d.** water **f.** carbon dioxide

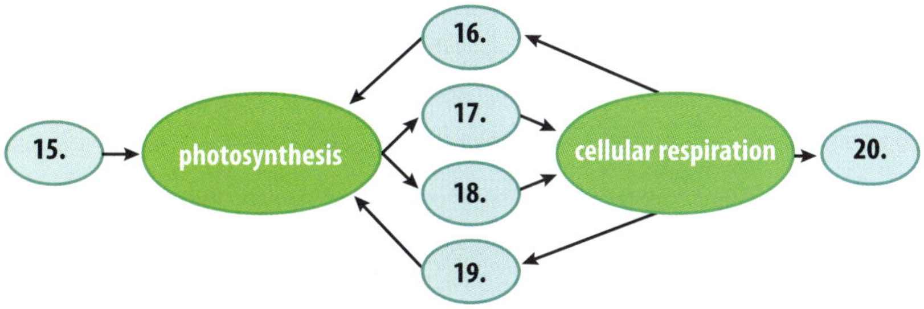

Critical Thinking

For Questions 21–23, indicate the name of the structure whose failure would cause the problem described.

21. A cell cannot produce proteins.

22. An algal cell cannot distribute proteins throughout the cell.

23. A bacterial cell dries up.

Four bags made of a membrane similar to a cellular membrane are filled with different concentrations of a salt solution. Each bag is placed in a beaker of water. Use this information and the table below to answer Questions 24–26.

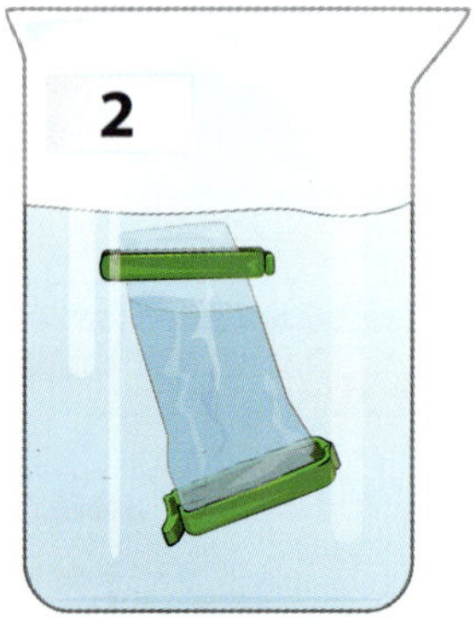

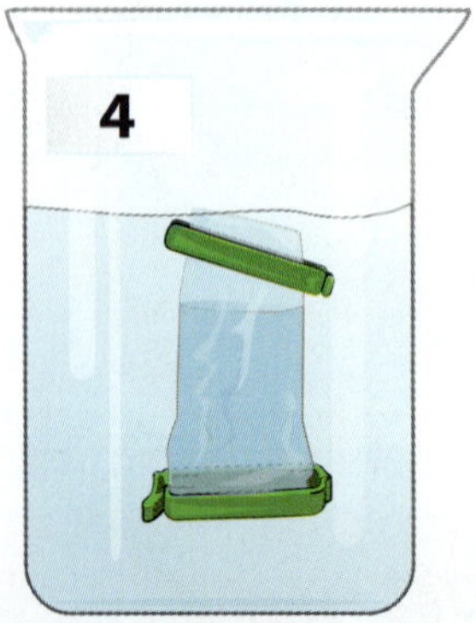

Table 2-1

Time to Bursting Through Osmosis

Time (min)	Volume (mL)			
	Bag 1	Bag 2	Bag 3	Bag 4
0	10	10	10	10
2	12	13	9	14
4	14	16	8	18
6	15	18	7	bag burst
8	16	19	7	
10	16	bag burst	7	

24. Which bag had the highest concentration of salt solution? Explain.

25. A new lab assistant bungled the procedure. He remembers pouring the water and salt solution but he doesn't remember which went into Bag 3 and which went into the beaker. What is the most likely explanation?

26. Another two bags of salt solution are placed in beakers of water. The concentration of the salt solution in Bag 1 was twice the concentration of the salt solution in Bag 2. During the first two minutes, the volume in Bag 2 increased by 2 mL. Which line in the graph below most likely describes the increase in volume in Bag 1 during that same two minutes? Explain.

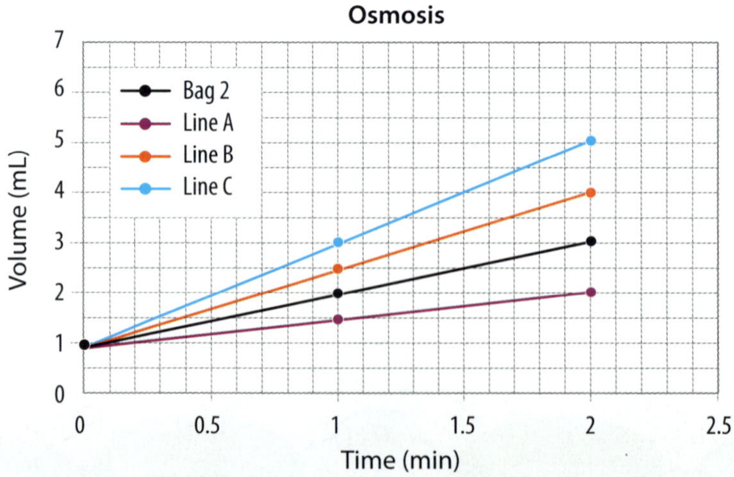

27. How does Genesis 9:6 apply to the issue of embryonic stem cell research?

INFORMATION IN THE CELL 3

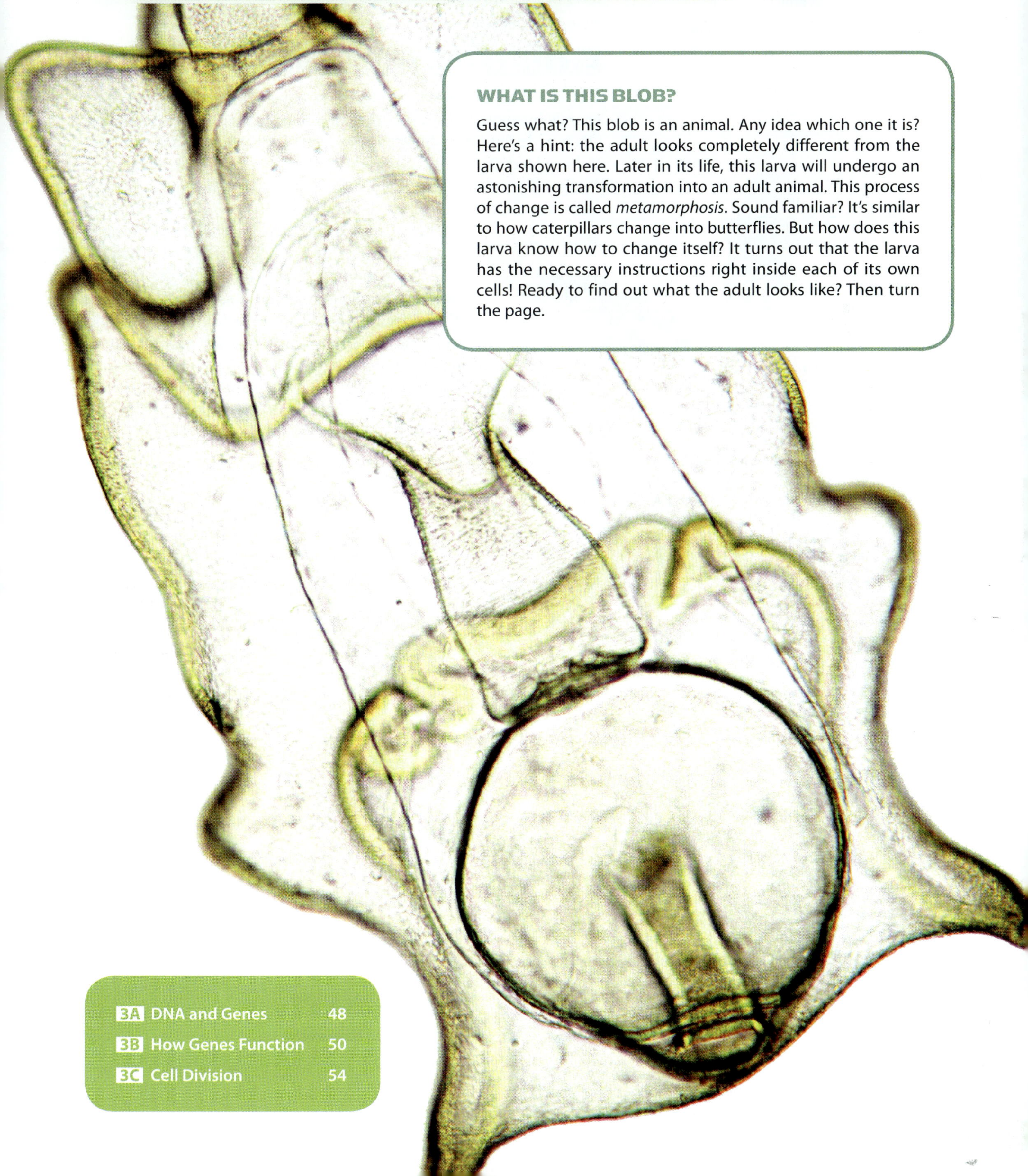

WHAT IS THIS BLOB?

Guess what? This blob is an animal. Any idea which one it is? Here's a hint: the adult looks completely different from the larva shown here. Later in its life, this larva will undergo an astonishing transformation into an adult animal. This process of change is called *metamorphosis*. Sound familiar? It's similar to how caterpillars change into butterflies. But how does this larva know how to change itself? It turns out that the larva has the necessary instructions right inside each of its own cells! Ready to find out what the adult looks like? Then turn the page.

3A Questions

- Why do cells need information?
- What does DNA look like?
- How does the structure of DNA hold information?

3A

DNA AND GENES

How do cells store information?

3.1 THE NEED FOR INFORMATION

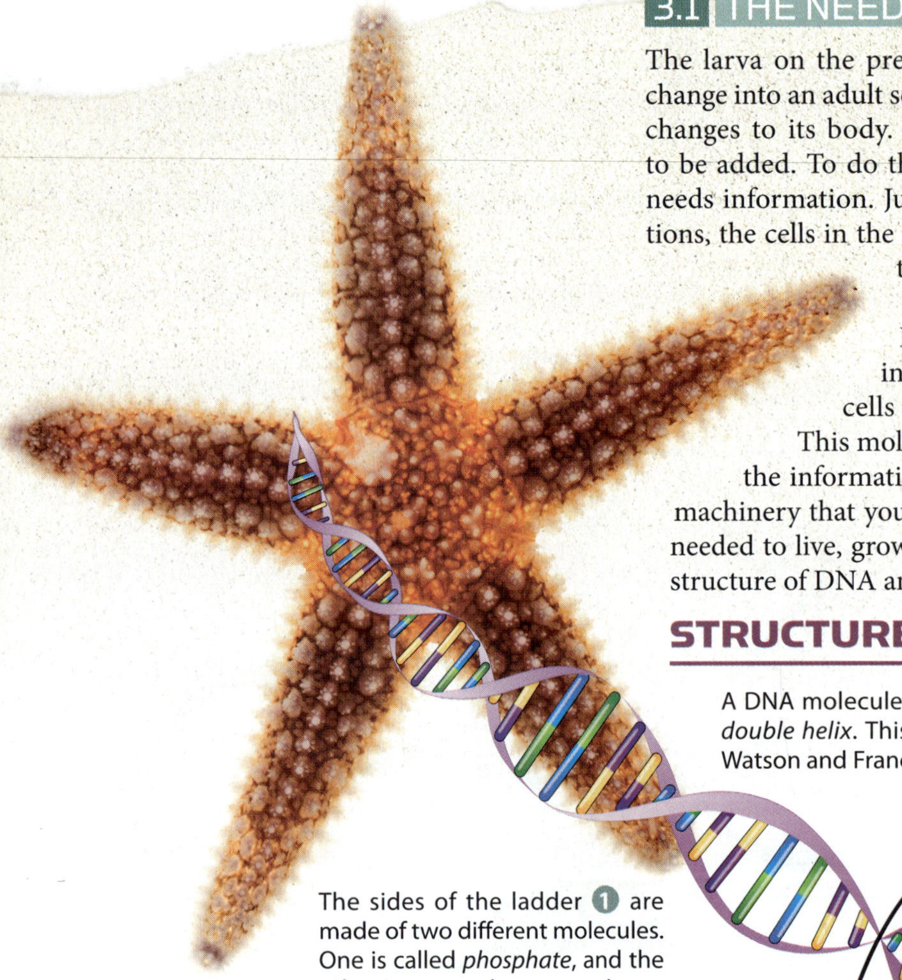

The larva on the previous page is a baby sea star! Surprised? To change into an adult sea star, the larva needs a plan for making major changes to its body. New types of cells with new functions have to be added. To do this, the sea star needs a certain something. It needs information. Just like people who build things need instructions, the cells in the sea star larva need instructions too. If it could think, the sea star larva might be asking itself, "What kinds of cells do I need? And how do I make them?" When God made the first living things, including sea stars, He built into their cells a special kind of information-rich molecule. This molecule, DNA, is a kind of *nucleic acid* that stores the information needed to make all of the complex cellular machinery that you learned about in Chapter 2—machinery cells needed to live, grow, and reproduce. Let's take a closer look at the structure of DNA and see how it stores information.

STRUCTURE OF DNA

A DNA molecule looks like a twisted ladder, a shape called a *double helix*. This structure was first suggested by James D. Watson and Francis H. C. Crick in 1953.

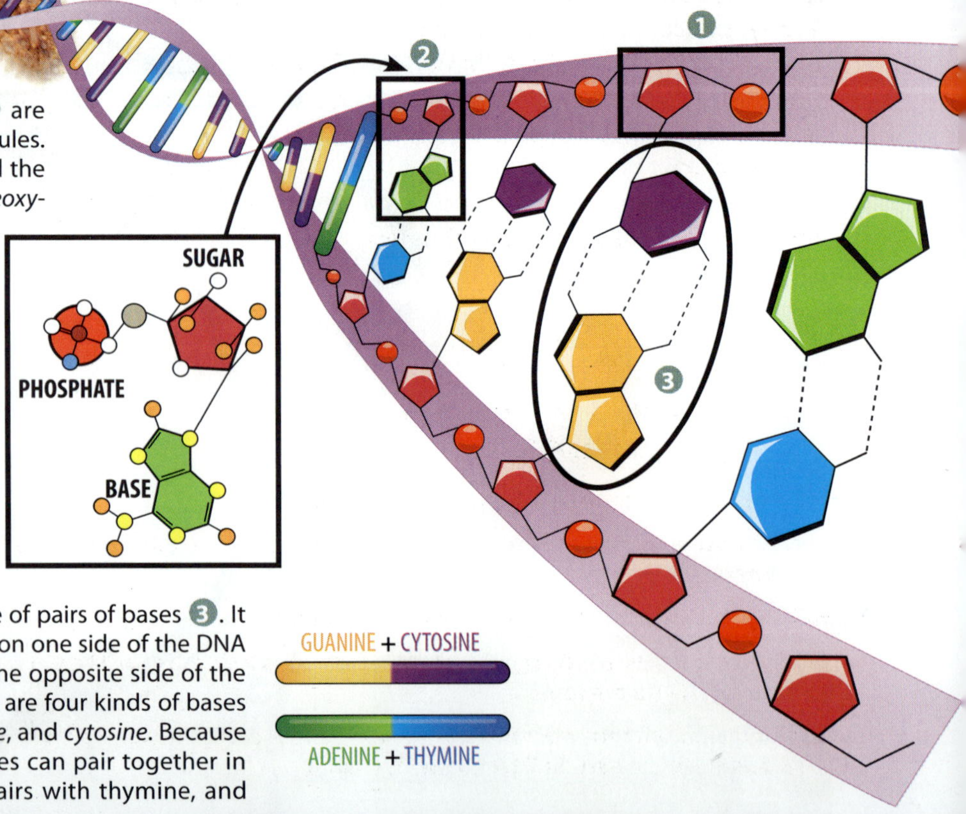

The sides of the ladder (1) are made of two different molecules. One is called *phosphate*, and the other is a sugar known as *deoxyribose* (dee AHK see RI BOS).

The long DNA molecule is made of many smaller subunits known as **nucleotides** (2). Each nucleotide consists of a sugar, a phosphate, and a base. The DNA in each of your cells consists of about 6 *billion* nucleotides!

The "rungs" of the DNA ladder are made of pairs of bases (3). It should come as no surprise that a base on one side of the DNA molecule coupled with its partner on the opposite side of the molecule is known as a *base pair*. There are four kinds of bases found in DNA: *adenine*, *thymine*, *guanine*, and *cytosine*. Because of their chemical structures, these bases can pair together in limited ways only—adenine always pairs with thymine, and cytosine always pairs with guanine.

3.2 THE HIDDEN MEANING OF DNA

Now that you've seen the structure of DNA, let's look at how that structure stores information. But first, think about the words that you're reading on this page. How do they give us information? Each letter or combination of letters in a word means something to us. If we read the letters *c*, *a*, and *t* in that order—*cat*—then that word has a meaning, in this case a small, furry animal often kept as a pet. If we scramble the letters in the word *cat* so that they spell *tca*, then the word loses its meaning—it provides no information. The information in a word depends both on which letters are used to make the word and in what order the letters are placed. Sentences work the same way—the words used and the order of the words determine the meaning of the sentence.

DNA works much like words on a page. Different nucleotides are arranged in differing orders to convey the information that cells need to build proteins. Each section of a DNA molecule that codes for a unique protein is called a **gene**. Which protein gets built depends on which nucleotides are present in the gene and what order they're in. One possible arrangement of nucleotides might produce one kind of protein. Scrambling the order of those same nucleotides might produce a different protein—or no protein at all. It's almost as simple as that!

3-1
Pondering the origins of design

3A REVIEW QUESTIONS

1. What is the information in DNA used for?
2. Describe the shape of DNA's double helix.
3. Name the four bases found in DNA nucleotides.
4. Only two possible base pairings exist within a DNA molecule. What are they?
5. A section of DNA that codes for one kind of protein is a(n) __________.
6. What two features of a gene determine what kind of protein will be built using its information?
7. If the order of the nucleotides within a gene is changed, will the information in the gene still be useful? Explain.

3B Questions

- How does a cell "read" DNA?
- How are proteins made?

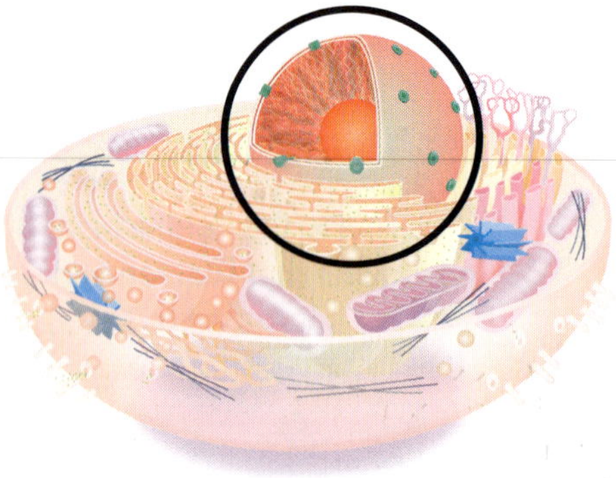

3B

HOW GENES FUNCTION

How do cells use information?

3.3 CODED MESSAGES

Transcription

Remember from Chapter 2 that DNA is found in the nucleus of a eukaryotic cell. But that's not where proteins are built. Proteins are built by ribosomes found in the cytoplasm. So how does a ribosome read the DNA in the nucleus? Actually, it doesn't! Just like a building contractor keeps a master set of plans in his office, a cell keeps a master set of DNA in its nucleus. A contractor sends copies of the master plans, called *working copies*, to the construction site for the builders to use. What the ribosomes need are working copies of the DNA that they can use to make proteins. The process cells use to do this is called **transcription**.

TRANSCRIPTION

MAKING A WORKING COPY

During transcription, an enzyme temporarily "unzips" part of the DNA molecule (1) in order to "read" a gene.

The enzyme reads the order of the nucleotides on the unzipped DNA strand. It uses that information to build a new but slightly different kind of nucleic acid (2) called **ribonucleic acid (RNA)**. The new molecule has only a single strand of nucleotides, contains the sugar ribose instead of deoxyribose, and in place of thymine has a different base called *uracil*. Since the RNA strand will carry its information out of the nucleus, it is called *messenger RNA* (mRNA).

(3) The mRNA-building enzyme adds nucleotides that pair with the original DNA nucleotides. For example, if a section of the original DNA reads cytosine-guanine-adenine, then the new mRNA section will be guanine-cytosine-uracil (remember, uracil replaces thymine in RNA).

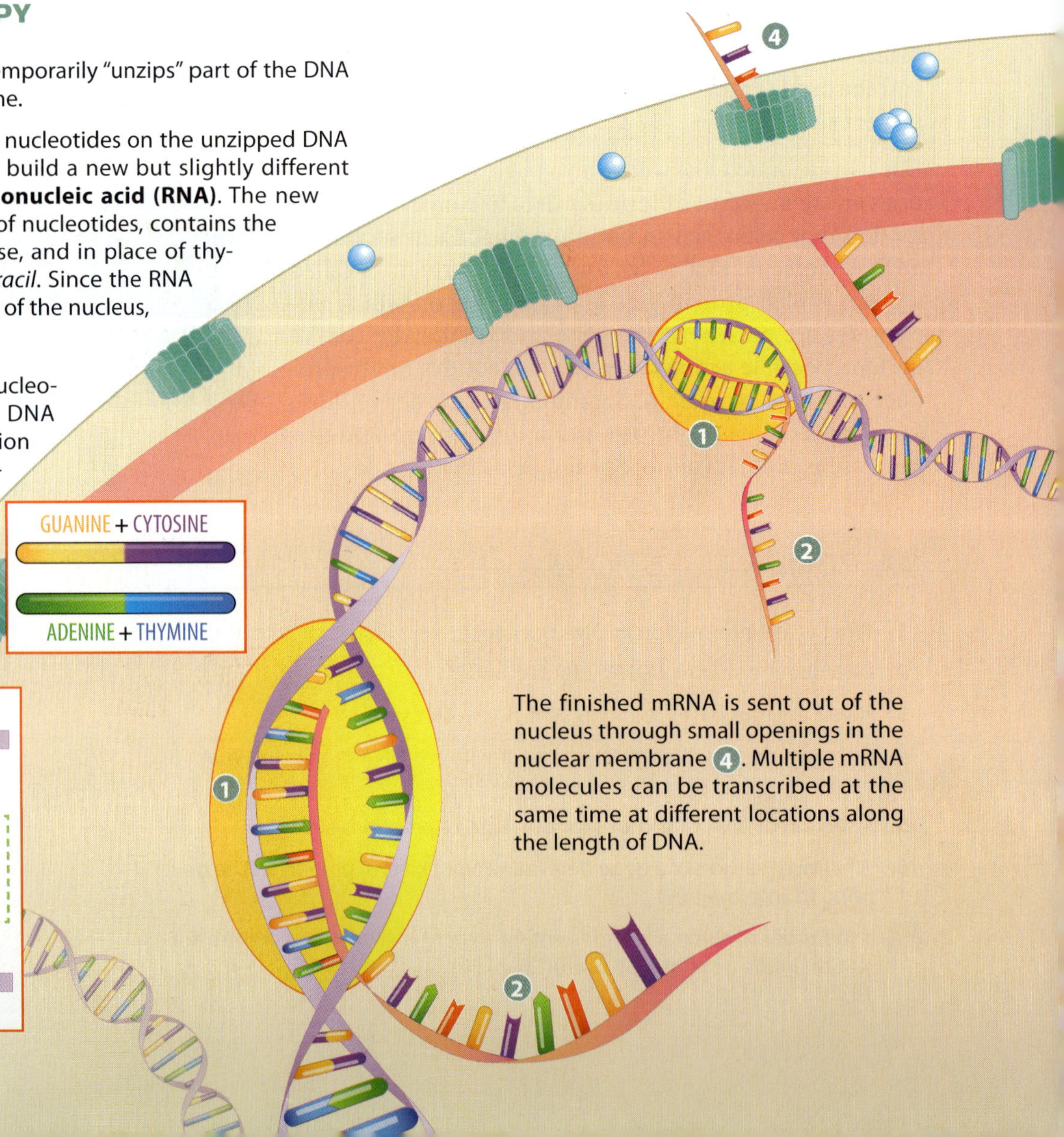

The finished mRNA is sent out of the nucleus through small openings in the nuclear membrane (4). Multiple mRNA molecules can be transcribed at the same time at different locations along the length of DNA.

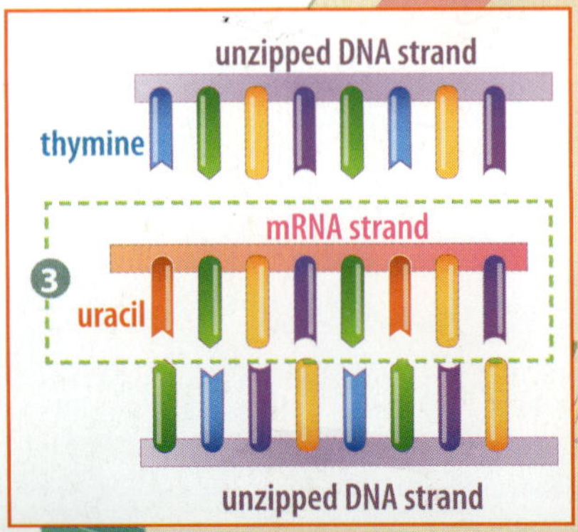

3.4 BUILDING A PROTEIN

Translation

DNA is a great molecule for storing information, but it's not useful for building the structures that cells need to live, grow, and reproduce. Most cellular structures are made of proteins. The genes in a DNA molecule are the instructions for building the different kinds of proteins that cells need. If DNA is like a coded message, then proteins are the decoded messages. Let's take a look at the process, called **translation**, of converting the DNA code into a protein.

TRANSLATION

The mRNA produced during transcription is read by a ribosome ①, which is also made of RNA, called *ribosomal RNA* (rRNA), along with some protein.

The "words" of the mRNA are sets of three nucleotides in a row called **codons** ②. Each "codes" for one of about twenty different **amino acids**, the molecules used to build proteins.

Amino acids are carried to the ribosome by yet another type of RNA called *transfer RNA* (tRNA) ③. Each has an **anticodon** on one end that matches a codon on the mRNA. At the other end is a site that holds the amino acid that is coded for by the mRNA codon. By finding the codons on the mRNA strand that match their anticodons, tRNAs line up their amino acids in the correct order to make the protein coded for by the mRNA.

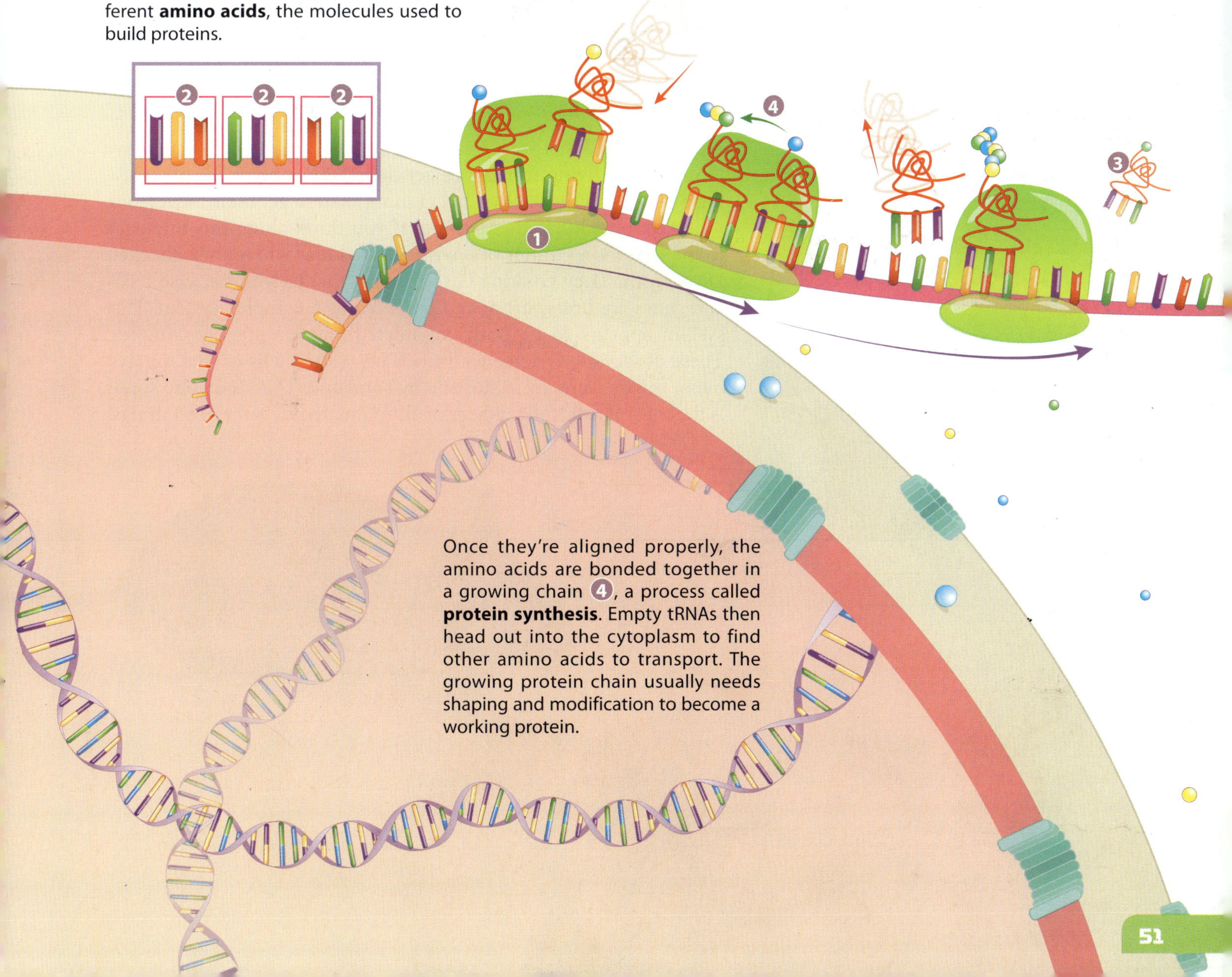

Once they're aligned properly, the amino acids are bonded together in a growing chain ④, a process called **protein synthesis**. Empty tRNAs then head out into the cytoplasm to find other amino acids to transport. The growing protein chain usually needs shaping and modification to become a working protein.

Engineered Systems

What do car factories make? Cars, obviously! But it's no accident that cars roll off the end of an assembly line at a car factory. The factory is *designed* to make cars! It's an example of an **engineered system**—a group of things that are planned, designed, and built to work together to do a certain job. The cars the factory makes are also engineered systems. Even parts within the car, like the engine or brakes, are engineered systems. Transcription and translation also appear to be engineered systems, as do the DNA and protein machines that carry out those processes. Engineered systems require engineers to design and build them. Car factories don't build themselves from nothing, and we should conclude that protein factories don't do so either.

Engineered systems often display another feature that shows that skill and knowledge were needed to make them: **irreducible complexity**. Irreducible complexity is a mouthful of syllables that simply mean that if one part of the system is removed or doesn't work, the entire system will not function. If a mousetrap, for example, is missing one of its parts, it won't catch mice. And what good is a mouse trap if it doesn't catch mice? Similarly, transcription and translation both require many parts working together to produce proteins. Take any of those structures out of the picture and the whole process will grind to a halt.

In the world that we humans can observe, only living things have DNA. Those living things can get their DNA only from other living things. And those other living things had to get their DNA from, well, yet some other living thing! If you think along this line back to the first living things, it's natural to wonder where they got their DNA from. They couldn't have made it themselves, because DNA and genes are both *part of* and *produced by* a complex system. That system appears to be both engineered and irreducibly complex. These observations agree with the Bible, which tells us that the first living things were, in fact, made by someone. That Someone surely is the most masterful engineer of all—God, the Creator of all things!

THE WONDER OF REGENERATION

Does every cell in every organism really have all the information needed to build the entire organism? Yes! In fact, some organisms can use this information to replace lost body parts. Sea stars are one of those organisms. Most sea stars can regrow an arm if it is lost. It's also common for an entire new sea star to grow from one or more arms if part of the sea star's central disc is included. A few kinds of sea stars can completely regenerate from just one arm, even without a part of the central disc. How does regeneration happen? Certain cells within the sea star, called *stem cells* (see page 42), have the ability to *differentiate* (change to become more specialized) into the different kinds of cells that a sea star needs. A human arm or leg cannot, of course, regenerate into a new body. That's because human adult stem cells are limited in their ability to differentiate, even though they have the instructions in their DNA to build any kind of cell. Because their ability to differentiate is limited, they cannot change into all the different kinds of cells that it would take to grow a complete human body.

3B REVIEW QUESTIONS

1. Copy the table below on your own paper. Compare transcription and translation by filling in the blanks of the table with the words *protein*, *nucleus*, *enzyme*, *ribosome*, *cytoplasm*, and *mRNA*.

	Transcription	**Translation**
Takes place in (cell location)		
Produces		
Nucleic acid "read" by		

2. Describe two ways that an mRNA molecule is different from a DNA molecule.
3. A group of three nucleotides that codes for an amino acid is called a(n) ______.
4. How does tRNA recognize where to place its amino acid in a protein being built?
5. Describe two ways that genes and their functions are similar to engineered systems.

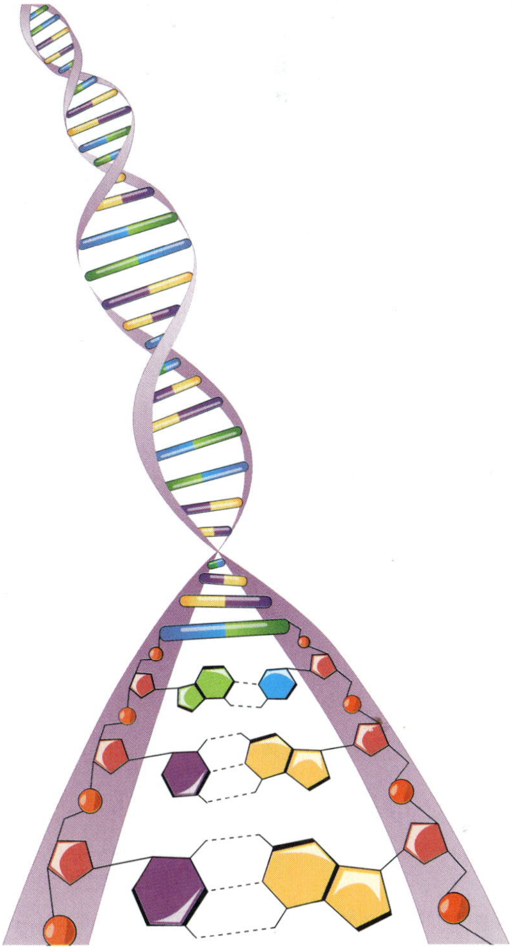

3C

CELL DIVISION

How do cells copy information?

3C Questions

- How do new cells get their own DNA?
- How do cells produce new cells?
- What is the difference between mitosis and meiosis?

3.5 THE CELL CYCLE

As most cells grow, they prepare to reproduce. The process of growth and reproduction is called the **cell cycle**. The cell cycle can be broken down into three stages: interphase, mitosis, and cytokinesis. How long the cell cycle takes to complete varies for different kinds of cells and organisms. It can be as short as several minutes or as long as a year.

Cells spend most of their lives in the **interphase** portion of the cell cycle, growing and carrying out their normal functions. When it's time to divide, the cell will need to make a copy of its DNA so that each new cell will have a set. We learned in the previous section how cells make proteins. Next we need to look at how the cell copies its DNA.

3.6 DNA REPLICATION

Scientists call the DNA-copying process **replication**. Knowing what a DNA molecule looks like helps us to understand how it makes copies of itself.

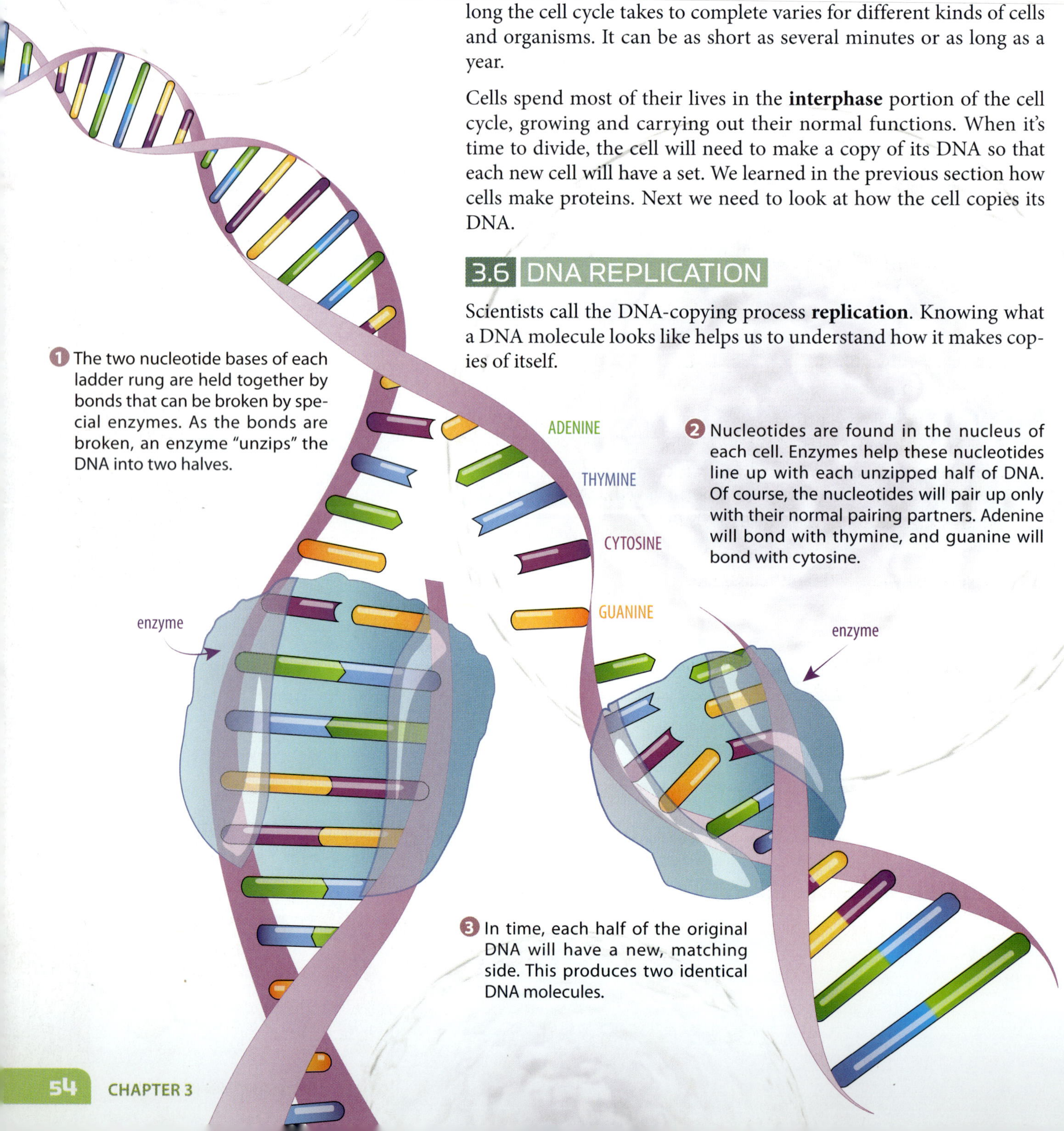

1 The two nucleotide bases of each ladder rung are held together by bonds that can be broken by special enzymes. As the bonds are broken, an enzyme "unzips" the DNA into two halves.

2 Nucleotides are found in the nucleus of each cell. Enzymes help these nucleotides line up with each unzipped half of DNA. Of course, the nucleotides will pair up only with their normal pairing partners. Adenine will bond with thymine, and guanine will bond with cytosine.

3 In time, each half of the original DNA will have a new, matching side. This produces two identical DNA molecules.

LIGHTS! CAMERA! ACTION! VISUALIZING DNA REPLICATION

How is DNA copied?

Have you ever thought of directing your own movie? Technology has made that possibility a reality for anyone with access to a smartphone. For this lab activity, you'll need a smartphone that has a stop-motion app loaded onto it. *Stop-motion animation* is a type of photography that consists of shooting a series of photos. In between the taking of each photo, the objects being photographed are moved slightly. When the photos are viewed quickly one after another, the objects appear to be moving.

E

Equipment

smartphone, tablet, or camera

craft supplies such as tape, colored paper, building blocks, clay, string, yarn, or other materials

tripod and mount

PROCEDURE

1. Obtain a smartphone, tablet, camera or some other device that has stop-motion video capability, familiarize yourself with how to use it, and mount it on a tripod.
2. Collect materials for depicting a strand of DNA and its nucleotides. There are lots of ways you could do this! Individual nucleotides can be indicated by paper cutouts, clay, felt, bits of yarn, toy building bricks, or even colored markers used on a whiteboard. Be creative!
3. Starting with a strand of DNA, use stop motion to show an enzyme unzipping the strand and new nucleotides joining the unzipped sides to create two new DNA strands.
4. Share your video with your classmates.

GOING FURTHER

1. Think about the video that you created. In what way(s) is it a model of the DNA replication process?
2. How could you improve the ability of your video model to explain the DNA replication process?

3.7 CELL DIVISION

After a cell is done copying its organelles and DNA, it's ready for the next phase of the cell cycle—**mitosis**. Mitosis ensures that each new *daughter cell* produced will have the same number and kinds of genes as those in the original *parent cell*. Multicellular organisms replace dead or worn-out cells by mitotic cell division. Some of your skin cells, for example, may go through this process several times a day to replace dead cells that get rubbed or washed off. Mitosis is also needed for growth. The trillions of cells that make up your body are the result of repeated cell divisions.

THE CELL CYCLE

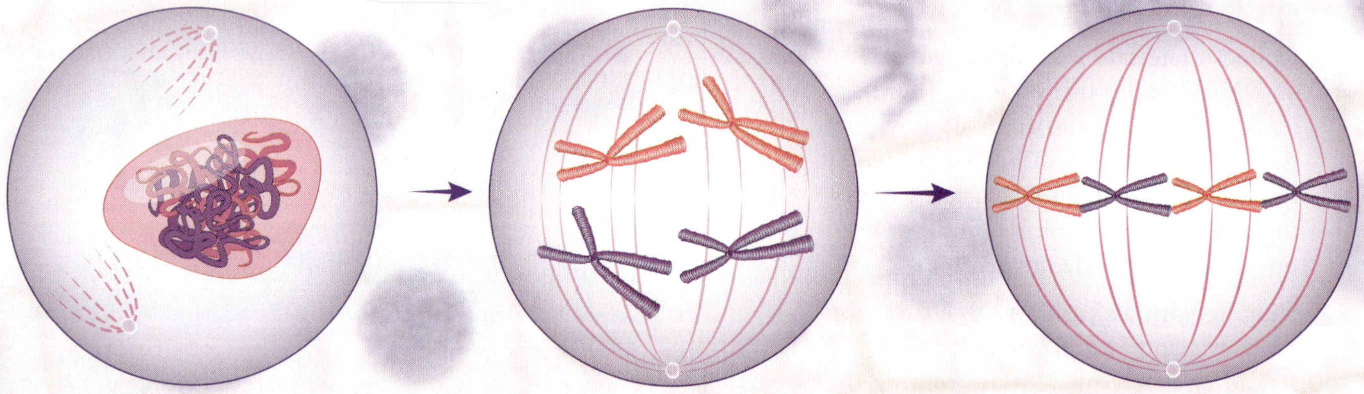

Normal cell growth and function take place during interphase. Toward the end of this phase, the cell copies its DNA as it prepares for mitosis. During this time, the DNA material, known as *chromatin*, exists as long, thin strands. Each separate strand is a chromosome. If you look at a cell under a microscope during this phase, the chromosomes look like a fuzzy, tangled mass inside the nucleus. It may look messy, but the chromosomes are actually highly organized structures.

In *prophase*, the membrane around the nucleus disappears and the chromatin of each chromosome begins to coil up. As the chromatin coils and compacts, it no longer looks like a fuzzy mass. Instead, the individual chromosomes become visible. At this point, each chromosome is attached to its copy. The original and its copy are each called *sister chromatids*. The two attached sister chromatids often appear X-shaped by the end of prophase. Also, during prophase a spindle of cytoskeletal fibers attaches to each chromosome and to each end of the cell. These fibers move the chromosomes during the next three phases.

In *metaphase*, the chromosomes are lined up at the center of the cell.

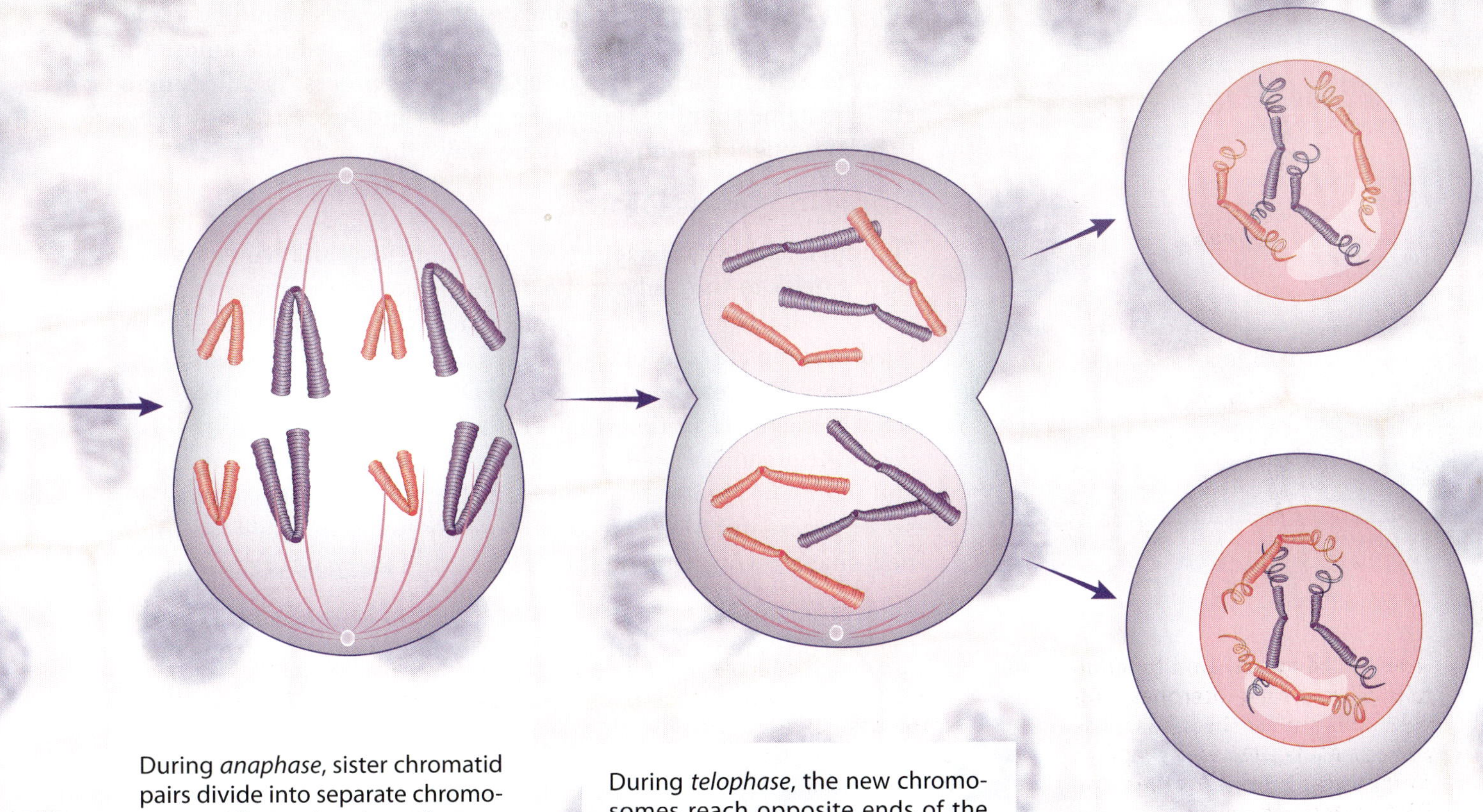

During *anaphase*, sister chromatid pairs divide into separate chromosomes. These *daughter chromosomes* are moved by the spindle fibers to opposite ends of the cell.

During *telophase*, the new chromosomes reach opposite ends of the cell and begin to uncoil. As they uncoil, the spindle disappears. New nuclear membranes form around each group of chromosomes. This results in two new nuclei, each with a complete set of chromosomes.

Cytokinesis (si toh kuh NEE sus) is the final stage of the cell cycle. This process gives each new daughter cell some cytoplasm and organelles. The parent cell pinches in between the two new nuclei until the cytoplasm is divided and two new cells are formed.

3.8 MEIOSIS

3-2
Planarians are a kind of free-living flatworm.

Many organisms, like amoebas, can reproduce using mitosis alone. A single amoeba goes through mitosis to produce two new amoebas. Each of the new amoebas has the correct number of chromosomes. Yeasts use a form of mitosis called *budding* to produce new yeast cells. In *regeneration* (see box on page 53), also a type of reproduction by mitosis, some organisms, like planarians, can be cut into pieces, and each piece will grow into a new organism. All of these are examples of **asexual reproduction**—reproduction that does not involve **gametes** (sperm and egg cells). But what about organisms that do reproduce using gametes, or **sexual reproduction**? If both an egg cell and a sperm cell that combines with it have the normal number of chromosomes, then the fertilized egg will have twice the amount of DNA needed. For most organisms, this would be a huge problem. What they need is a way to reduce by half the amount of DNA present in each sperm or egg cell. This process is called **meiosis**. The name sounds kind of like mitosis and suggests that the two processes might be similar. In many ways, they are!

Haploid Versus Diploid

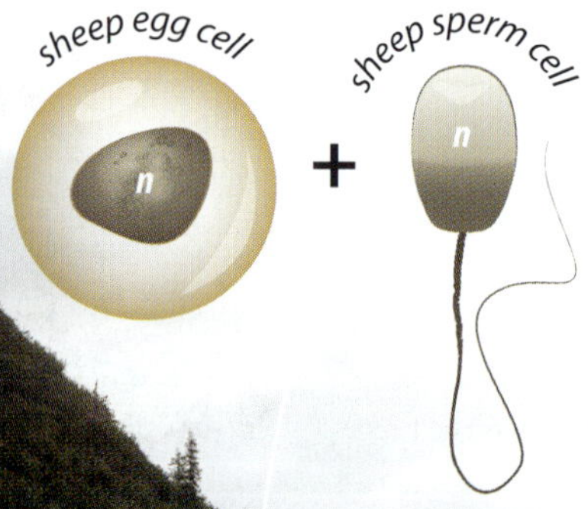

To understand what's happening during meiosis and why it's necessary, it helps to think about the number of chromosomes in a cell at different times in its life. Remember back when we said that each cell needs a complete copy of the organism's DNA? Most organisms don't only have just one complete set of DNA—they actually have two! That is, each cell in the organism has two complete sets of all the genes the organism needs. Scientists call this condition being *diploid* and represent it with the symbol $2n$. The n stands for one complete set of genes, and $2n$ means that there are two sets. You'll see in the next chapter why being diploid is important for most organisms.

Since it is normal for most cells to have two sets of genes, copying them to prepare for mitosis temporarily results in having four sets of genes, or $4n$. Mitosis reduces the number back down to $2n$ for each daughter cell. But that's still too many genes for a sperm cell or egg

cell, so mitosis won't work for making sex cells. They need a different process, and that's where meiosis comes in. Meiosis reduces the normal $2n$ number of genes down to just n. Sperm cells and egg cells produced by meiosis will each have only one set of genes (n) instead of the usual two found in most other cells. Scientists describe having one set of genes as being *haploid*. Later, when a haploid sperm cell (n) and a haploid egg cell (n) unite, the newly fertilized cell is once again diploid ($2n$).

Homologous Chromosomes

Each chromosome in a diploid cell has a matching partner with the same genes on it. These matching pairs of chromosomes are called **homologous** (huh MAH luh gus) **chromosomes**. It's like having two copies of a software program on a computer. The two programs each have the same number of files, and each file has the same set of instructions in it. So it is with a diploid cell's two sets of genes ($2n$). Each of the two sets has the same number of chromosomes. Each chromosome within one set has genes for the same types of traits as its partner in the other set. Thinking of genes and chromosomes this way will help you understand what happens during meiosis.

It should be noted, though, that the genes on homologous chromosomes may not be exactly alike. Think about the software program again. It might contain a line of code that codes for a yellow icon on a computer screen. A slight change in the code might produce a red icon. Genes function in much the same way. One gene on a chromosome may code for brown eye color. Its partner on the other chromosome within the homologous pair may code for blue eyes. Both genes code for the same trait, eye color, but each is slightly different from the other. Each slightly different version of a gene coding for the same trait is called an *allele*. You'll learn about the importance of alleles in Chapter 4.

MEIOSIS

1. Meiosis begins much like mitosis, with the number of genes in the parent cell temporarily doubled (4*n*).

2. Meiosis consists of two cell divisions instead of just one. The two divisions are called *meiosis I* and *meiosis II*. During meiosis I, sister chromatids don't separate. Instead, each daughter cell will receive one chromosome from each homologous pair of chromosomes. Since each chromosome at this point still consists of two attached sister chromatids, the daughter cells from the first division are diploid (2*n*).

3. During meiosis II, the sister chromatids in each of the diploid daughter cells from the first division will separate. Each of the daughter cells produced by the second division will receive one sister chromatid from each newly separated pair. So instead of yielding just two diploid daughter cells as in mitosis, meiosis produces a total of four haploid daughter cells. These haploid daughter cells are called *gametes*. Female gametes are *eggs*. Male gametes are *sperm*. Keep in mind that even though each gamete is haploid, it still contains one complete set of the organism's genes.

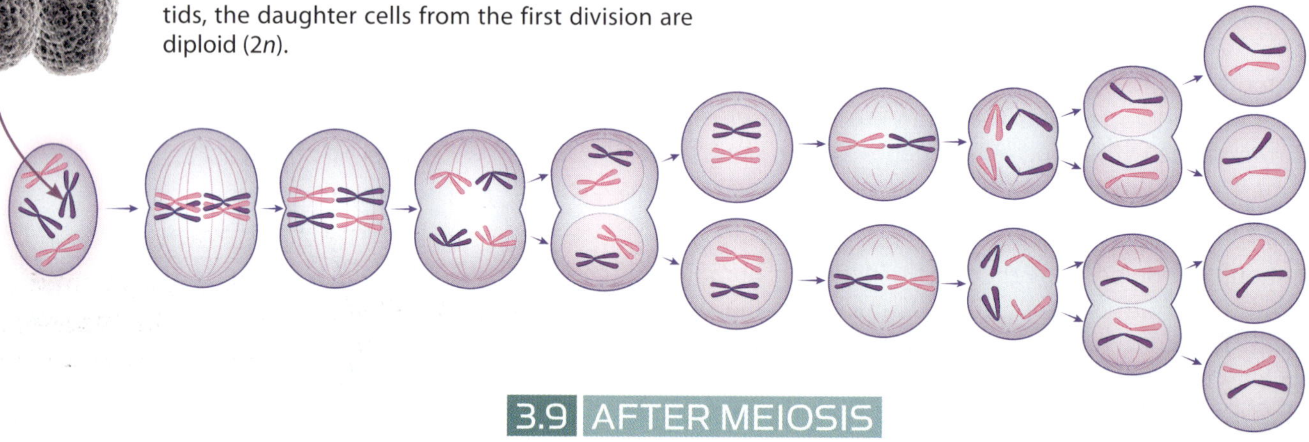

3.9 AFTER MEIOSIS

During sexual reproduction, the sperm and the egg each contribute one-half of each pair of chromosomes. For instance, the sperm and egg cells of the maned wolf each have 37 chromosomes. This is one-half of the chromosomes that maned wolf cells normally have. When a sperm from a male and an egg from a female unite, the new cell that forms has 74 chromosomes (37 pairs).

3-3
Not a true wolf, the maned wolf is South America's largest canid, the group that includes wolves, foxes, and domestic dogs.

This joining of gametes is called **fertilization** (FUR tul ih ZAY shun). Fertilization occurs when two haploid gametes (one egg and one sperm) unite and form a diploid cell. The diploid cell that results from fertilization is called a **zygote**. Zygotes contain two of every kind of chromosome typically found in the organism, so they have two copies of each gene, just like their parents. In most diploid organisms, each parent provides one-half of each pair of genes. In multicellular organisms, the zygote and its daughter cells then go through mitosis millions of times to grow into an adult.

A major difference between asexual and sexual reproduction is seen in the offspring. Asexual reproduction yields offspring that are identical to the parent. Because the reproduction happened by mitosis, the offspring's genetic information is the same as that of the parents. There is no variety. Sexual reproduction, on the other hand, produces offspring that are different from both parents. Each parent provides half of the offspring's genetic material. As a result, the offspring are similar to, yet different from, both the mother and the father.

3.10 EVIDENCE OF DESIGN

Remember the F-22s you read about in Chapter 2? Imagine that you were to stumble across an airplane while hiking in the wilderness. Would you think to yourself, "Wow! It's amazing how the forces of nature put this airplane together out here in the middle of nowhere"? Probably you wouldn't! Instead, if you're like most people, when you see something that looks man-made you think, "Wow! I wonder who made this and how it got here." That's because all of us recognize that something like an airplane is an orderly structure—its many parts are put together in a planned way. Information was required to build it. It would be foolish to think that an airplane somehow assembled itself completely by chance. It's natural to assume that an airplane must have had a builder.

3-4
Made by nature?

Now imagine that tucked away inside your mystery airplane is a complete set of blueprints and diagrams showing how the plane is put together. Now you would have no doubt that some intelligent person designed that airplane. As complex as an airplane is, living things are even more so. The amount of information tucked away in the DNA of living things is staggering. Human DNA, as you read earlier, contains about 6 *billion* nucleotide pairs. One species of Japanese flower has *fifty times* as much DNA as a human! All of these billions of nucleotides are placed in a particular order so that particular proteins can be built. The possibility that all this order could result from chance alone seems very unlikely. This *argument from design* is one of the strongest evidences in favor of a wise and powerful Creator. Of course, this Creator is exactly what we find in the first chapter of Genesis. Christians can take comfort in the fact that what we observe in God's created world agrees with God's written word—just as we would expect it to do!

3C REVIEW QUESTIONS

1. Why do cells need to copy their DNA before mitosis?
2. In which phase of the cell cycle do cells spend most of their existence?
3. (True or False) During replication, a DNA molecule is disassembled into individual nucleotides that are then used to build two new DNA molecules.
4. Place the following phases of mitosis in the correct order: *telophase*, *metaphase*, *anaphase*, *prophase*.

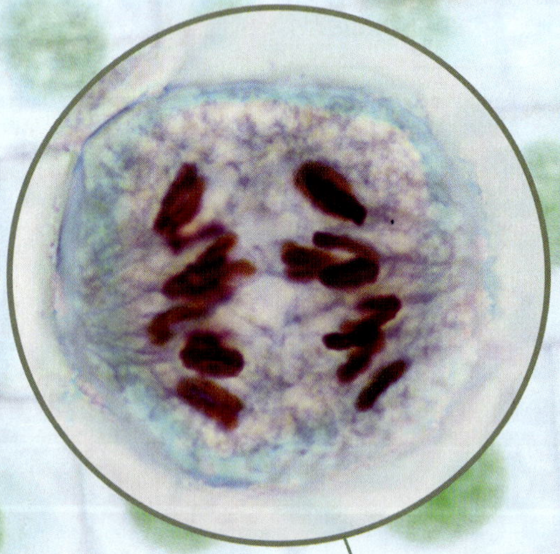

5. Which phase of mitosis is shown in the photo on the left?
6. Create a T-chart comparing mitosis and meiosis. Place "Mitosis" and "Meiosis" at the top of the chart. Under each term, identify how many cell divisions take place, how many daughter cells are produced, whether the daughter cells are haploid or diploid, and whether the process is used for asexual reproduction or to produce gametes. (For instructions on making a T-chart, see Appendix B.)
7. How does DNA give evidence of having been designed by God?

CHAPTER SUMMARY

3A DNA AND GENES

- Cells need information in order to live, grow, and reproduce.
- Cells store information in DNA, a kind of nucleic acid.
- DNA exists as a double helix, a ladder-shaped molecule.
- DNA is made of individual nucleotides. Each nucleotide consists of a sugar, a phosphate, and one of four bases: adenine, thymine, guanine, or cytosine.
- A gene is a section of DNA that codes for a protein.
- The kind of protein coded for by a gene depends on the order of the nucleotides within the gene.

3A TERMS

nucleotide	48
gene	49

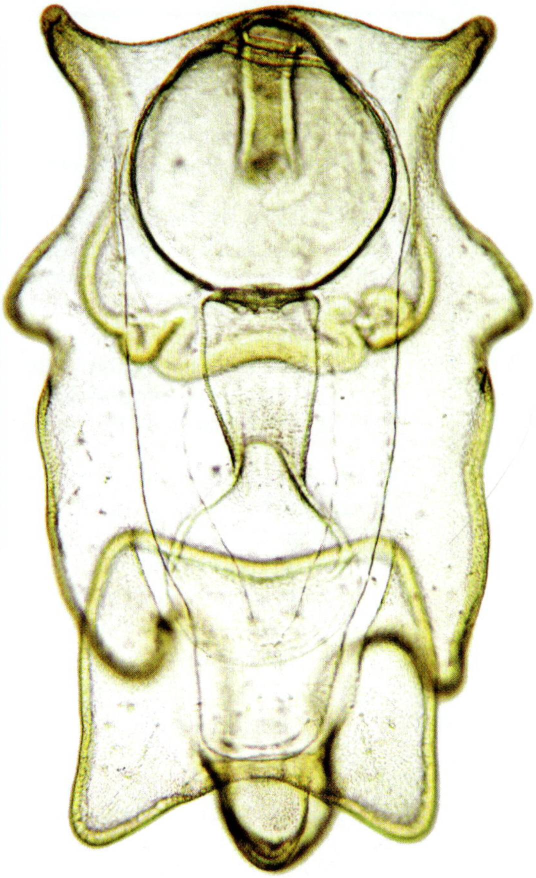

3B TERMS

transcription	50
ribonucleic acid (RNA)	50
translation	51
codon	51
amino acid	51
anticodon	51
protein synthesis	51
engineered system	52
irreducible complexity	52

3B HOW GENES FUNCTION

- The information in a gene is transcribed into mRNA before being sent out of the nucleus.
- During transcription, a DNA molecule is temporarily unzipped by an enzyme. The enzyme reads the DNA to determine the order of nucleotides for the mRNA being made.
- Messenger RNA differs from DNA in being only a single strand instead of a double strand. In mRNA, thymine is replaced by uracil.
- Finished mRNA molecules exit the nucleus through pores in the nuclear membrane.
- During translation, ribosomes read the information in mRNA and use it to build proteins.
- Amino acids are shuttled to the ribosome by tRNAs.
- Transfer RNAs place their amino acids in the correct sequence by matching their anticodon to the proper codon in the mRNA.
- Amino acids are bonded together during protein synthesis.
- The building of proteins using the information in DNA shows evidence of being an engineered system, including being irreducibly complex.

CHAPTER 3

CHAPTER SUMMARY

3C CELL DIVISION

- The cell cycle is the cycle of growth and reproduction in cells.
- Interphase is the longest part of the cell cycle. During interphase, the cell is growing and functioning normally. Toward the end of interphase, the cell makes copies of its DNA. At the end of interphase, the DNA in a cell is temporarily doubled in preparation for mitosis.
- During DNA replication, enzymes unzip the two sides of a DNA molecule. Other enzymes bring in nucleotides to build new, matching sides for each of the two original sides. This makes two identical molecules.
- Mitosis is the part of cell division in which the nucleus is divided.
- Mitosis consists of four phases: prophase, metaphase, anaphase, and telophase.
- During mitosis, the sister chromatids within each chromosome separate and each daughter cell receives an identical set of chromosomes.
- During the last phase of the cell cycle, cytokinesis, the parent cell's cytoplasm is divided between the two new daughter cells.
- Meiosis is the process that produces haploid egg and sperm cells.

- Meiosis includes two cell divisions. The first division (meiosis I) separates homologous chromosomes. The second division (meiosis II) separates sister chromatids.
- The huge amount of information found in the DNA of organisms strongly supports the existence of an intelligent Creator.

3C TERMS

cell cycle	54
interphase	54
replication	54
mitosis	56
cytokinesis	57
asexual reproduction	58
gamete	58
sexual reproduction	58
meiosis	58
homologous chromosomes	59
fertilization	61
zygote	61

CHAPTER REVIEW QUESTIONS

Recalling Facts

1. What are the three parts of a nucleotide?
2. In DNA, what base pairs with guanine?
3. Describe the shape of a DNA molecule.
4. What is a gene?
5. During transcription, the information in a DNA molecule is copied to produce ______ molecule.
   - **a.** an mRNA
   - **b.** a tRNA
   - **c.** an rRNA
   - **d.** another DNA
6. In RNA, what base bonds with adenine?
7. A codon codes for a(n)
   - **a.** amino acid.
   - **b.** protein.
   - **c.** nucleic acid.
   - **d.** base.
8. What determines the order of amino acids in a protein?
9. What is the function of tRNA?
10. The process where one DNA molecule forms two identical DNA molecules is ______.
11. How do mitosis and cytokinesis differ?
12. In which phase of mitosis do the sister chromatids separate into two different chromosomes?

13. How do the sister chromatids reach opposite ends of the cell during mitosis?

14. During which part of the cell cycle is cytoplasm divided between the new daughter cells?

15. What is the purpose of meiosis?

16. (True or False) If there is a gene for eye color on one chromosome of a homologous pair, then the other chromosome of that pair also contains a gene for eye color.

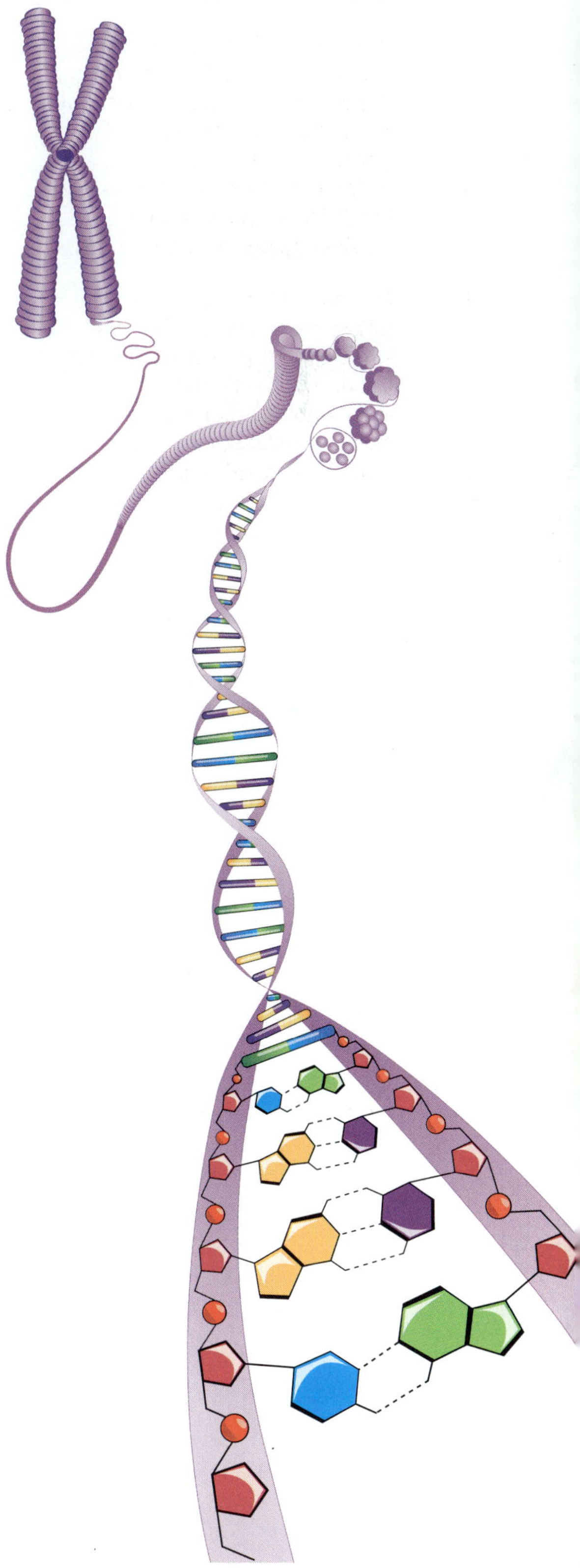

Understanding Concepts

17. Draw or build a model of DNA. Label these parts on your model: nucleotide, sugar, phosphate, adenine, thymine, guanine, cytosine.

18. How do cells and DNA demonstrate irreducible complexity?

19. Construct a concept map of the cell cycle using the following terms or actions: cell cycle, interphase, DNA is duplicated, sister chromatids, mitosis, chromosomes are divided, prophase, metaphase, anaphase, telophase, cytokinesis, cytoplasm is divided, daughter cells. You will need to come up with your own linking words. (For information on how to make a concept map, see Appendix B.)

20. Suppose a cell had 20 chromosomes at the beginning of its cell cycle. After mitosis and cytokinesis were completed, each newly formed cell would have ______ chromosomes.

21. The body cells of a rabbit contain 44 chromosomes. How many chromosomes would a rabbit egg cell contain?

    a. 11
    b. 22
    c. 44
    d. 88

Critical Thinking

22. How are DNA replication and transcription similar? How are they different?

23. A cell is undergoing transcription of a certain gene. If the gene's DNA base code is thymine—guanine—guanine—adenine—cytosine, what will be the correct order of bases on the resulting mRNA molecule?

    a. adenine—cytosine—cytosine—uracil—guanine
    b. thymine—guanine—guanine—adenine—cytosine
    c. adenine—cytosine—cytosine—thymine—guanine
    d. uracil—guanine—guanine—adenine—cytosine

24. What would happen if a ribosome accidentally skipped just one base when decoding the mRNA's codons?

25. How do DNA and genes show evidence of a Designer?

26. A diploid donkey cell has 62 chromosomes. How many chromosomes would this cell have after mitosis and after meiosis?

    a. 62 after mitosis and 31 after meiosis
    b. 62 after mitosis and 62 after meiosis
    c. 31 after mitosis and 62 after meiosis
    d. 31 after mitosis and 31 after meiosis

27. How do asexual reproduction and sexual reproduction differ?

ETHICS Cloning

When you read the word "clone," perhaps you think of sci-fi movies with armies of look-alike soldiers. But just what is cloning? A *clone* is an offspring that has the same genetic information as its parent(s). Cloning is actually nothing new. It's been happening ever since God created the first living things. Asexual reproduction, which you read about earlier, produces *natural clones*. Many kinds of microorganisms, plants, and insects reproduce this way. And humans have been using cloning for centuries. Many of the fruits that you eat, for example, are plant clones produced by a technique called *grafting*.

But Dolly the sheep was different. Back in 1996, Dolly was an international sensation. She was the world's first mammal successfully cloned from an adult, non-gamete cell (called a *somatic cell*). Dolly was a type of *artificial clone*. Scientists produced her in a laboratory by removing the nucleus of a mature cell from one sheep and inserting it into a cell from another sheep. The new cell was implanted into the womb of another sheep to grow and eventually be born. This technique is called *nuclear transfer*. In the decades following Dolly's birth, scientists have used nuclear transfer to clone many other kinds of mammals.

Why is this ordinary-looking sheep stuffed and on display at the National Museum of Scotland? It's because she's no ordinary sheep—she's Dolly!

28. Do a Web search using the key term "cloning." Then write a brief summary of what you learned about this issue.

29. What seem to be the most important ethical issues regarding artificial cloning?

30. Does the Bible have anything to say about this issue?

31. What are the acceptable and unacceptable options of artificial cloning?

32. What might be the consequences of adopting the acceptable and/or unacceptable options?

33. What views or actions should you take with regard to cloning?

GENETICS 4

THE GENE GUN

Since the beginning of time, mankind has worked the land to raise useful plants. For centuries farmers have been breeding plants and animals. But what if you could make a plant with certain features by literally shooting DNA at it? That's just what John Sanford and his colleagues at Cornell University did. They coated tiny metal particles with DNA and used a gun to shoot the particles into the nuclei of its cells. Some of the DNA was incorporated into the plant's chromosomes. This so-called gene gun was used to create some of the world's first genetically modified plants. How does this fit into God's commands to the human race?

4A Questions

- What were Mendel's experiments and principles of inheritance?
- What is the difference between dominant and recessive alleles?
- What is the relationship between genotype and phenotype?
- How can you use a Punnett square to predict the outcome of a cross?

4-1
Gregor Mendel, an Austrian monk, scientist, and educator, is called the Father of Genetics. Interestingly, he wrote more about meteorology, but his work on genetics has had a greater impact on the development of science.

4A

PRINCIPLES OF HEREDITY

How are traits passed from one generation to the next?

4.1 MENDEL'S EXPERIMENTS

God's command to the human race to wisely use His world makes it important for us to work to make it more useful. But caring for this world is a daunting task. We must consider the consequences of our actions when choosing how to rule the earth wisely. Understanding genetics and exploring the potential of genetically modifying organisms is an important goal of life science. And it all started with Gregor Mendel.

Farmers had understood for centuries that they could breed plants or animals and sometimes get the characteristics they wanted. In the early 1860s, while the United States was experiencing the Civil War, an Austrian monk named Gregor Mendel was tending his pea plants in the monastery garden. He was curious about how characteristics were passed down from one generation to the next. Mendel used pea plants because their **characters**, like seed color and plant size, were easy to observe. Many of these characters had two or more forms, called **traits**. For instance, some peas had the trait of yellow pea pods while most peas had the trait of green pea pods. It was easy to breed or *cross* different plants and observe which traits the offspring received. But no one had ever done any scientific inquiry on how these traits were passed down.

BREEDING PEAS

1 ***Purebred.*** Mendel started with purebred plants. A **purebred** plant is a plant that always produces descendants with the same trait. For example, when Mendel planted peas (seeds) from purebred plants with green pea pods, the peas always grew into plants that produced green pea pods. The same thing happened with pea plants that produced yellow pea pods.

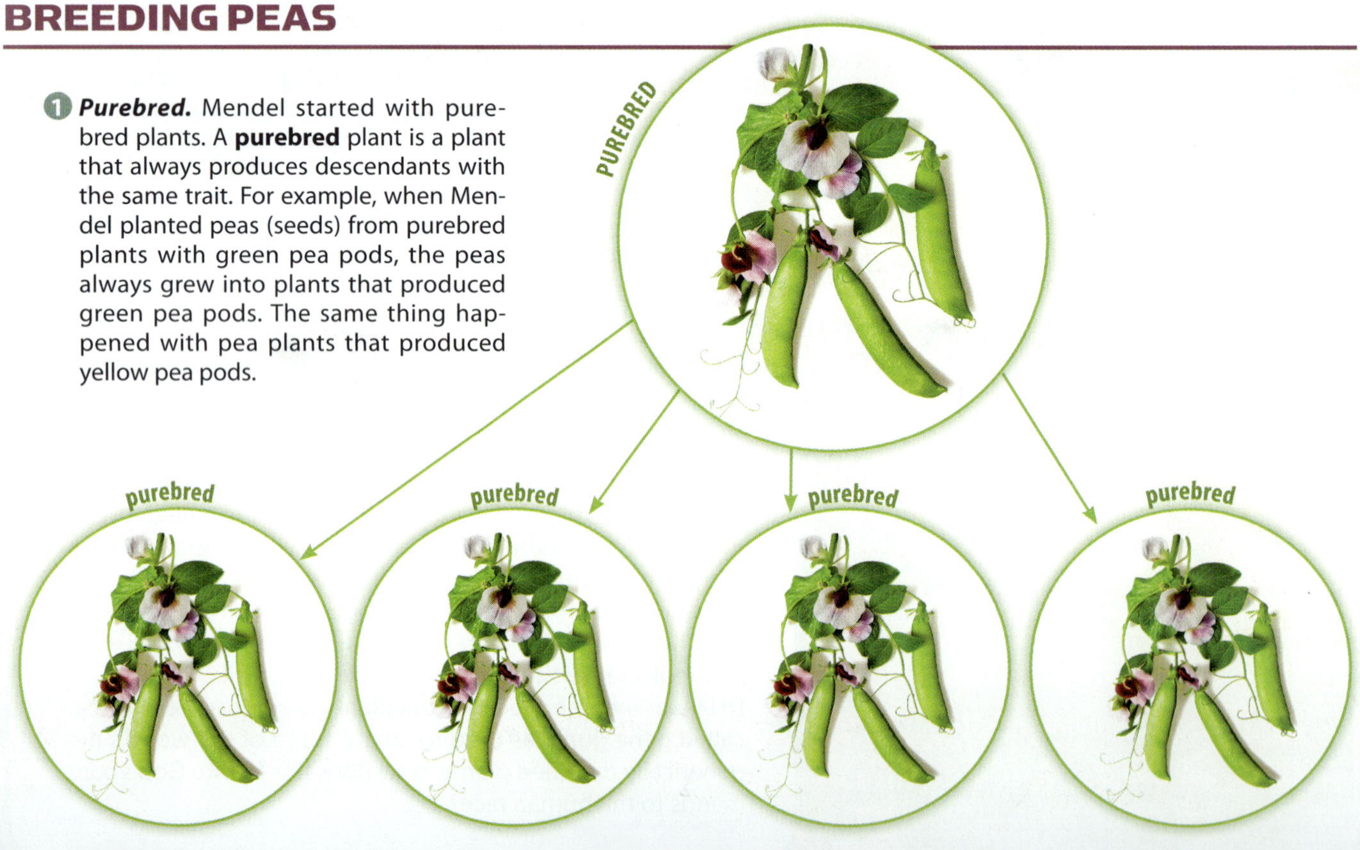

❷ ***Hybrid.*** Then Mendel started crossing his peas. He crossed purebred peas having green pods with purebred ones having yellow pods. Each offspring had a mix of alleles (see page 59) from its parents. An offspring with a mix of alleles for a specific character is called a **hybrid**. The first-generation hybrid offspring all had green pods. Mendel concluded that green pods is a dominant trait for pea plants. **Dominant traits** are the traits that appear in a hybrid. **Recessive traits** are the traits that are hidden in a hybrid. Mendel found that every first-generation hybrid showed only the dominant traits for all of the pea plant characters that he studied.

❸ ***Hybrid Crosses.*** Mendel continued to cross his pea plants. He tried different combinations of purebreds and hybrids, and he did multiple generations of hybrids. He kept track of the ratios of plants that had pods of different colors.

The characters that Mendel observed were controlled by only one gene, and each gene had only two alleles. The dominant trait was always expressed in a hybrid plant. This is called **simple dominance**. Mendel examined many characters in pea plants, including their height, seed shape and color, pod shape, and flower position. In eight years, he experimented with 30,000 plants! Take a look at the results that he got in the table below.

Table 1

Results of Hybrid Crosses

Flower position	 651 axial position	 207 terminal position
Height	 787 tall plants	 277 short plants
Pod shape	 882 inflated pods	 299 pinched pods
Pod color	 428 green pods	 152 yellow pods
Seed shape	 5474 round seeds	 1850 wrinkled seeds
Seed color	 6022 yellow seeds	 2001 green seeds
Seed coat color	 705 gray-brown	 224 white

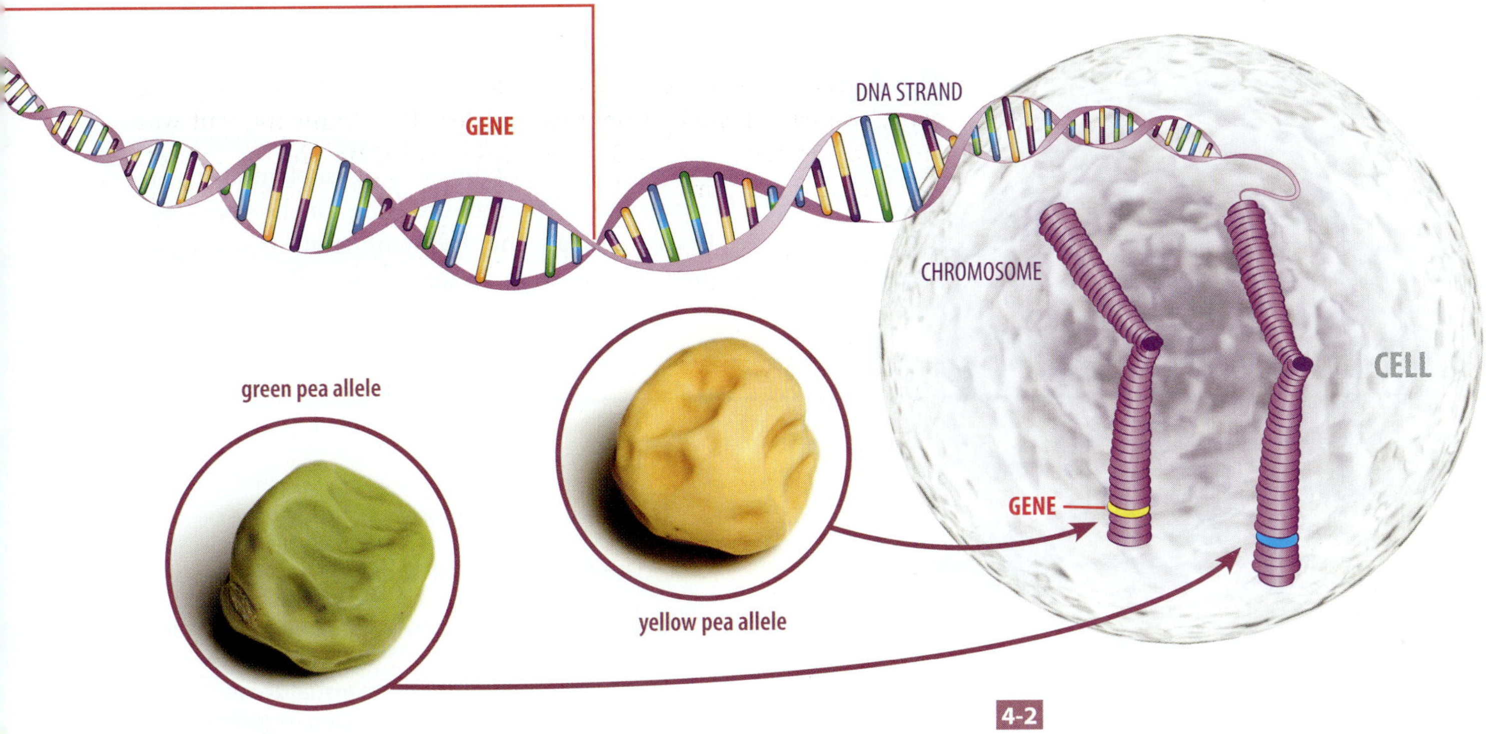

4-2
Cells have chromosomes made up of coiled DNA. Sections of this DNA code for specific characters. These sections of DNA are genes.

Mendel's research was the beginning of the science of **heredity**—the study of how traits are passed from one generation to the next. Traits are determined by genes. As you saw in Chapter 3, genes can exist in different forms known as *alleles*. Different alleles result in different traits. So all pea plants have the gene that codes for pod color. Some have only the allele that results in green pods, some have only the allele that codes for yellow pods, and some have one of each allele.

Later scientists who discovered and repeated Mendel's work summarized his observations to formulate the three principles of heredity.

1. In a cross of purebred parents, all offspring will show the dominant trait.
2. For any one character, the two alleles for a parent's gene separate and then combine with the other parent's alleles to determine the offspring's trait.

   For example, a hybrid plant with green pea pods has one green pod allele and one yellow pod allele. But each of that plant's offspring receives only one of those alleles. Some offspring will receive a green pod allele, and some offspring will receive a yellow pod allele. Each offspring also receives a second allele from the other parent.
3. A parent passes each allele to its offspring independently of all other alleles.

   So a pea plant that is hybrid for both the pod color character and height character might pass the green pod allele and the tall allele to an offspring. Or it might pass the green pod allele and the short allele. Whichever pod color allele the offspring receives has no effect on which height allele it will receive.

Although many scientists of Mendel's day rejected his discoveries, scientists rediscovered and confirmed his findings around 1900. A few decades after this rediscovery, a British scientist named Reginald Punnett created a tool now known as a *Punnett square*, which uses the alleles of the parents to predict the likelihood of offspring displaying a particular trait.

4.2 PUNNETT SQUARES

The Punnett square is a grid that uses letters to represent alleles. For example, we could track the allele for tallness in a pea plant using the letter *T*. We use a lowercase *t* to represent the allele for shortness. We use a capital letter (*T*) to indicate the dominant trait, or taller pea plants. Using a Punnett square, you can quickly see all the possible outcomes of a cross.

TRACKING TALLNESS IN PEA PLANTS

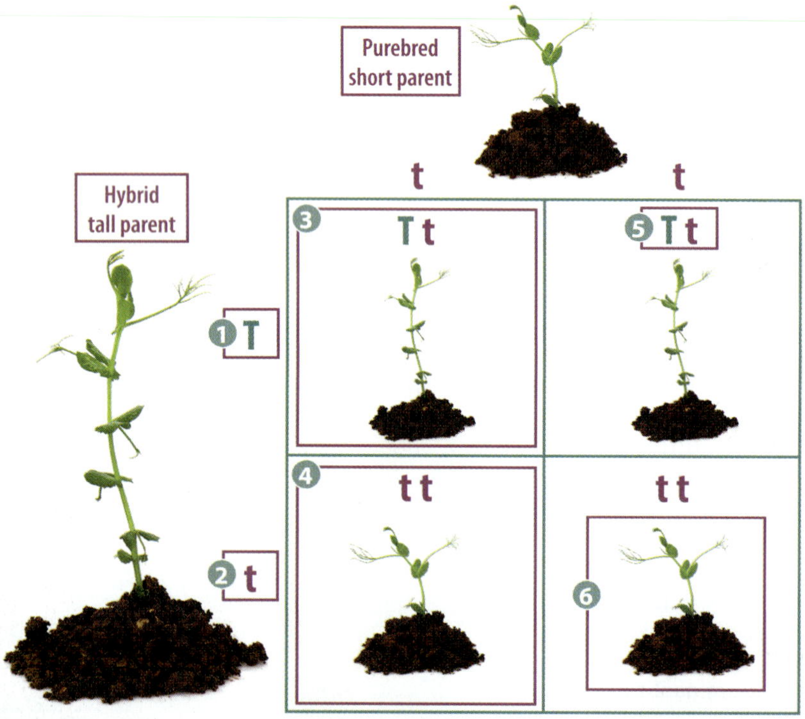

1. Dominant allele, tall
2. Recessive allele, short
3. Offspring, *hybrid* tall (2 different alleles)
4. Offspring, *purebred* short (only 1 kind of allele)
5. The letters that help you visualize an organism's genetic makeup are called its **genotype**.
6. The trait that an organism expresses as determined by its genotype is called its **phenotype**.

Each box represents a 25% chance of a particular genotype and phenotype. So any offspring of one hybrid tall pea plant and one purebred short pea plant has an equal chance of being tall or short since half of the four squares have a genotype that produces a tall phenotype and half have one that produces a short phenotype.

If you were to continue to cross tall and short pea plants, you could track the genotypes and phenotypes of the offspring for several generations. There can be three different combinations of alleles or genotypes: *TT*, *Tt*, or *tt*. But there are only two possible phenotypes, tall and short. The lab activity on the facing page takes a closer look at the predictive power of Punnett squares.

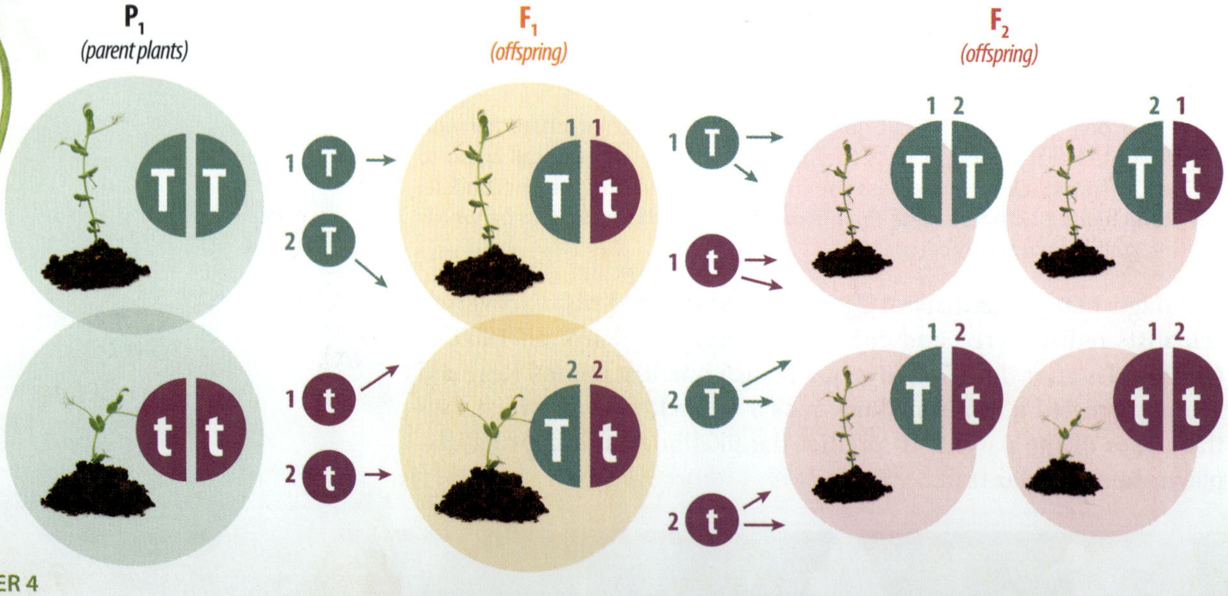

GENETICS: USING YOUR MARBLES

Genetics is all about probability. When we breed two organisms, a variety of traits may be possible. It's like rolling dice! Which traits we get is based on probability. A Punnett square can't tell you which genotype an offspring will have. It can tell you only how likely an individual is to have a certain genotype.

In this lab activity, you will model heredity using marbles. First you will predict the probability of a certain genotype using a Punnett square. Then you will use the marbles to see how close your prediction was.

Equipment

red marbles (or other color) (30)

white marbles (or contrasting color) (30)

small paper bags (2)

PROCEDURE

❶ Select 30 red marbles and 30 white marbles. Red marbles represent the dominant allele and are represented by *R*. They result in a red phenotype. White marbles represent the recessive allele and are represented by *r*. They result in a white phenotype.

1. Which genotypes produce a red phenotype?

❷ Place 15 marbles of each color into a small paper bag. Label the bag "male." Place the remaining 30 marbles (15 of each color) in a second bag labeled "female." This will represent both parents being hybrids for this character.

2. What are the genotypes of the "parents"?
3. What are the phenotypes of the "parents"?
4. Create a Punnett square for this cross.
5. On the basis of your Punnett square, predict what percentage of the "offspring" will have the dominant color.

❸ Without looking, remove a marble from each bag. Record the genotype and phenotype of the "offspring." Replace the marbles into their original bags.

❹ Repeat Step ❸ fifteen more times.

6. Now create a fraction with the total population in the denominator. Put the number of individuals with the dominant trait in the numerator. What percentage of the offspring have the dominant color?
7. How does this compare with what you predicted in Question 5?

4A REVIEW QUESTIONS

1. How did Mendel study heredity?
2. How do you know which traits are dominant in an organism like a pea plant?
3. Write down the dominant traits for pea plants that Mendel studied (see Table 1 on page 70).
4. Write the three principles of heredity in your own words.
5. How do genotypes and phenotypes relate?
6. A pea plant with yellow pods has a genotype of *gg*. Is it purebred or a hybrid? Is the yellow pod trait dominant or recessive?

For Questions 7 and 8, a purebred pea plant with green pods (*PP*) has been crossed with a purebred pea plant with yellow pods (*pp*).

7. What is the genotype and phenotype of the first-generation offspring?
8. What is the probability that an offspring will have this phenotype?
9. Using Table 1, determine the percentage of tall plants produced by a hybrid cross.
10. Create a Punnett square for a hybrid cross of two tall parents with the genotype *Tt*.
11. Using the Punnett square in Question 10, determine the percentage of offspring from the hybrid cross that should be tall. How does this compare with the percentage that you calculated in Question 9?

For Questions 12 and 13, a purebred pea plant that bears yellow peas (*YY*) has been crossed with a hybrid pea plant that bears yellow peas (*Yy*).

12. Create a Punnett square for the cross.
13. What phenotypes will the offspring have?

4B Questions
- What is incomplete dominance?
- What is codominance?
- How are the traits of characters with more than two traits passed to offspring?
- How can you predict the outcomes of inheritance patterns that don't follow Mendel's principles?

4B PATTERNS OF INHERITANCE

Why don't all traits follow Mendel's principles of heredity?

4.3 INCOMPLETE DOMINANCE

As scientists rediscovered Mendel's experiments in the early twentieth century, they began to notice that some organisms didn't follow the principles of heredity. One such pesky organism is the snapdragon. In a cross of red snapdragons (C^rC^r) and white snapdragons (C^wC^w), Mendel's principles would predict that the first generation of offspring should all be either red or white depending on which allele was dominant. (C stands for the color of the flower, either red (r) or white (w).) Instead, all the first-generation offspring are pink (C^wC^r)! If some of these pink snapdragons are crossed, about half of the second generation are pink, about one-fourth are red, and about one-fourth are white. What happened?

4-3
Snapdragons do not follow Mendel's principles of heredity.

Snapdragon flower color is an example of **incomplete dominance**. There is no "hidden" trait; both are blended together. Hybrid offspring have a phenotype that blends the two parents' phenotypes. So red and white snapdragons produce pink offspring.

4.4 CODOMINANCE

Snapdragons aren't the only organisms that don't follow Mendel's principles. Camellias don't either. If a camellia with red flowers is crossed with a camellia with white flowers, you would expect the offspring to have pink flowers, right? Actually, they don't. The offspring will have flowers that are both red *and* white. That's because camellia flower color is an example of **codominance**. The phenotypes of both parents are there, but they aren't blended. If two of these multicolored offspring are crossed, about half of the offspring will also be multicolored. About another fourth of the offspring will be red, and the other fourth will be white.

The Punnett square for codominance looks a little different from the ones for simple dominance that you created for pea plants. Since both alleles for petal color are dominant, we can represent them both with capital letters. When a white camellia (C^WC^W) is pollinated with a red camellia (C^RC^R), it produces a plant with variegated blooms (C^WC^R).

4.5 MULTIPLE ALLELES

All the characters that we've studied so far have had genes with only two alleles. The gene for pea plant height had just two alleles, *T* and *t*. But some genes have more than two alleles. Each individual organism will still have only two alleles for a gene, one from each parent. But there are more than two alleles for the gene present in the population of organisms.

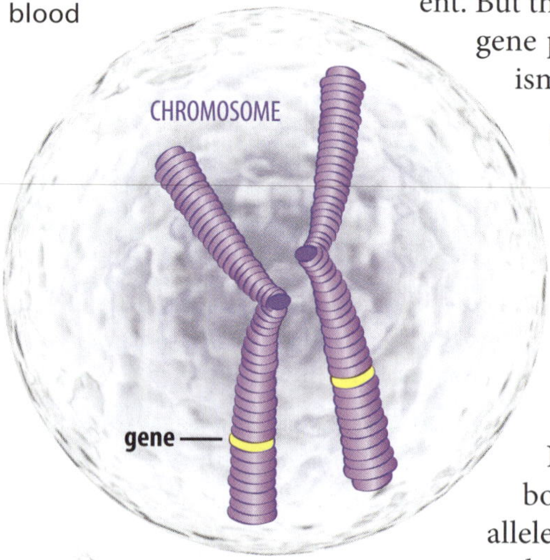

One example of multiple alleles in humans is the gene that controls the type of blood that people have. Some cells in human blood carry molecules called *antigens*. Your body recognizes the shape of its own antigens, but it attacks antigens with a different shape.

Notice that when a person inherits both I^A and I^B alleles (see far left), both alleles are expressed. The I^A and I^B alleles are codominant. Both are dominant to the *i* allele.

The gene that determines the shape of antigens has three different alleles. The presence and shape of the antigen determine what blood type a person has.

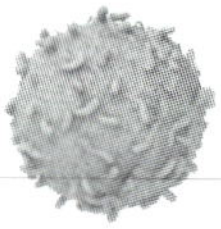

Type O blood (ii). The allele *i* produces no antigens.

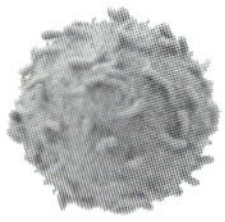

Type A blood (I^AI^A or I^Ai). The allele I^A produces antigen A.

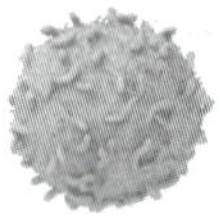

Type B blood (I^BI^B or I^Bi). The allele I^B produces antigen B.

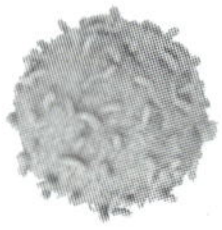

Type AB blood (I^AI^B). Having both I^A and I^B alleles produces both antigen A and antigen B.

4.6 POLYGENIC INHERITANCE

If incomplete dominance and codominance were the only exceptions to Mendel's principles, geneticists would have an easy time. But there are lots of other exceptions. Another variation is **polygenic inheritance**. This occurs when two or more genes in different places on a chromosome or on different chromosomes interact to control a single character. Scientists believe that many characters, like skin color, height, and intelligence, are controlled by polygenic inheritance. Scientists think that height, for example, is influenced by 424 genes with 697 alleles! Because so many genes influence human height, we see a wide variety of heights.

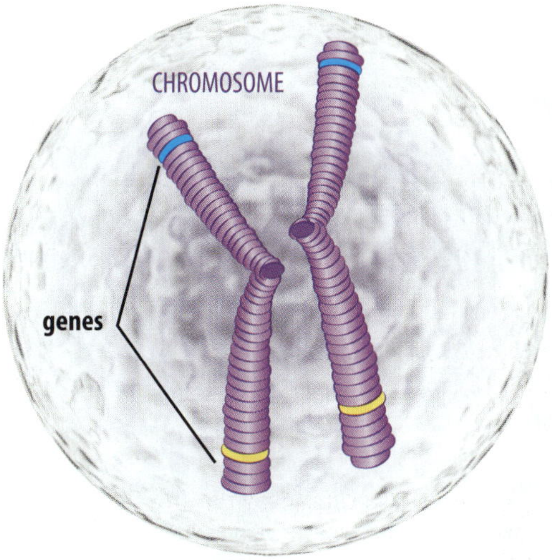

4-4
Polygenic inheritance happens when two or more genes interact to control a single character. This is part of what makes people so diverse!

But organisms are not just programmed by their genetics. Environment plays a role too. Things like weather, soil, and other organisms affect how tall a plant grows. Your health and habits also affect how tall you grow.

Now you've learned about the main types of inheritance. Believe it or not, there are actually a lot more types! Even now, over 150 years after Mendel published his research on pea inheritance, scientists are still learning new things about genetics. This complexity is no accident. God has fashioned a creation with built-in diversity. He knew you while you were still developing in your mother's womb, and He knows every gene in your cells (Ps. 139:14–16). For some people who have genetic disorders, the probabilistic nature of genetics can feel like some kind of cruel twist of fate. This isn't true! God's creation was very good (Gen. 1:31), but the Fall affected all things for the worse, even our genes. The biblical hope for a new body at the resurrection is magnified for those suffering deeply from genetic disorders (Rom. 8:22–23).

4-5

A roan calf

4B REVIEW QUESTIONS

1. What does it mean for a character to show incomplete dominance?
2. A red tulip is cross-pollinated with a white tulip. If tulips exhibit incomplete dominance, describe the color of the offspring.
3. What does it mean for a character to show codominance?
4. A white bull was crossed with a red cow. The offspring was a roan calf (see Figure 4-5). What type of inheritance pattern does hair color in cattle exhibit: simple dominance, incomplete dominance, or codominance? Explain your answer.
5. Copy the following table on your paper and fill in the cells to compare the number of genes and alleles involved in the different kinds of inheritance.

Example (Inheritance Pattern)	Number of genes involved	Number of alleles involved
Pea flower color (simple)		
Snapdragon flower color (incomplete)		
Camellia color (codominance)		
Blood type (multiple alleles)		
Human height (polygenic inheritance)		

6. A mythical red griffin is paired with a yellow griffin. Griffin color inheritance is an example of incomplete dominance. Describe the color of the offspring.

4C Questions

- What is genetic drift?
- What is natural selection?
- How do genetic drift and natural selection affect a population?
- How much genetic change can occur in a population?
- How can you predict the results of changes in a population?
- Do creationists believe in natural selection?

4C GENETICS AND POPULATIONS

Can the environment cause changes in a population?

4.7 GENETIC DRIFT

At a farm near you, black, white, and black-and-white (erminette) chickens live together in a coop. Inheritance of feather color in chickens is an example of codominance (see page 75). We will use the letter *C* (for *color*) to represent the alleles involved in feather color inheritance. Geneticists are interested in how black chickens (C^BC^B) and white chickens (C^WC^W) cross to form erminette chickens (C^BC^W).

But instead of focusing on the individual chickens, we could also take a look at how the flock in the coop changes over the years. Some are born, some have chicks, and some die. *Population geneticists* are interested not in individuals, but in the ratio of the black alleles (C^B) to the white alleles (C^W) within the population of chickens in our coop. They want to know whether that ratio is changing over time, and, if so, what is causing the change. The ratio of alleles in a population changes randomly over time, in a process called **genetic drift**. Let's look at how the ratio of alleles in a flock of chickens could change over time.

CHANGES IN THE COOP

Right Now

1 white chicken (C^WC^W)

1 black chicken (C^BC^B)

50% C^B, 50% C^W

1 Year Later

3 white chickens (C^WC^W)

2 black chickens (C^BC^B)

3 erminette chickens (C^WC^B)

56% C^W, 44% C^B

2 Years Later

3 white chickens (C^WC^W)

1 black chicken (C^BC^B)

9 erminette chickens (C^WC^B)

58% C^W, 42% C^B

Notice a few things about our changing chicken flock. First, the ratio of alleles doesn't necessarily reflect the ratio of the feather colors of chickens. Second, it looks like the white alleles have become slightly more common than the black alleles.

Why would a population change like this? Well, some chickens have chicks, and some chickens die. Some chickens have more chicks than others do. There are two key characteristics of genetic drift. First, it is completely random and unpredictable. We don't know how the chicken coop's population will change over time. Second, the effects of genetic drift are most felt on smaller populations. This makes the chicken coop a good example. Genetic drift is more likely to occur in a smaller population than in a larger one.

4.8 NATURAL SELECTION

At the beginning of the chapter, we saw that Mendel bred pea plants to explore how they passed on their traits. People have been breeding animals and plants for specific traits for centuries. When farmers breed organisms, they are choosing which organisms to allow to reproduce. This isn't a random process. They are purposefully trying to change the ratio of the alleles in a population to favor a certain phenotype. Breeding is also called *artificial selection*.

This can happen in nature, too, but there is no human to guide the process. For example, foxes can have a variety of ear sizes. For mammals that live in hot environments, large ears are a great way to lose heat and cool down. It's likely that when a population of foxes moved into the desert after the Flood, some individuals had larger ears than others. The ones with larger ears could release heat better than those with smaller ears. These larger-eared foxes likely survived better and had more offspring than smaller-eared foxes. An inherited trait that helps an animal survive to reproduce, like large ears in the desert, is called an **adaptation**. Over time, individuals with small ears became rare, and eventually the entire population had large ears. And in fact, this is exactly what we see with the kit fox of the American southwest and the fennec fox of the Sahara.

4-6

The fennec fox (two are shown below), the world's smallest fox, lives in the Sahara Desert. It cools off through its large ears. Compare this with the arctic fox (facing page) that lives in cold climates.

Contrast the fennec fox with the Arctic fox, which lives in very cold places. The Arctic fox needs small ears to keep in its body heat. In the original population that moved to the Arctic after the Flood, individual foxes with larger ears probably lost too much heat and could not survive to reproduce as well as foxes with smaller ears. Eventually, all Arctic foxes had small ears.

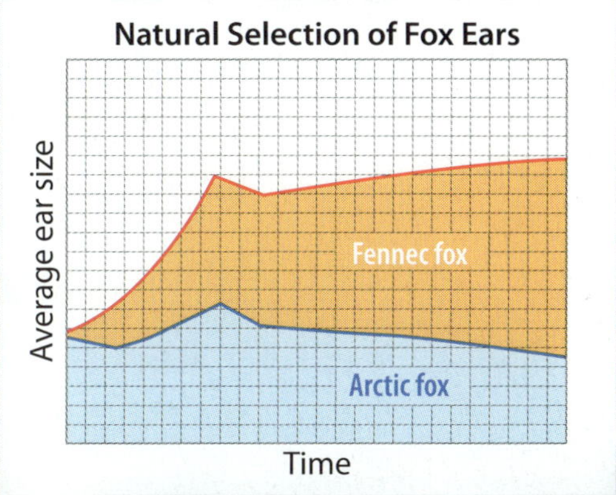

Notice that the change in the populations of foxes that must have occurred over time was driven by the environment, not genetic drift. This kind of change is called natural selection. **Natural selection** is the process through which organisms with certain adaptations survive to pass on their traits to a greater number of offspring than other organisms do. Three things have to be true in order for natural selection to occur.

1. There is variation in a population.
2. This variation is passed down from one generation to another.
3. The variation makes an organism more likely to survive and reproduce.

The first two are also true for genetic drift, but the last is true only of natural selection.

4.9 NATURAL SELECTION AND WORLDVIEW

Now, wait a second! Isn't natural selection the key to evolution? Evolutionists often base their arguments for evolution on examples of natural selection like the fox ears. Natural selection was a central idea in the work of Charles Darwin. He argued that it was the means by which all creatures descended from an original common ancestor. But the fact that Darwin proposed natural selection as a means for driving evolution doesn't make it a bad idea.

We observe natural selection in the real world. Populations do change when the environment favors certain traits. The real difference is the way that one interprets these observations. So creationists and evolutionists agree on the effects of natural selection. They just disagree on how far natural selection can go.

For example, animals from the dog kind that Noah took on the Ark multiplied and moved into different parts of the world. Foxes came from this kind. Natural selection eliminated alleles for traits that made the dogs less likely to survive in their habitats. But of course, there are lots of different habitats, so the different populations of the dog kind began to look different from each other. But to this day they are still all dogs! Natural selection can't create new genetic information; it can only *remove genetic information from a population.* Natural selection affects existing traits that God programmed inside the cell to produce the amazing variety that we see in living things.

4C REVIEW QUESTIONS

1. Define *genetic drift*.
2. (True or False) Genetic drift is measured by the ratios of organisms with a specific phenotype.
3. Define *natural selection*.
4. Compare genetic drift and natural selection.
5. (True or False) Natural selection can remove but not add genetic information.
6. A population of deer mice living in the dark soil of Nebraska move onto the sand dunes deposited by glaciers. While most of the mice are dark, some are light and blend in well with the light-colored sand. Predict which color of mouse will be more common in the sand dunes after fifty years. Explain your answer.
7. How do creationists and evolutionists differ in their views of natural selection?

4-7
Which mouse will become more common in an area with light-colored sand?

ETHICS Genetic Engineering

The gene gun mentioned in the Chapter opener is one tool used in genetic engineering. Genetic engineering is the process of deliberately manipulating the genes in an organism in ways other than natural processes like reproduction. It involves changing an organism's DNA by inserting new genetic material. Genetic engineering is a blossoming field of biology with real potential to solve significant real-world problems. It has been extensively applied to enhancing food production by modifying the genes of plants.

Scientists are convinced that genetically engineered food crops (called GMOs—genetically modified organisms) are safe to eat after extensive testing. However, some people have become very concerned about GMOs. They fear that these food crops may cause long-term health problems. But GMOs have been used in food for over a decade; so far, no evidence of ill effect has been discovered.

As the human population grows, we must think about the amount of food needed to keep everyone healthy. Genetic engineering may be the answer to this potential problem.

Use the Ethics box above to answer Questions 8–13.

8. Do a web search using the key term "genetic engineering." Write a brief summary of what you have learned.
9. What seems to be the most important ethical issue regarding genetic engineering?
10. Does the Bible have anything to say on this issue?
11. What are acceptable and unacceptable options on this issue?
12. What are the possible consequences of the acceptable options?
13. What views or actions should you take with regard to this issue?

CHAPTER 4

CHAPTER SUMMARY

4A PRINCIPLES OF HEREDITY

- Gregor Mendel studied inheritance with pea plants, which led to three principles of heredity.
- The phenotype for the dominant allele is expressed in a hybrid organism.
- An organism's genotype is the combination of alleles that it has inherited from its parents.
- An organism's phenotype consists of the traits coded for by the alleles in its genotype.
- A Punnett square is a tool that geneticists use to predict the likelihood that an offspring will express specific traits passed down from its parents.

4A TERMS

4B PATTERNS OF INHERITANCE

- A character demonstrating incomplete dominance blends together traits from the parents.
- A character demonstrating codominance expresses both traits at the same time.
- In multiple allele inheritance, one gene has more than two possible alleles.
- More than one gene is responsible for one phenotype in polygenic inheritance.

4B TERMS

CHAPTER SUMMARY

4C GENETICS AND POPULATIONS

- Genetic drift is the random change in the ratio of alleles in a population over time.
- Natural selection is the process through which organisms with adaptations that increase their ability to survive and reproduce pass on their traits to their offspring.
- Genetic drift and natural selection affect a population by decreasing the amount of its genetic information over time.
- Genetic change in populations is limited by the information present in the population's genotype.
- Creationists and evolutionists alike observe natural selection at work in the world, but they view it using different stories and worldviews.

4C TERMS	
genetic drift	79
adaptation	80
natural selection	81

CHAPTER REVIEW QUESTIONS

Recalling Facts

1. Summarize Mendel's three principles of inheritance.
2. Two pea plants with yellow peas were crossed, and the seeds were planted. Of the 100 offspring, 73 had yellow peas, and the other 27 had green peas. Which allele is dominant?
3. Identify each of the results of these purebred crosses as an example of simple dominance, incomplete dominance, or codominance.
   a. A purebred pea plant with wrinkled seeds is pollinated with a purebred pea plant with round seeds. All of the offspring have round seeds.
   b. A mom with type A blood and a dad with type B blood have a child with type AB blood.
   c. A purebred white poppy is pollinated by a purebred red poppy. All of the offspring have pink blossoms.
   d. A purebred white sheep is bred with a purebred black sheep to produce a white lamb.
4. A white albino rabbit lacks fur color. It carries the alleles *rr*. What is its genotype and phenotype?
5. Human blood type involves what kind of inheritance?
6. Labrador retriever coat color is determined by two different genes. What type of inheritance is this?

7. What do we call random changes to the ratio of alleles in a population?
8. What do we call the change in the ratio of alleles caused by the environment?

Understanding Concepts

9. Wrinkled pea pods are the result of a dominant allele. What is the genotype of non-wrinkled pea pods? Choose your own letter to represent the genotype.
10. In a simple dominant trait, will the genotypes *JJ* and *Jj* produce different phenotypes? Explain.
11. In pea flowers, purple is dominant to white. A purebred plant with white flowers (genotype *pp*) is crossed with a purebred plant with purple flowers (genotype *PP*). What is the genotype and phenotype of the first-generation offspring?
12. Draw a Punnett square to represent the cross of two hybrid pea plants with yellow pea pods (genotype *Yy*). What are the genotypes and phenotypes of the offspring? (The recessive allele codes for green pods.)
13. A plant with blue flowers is crossed with a plant with purple flowers. Flower color for this plant is a codominant trait. What will the offspring's flowers look like?
14. The lynxes of an area are split into two populations by a river. The population on the north side of the river experiences much colder weather than the population on the south side. Both populations contain individuals with thicker coats of fur and individuals with thinner coats of fur. Predict how the coats of fur of the two populations will differ after many years.
15. What limits natural selection?
16. Why can natural selection not produce cats from the dog kind?
17. Explain the difference between the creationist's and evolutionist's views of natural selection.

4-8
Lynx are shy cats, easily identified by black tufts of hair on top of their ears.

Critical Thinking

Bb	**Bb**
Bb	**Bb**

18. Look at the Punnett square on the right. Work backward to find the genotypes of the parents.
19. A hybrid pea plant with purple flowers is crossed with a purebred pea plant with white flowers. Some of the first-generation offspring have purple flowers, and some have white flowers. Two of the first-generation offspring with purple flowers are crossed. What is the percent probability that a second-generation offspring will be purple? What is the percent probability that it will be white? Include Punnett squares in your answer for these two crosses. Use the letter *B* for the alleles.
20. A group of scientists discover a species of plant with flowers that exhibit both red and white colors. They cross two of the plants, and all the first-generation offspring are also red and white. They cross two of the first-generation offspring. All the second-generation offspring are red and white. Is this an example of codominance? Explain.
21. In a population of gray squirrels, some individuals are larger than others. How is genetic drift likely to affect this population without any pressure from the environment?

22. In the same population of gray squirrels, larger individuals are stronger and able to escape from hawks more effectively. On the basis of this information alone, predict how natural selection is likely to affect this population.

23. Suggest some possible reasons why it would be harder for larger individual squirrels to survive to reproduce.

Use the Case Study below to answer Questions 24–26.

CASE STUDY COLORBLIND ISLAND

Some people on the South Pacific island of Pingelap cannot see color. This condition, called *achromatopsia*, is extremely rare in the rest of the world. But for every ten people born on this island, one has achromatopsia. Compare this with the less than .005% of the US population who have the condition. How did this recessive trait become so common on this tiny island? It all started with a storm.

In 1775 a powerful typhoon swept over the island. Fewer than two dozen people on the island, most of them women, survived the storm and famine that followed. The chieftain of the island survived and married many wives. He is believed to have been a carrier of the recessive allele. With so few people left, many of the next few generations married close relatives, making it more likely that their children would receive two copies of the recessive allele. Today, complete colorblindness is much more common on Pingelap than anywhere else in the world. All the Pingelap residents who are colorblind are descended from the ruler who survived the typhoon.

This shift in the population is an example of a *bottleneck effect*. A much larger population is reduced to a smaller population with less genetic variation. This results in the ratios of alleles changing as the smaller group starts a new population.

24. Is a bottleneck effect an example of natural selection or genetic drift? Explain your answer.

25. Nanmwarki Mwanenihsed, the ruler of the island during the typhoon, was probably not colorblind. In fact, colorblindness did not appear on the island until several generations later. Why were his children not colorblind?

26. If the typhoon had not killed most of the island's population, would the number of colorblind people be as high? Explain.

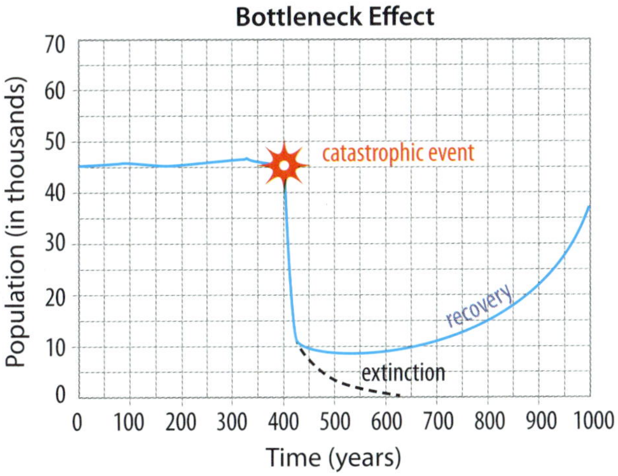

5 CHANGE IN NATURE

DESCENDED FROM DINOSAURS?

Many scientists believe that this blue heron and all other birds evolved from dinosaurs. This theory has been so talked about in books, magazines, movies, and other media over the years that the general public believes it to be true. But is it? Many people don't even stop to think about questions like this. The idea that living things have evolved over millions of years has become deeply ingrained in our culture. But does the theory of **evolution** do the best job of explaining how we got all the variety of living things that we observe today? Is there a better way to explain the diversity of plants and animals on our planet? Creationists believe there is, and their explanation begins with an eyewitness testimony—the testimony of God's Word.

5A

EVIDENCE OF CHANGE

Do living things change?

Where do we come from? How did we get here? These are questions that people have wondered about for thousands of years. Many people believe that living things have *evolved*—that simpler organisms have slowly changed into new and more complex organisms over millions of years. The Bible, though, teaches that God *created* all living things, and that He took only six days to do it. Judging from records in the Bible, the earth is only about 7000 years old. Which model—the evolution model or the Creation model—better explains the evidence that we observe today? We'll look at each model's ability to explain the evidence later. First, though, we need to examine another question: Is there, in fact, evidence that living things *can* change over time?

5A Questions

- What evidence is there that living things change over time?
- What are fossils?
- How does a person's worldview affect how he constructs a history of change in life on Earth?

5.1 CHANGE IN THE PAST

Bones, teeth, shells—all over the world, people have found the remains of plants and animals that lived long ago. These preserved remains of plants and animals are called **fossils**. Fossils come in many forms. Some are complete organisms preserved in ice, tar, amber, or other materials. Other fossils are shells or bones that have not yet decayed.

By studying the fossil record, we know that there were many kinds of plants and animals in the past. Many of them are no longer around today. Dinosaurs like *T. rex* were once common but now are known only from their fossilized remains. We can also see that some kinds of organisms are less common now than in the past. Fossilized redwood trees, for example, are found throughout much of the Northern Hemisphere, but the current range of living redwoods is only a small fraction of that. All of this points to a world in the past that was different from the world we know today.

5-1
Evolutionists believe that birds evolved from dinosaurs similar to this *Microraptor*.

FOSSILS: A RECORD IN THE ROCKS

Mineralized Fossil. If an organism is buried by wet sediment, chemicals in the water can leach into the small spaces within the organism's tissues. Some of these chemicals can eventually harden. Slowly, the organism's tissues are replaced by an exact copy made of stone. This piece of petrified wood is a kind of mineralized fossil.

Carbonized Fossil. Carbonized fossils are formed when a buried organism decays, leaving behind only a thin film of carbon. This carbon film is like a printed image of the organism. Many fossilized plants exist as this kind of fossil.

Molds and Casts. Sometimes, buried organisms decay completely without becoming mineralized, leaving behind an empty space. The empty space, called a *mold*, is in the shape of the decayed organism. In such cases, more sediment may eventually fill the space and harden, creating a *cast*. The cast is a copy of the original organism. Molds and casts are often found still joined together. This fossil of a trilobite is the cast portion of a mold and cast fossil.

Trace Fossils. A trace fossil is evidence that an organism leaves behind that is not part of the organism itself. Fossilized footprints, imprints, burrows, and bite marks are all examples of trace fossils.

Original Material. In some cases, a fossil may consist of all or part of the original organism—not a mold, a cast, or mineralized remains. One type of this fossil is *amber*. Amber is fossilized tree resin. If an insect or animal is trapped in sticky tree sap and the sap gets buried in sediment, over time the resin can be transformed into amber. The insect or animal remains encased in the hardened amber.

5.2 CHANGE IN THE PRESENT

As you learned in Chapter 4, groups of organisms can change even without the help of humans. Genetic drift and natural selection can alter the ratios of traits within a population. We've seen before how populations can develop adaptations that make them more likely to survive in order to reproduce.

The evidence shows that living things have changed in the past, and we see that living things are changing in the present too. These are things that we can easily observe. But what are we to make of all this evidence? Could complex animals really have evolved step by step from single cells over millions of years? Evolutionists and creationists strongly disagree on the answer to that question. Why is that so? It goes back to the issue of worldview.

5.3 WORLDVIEW AND THE HISTORY OF CHANGE

Evolutionists and creationists don't agree about the meaning of change in living things because they start with different beliefs about the story of the world. Evolutionists view the world through the lens of naturalism. Most of them do not believe that God created the world. They do not believe that living things were carefully engineered by a wise and powerful Creator. They believe that all the living things that we can observe are the result of entirely natural and random processes taking place over many millions of years.

Christians view the world differently. They believe that God created living things, just as the Bible tells us in Genesis. The Bible describes God as all-knowing and all-powerful. The God of the Bible is clearly capable of creating the amazingly complex plants and animals that we see in the world today.

So which of these two beliefs is right? Science cannot *prove* that one or the other is correct since the world's past is no longer observable. What we can do, though, is examine how each of these belief systems—evolutionism and creationism—explains the evidence that we observe today.

5A REVIEW QUESTIONS

1. Identify each of the following fossils as either a mineralized fossil, a carbonized fossil, a mold and cast, a trace fossil, or original material.

a.

b.

c.

2. How does the fossil record show that life has changed on Earth?
3. Why do evolutionists and creationists view the history of change in living things differently?
4. A friend tells you that scientists have proved the theory of evolution. How might you respond?

5B

INTERPRETING EVIDENCE WITH EVOLUTION'S STORY

How do evolutionists explain change in living things?

5B Questions

- What is Darwinism?
- What is the role of genetics according to the modern synthesis?
- How do evolutionists interpret the evidence for change in living things?
- What are some of the problems with the evolutionary story?

5.4 A SHORT HISTORY OF DARWINISM

A Theory Takes Shape

Born in England in 1809, Charles Darwin had shown a keen interest in collecting and life science even as a boy. His father expected him to become a doctor, but he did poorly in medical school. Darwin's father next arranged for him to study for the ministry, but by 1831 he had lost interest in this as well. In August of that year, one

of his professors informed him of an opening aboard HMS *Beagle*. The ship would be making a voyage of discovery and needed a naturalist, someone with knowledge of the life sciences. Darwin's father opposed the idea at first but later agreed to fund his son's passage. The *Beagle* sailed in December. Darwin did not set foot in England again for almost five years.

In nineteenth-century England, most people believed that the earth was only a few thousand years old. They also believed in the *fixity of species*, the idea that species do not change over time. But by this time many scientists were beginning to doubt these beliefs. New theories discussed how living things might change over time. One such theory was the *theory of acquired characteristics*. This theory proposed that organisms could acquire new traits during their lifetimes and pass them on to their offspring. Darwin's own grandfather had written about an idea like this one in 1796.

Another idea that gained popularity during Darwin's early years was the concept of **uniformitarianism**. Uniformitarianism suggests that all of Earth's physical processes, such as erosion and mountain building, happen at the same rate in the present time as in the past. Because these processes happen slowly today, many scientists assume that vast amounts of time were needed in the past to form Earth's present physical features. Thus, uniformitarianism suggests that the earth could be millions, even billions, of years old.

Darwin was familiar with many of these new ideas before he sailed from England. While at sea he read more books on natural history from the ship's library. The ship's captain gave him a copy of Charles Lyell's *Principles of Geology*. Lyell's ideas about uniformitarianism greatly influenced Darwin's thinking.

5-2

Uniformitarian geologists believe that rock layers like these were formed from sediments that accumulated over millions of years.

The *Beagle*'s main duty was marine charting, a slow task. This gave Darwin plenty of time to spend on shore. There he gathered animals, plants, and fossil specimens and made many notes on local geology. By the end of the voyage, Darwin had filled fourteen field notebooks with observations. When he returned to England, his collections were given to experts who could identify the plants and animals that he had sampled.

Darwin had been impressed by the animals on the Galápagos (guh LAH puh gus) Islands off the coast of Ecuador. He had noticed that the mockingbirds on some islands varied slightly from those on the South American mainland. Later he also learned that some of the Galápagos birds that he thought were unrelated species were in fact each a type of ground finch. Like the difference between mainland and island mockingbirds, the finches also differed slightly from island to island. For example, the finches on one island had large, strong beaks for cracking the large seeds that grew on that island. But the finches on another island had smaller beaks for cracking the smaller seeds found on that island.

5-3

Sketches of four species of Galápagos finches drawn by John Gould, the zoologist who identified many of Darwin's bird specimens from the *Beagle* voyage

Darwin's Theory of Natural Selection

In 1859, Darwin published his book *On the Origin of Species by Means of Natural Selection*. In this book he described the **theory of natural selection** and gave examples from what he saw during his journey. The main points of his theory are the following:

» In every species more young are born than can survive.

» The individuals in a species compete for the things needed to survive, such as food, shelter, and mates.

» Some individuals have traits that make them better able to compete for those things than others. Those individuals will survive and reproduce. This process is now sometimes called the *survival of the fittest*.

» Individuals that reproduce pass on their traits.

» With each new generation, the traits that better allow survival become more common.

Darwin's book sold very well and his theory became well-known. But Darwin did not know how new traits might be formed in a population of organisms. In 1859, the inner workings of the cell were still very poorly understood.

The Theory Today

During the twentieth century, scientists learned how organisms pass traits from one generation to the next. They discovered the role of DNA and genes. And they also began to see how population genetics affects the changes that can happen in a group of organisms. All these new findings were woven into Darwin's theory of natural selection. Scientists understood that for natural selection to work, new traits had to be produced. They also knew that since genes produce traits, then it is the genes that must change. And since genes are sections of DNA, the changes must occur within the structure of the DNA molecule. These random changes in DNA are called **mutations**. According to the new version of the theory, if the mutations help an individual survive and reproduce, then they will be passed on to the individual's offspring. Today, this blend of mutation and natural selection is often called the **modern synthesis**.

5.5 CONSTRUCTING THE OLD-EARTH STORY

As we've noted, the story of evolution begins with an old earth. But as you'll soon see, evolution not only assumes that the earth is old—the theory *requires* the earth to be very old. Let's look at how the theory of evolution attempts to explain the history of change on Earth.

Fossil-bearing sedimentary rocks (rocks formed from sediments such as sand, silt, or mud) are found all over the world. Evolutionists believe that these layers of sediment were laid down over many years in streams, lakes, and oceans. The present rates at which sediments are laid down and turned into rock are indeed very slow. At these rates, it would take many millions of years to form the rock layers that we see today. But these rocks do not have labels that say how old they are! Scientists attempt to find out the ages of rocks by a process called **dating**.

There are two ways to do this. In the first, *relative dating*, rock layers are compared to other layers near them. According to uniformitarian geology, the oldest rocks in a layered column of rocks must be those at the bottom of the column. Younger rocks are found at the top of the column. This assumption affects how fossils are dated. If a fossil is found in a certain rock layer, and a second fossil is found in a lower layer, then the second fossil is assumed to be older than the first.

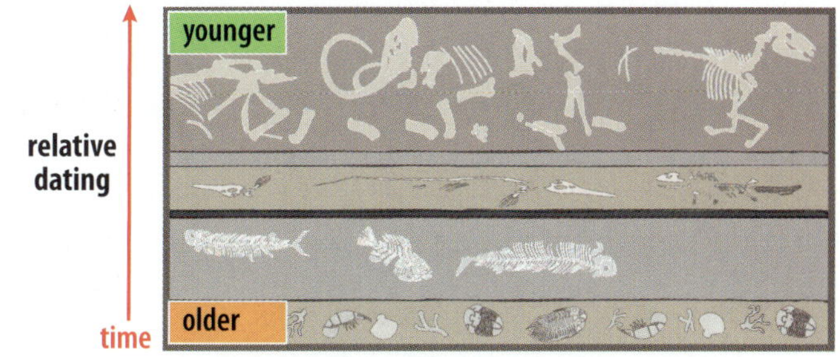

The second type of dating is known as *absolute dating*. Absolute dating relies on the known rates of radioactive decay of certain elements found in igneous rocks. Unstable *isotopes* (radioactive variants) of these elements slowly decay into stable (nonradioactive) isotopes. Scientists can measure the amounts of the unstable and stable isotopes in an igneous rock layer and use this information to calculate its age. If a sedimentary rock layer that contains fossils is found below the igneous layer, scientists then assume that those fossils are older than the calculated age of the igneous rock layer.

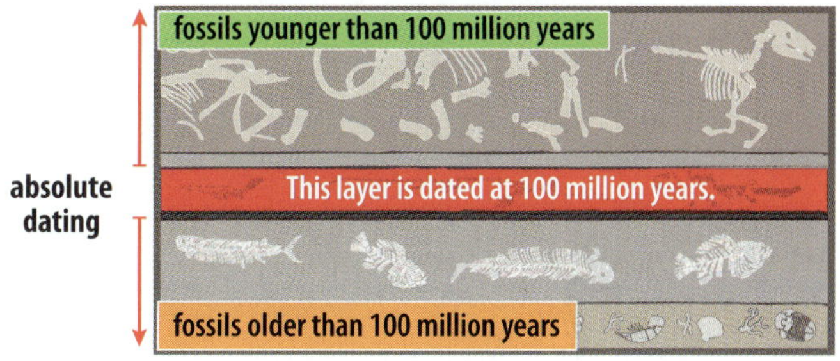

A type of absolute dating that involves measuring isotopes in the fossils themselves is called *radiocarbon dating*, or *carbon-14 dating*. Most of the carbon on Earth exists as carbon-12. But a very small fraction of carbon exists as radioactive carbon-14, which decays into carbon-12. Plants and animals take in carbon while they are alive, including some carbon-14. When they die, they no longer take in carbon. Scientists can measure the amount of carbon-14 left in a fossil and then calculate the fossils age. This process is used only to date fossils that are presumed to be very "young," since carbon-14 decays rapidly compared with other isotopes. It is believed that carbon-14 dating can be used to date fossils only up to 100,000 years old.

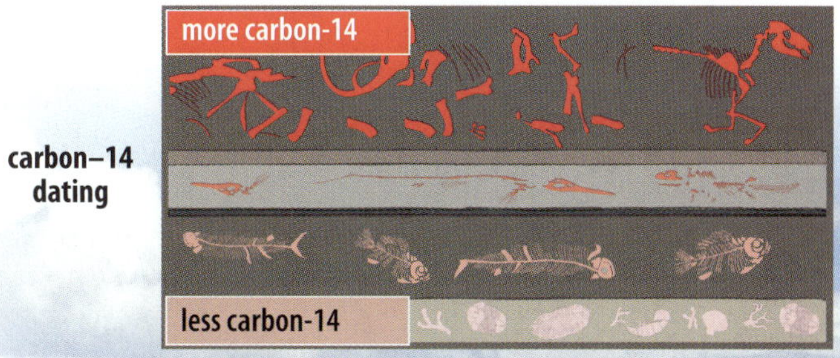

5.6 LOOKING AT THE EVIDENCE THROUGH AN EVOLUTIONIST'S LENS

The Evolutionist View of the Fossil Record

Evolutionists assume that the change from one kind of organism to another is very slow. To get from a simple one-celled organism, like a bacterium, to a complex multicellular plant or animal would take many millions of years. Evolutionists believe that the fossil record provides evidence for this slow process. In general, simpler organisms are found in the lower, older layers of sedimentary rock. More complex organisms are found in younger layers of rock. Evolutionists have argued that many kinds of organisms in the fossil record exist in a series of forms. They say that these forms show evolution in action. The figure at left is a branching diagram of one such series—the supposed evolution of horses. Each point where the diagram branches represents a presumed **common ancestor**.

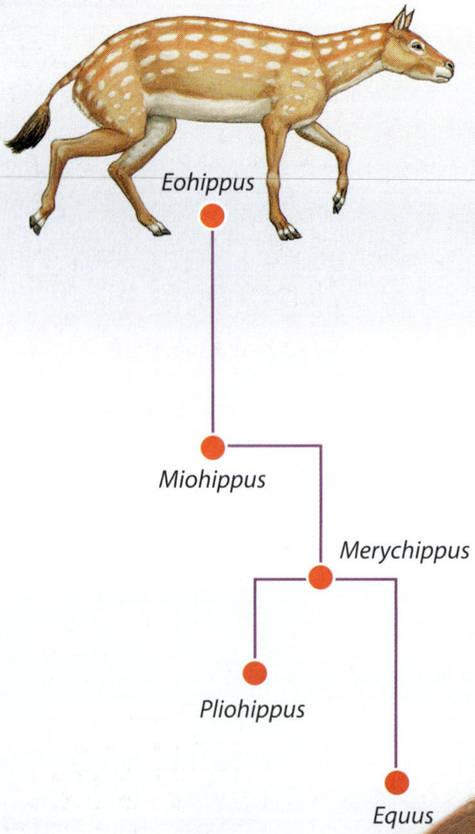

Many kinds of plants and animals found in the fossil record are no longer living today. There is evidence in the fossil record that many of these organisms became extinct at about the same time. This is often called a *mass extinction* event. Evolutionists are not certain about what could have caused such events. But they also believe that mass extinctions made new opportunities for evolution. Plants and animals that survived the event had opportunities to expand into new kinds of habitats. Evolutionists believe that this allowed natural selection to produce new kinds of organisms.

The Evolutionist View of Genetics

Evolutionists believe that the presence of DNA in all organisms supports their theory. They believe that the more similar the genes of two different species, the more closely related they are. As you read earlier, for evolution to be true, organisms must develop new traits. Evolutionists believe that the information needed to produce new traits is caused by mutations in DNA. But a mutation is in fact a fairly rare event. Most cells do not experience a mutation during replication. And most mutations happen in body cells, where they cannot be passed on to offspring. Only mutations in the DNA of gametes can be passed on to offspring.

How many genes must change to produce a new kind of organism? Evolutionists believe that dogs and cats evolved from an ancient common ancestor. Some of their genes are quite similar. For example, both have genes for the blood protein hemoglobin, with only slight differences. Many other genes may also differ only slightly. But thousands of their genes are *very* different. For just one of these animals to have evolved, millions of mutations would have been required. Because mutations are rare and many are needed for evolution to be possible, a lot of time is needed. This is another reason why evolutionists believe that the earth must be very old.

5-4

How many generations would it take to get from one pair of original animals to a variety of animals?

close relatives *very distant relatives*

OTHER PIECES OF THE EVOLUTIONARY STORY

1 ***Homologous Structures***. Many animals share certain features. For instance, most vertebrates have limbs. If you strip away the fur, skin, and muscle, their bones look very similar. An evolutionary biologist calls these similarities **homologous structures**. Evolutionists believe that organisms that share similar structures must also share a common ancestor.

2 ***Vestigial Structures***. A **vestigial structure** is one that, according to evolutionists, no longer has a function. To them, the structure may have served a purpose in the past, but the organism has since evolved to a point where the structure is no longer needed. One such structure is the pelvic spur of a boa (see inset), thought to be the remnant of hind legs.

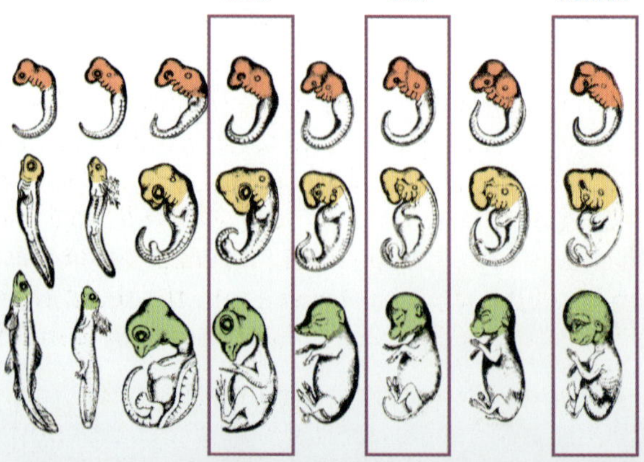

3 ***Embryology***. The embryos of many vertebrate animals resemble each other in their very early stages of development. They all have, for example, structures called *pharyngeal pouches*. These pouches later develop into parts of the face and neck. Evolutionists believe that these similarities in embryos suggest that all vertebrate animals had a common ancestor.

5.7 CRACKS IN THE EVOLUTIONARY STORY

Evolution is one of the most widely accepted theories within the world of science. Some scientists claim that an understanding of evolution is vital for studying every other area of biology. And it is no secret that evolutionists are often openly hostile toward and scornful of creationists. But is the case for evolution as solid as its supporters would have us believe? On closer examination, many of evolution's basic assumptions don't hold up well.

A Rocky Record?

Take the issue of dating rocks. Though there are several ways to date the ages of rocks, all of them depend on assuming certain things to be true. One assumption is uniformitarian in nature. It supposes that the rates at which unstable isotopes decay has remained constant throughout history. Another assumption is that when the rocks first formed, only the unstable isotopes were present and none of the stable isotopes. Are these assumptions valid? There is no way for us to know. In most cases, humans were not around when the rocks were formed. No one was there to measure the amounts of isotopes in them! In addition, it is quite common for absolute dating methods to yield ages that are thought (and sometimes known) to be too old or too young. What do scientists do with those age estimates? They simply discard them!

Carbon-14 dating presents a special problem for evolutionists. Many fossils thought to be millions of years old have been proved to contain measurable amounts of carbon-14. This suggests that they are far younger than believed and much too young to fit neatly into the evolutionary story. This is just one of the many problems with the evolutionary view of the fossil record.

Perhaps the biggest problem is the lack of transitional forms (sometimes called missing links). If evolution is true, then there have been many millions of organisms in the past with body parts that are intermediate between early ancestral forms and later fully developed forms. Yet there is little, if any, evidence of such organisms in the fossil record. For example, many evolutionists believe that jaws first evolved in fish. They believe that the jaws of these fish evolved from the gill arches of jawless fish. But there are no fossils of such partially jawed fish. In fact, partially formed jaws would give a fish no competitive advantage. A fish with such jaws would likely not survive to pass on its trait. And over and over again in the fossil record we see distinct kinds of animals with fully formed and fully functional features.

5-5

The coelacanth is likely the most well-known example of a living fossil.

There's a further problem with the fossil record. It doesn't just contain organisms that are different from those we observe today. It also contains *many* preserved plants and animals that are nearly identical to those we see today! Evolutionists do not have an answer for why these *living fossils* appear suddenly in the fossil record and then remain mostly unchanged until today.

One of the most famous living fossils is the coelacanth. It was long believed by evolutionists that coelacanths became extinct along with the dinosaurs about 65 million years ago. They also believed that coelacanths lived in shallow seas and used their lobed fins to help push themselves along the sea bottom. This, they thought, may have been a step in the evolution of legs and walking on land. Imagine their surprise when a living coelacanth was caught off the coast of South Africa in 1938! Afterward they learned that coelacanths live in the deep ocean, not in shallow water, and use their fins only for swimming, just like other fish.

Evolution in Action?

Scientists estimate that there are nearly nine million species on Earth today—give or take a few million. Several hundred thousand more have been identified in the fossil record. Evolutionists believe that all of these organisms arose from one common ancestor. All these changes from one kind of organism to another were supposedly caused by changes—mutations—in DNA.

But is this kind of change what we observe happening today? Far from it! What really happens when random changes happen in an organism's DNA? Let's imagine something like this happening to a finely engineered car. Suppose a random mistake is made in copying the blueprints for the car, and the result is one less bolt holding on a bumper. Would the car still work? Of course it would! But what if the one random mistake caused there to be no connection between the fuel tank and the engine? The resulting car would not be very good for getting around! Mutations are kind of like that. Some may be fairly harmless, but others may be fatal. And each nonfatal mutation has a chance of being passed on to the next generation.

The random mutations that accumulate within the DNA of a population of organisms increase its **genetic load**. Every generation of that organism is like a car rolling off an assembly line with more random mistakes in its construction than the car before. How many cars can be made before the plant produces only useless, nonfunctioning vehicles? Similarly, in every generation that passes, we see more and more mutations piling up within the DNA of all living things—including humans—making them less fit for long-term survival. Given enough time, the genetic load of every kind of organism will result in its eventual extinction. Random mutations do not add information—they degrade it.

5-6
The small size of the population of lions in Africa's Ngorongoro Crater puts them at risk for an increased genetic load.

Problems, Problems, Problems!

The other evidences for evolution that you have read about all have their own problems. The idea that homologous structures prove common ancestry is based on an assumption. Evolutionists assume that if some animals *did* have a common ancestor, then they would have homologous structures. But there is no way to confirm this assumption. One could also suppose that if some animals had a *common designer*, then that designer would engineer them to have homologous structures. Is there a reason why God should have to create unique structures for every kind of organism to prove He is the Creator? Of course not! In fact, the existence of homologous structures that perform a range of tasks—walking, flying, steering, grasping—might even suggest that the designer is very inventive!

And do vestigial structures really support the idea of evolution? Or do they simply show how much we still have to learn about God's creation? At one time, there were over 180 human structures that evolutionists considered to be vestigial. Examples include the tonsils, appendix, pituitary gland, and thymus gland. Today we know that each of these structures has a valid function within the human body. With further research, it may be that we will discover the purpose for every human structure that evolutionists once thought was an evolutionary "leftover."

And finally, the fact that some embryos look alike during their early development is not as helpful as it first appears. There are major differences between the body structures of fish and frogs for instance. That means that there are also major differences in the genes that code for those structures. Those genes are present in each animal from the time it begins life as a fertilized egg. So from the moment of conception, and long before they are ever embryos, fish and frogs are just as different as they will be once they are adults.

5B REVIEW QUESTIONS

1. Describe two early nineteenth-century ideas that influenced Darwin's thinking, leading up to his theory of natural selection.
2. Which of the following is *not* one of the main points of Darwin's theory of natural selection?
    a. The individuals in a species compete for the things needed to survive, such as food, shelter, and mates.
    b. Individuals evolve the traits that they need to adapt to their surroundings.
    c. Only the individuals with the traits best suited to survive will be able to reproduce and pass on their traits.
3. How did the modern synthesis modify the theory of natural selection?
4. What is the difference between relative dating and absolute dating?
5. Which fossils in the image at left would evolutionists assume are older, those in layer A or layer B?

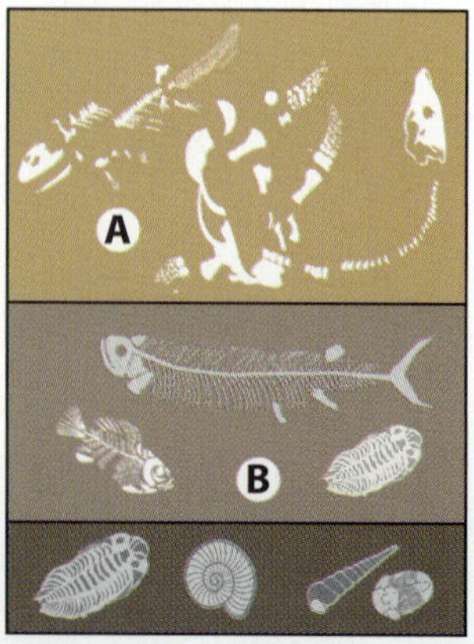

6. Why does evolutionary theory absolutely require the earth to be very old?

7. What do scientists assume to be true when calculating the absolute ages of rocks? Is their assumption a valid one?
8. Why would the discovery of one or even a few transitional-form fossils be insufficient evidence that evolution works?
9. How does the concept of genetic load provide evidence against evolution?

LAB EXERCISE

TREE TROUBLE: INTERPRETING EVOLUTIONARY FAMILY TREES

How do evolutionists try to show evolutionary relationships?

Scientists can picture the way they believe evolution occurred by using a type of branching diagram that we'll call an evolutionary family tree. The branches on the tree each represent a kind of organism. The places where the branches connect, called *nodes*, represent common ancestors. So from each node you'll see branches emerging that represent all the organisms thought to have arisen from that common ancestor.

Scientists who support evolution do not always agree on the order in which various organisms evolved from supposedly older organisms. Because of this, there are almost as many evolutionary trees as there are opinions on the order in which evolution took place! And because ideas about this order of events are constantly changing, the trees are frequently changed as well.

PROCEDURE

Shown at right are four evolutionary family trees for vertebrates (animals with backbones). Look for differences in evolutionary paths and in common ancestors on these trees. Then answer the questions that follow.

1. Describe some of the differences between the four trees.
2. Which tree(s) show(s) birds as being most closely related to mammals? How can you tell?
3. Which tree(s) show(s) bony fish and cartilaginous fish as having a common ancestor?
4. Why are no organisms named at the nodes on each tree?
5. How are evolutionary family trees an example of modeling?

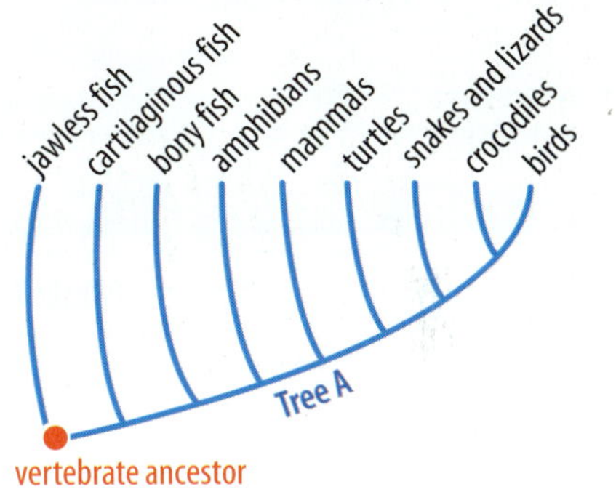

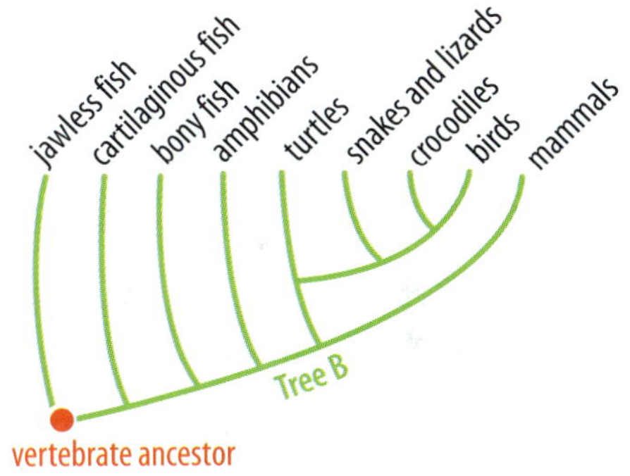

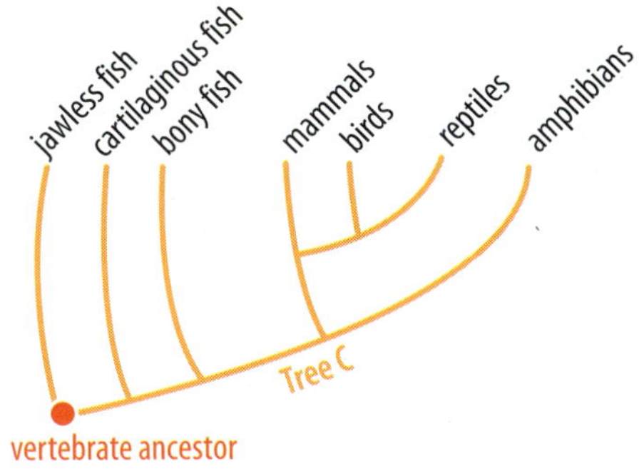

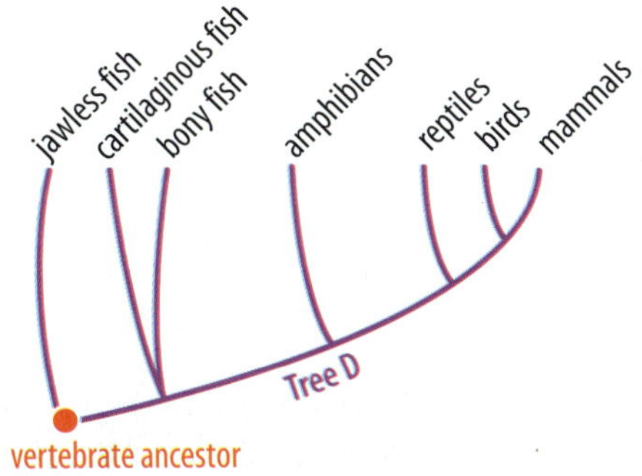

5C Questions

- What does the Bible say about Creation?
- How do creationists interpret the evidence for change in living things?
- Does the evidence for change in living things fit the Bible's story?
- Is it possible to harmonize evolution with the Bible?

5-7

Change happens only within created kinds.

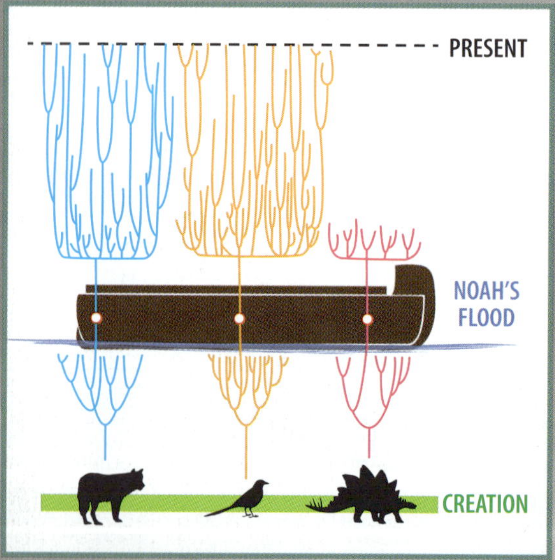

5C

INTERPRETING EVIDENCE WITH THE BIBLE'S STORY

How do creationists explain change in living things?

5.8 THE TESTIMONY OF SCRIPTURE

To understand the creationist view of changing life on Earth, we must keep in mind the Creation story. In Chapter 1 we presented an overview of the Bible's "big picture"—Creation, Fall, and Redemption. Now we need to focus in on a few parts of that story, namely Creation Week and the Flood. Does the evidence for change we've looked at so far fit the Bible's story?

The Biblical Kind

There are two important things to take note of in Genesis 1. The first is the repeated use of the word **kind**. Each time the story describes God's creation of a type of organism, whether plant or animal, the Bible says that it was created "after his kind" or "after their kind." This tells us that God created types of organisms that were distinct from each other. Plants were plants, birds were birds, and beasts of the field were beasts of the field. Although it does not tell us how many kinds of plants and animals were created, the Bible assures us that every created kind existed in the very beginning. There is no evidence anywhere in the Bible that one kind of organism has ever changed or can ever change into a different kind of organism. Darwin's finches, for example, have adapted to eating different sorts of food on different islands, but they're still finches. Be careful not to equate the biblical kind with the modern concept of species! There are a great many species on Earth, but many of them can interbreed with each other and produce fertile offspring. These similar species would be grouped together in a biblical kind.

The second thing to note is that each kind of organism was created fully formed. When God called for plants on Day 3, plants came forth. When He called for cattle and creeping things on Day 6, that's exactly what the earth produced. God did not call forth living things as unicellular life that then needed to evolve into plants and animals. They have existed as plants and animals ever since they were first created.

The Fall and the Flood

The Bible tells us that God's original creation was "very good." There was no death or disease. All of God's creatures had what they needed to live and thrive. But mankind sinned against God and brought the earth under a curse. Sickness and death entered into the world. Even dinosaur fossils show signs of growth defects, arthritis, and bone fractures.

After Adam and Eve were banished from Eden, mankind's wickedness increased to the point that God chose to bring a terrible judgment on the earth—the Flood. This global disaster covered the entire earth in water and lasted about a year. All the air-breathing animals that lived on land perished, along with most of the people. Only Noah's family and the animals onboard the Ark survived.

5-8
Same kind, different species

5-9
Since the Flood was a worldwide disaster, we should expect to see evidence of it in the fossil record.

These two events—God's creation of distinct kinds of life and the Genesis Flood—are significant to the creationist view of the history of life on Earth. If God created distinct kinds of life, then we should find *no* evidence that one kind of life has changed into another. If there really was a global flood, then we should be able to find evidence for it all over the world. And if that flood destroyed almost all life on Earth, we should also be able to find evidence of the large numbers of plants and animals that died in that event. So what do we actually find when we look at the evidence?

5.9 LOOKING AT THE EVIDENCE THROUGH THE BIBLE'S LENS

The Biblical View of the Fossil Record

The fossil record is the creationist's friend, for it clearly shows three things. First, most fossils are found in sedimentary rock. This kind of rock is formed from layers of *waterborne* sediment, such as sand, silt, or clay. Vast areas of the earth are covered by this sort of rock. Many of the fossilized animals found in sedimentary rock also show strong evidence of having died in watery surroundings. They are often found in what is known as the "dinosaur death pose." In this position, the animal's neck and tail are severely arched over its back. This posture is seen today in animals that have drowned.

5-10

Many fossils show evidence of death by drowning.

5-11

The preservation of fine detail in a fossil requires rapid burial.

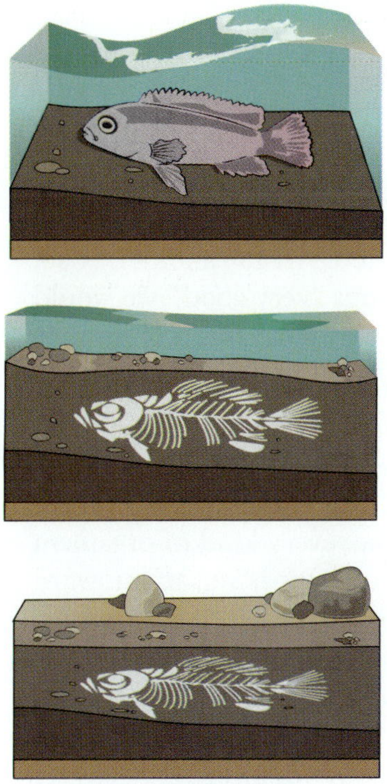

Second, it must be noted that in order for fossils to form, an organism must be buried *quickly*. Evolutionist stories of fossil formation often describe an organism dying in a river or lake, then slowly being covered by sediment. But that is not what happens today. Animals that die in water are usually quickly scavenged. What remains afterward normally decays away long before it can be covered in sediment. Only organisms that are quickly buried are protected from the effects of scavengers and decay.

Finally, the fossil record shows that fossils are often found in what are known as *fossil graveyards*—large numbers of fossils found together in one area. These graveyards suggest that many organisms died suddenly and were quickly buried.

Billions of organisms, all fully formed and distinctly different, quickly buried in waterborne sediment. This is *exactly* what we would expect to see if the Genesis Flood were the worldwide event that God's Word says it was. The very rocks of Earth give testimony of the truthfulness of the Bible's story.

THE CAMBRIAN EXPLOSION: LIFE, SUDDENLY

In Subsection 5.7, we saw that the fossil record lacks evidence of transitional forms. One instance of this has proved especially troublesome for evolutionists, including Charles Darwin. Scientists call it the *Cambrian Explosion*. The Cambrian Explosion describes the sudden appearance in the fossil record of many kinds of multicellular marine animals. Almost all of the major animal groups that we see today are present among the Cambrian fossils. There are no transitional forms between these Cambrian animals and the Precambrian life forms found in "older" rock layers. So far, evolutionists have not been able to satisfactorily explain how these animals could have so quickly evolved in such a short amount of time. Their theories about rapid evolution run into many of the same problems facing theories of gradual evolution.

Creationists believe that the Cambrian animals were formed by God during Creation Week, along with every other form of life. The Cambrian fossils are mostly the remains of small and not very mobile animals. They would have been among the first to perish during the early stages of the Flood.

The Creationist's View of Genetics

If the Bible's story of Creation is true, what should we expect to see with regard to genetics? First, we would expect to see that each created kind of organism has at least some DNA that is unique to that organism. And since God created living things to reproduce after their kind, we should not expect to find evidence that the DNA of one kind of organism can change into the DNA for another. Also, since God created the earth and its creatures "very good," we would expect any changes in their DNA codes (mutations) to be mostly harmful. Is this, in fact, what we observe in nature?

5-12

Albinism is an example of loss of genetic information. Albino animals are less likely to survive in the wild.

As was pointed out earlier in Subsection 5.7, the harmful character of mutations is exactly what we observe in nature. Everywhere that scientists look in the world of living things, there seems to be a downward trend. With the passage of time, every kind of organism is experiencing an increasing genetic load. This is not the upward march of progress that we would expect to see if evolution were true. Rather, it is the kind of increasing disorder we would expect as the result of God having placed the earth under a curse for Adam's sin. In the meantime, God has placed mechanisms within every cell that try to minimize the damage done to DNA by mutations. But this design only slows the rate of harmful change; it doesn't prevent it completely.

We saw in Chapter 3 that DNA is an incredibly rich information-storage molecule. Every plant and animal appears to be divinely engineered, right down to the information found in its DNA. God built a wondrous amount of variation within the DNA of the first created kinds. This allowed them to adapt quickly and fill the earth, especially after the Flood, producing the many species that we see today. But the change from one *kind* of organism into a different *kind* requires new information. This new information must be added to the DNA of an organism. But we do not see such addition of new information happening in nature today. Even some evolutionists admit that they are hard-pressed to come up with even *one* known example of such an addition.

In Chapter 4, we saw that small, isolated populations of organisms are more prone to genetic drift. Yet in the real world, scientists often work as hard as they can to *prevent* such isolated populations! Why is that? It's because even evolutionists can see that most mutations are harmful and that they are more likely to become established in small populations. That's why scientists try to prevent small, isolated groups of organisms. Scientists know that such populations are more likely to become extinct than large, well-connected populations.

GLOBAL SPECIES MANAGEMENT PROGRAMS

China's blue-crowned laughingthrush is believed to be critically endangered. Only about 250 are thought to exist in the wild. But this rare bird has an ally in its fight against extinction. The blue-crowned laughingthrush is one of the species being aided by a Global Species Management Plan run by the World Association of Zoos and Aquariums (WAZA). Zoos taking part in this program make every effort to ensure that the laughingthrushes in their care remain as genetically diverse as possible. This involves detailed record keeping and the exchange of birds between WAZA member zoos. This is just one example of many similar programs now in operation all around the world. All of these zoos are working together to make sure that their laughingthrushes don't continue to exist as small, isolated populations. Yet that is what evolutionary theory says must happen in order for new traits to become established in a population. In real practice, the biologists who run these programs confirm what creationist theory predicts—that small, isolated populations lead to a loss of genetic information.

5.10 CAN THE BIBLE AND EVOLUTION WORK TOGETHER?

The church of the late 1800s failed to stop the rising tide of Darwinism. Many Bible scholars and laypersons alike began to question basic church doctrines. They began to doubt the existence of miracles, the virgin birth, and even the deity of Christ. People started thinking that the stories of Creation and the Flood might not be real history but only fables. Such stories could tell us something about God's *nature*, but they don't describe things that God actually *did*.

At the same time, trust in the abilities of science and technology was growing. People marveled at the invention of steamships, railroads, telegraphs, and electric lighting. The great scientists and inventors of the time were often as famous as movie stars are today. As mankind's understanding of the natural world rapidly grew, the possibilities for the future seemed endless.

Faced with this growing faith in science, some theologians tried to reconcile naturalism with the Bible. Some simply squeezed the millions of years needed for evolution somewhere into the Bible's story. They did so in different ways. Some said that the millions of years happened before Creation Week. Others said that the millions of years took place in between each of the six days of Creation. Still others concluded that the days of Creation Week were really vast ages themselves instead of real twenty-four-hour days.

But attempts to harmonize Darwinism and the Bible run into problems. First, they ignore the plain meaning of the text. The Bible says that the earth was created first, then the sun. Evolution claims that it's the other way around. The Bible also says that plants were created before the sun. That would clearly not work if plants lived for millions of years before the sun was created! Genesis is written in a way clearly meant to be understood as real history. Jesus also quoted Genesis as though it described real people and events. So too did the authors of the Bible. The Creation story is not a myth—it's a vital key to properly understanding mankind's sin and need for salvation. And there is danger in rejecting Genesis as real history. After all, if the Bible doesn't tell the true story of the beginning of the world, how can it be trusted when it speaks about sin, salvation, and eternal reward and punishment?

Ignoring Genesis as real history presents a second serious problem. The Bible teaches that death hasn't always existed. It entered into the world as a result of Adam's sin and through Adam to the rest of humankind. Christ's death paid the penalty for sin, and His resurrection is the guarantee of eternal life for those who believe. In time, God will create a new heaven and a new earth in which there will be no more death. But if evolution is true, then death has always existed, even before there were any people. In evolution's story, death is normal—and even necessary. And there is no goal or purpose in evolution's story. In fact, the naturalistic worldview teaches that all life will ultimately cease to exist.

Naturalism gives no meaning to life. Its worldview is at complete odds with biblical teaching. God's Word reveals that all of creation has a purpose—to glorify God. We see too in the Bible that redeemed mankind has a special purpose as well—to live for eternity in perfect, unbroken fellowship with our Creator. Evolution offers only a random, impersonal, and ultimately pointless process. What a difference between that bleak picture and the present purpose and future hope of eternal life that all believers have in Christ!

5C REVIEW QUESTIONS

1. Was it necessary for God to create every species in order to have the diversity of life that we see today? Explain.
2. Describe two ways that the fossil record affirms the truth of Scripture.
3. Though genetic load poses risks for small populations today, why might this not have been a problem for the animals exiting the Ark after the Flood?
4. Give two ways that the fossil record shows evidence of animals dying in a universal flood.
5. How would a creationist explain the role of mutations in living things?
6. Describe two ways that people have tried to reconcile evolution with the Bible's story of Creation.
7. What is one of the major problems arising out of a person's rejection of Genesis as real history?

ETHICS Eugenics

The popularity of Darwin's theory of natural selection had an unpleasant consequence in the late nineteenth and early twentieth centuries. As people began to abandon the Bible's story of Creation and mankind's special place within it, they began to view humans as simply highly evolved life. This led naturally to the belief that some races of humans were also "more evolved" than others. This belief was widespread in Western Europe, whose citizens considered themselves the most advanced nations on Earth. Some people even took this belief to a terrible, but logical, conclusion. If humans were merely evolved animals, and evolution relied on the "survival of the fittest," then why not give evolution a helping hand? Many Western European and American whites began to actively promote the concept of *eugenics*. Eugenicists believed that only the most "fit" humans should be allowed to reproduce. They worked to enact policies that would discourage or prevent "inferior" persons from having children. The German dictator Adolf Hitler learned from American eugenicists and was able to apply these principles in what we know as the Holocaust. During World War II he began a program to permanently rid Germany and its territories of *Untermenschen* ("subhumans"). In the end, over six million Jews were killed, along with many more Gypsies, Slavs, disabled persons, and others deemed unfit by the Nazis.

Use the Ethics box above to answer Questions 8–15.

8. Do a web search using the keyword "eugenics." Write a brief summary of what you learn about this issue.
9. What seemed to be the most important ethical issues regarding eugenics?
10. What types of people were most likely to be affected by decisions made on the basis of eugenics?
11. What does the Bible have to say on this topic? (*Hint*: Don't try to find the word *eugenics* in the Bible. It's not there! Instead, think about what the Bible has to say about how we treat others.)
12. On the basis of your answer to Question 11, what are acceptable and unacceptable options for believers?
13. What might be the consequences of adopting the options you described in Question 12?
14. Is there evidence that eugenics is still an issue today?
15. What view (or actions) should you take regarding eugenics?

CHAPTER SUMMARY

5A EVIDENCE OF CHANGE

- Both the fossil record and the field of genetics give evidence that living things can change.
- Fossils are the preserved remains of plants and animals. They can exist in several forms, the most well-known of which are mineralized fossils.
- A person's worldview affects the way in which he interprets the evidence for change in living things. Evolutionists and creationists begin with different assumptions about how things came into existence.

5A TERMS

evolution	88
fossil	89

5B INTERPRETING EVIDENCE WITH EVOLUTION'S STORY

- Charles Darwin published his book on natural selection in 1859.
- Darwin's theory was based on observations that he made during his voyage on HMS *Beagle*, coupled with his belief in uniformitarianism.
- The theory of natural selection states that organisms with traits that give them an advantage for survival are the organisms most likely to breed successfully and pass those traits on to their offspring.
- Modern evolutionary theory, or the modern synthesis, combines Darwinian natural selection with mutations and genetics.
- Scientists date fossils either by relative dating or by absolute dating. These dating methods rely on certain assumptions that cannot be proved true.
- Evolutionists view the fossil record as evidence of gradual change in living things that happened over millions of years. But the transitional forms that such a process requires are lacking in the fossil record.
- Evolutionists believe that mutations produce new traits in organisms. These can then be acted on by natural selection. They believe this in spite of the evidence that mutations increase the genetic load of living things.
- Evolutionists also cite evidence from homologous structures, vestigial structures, and embryology as additional support of evolution.

5B TERMS

uniformitarianism	93
theory of natural selection	95
mutation	96
modern synthesis	96
dating	96
common ancestor	98
homologous structures	100
vestigial structure	100
transitional form	101
genetic load	103

CHAPTER 5

CHAPTER SUMMARY

5C INTERPRETING EVIDENCE WITH THE BIBLE'S STORY

- The Bible teaches that living things were created by God only a few thousand years ago and that living things were created as distinct kinds. It also teaches that the creation was afflicted by death and suffering as part of the Curse after mankind fell into sin.
- The Bible teaches that the earth experienced a worldwide flood. The Genesis Flood killed all humans and land-dwelling animals that breathed air, except those humans and animals on the Ark.
- Creationists view the fossil record as evidence of the worldwide Flood. Fossil formation requires rapid burial in waterborne sediment, and many fossilized animals show evidence of death by drowning.
- The fossil record lacks transitional forms. Instead, it shows distinct kinds of organisms, which accords with the Genesis account of Creation.
- Creationists do not view mutations as agents capable of changing one kind of organism into a different kind. Instead, mutations are evidence of the increasing disorder in the creation as a result of God's curse.
- Attempts to harmonize evolution with the Bible generally fall into two types. The first attempts to fit the millions of years needed for evolution somewhere into the Bible's story. The second claims that the Bible's story is not real history.

5C TERMS	
kind	106

CHAPTER REVIEW QUESTIONS

Recalling Facts

1. Give a simple definition for biological evolution in your own words.
2. A fossil is the ______ of a plant or animal.
3. (True or False) Only the hard parts of animals, such as bones or teeth, can become fossils.
4. Describe two theories that contributed to Darwin's thinking leading up to the theory of evolution.
5. In addition to his observations of local wildlife, what else did Darwin observe, collect, or take note of during his voyage aboard HMS *Beagle*?
6. What does the expression "survival of the fittest" mean?
7. According to the modern synthesis, new traits are produced by ______.
8. Absolute dating measures the ratio of stable and unstable ______ to determine the ages of rock layers.
9. Evolutionists believe that similar organisms evolved from a(n) ______.
10. The foreleg bones of frogs and foxes are similar. What would evolutionists call this an example of?
11. (True or False) The fossil record contains many kinds of organisms that are still alive and little changed today.

12. What two events in Genesis have a great impact on how creationists view the history of life on Earth?

13. What do all attempts to harmonize evolution with the Bible have in common?

Understanding Concepts

14. Which of the following is *not* an example of a fossil?
    - **a.** ground sloth bones found in a tar pit
    - **b.** bird eggshell fragments found in an abandoned nest
    - **c.** an ant found in a piece of amber
    - **d.** a hadrosaur track found in a layer of rock

15. Describe how relative dating is used to date fossils.

16. If a fossil is found in a layer of rock above another layer whose age is "known," is the fossil considered to be older or younger than the layer of known age?

17. Why can radiocarbon dating be used to determine the ages of "young" fossils only?

18. Describe three difficulties with the evolutionist view of the fossil record.

19. Why is the concept of genetic load a serious problem for evolution?

20. What evidence suggests to creationists that the fossil record was formed quickly during the Flood rather than over millions of years?

21. According to a creationist view, do mutations add increasing order or disorder to an organism's DNA? Explain.

Critical Thinking

22. Why do evolutionists and creationists look at the same fossil evidence but draw different conclusions about how it was formed?

23. Why is it a mistake to say that God created every species of living thing during Creation Week of Genesis?

Use the information below to answer Questions 24 and 25.

The peppered moth is often used in textbooks as an example of natural selection in action. The story is usually told that peppered moths are normally white and rest on tree trunks during the day. Their white coloring serves as camouflage from hungry birds. In the polluted air of England during the Industrial Revolution, tree trunks became darkened with soot. Against that backdrop, the white moths were more easily seen and eaten by birds. The occasional darker-colored peppered moth could now survive and pass on its dark coloration to its offspring. Eventually, most of the peppered moths were of the dark-colored form.

24. Would creationists agree that this is an example of natural selection? Explain.

25. Why is the case of the peppered moth *not* a good example to use in the defense of evolution?

26. The Bible teaches that death entered into the world because of Adam's sin (Rom. 5:12). How is this teaching affected by a belief in biological evolution?

27. Does a belief in evolution provide a solid basis for moral behavior? Explain.

unit 2 MICROORGANISMS AND PLANTS

FROM A PLANT GENETICIST TO YOU

I loved science as a youth—I delighted in learning how things work. In middle school I wanted to be an inventor and I tried building my own helicopter (I had some technical difficulties!). I ended up studying plant science in college—I wanted to help feed a hungry world.

By God's grace I got my PhD in plant genetics, and then I invented something called the gene gun, which had a significant impact on world food production. So God really blessed my efforts and also the efforts of my many excellent partners in research!

Even while these exciting things were happening, my wife and I were coming to Jesus and being saved. Sometime after this, I redirected my scientific research toward showing that "good science affirms Scripture." I have been doing this for the last seventeen years. I believe God blessed my gene gun work such that I could do the work I am now doing.

This last part of my scientific career has been very challenging—it is hard to swim against the current. But it has really been worth it; God has given fruit to my efforts. I have seen my work encourage the faith of many people.

Dr. John Sanford

6 BACTERIA AND VIRUSES

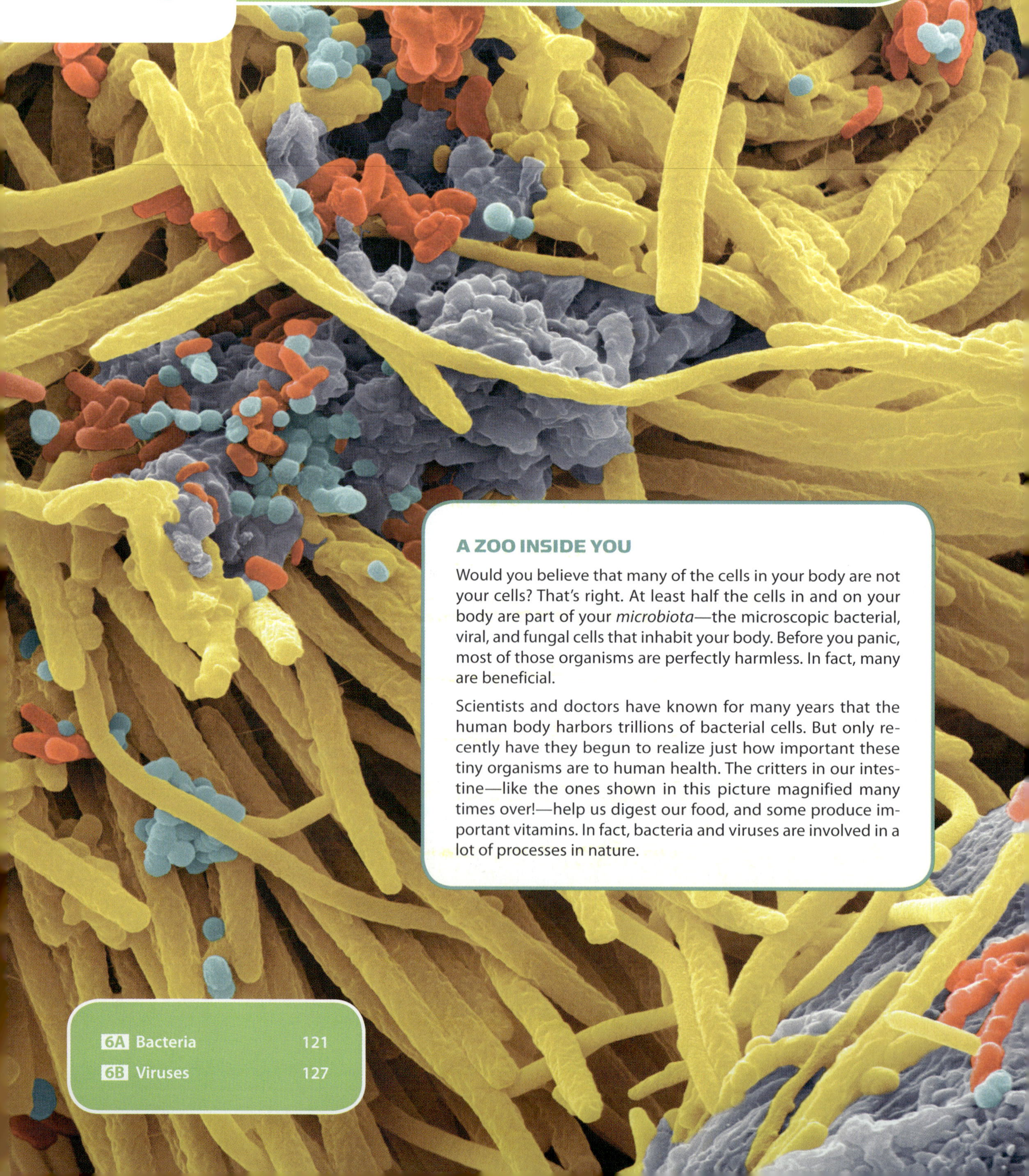

A ZOO INSIDE YOU

Would you believe that many of the cells in your body are not your cells? That's right. At least half the cells in and on your body are part of your *microbiota*—the microscopic bacterial, viral, and fungal cells that inhabit your body. Before you panic, most of those organisms are perfectly harmless. In fact, many are beneficial.

Scientists and doctors have known for many years that the human body harbors trillions of bacterial cells. But only recently have they begun to realize just how important these tiny organisms are to human health. The critters in our intestine—like the ones shown in this picture magnified many times over!—help us digest our food, and some produce important vitamins. In fact, bacteria and viruses are involved in a lot of processes in nature.

6A

BACTERIA

How do prokaryotes function without membrane-bound organelles?

6.1 BACTERIAL STRUCTURE

Your body isn't the only thing with lots of bacteria. They are, almost literally, everywhere—in the air, in soil, and in water. In fact, some estimate that a single cup of soil contains more bacteria than there are people on Earth (that's over 7 billion!). Although bacteria are an extremely diverse group that live in so many different environments, most of them have certain structures in common.

6A Questions

- What structures do bacteria have?
- What are the differences between archaea and bacteria?
- How do bacteria reproduce?
- How do bacteria affect the environment?
- Is antibiotic resistance an example of biological evolution?

BACTERIAL STRUCTURES

1. Because a bacterium lacks a nucleus, its DNA, which forms a circular chromosome, is suspended in the cytosol.
2. Many bacteria have one or more **plasmids**, small, often circular strands of DNA that are separate from the main chromosome.
3. Bacteria have ribosomes that produce the proteins they need to survive.
4. Like all cells, bacteria have a cell membrane, and many have a cell wall.
5. Some bacteria have a slimy coating called a *capsule* that covers their cell walls. The capsule prevents the bacterium from drying out.
6. Some bacteria have whip-like flagella, which the bacteria use to move, with some traveling at very fast speeds.

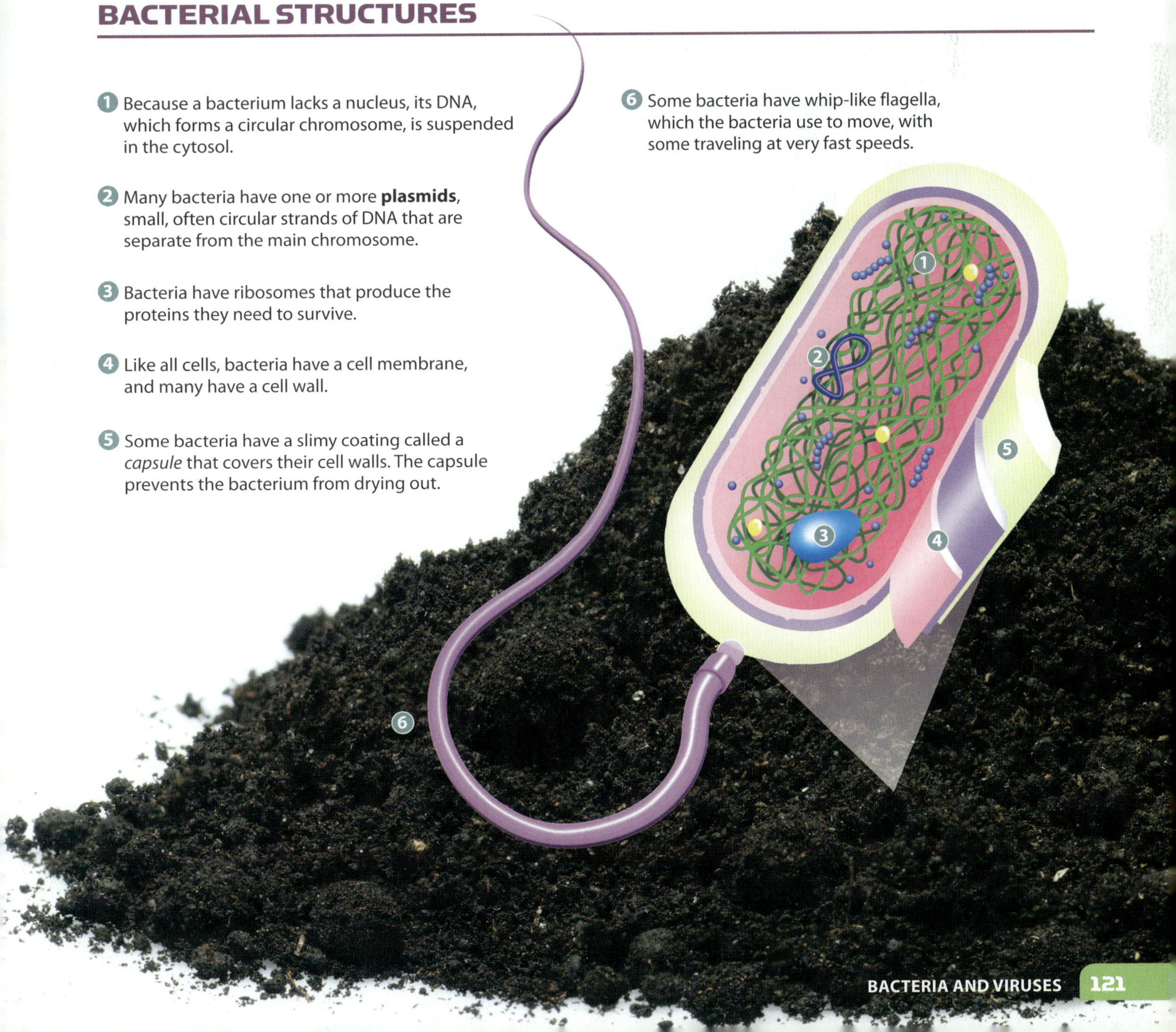

6.2 BACTERIAL TYPES

The hot, rainbow-colored waters of the Grand Prismatic Spring in Yellowstone National Park shown in this page spread would be deadly for a human. But they are teeming with microscopic life. Many bacteria live there. The water becomes cooler as it moves away from the center. Bacteria living in the different temperature areas of the spring produce the different colors. They share the water with other prokaryotes classified in the kingdom Archaea. Although like bacteria in many ways, archaea are also quite different. Their cell walls lack a molecule called *peptidoglycan* that is found in bacteria, and some of their genes are more similar to eukaryotes than to bacteria. Although archaea are most famous for living in extreme places such as hot springs, they also live in more ordinary places such as the ocean, swamps, and even the human body.

Archaea are a small group with only a few hundred recognized species. Most prokaryotes are classified in kingdom Bacteria. They lack membrane-bound organelles like a Golgi apparatus or endoplasmic reticulum. But even without these, they can still accomplish everything they need to in order to survive. Although they live in many very different places, they come in three basic shapes: spherical, rod shaped, and spiral shaped.

6.3 BACTERIAL REPRODUCTION

Like all living organisms, bacteria reproduce. Normally bacteria reproduce by **binary fission**. The DNA in the chromosome and any plasmids is replicated. The cell membrane then pinches together to form two bacterial cells. The two bacteria are identical to each other and to the original bacterial cell. Some bacteria can reproduce every twenty minutes.

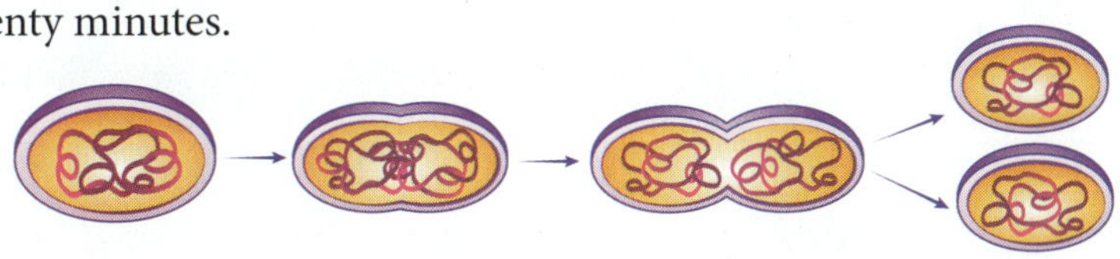

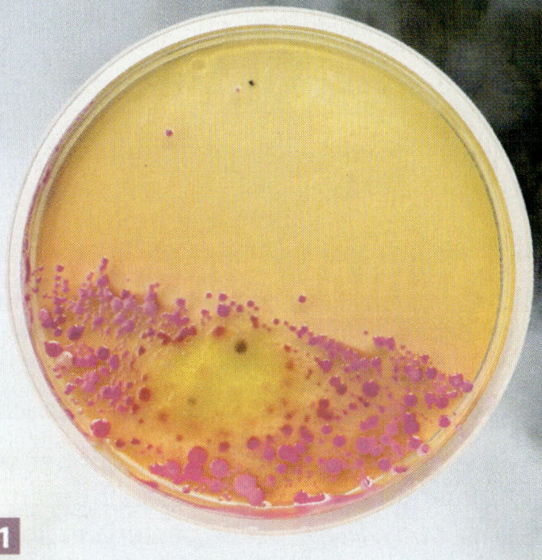

6-1

Bacterial colonies are rarely larger than half a pea.

Let's think about this. In twenty minutes, one bacterium can produce two bacteria. In another twenty minutes, those two bacteria divide to produce a total of four bacteria. So in an hour, one bacterium can become eight bacteria.

If this continued, at the end of a day one bacterium would become nearly 5 sextillion (that's 5 followed by 21 zeros!) bacteria. A lump of bacteria that large would weigh 4.5 metric tons.

But if bacteria really reproduce that quickly, why haven't they engulfed the earth? The reason the earth hasn't been destroyed by bacteria is that the conditions must be exactly right for bacteria to reproduce so quickly. And the perfect conditions rarely exist for long. Bacteria usually don't have enough food, moisture, and oxygen, and as bacteria reproduce, the amount of resources available for each bacterium decreases. So you don't need to worry about the earth being swallowed by a giant glob of bacteria.

Sometimes the conditions are so poor that bacteria struggle to survive. But some bacteria escape these conditions by forming **endospores**. The bacterium's DNA and ribosomes are enclosed in a tough case that can survive heat, cold, and a host of other conditions that would normally kill a bacterium. When the conditions become more favorable, the endospore opens up to produce a normal bacterium once again. Some endospores can survive for thousands of years.

6.4 BACTERIA IN THE ENVIRONMENT

Bacteria are so common that it's hard to find a process that they are not involved in. And while we often think of bacteria as something to be avoided or killed, many of them are quite helpful. Let's take a look at some of the most important beneficial roles of bacteria.

BACTERIAL FUNCTIONS

1. As you learned in the Chapter opener, you have as many nonhuman cells as human cells in your body. Many of the bacteria in your microbiota keep other, less friendly bacteria from starting to grow in or on your body. Other bacteria in your body take a more active role in your health.

2. Plants need nitrogen to grow, but they can't use the nitrogen gas in the atmosphere. Instead, the atmospheric nitrogen must be transformed into a form that plants can use. Bacteria that live in the soil are especially important to this important task of *fixing* nitrogen. Some of these bacteria live in the roots of plants such as beans and clover.

3. The most common nutrient on Earth is cellulose. Too bad people and most animals can't digest it! But some bacteria can. Certain mammals called *ruminants* have a special chamber in their stomach that contains these bacteria. The bacteria break down the cellulose into other nutrients that the ruminant can digest. Ruminants include cattle, sheep, and goats.

❹ Plants and animals use material from their environments to build their body systems. If those nutrients remained trapped in their bodies after they died, the world would soon run out of nutrients. However, bacteria prevent that from happening. Bacteria called *decomposers* are responsible for recycling matter from waste materials and the remains of plants and animals into the soil where they can be reused.

❺ Bacteria are also essential to making many foods, including cheese, green olives, yogurt, pickles, and salami. In each of these foods, a harmless bacterium is used to ferment the food, and the fermentation changes the food so that harmful bacteria cannot grow on it as quickly. Often the changes produce different tastes and textures that people enjoy.

6.5 BACTERIA AND A BIBLICAL WORLDVIEW

Of course, not all bacteria are beneficial. At the end of the Creation Week, God said that everything He had made was "very good." So we know that bacteria weren't harmful then. But when Adam and Eve disobeyed God, the whole earth was cursed (Gen. 3:17–19). Part of the Curse included bacteria becoming the cause of diseases such as tuberculosis, strep throat, and food poisoning.

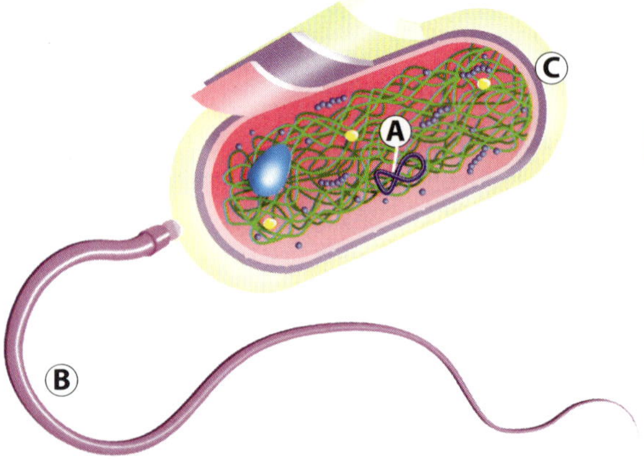

6A REVIEW QUESTIONS

1. Identify the three organelles labeled in the image on the left.
2. Name one of the differences between prokaryotes in kingdom Archaea and those in kingdom Bacteria.
3. Place the steps of binary fission in order: two bacteria; one bacterium; cell membrane pinches in two; DNA replicates.
4. Explain the importance of decomposers.

CASE STUDY THE SUPERBUGS

Jesus said that the second great commandment was to "love thy neighbour as thyself" (Matt. 22:39). Finding new ways to fight diseases is one way to do this. One of the greatest discoveries in this fight occurred in 1928 when Alexander Fleming discovered the first antibiotic, called *penicillin*. By the end of World War II, penicillin was being mass produced and was being used to treat bacterial infections that had previously been fatal. In the years that followed, many more antibiotics were discovered, increasing the number of bacteria that could be fought with antibiotics.

But some bacteria become resistant to one or more types of antibiotics. Resistant bacteria are a serious problem in hospitals. Evolutionists often mention resistant bacteria as an example of evolution. But is that an accurate description?

Remember plasmids, the small, circular pieces of DNA that some bacteria have? Some plasmids make bacteria resistant to an antibiotic. This type of resistance is especially troublesome because bacteria can share their plasmids through a process called **conjugation** (see image below). In conjugation, one bacterium connects to another bacterium. It then shares its plasmid with the other bacteria. This can make both bacteria resistant to the antibiotic. If the second bacterium is a harmful bacterium, it will be much harder to treat if it infects someone. But this isn't evolution. It's just DNA being passed between bacteria.

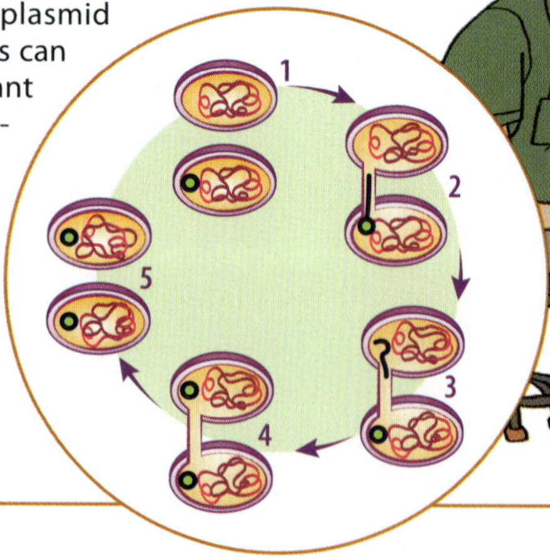

Other bacteria become resistant to antibiotics through mutation. Most antibiotics target a specific structure in the bacteria. Sometimes these structures mutate so that the antibiotic cannot attack it as well. The mutated structure almost never works as well as the original structure did. In fact, normally the bacterium with the mutated structure wouldn't survive. But when antibiotics kill all the normal bacteria, only the mutated ones remain. This isn't really an example of evolution. Instead it's an example of natural selection. The bacterium isn't evolving into something else. It's just a mutated bacterium that happens to be able to survive an antibiotic attack.

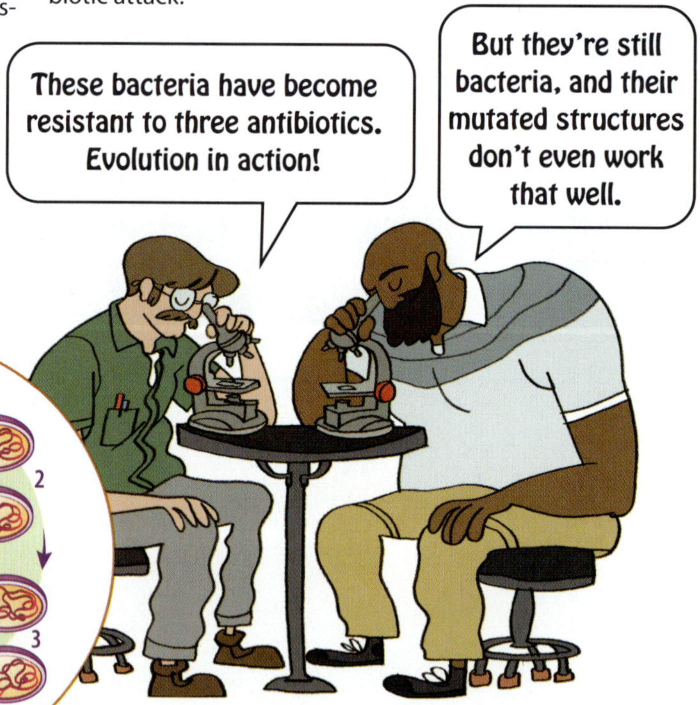

Use the Case Study above to answer Questions 5 and 6.

5. A scientist treats several colonies of bacteria in a petri dish with an antibiotic. A few are resistant to that antibiotic. Soon all the bacteria in the petri dish are resistant. The scientist suggests that this is an example of evolution. Why would you disagree?
6. Explain why antibiotic resistance is not an example of evolution.

6B VIRUSES

Are viruses alive?

6B Questions
- What structures do viruses have?
- Why are viruses classified differently than living organisms are?
- What is the difference between an active virus and a latent virus?
- How do viruses function in the environment?

6.6 VIRUS STRUCTURE AND FUNCTION

You have a big game, a recital, or a family trip coming up, but the night before the big day you start feeling not quite right. After a restless night, you awake in the morning with a sore throat, a fever, and a stuffed-up nose. Instead of a fun day, you get to spend the whole day in bed with a box of tissues and lots of liquids.

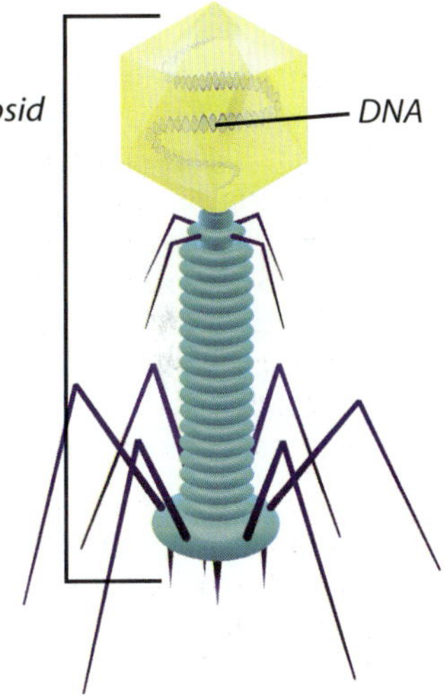

Have you ever had an experience like this? If so, then you've suffered a viral attack. Compared with a bacterium, a **virus** is pretty simple. It consists of only a nucleic acid, either DNA or RNA, surrounded by a protein coat called a **capsid**. Viruses have no cytosol, no organelles, and no cell membrane. They are not cells, and scientists debate whether they are even alive. As a result, scientists classify viruses using a different system than the one they use for living organisms.

Most viruses are very small; in fact, most cannot be seen even with a light microscope. When viewed with an electron microscope, viruses look like spheres, rods, polygons, or a combination of geometric shapes.

Unlike living cells, viruses cannot reproduce on their own. Instead, a virus must infect a cell in another organism, called the **host** cell.

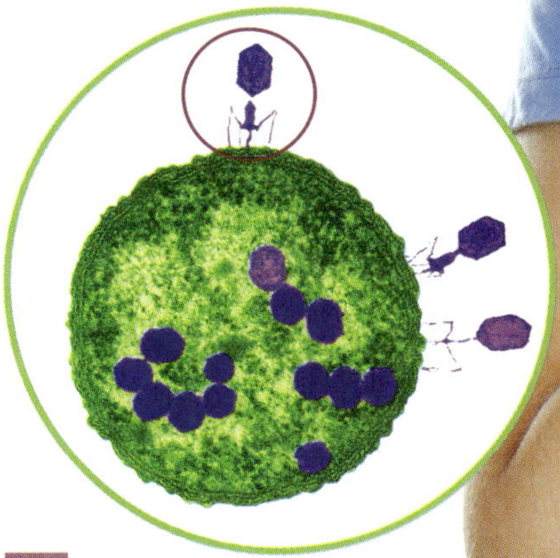

6-2
The small purple structures on the outside of the bacterium are viruses seen through an electron microscope.

VIRAL REPRODUCTION

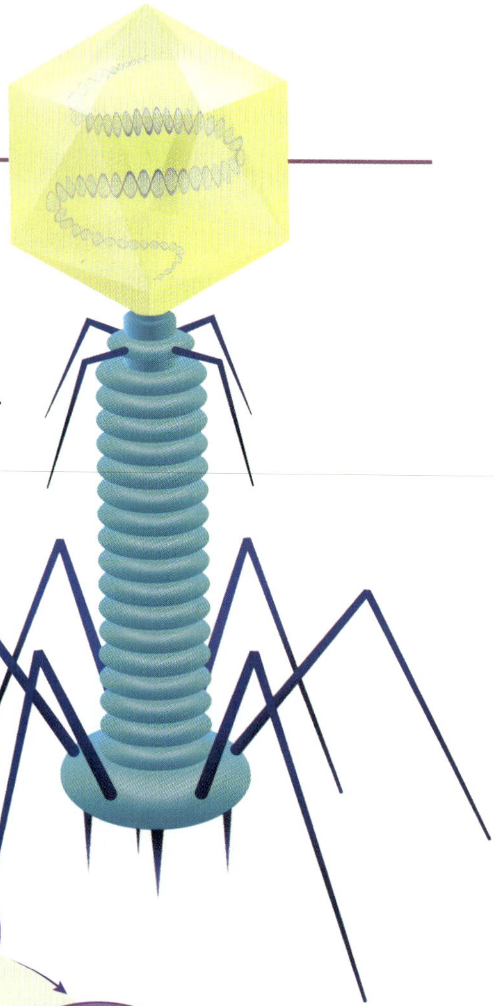

1. The virus attaches to the host cell.
2. The virus injects its nucleic acid into the host.
3. The viral nucleic acid uses the cell's own cellular machinery to make more viruses.
4. The viruses then destroy the cell by bursting it open and escaping to infect new cells. Viruses reproducing in this manner are called *active viruses*.
5. But some viruses do not kill the host cell immediately. Instead the viral DNA remains inactive for months or even years. This type of virus is called a *latent virus*.
6. Each time the host cell goes through mitosis, the viral DNA is replicated and passed to the daughter cells.

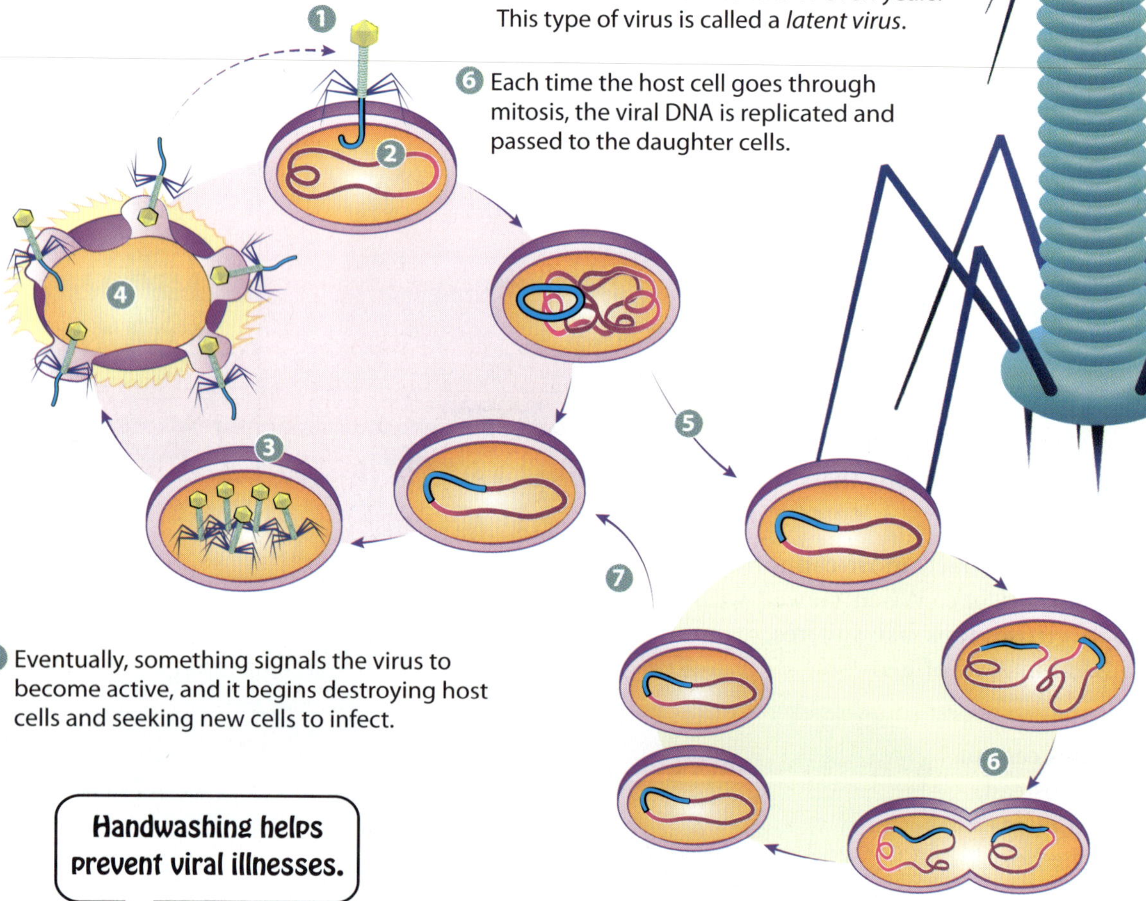

7. Eventually, something signals the virus to become active, and it begins destroying host cells and seeking new cells to infect.

Cold sores are caused by a latent virus. These viruses sometimes remain latent for years, but when the body is stressed, such as during a cold, they become active. The destroyed host cells form a cold sore.

6.7 FIGHTING VIRAL INFECTIONS

There are several things that you can do to limit the number of viruses that infect you. One is to practice good hygiene. Covering your coughs and sneezes can help prevent viruses from spreading to others. Washing your hands often and thoroughly can help prevent you from becoming sick in the first place. In fact, it's one of the most effective ways to avoid illness.

Healthy living can also help you beat germs. Eating a balanced, nutritious diet and exercising regularly may limit the number of illnesses you suffer. It will definitely help you recover more quickly when you do get sick.

Finally, you can avoid some of the most serious viral diseases by being vaccinated against them. A *vaccine* is an injection of a weakened or dead virus. This stimulates the body to fight the weakened virus in the vaccine. If you are ever attacked by an unweakened form of that virus, your body will already know how to defend itself against it.

LAB EXERCISE

HANDWASHING: IT'S A GOOD PRACTICE

How long does it take to wash off germs?

Equipment

- black light
- germ simulation kit
- soap
- stopwatch

E

You've heard from adults since before you were in school, "Be sure to wash your hands." Is it really that important?

This is your chance to investigate. Since it would be dangerous to spread real bacteria or viruses on your hands, you will use germ simulation gel. It's actually really cool—it glows under a black light!

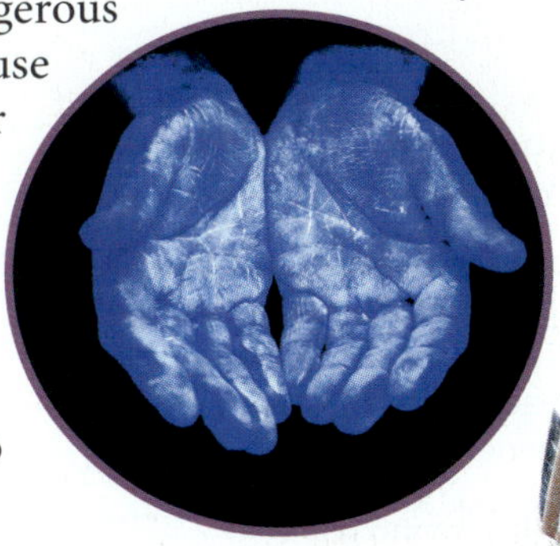

PROCEDURE

❶ Follow the instructions on the bottle for applying the germ simulation gel to both of your hands. Be sure to cover both the fronts and backs of both hands.

❷ Look at your hands under the black light.

1. How long do you think it will take to wash off the germ simulation gel?

❸ Wash your hands with soap for five seconds. Look at them under the black light. Record on separate paper whether you see any glowing gel.

❹ Repeat Step ❸ three more times. Be sure to keep track of how many total seconds you have washed using a stopwatch.

2. Are your hands clean (no glowing gel remaining)?

❺ If you answered "yes" to Question 2, continue to Question 3. If you answered "no," repeat Steps ❸ and ❹ until your hands are clean, and then continue to Question 3.

3. How many seconds did it take to thoroughly clean your hands?
4. What areas of your hands took longer to clean?
5. Why do you think these areas took longer to clean?
6. On the basis of what you have observed, write some guidelines for how people should wash their hands.

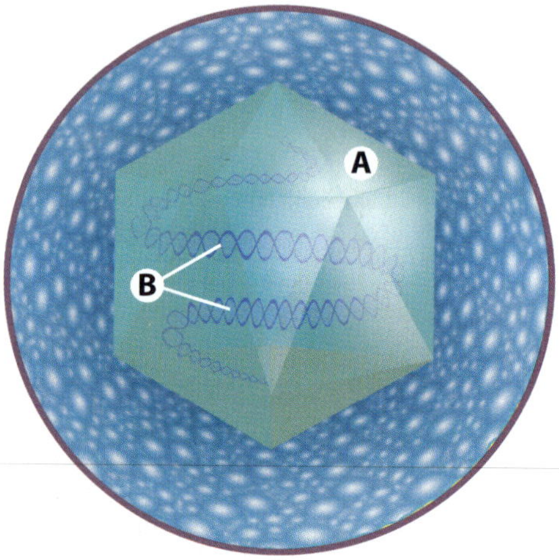

6B REVIEW QUESTIONS

1. Identify Structures A and B in the virus shown at left.
2. Why do scientists use a different system to classify viruses?
3. Give evidence that viruses are or are not alive using the characteristics of life (see page 9).
4. Does the diagram below show an active or a latent virus?

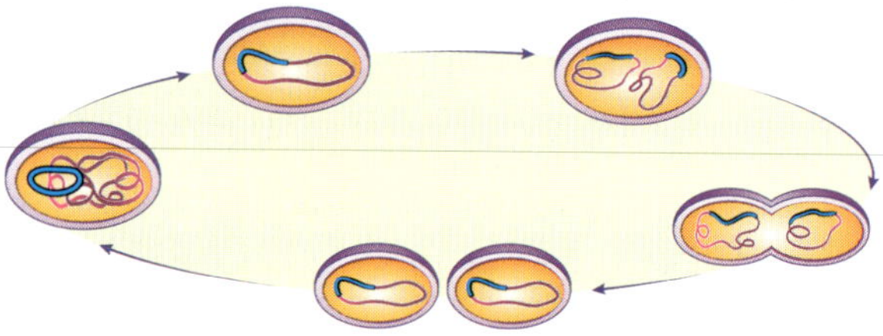

5. What is one way to help avoid getting a viral disease?

ETHICS Gene Therapy

We usually think of viruses as bad things. But what if instead of causing diseases, viruses could be used to cure them? That is the dream of gene therapy.

Some diseases are caused by mutated genes. Gene therapy seeks to use viruses to repair these defective genes. The viruses are changed so that instead of injecting their own nucleic acids into the cell, they inject DNA with the corrected gene into the cell.

However, gene therapy is not that simple. Many genetic diseases are caused by genes expressed throughout the body, so it is very hard to get the viruses to all the cells that need the new gene. Also, there is concern about the safety of gene therapy. Some gene therapy treatments are now being tested to see whether they are safe and can cure disease.

Use the Ethics box above to answer Questions 6–12.

6. Do a web search using the keywords "gene therapy." Write a brief summary of what you learn about this issue.
7. What are some diseases that gene therapy may be effective in treating?
8. What does the Bible say about the issue? (While Scripture doesn't deal with gene therapy itself, it does have a lot to say about respecting human life.)
9. On the basis of your answers to Questions 7 and 8, what are the acceptable and unacceptable options?
10. What may be the consequences of the acceptable options you identified in Question 9?
11. What view or action should I take regarding gene therapy?
12. Write a paragraph summarizing the key ideas from answering the five ethics questions on gene therapy. Be sure to include biblical principles and outcomes in your paragraph.

CHAPTER SUMMARY

6A BACTERIA

- Bacteria lack nuclei and membrane-bound organelles.
- Bacteria reproduce by binary fission and endospores.
- Bacteria promote good health, create usable forms of nitrogen for plants, decompose wastes and dead bodies, and help create foods.
- Bacteria can cause disease.

6A TERMS	
plasmid	121
binary fission	123
endospore	123
conjugation	126

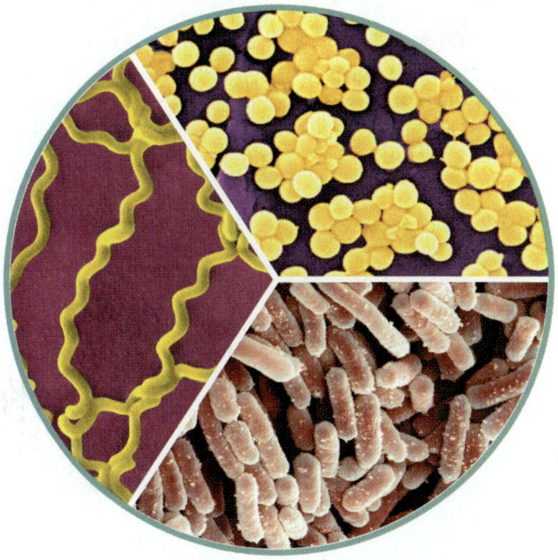

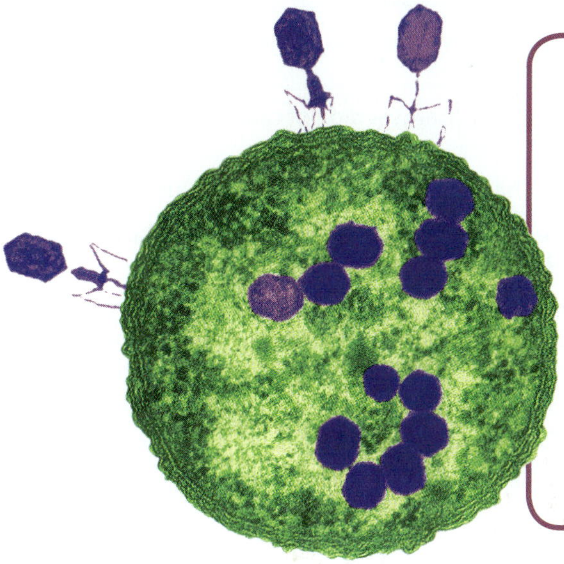

6B VIRUSES

- Viruses are not made of cells. Some scientists consider them nonliving.
- A virus consists of a protein coat and nucleic acid.
- An active virus injects its nucleic acid into a cell. The virus then uses the cell to produce more viruses until the cell bursts.
- Some viruses remain latent inside the cell for a period of time before becoming active.
- Viral illnesses can be prevented or overcome more easily by good hygiene, healthy living, and vaccinations.

6B TERMS	
virus	127
capsid	127
host	127

CHAPTER REVIEW QUESTIONS

Recalling Facts

1. Some bacteria have a small circular strand of DNA separate from their main chromosome called a(n) ______.
2. What is the purpose of a bacterium's flagellum?
3. What are the two groups of prokaryotes?
4. When would a bacterium likely form an endospore?
5. What do we call the protective protein coat of a virus?
6. What do viruses have instead of a nucleus?
7. What occurs when a cold sore erupts in the same place where another cold sore occurred three years before?

6-3

Thermophiles are heat-loving archaea that thrive in the boiling temperatures of hot springs.

Understanding Concepts

8. Scientists discover something that is very small and has DNA and ribosomes. Is it a small bacterium or a virus? Explain.
9. Explain what occurs in binary fission.
10. Explain how nitrogen-fixing bacteria help plants.
11. How do decomposer bacteria benefit the environment?
12. Explain the role of conjugation in antibiotic bacteria.
13. Why do some scientists believe that viruses are not alive?
14. Explain how a virus makes a new virus.
15. Explain how a latent virus is different from an active virus.
16. Why is it important to wash your hands?

Critical Thinking

17. Sketch a bacterium and draw and label the following four parts: cell wall, chromosome, flagellum, and ribosomes.
18. What is the difference between organisms in kingdoms Archaea and Bacteria?
19. How do we know that bacteria and viruses did not cause diseases when God created them?
20. Why are antibiotic-resistant bacteria not an evidence of evolution?
21. Some scientists believe that viruses are alive. If they are correct, how will this affect the cell theory?
22. In 2017, scientists announced that they had successfully used gene therapy to correct a genetic heart defect in a human embryo. This embryo was created in a laboratory and then destroyed when the study was completed. From a biblical worldview, is this study ethically acceptable? Explain.
23. What might be some benefits of gene therapy?

6-4

Compost piles depend on the action of decomposer bacteria and other organisms.

7 PROTISTS AND FUNGI

WEIRD WILDLIFE

Take a look at this photo. What do you suppose this is? At first glance you might think that this is a kind of weird moss, but it isn't. It's not any kind of plant. It's not an animal either. Here are some hints: It grows on rotting wood or mulch. Although it looks harmless, it is actually a dangerous predator—if you are a bacterium! And though it looks like it's growing in one place, it's not. It can move around, though very slowly. And here's the kicker—it's only one cell! Hmmm … grows on rotting wood but isn't moss … moves and eats but isn't an animal … easily seen with the naked eye but is only one cell. Are you puzzled? You aren't alone! Scientists have wrestled for years with how to classify organisms like this one. So for now they've placed them into a group all their own—the *slime molds*. Nice name, eh?

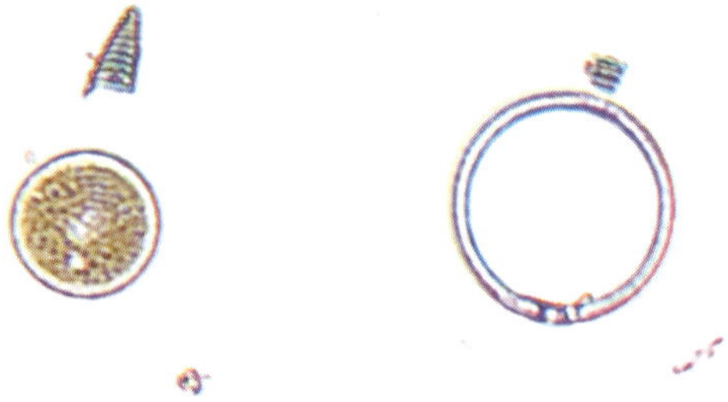

7A PROTISTS

How do protists live and function?

7A Questions

- What are protists?
- How are protists classified?
- How do protists move, get energy, and reproduce?
- What role do protists serve in the environment?
- Are protists lower life forms?

7-1
These dazzling geometric forms are the shells of microscopic diatoms.

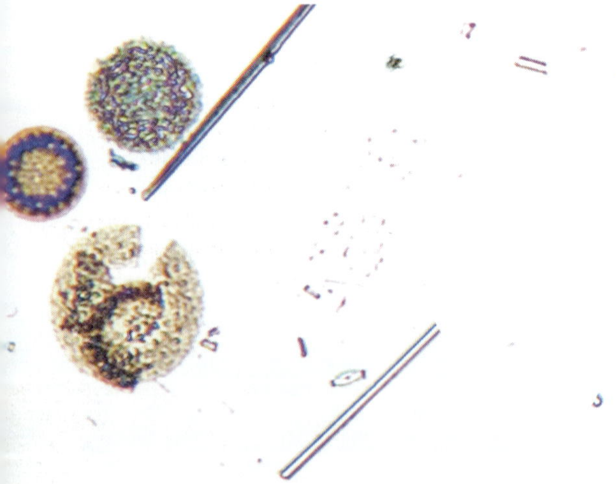

7.1 INVISIBLE WONDERS

Slime molds are just one group of organisms from among many known as protists. A **protist** is any organism that is not an animal, plant, bacterium, or fungus. Protists come in a wide variety of shapes, forms, and sizes, but they all have one thing in common: every protist is a eukaryote and has both a membrane-bound nucleus and membrane-bound organelles. Beyond that, there is a lot of diversity within the world of protists. Some are animal-like, some are plantlike, and some are fungus-like. Some make their own food while others feed on other organisms. Some do both! In fact, protists can move, obtain food, and reproduce in as many different ways as are found in all the other kingdoms combined.

You may not realize it, but you share this planet with trillions of protists, most of which are so small they can't be seen clearly without the aid of a microscope. They can be found on the ground you walk on, the water in a nearby pond, and just about anyplace else that offers moisture and a hint of food. They have been found in extreme environments from the icy poles to the bottom of the ocean. Some can even live inside your body. But though protists are small, they play a significant role in God's world—for good and for bad. Protists are an important food source for many creatures, and some plantlike protists produce much of Earth's oxygen. Other protists are destructive. They can pollute lakes and rivers, poison shellfish, and cause disease in humans. They can affect your life in ways you may not even know about. Let's take a closer look at these tiny organisms.

PROTIST PORTRAITS

Classifying the many thousands of species of protists is a very difficult and ongoing task—they are just too diverse and wacky! The groups you see on these two pages show one of several ways that scientists have attempted to classify some protists. These groups are based mainly on similarities in structure and mode of living.

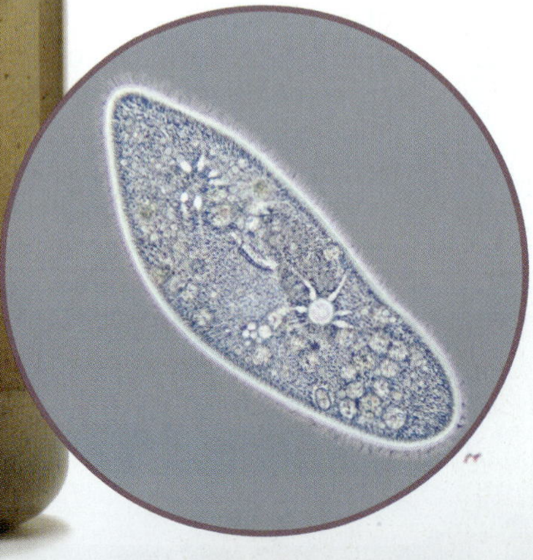

Ciliates

Named for their hair-like *cilia*, ciliates like this *paramecium* are common throughout the waters of the world. A paramecium has between 10,000 and 14,000 cilia!

Amoeboid Protists

Amoeboid protists are animal-like protists capable of an oozing kind of movement using pseudopods. Some amoeboid protists have shells and others don't. *Amoeba proteus* is one of the most commonly studied protists in biology classrooms.

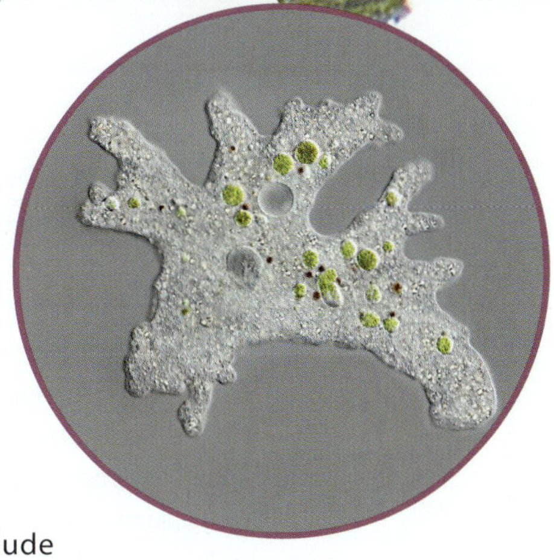

Plantlike Protists

Plantlike protists include many photosynthetic varieties that are sometimes referred to by the catchall term *algae*. Important members of this group include diatoms (see image in Chapter opener) and coccolithophores (shown on the left)

Microsporidians

All microsporidians are animal parasites, and most of their hosts are insects. This makes them useful control agents to fight unwanted insect pests. Microsporidians are among the smallest of eukaryotic cells.

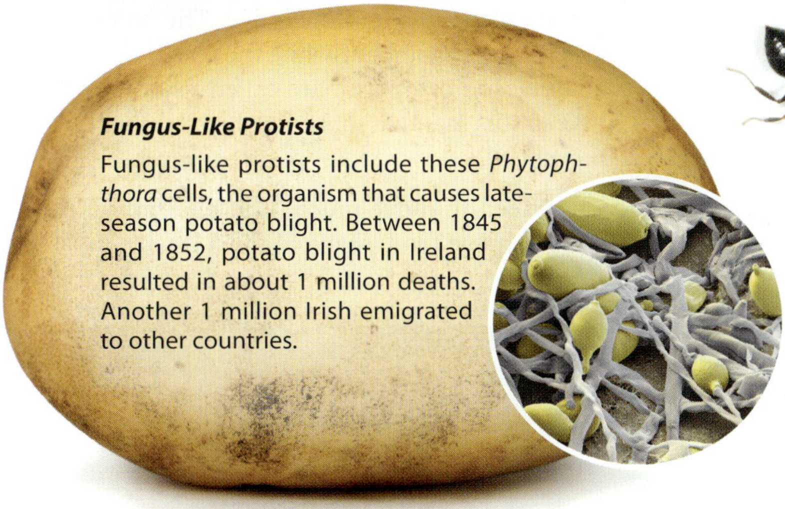

Fungus-Like Protists

Fungus-like protists include these *Phytophthora* cells, the organism that causes late-season potato blight. Between 1845 and 1852, potato blight in Ireland resulted in about 1 million deaths. Another 1 million Irish emigrated to other countries.

Slime Molds

Slime molds, like the *Stemonitis axifera* shown below and the *Fuligo septica* at the beginning of the chapter, exist most of the time as small, individual cells. But at other times, such as when food is scarce, the cells signal each other to clump together into large masses that move and act as one body. In some species, this mass remains a colony of individual cells. In others, the cells fuse together into an enormous single cell with many nuclei.

7.2 MOVEMENT, ENERGY, AND REPRODUCTION

Movement of Protists

Many protists are capable of movement in one of three ways. One method of moving involves whip-like flagella, which are whipped or flexed back and forth to propel the cell. *Giardia lamblia*, an organism that can cause abdominal pain and vomiting in persons who drink contaminated water, moves itself with flagella.

A second method involves smaller but more numerous, hair-like **cilia** (s. cilium). The cilia move back and forth like tiny oars, propelling the cell. They are often coordinated to move in waves. A paramecium moves itself in this manner and can do so quite quickly. When pond water is observed under a microscope, paramecia can often be seen zipping through the field of view. Some protists even have bundles of cilia fused into structures called *cirri*; they use these like legs for crawling along on surfaces.

A third way of moving involves a cell forming a bulge in its cell membrane. Its cytoplasm then streams into the bulge. The bulge may attach to a surface and serve as an anchor while the rest of the cell pulls itself forward. The bulges of cytoplasm are called **pseudopods**. Because amoebas move in this manner, this type of movement has been called *amoeboid movement*.

Some protists do not need to move much, if at all. Photosynthetic protists do not need to pursue food. Many of them often simply drift along on ocean currents.

7-2
Giardia lamblia

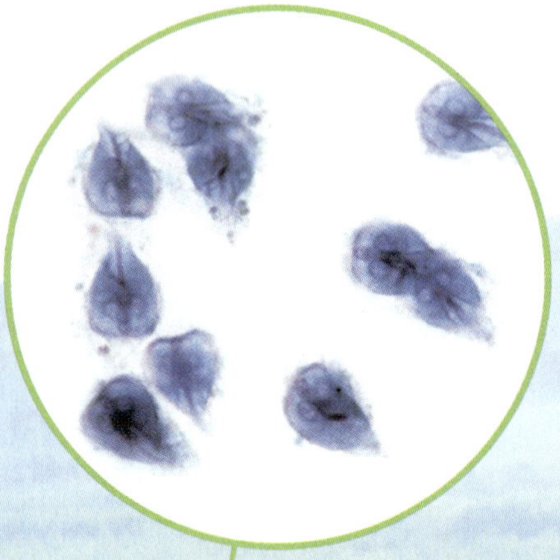

Nutrition of Protists

How do protists get energy? Some protists eat other protists, some eat bacteria, and some eat whatever debris is floating their way. Others can make their own food through photosynthesis. Some can do both, photosynthesizing when light and nutrients are abundant and switching to eating when they are not.

Protists that eat other organisms or particles of food are **consumers**; they rely on other organisms for their energy source. The paramecium, for example, has an *oral groove* for its mouth. The groove is lined with cilia that sweep food through the opening. Many protozoans form pockets in their cell membranes that hold trapped food. These pockets pinch off from the membrane toward the inside and form *food vacuoles*. Paramecia form food vacuoles at the inside end of their oral grooves, but they may form them at other places as well.

Once the food has been engulfed in a food vacuole, digestion with enzymes begins. In time the digested materials diffuse through the membrane of the vacuole into the cytoplasm. The indigestible wastes that remain in the vacuole are transported out of the cell. Unlike paramecia, amoebas do not have a permanent food opening. Instead, they surround and engulf food with their slow amoeboid movements.

Protists that perform photosynthesis are producers; they can make their own food in the form of sugars. Some of these photosynthetic protists are often called *algae*, though that term can also refer to many organisms that are presently classified as plants by some scientists.

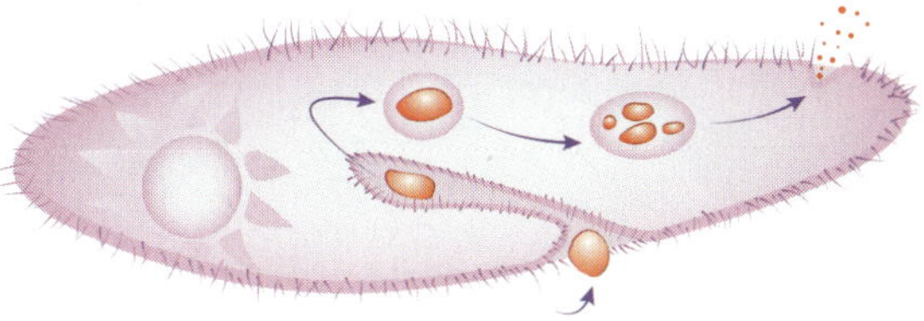

7-3

A paramecium uses its cilia to sweep food into its oral groove (above), where vacuoles engulf the food. An amoeba surrounds and engulfs its food with pseudopods (below).

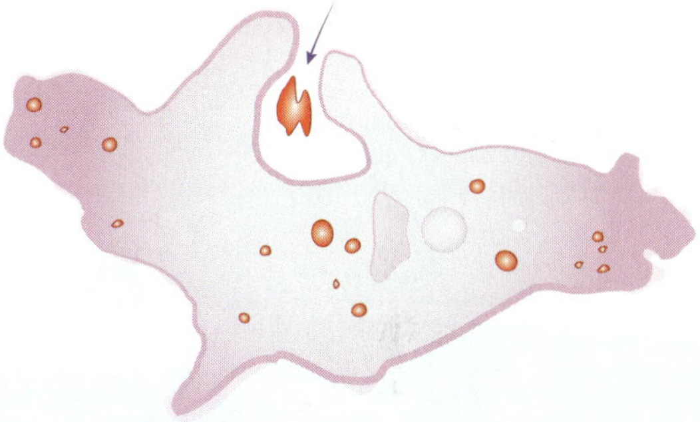

Paramecia conjugating

Reproduction of Protists

Most protists reproduce asexually by mitosis. For instance, under normal conditions an amoeba can divide about every three days. Many protists also reproduce sexually. They form gametes that later fuse together to make a new organism. Another type of sexual reproduction in protists is **conjugation** (KAHN juh GAY shun). Paramecia usually reproduce asexually by mitosis, but they can also conjugate. During conjugation, they join their oral grooves together and exchange DNA. Each paramecium then has a new combination of genes. After the exchange, the two paramecia separate.

Sexual reproduction in protists sometimes goes beyond just simply producing gametes. *Plasmodium falciparum*, the protist that causes malaria in humans, has a multistage life cycle; some of these stages reproduce sexually, while others reproduce asexually.

7.3 PROTISTS IN THE ENVIRONMENT

Though very small, protists are important to the lives of other organisms. Some of Earth's most vital protists are those that form part of the ocean's **plankton**. Plankton are tiny organisms that float near the ocean's surface. Directly or indirectly, plankton is the food source for most marine animals. Photosynthetic protists are also very important as a source of oxygen. As they go through photosynthesis, oxygen is released as a byproduct.

Protists, or their remains, have important economic uses. Certain commonly used minerals, such as chalk, limestone, and diatomaceous (DYE ah tah MAE shus) earth are largely made of the hard remains of protists. These minerals are used in products as diverse as building materials, soil conditioners, toothpaste, and polishing agents.

Protists can also cause serious diseases and other harmful events. Malaria, amoebic dysentery, and African sleeping sickness are human diseases caused by protists. When ocean conditions are just right, a certain species of a group of protists called *dinoflagellates* (DYE no FLAJ uh lits) can overpopulate an area of ocean water, giving it a reddish color. These *red tides* are deadly to fish because of a toxin that the dinoflagellates produce. Shellfish harvested during and shortly after a red tide are poisonous to humans because the shellfish concentrate the toxin in their bodies.

7-4
A red tide near Tsitsikamma National Park, South Africa, was caused by the dinoflagellate *Karenia brevis*.

7.4 ARE PROTISTS SIMPLE LIFE FORMS?

There is a tendency among those who favor evolution to think in terms of "simple" and "complex." Multicellular plants and animals are obviously complex, made of billions of cells with highly specialized functions. Because these organisms supposedly evolved from single-celled protists, some people regard protists as simple life forms by comparison. However, as we saw in Chapter 5, there is no biological or biblical evidence that confirms that such evolution takes place. And in Chapter 2 we learned that cells are incredibly complex structures. In multicellular organisms, cells are specialized to do different things, like help an organism move, digest food, or provide structure. Protists can do all these things with only *one* cell!

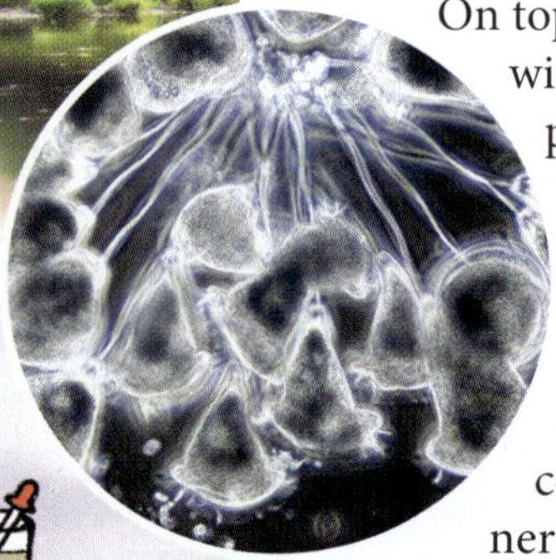

On top of all that structural complexity, protists also display a wide range of behaviors. Various protists can hunt, avoid predators, or build protective shelters for themselves. Some can respond to light, chemicals, or touch. Others can switch their way of getting energy from foraging to photosynthesis. We learned in the Chapter opener that some protists can even go from being lone, single cells to bunching together to become a kind of "super cell." And they do all these things *without* any brains or nervous tissue! Scientists still have much to learn about how these "lower" life forms perform their many complex behaviors. There is really nothing simple about them at all!

In truth, protists are not any more and certainly not any less equipped for survival than other forms of life. They are more evidence that God has carefully designed each kind of organism with the special structures and abilities it needs to thrive in His world.

7A REVIEW QUESTIONS

1. What is the *one* characteristic that all protists have in common?
2. Why are protists difficult to classify?
3. Classify the protist on the right as either a ciliate, an amoeboid protist, or a plantlike protist. Explain your reasoning.
4. List two protists that can cause human disease and the diseases they cause.
5. How is movement with cilia different from movement with flagella?
6. Describe amoeboid movement.
7. (True or False) Some protists are both consumers and producers.
8. How do protists digest their food?
9. How do most protists reproduce?
10. Describe the process of conjugation.
11. Describe two important functions of protists as part of the ocean's plankton.
12. What evidence suggests that protists are not "lower" life forms compared with other organisms?

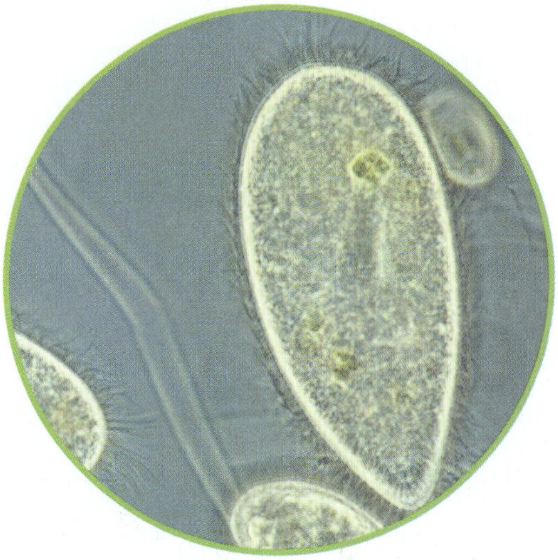

CASE STUDY PLANT OR ANIMAL?

CLASSIFYING ISN'T ALWAYS EASY!

Take a look at these protists. They're one example of a group of protists called *euglenoids*. They're green, so they sort of look like plants. And, like plants, they can photosynthesize. But they also have flagella, which means that they can move around like animals. And to make the riddle of classifying them even harder, euglenoids can also consume food like animals do. So what are they—plants or animals?

When Linnaeus first created his classification system, it had three kingdoms—plants, animals, and minerals (no, he didn't think minerals were alive, and yes, we no longer use his system to classify minerals). Scientists were stumped by euglenoids, though, since they had both plantlike and animal-like characteristics. Which kingdom should they be placed in?

Finally, in the 1860s, scientists created a fourth kingdom, *Protista*, into which they lumped all the "primitive" single-celled life forms, including euglenoids. That simplified matters for a while, but scientists realized that it wasn't an ideal solution. To this day, scientists are still wrestling with the issue of how to classify the many different kinds of euglenoids and other protists.

Use the Case Study above to answer Questions 13–14.

13. Why were euglenoids difficult for scientists to classify?
14. How did scientists solve this problem?

7B

FUNGI

7B Questions

- What are fungi?
- What are the parts of a fungus?
- How do fungi get energy and reproduce?
- How are fungi classified?
- What roles do fungi serve in the environment?

How do fungi live and function?

7.5 CLASSIFYING FUNGI

When you hear the word "fungus," what do you think of? Something slimy and unpleasant? How about when you hear the word "mushroom," which, by the way, is a fungus? When you hear the latter, there's a very good chance that the first sort of image that pops into your mind looks something like the one in the background image on this page spread. The white mushroom, *Agaricus bisporus*, is one of the most commonly grown and eaten mushrooms in the world. Do you know what the difference is between a white mushroom and the more expensive portobello mushroom? Only the color! They are two varieties of the same species of mushroom.

All fungi produce spores—some species can produce up to five different kinds. **Spores** are involved in reproduction and allow fungi to survive through unfavorable growth conditions. The tiny spores are easily transported by air. When conditions are again favorable, spores begin to grow, producing new fungi. Spores are produced in several different ways. As we'll see next, these ways of producing spores are used to classify the major groups of fungi.

7-5
White mushrooms

CLASSIFYING FUNGI

Sac Fungi. Sac fungi produce spores in microscopic, sac-like structures, hence their name. Sac fungi include delicacies like morels and truffles, along with agents of plant disease like ergot and powdery mildew. *Penicillium*, the fungus from which the antibiotic penicillin was extracted, is a sac fungus. Did you have toast for breakfast? Baker's yeast—used to make bread—is also a sac fungus.

Club Fungi. Club fungi are what we picture when we hear the word "mushroom," like the *A. bisporus* on your pizza. As their name suggests, the spore-producing structures of club fungi are club-shaped. In addition to mushrooms, familiar club fungi include puffballs and shelf fungi. As these scarlet cups demonstrate, some fungi can even be beautiful.

Zygote Fungi. Have you ever reached for that last slice of bread in a bag and been greeted by something like this? Then you have had a close encounter with black bread mold!

Chytrid Fungi. Chytrid fungi produce spores in pot-shaped structures. They include the fungus that has killed millions of frogs in recent years. But most chytrid fungi are decomposers, serving the environment by recycling organic matter.

Imperfect Fungi. The imperfect fungi are those with no known form of sexual reproduction, even if they otherwise fit into one of the other groups. One of the most common imperfect fungi is *Aspergillus niger*, a frequent contaminant of fruits and vegetables.

7.6 STRUCTURE OF FUNGI

A **fungus** is any member of kingdom Fungi. Fungi are not considered animals because fungal cells have cell walls, while animal cells do not have true cell walls. Plant cells have cell walls, too, but fungi aren't considered plants either. That's because plant cell walls are made of cellulose, but fungal cell walls are made of *chitin*. The same substance covers the bodies of insects and crustaceans. And whereas plants can make their own food through photosynthesis, fungi cannot.

One fungus may appear different from another, but the cells of different fungi are quite similar. Most fungal cells are long cells bound together to form **hyphae** (HY fee) (s. hypha). These filaments are connected end to end and may be grouped with others to form larger structures. The stalk of a mushroom, called the **stipe**, is an example of hyphae grouped together to form a much larger and more complex structure. Mushrooms themselves are a type of **fruiting body**, the spore-producing part of a fungus. The part of a fungus that you can see, usually the fruiting body, is often just a small part of the entire organism. The mushrooms in the fairy ring in Figure 7-6, for example, are growing at the spreading outer edge of a much larger mass of underground hyphae. Their spores are produced on the tips of special cells lining the **gills** on the underside of each mushroom's **cap**.

7-6
This fairy ring is the result of fungal hyphae spreading out through the soil seeking nutrients.

7-7
Mushrooms, like other fungi, are made of hyphae and produce spores.

7.7 FUNGI NUTRITION

Since fungi do not have chloroplasts, they must obtain energy from material around them. They are consumers. Unlike protists, they cannot engulf food or sweep it into vacuoles. Instead, most fungi secrete digestive enzymes into the area around them. These enzymes digest the food, and the small food molecules are then absorbed into the fungal hyphae. If the fungus obtains food from dead organic matter, the fungus is called a **saprophyte**. If it obtains food from a living organism (known as a host), the fungus is called a **parasite**.

Some fungi live in association with the roots of plants. These fungi colonize the roots of host plants. The fungi absorb nutrients from the soil much better than plant roots alone can. In exchange for nutrients, the plant provides the fungus with energy-rich sugar molecules. Plants living together with fungi usually grow better than identical plants without them. An association like this in which both species benefit from living together is a form of **symbiosis**.

There is another way that some fungi may obtain energy. They live together with algae in such an intertwined condition that they appear as a single organism called a **lichen** (LYE kuhn). The fungus obtains energy from living and dead algal cells. In return, the alga receives protection, water, and perhaps some minerals from the fungus. This is another example of a symbiotic relationship.

7.8 FUNGI REPRODUCTION

Most fungi can reproduce both sexually and asexually, but not all. This is another way to classify fungi. Fungi that can reproduce both ways are called **perfect fungi**. Those that can reproduce only asexually are called **imperfect fungi**.

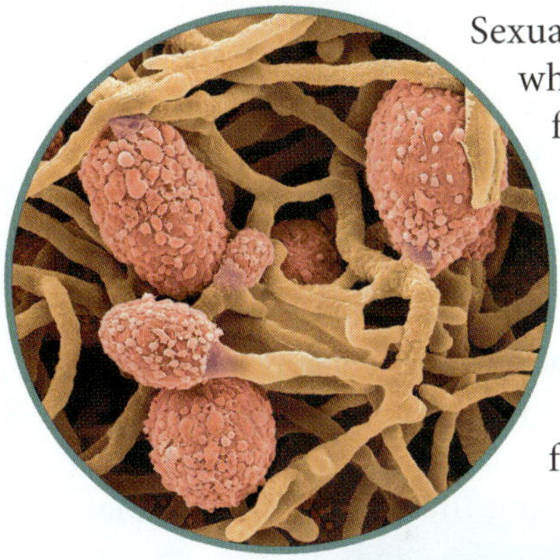

Sexual reproduction in fungi occurs mainly when growing conditions become unfavorable. Hyphae from two fungi that come in contact with each other first join together. These joined hyphae form special reproductive structures that produce spores. The spores produced by sexual reproduction have DNA that is different from that of the parent fungi.

7-8
By digesting the cellulose in this wood, these shelf fungi are helping to return nutrients to the soil.

7.9 FUNGI IN THE ENVIRONMENT

Good Fungi

You most likely go days without thinking too much about fungi—if you ever think about them at all. But fungi are incredibly important, both to the environment and to human trade and commerce.

Along with bacteria, fungi are the main **decomposers** of the world. Fungi digest dead plant and animal material and recycle its nutrients back into the environment. Plants can use these nutrients to produce new materials. Without fungi, dead organic matter would just pile up and sit there, which is hardly a pleasant thought!

Fungi are of great commercial value. The edible mushroom market alone is a multimillion-dollar industry. Cheeses, such as blue cheese, are often flavored or cured with fungi. The carbon dioxide produced by yeast—a fungus—causes baked goods to rise. Harvested cacao beans are fermented by a fungus during the making of chocolate.

Chemicals such as alcohol and acetic acid are produced by fungi. Antibiotics were first derived from a mold, *Penicillium*. Other drugs extracted from fungi have been used to treat cancer, high cholesterol, diabetes, malaria, and even fungal infections!

Lichens serve a vital role in environments that are too cold for plants to survive, such as tundras. They are often the first link in the Arctic food chain. Lichens are also decomposers—not of wood, but of stone. Lichens that grow on rocks secrete chemicals that slowly break down the rock. This is why stone structures such as grave markers, bridges, and even Mount Rushmore must now and then be carefully cleaned of lichens.

Bad Fungi

We read earlier about how some fungi work with plants in beneficial ways, but fungi can also be major plant pests. Fungal plant diseases can cause crop losses worth billions of dollars each year. Black stem rust, a fungal disease, does serious damage to the world's wheat harvest each year. Within the last twenty years, new types of black stem rust have appeared against which wheat has no natural resistance. And this is only one of several fungal diseases that attack crops. Some of the famines reported in the Bible (e.g., in Genesis 41 and Ruth 1) could have been caused by fungal plant diseases.

Fungi can cause human disease as well. *Ringworm* (athlete's foot) and *thrush* are easily treated fungal infections. However, some fungi can cause dangerous infections. *Valley fever* is a fungal disease common in some parts of the American southwest. The spores of the fungus that causes the disease reside in dry soil. When disturbed, perhaps by farming or construction, the spores become airborne and can be inhaled. Most people who are exposed to the fungus don't get sick, but a few do. They experience flu-like symptoms and normally get better in a few weeks. In severe cases, infections in the lungs, nervous system, or bones will require special treatment.

Not all molds are strictly good or bad—a type of mold similar to black bread mold is used to make *tempeh*, a type of fermented soybean cake popular in Indonesia. And the black mold we read about earlier that can ruin fruit, *A. niger*, is used to make a kind of Chinese tea called *pu-erh*.

The many negative and positive aspects of fungi in our world today are reminders of the goodness of God's original creation, the effects of the Fall and the Curse, and God's continued care for His world.

7-10
Pu-erh—bad fungi gone good?

7-9
Breaking down stone is good—unless it happens to be a stone monument.

LAB EXERCISE

DON'T EAT THE MUSHROOMS

How do I know which mushrooms are poisonous?

Ever spot a mushroom growing in the woods? Some people harvest wild mushrooms for eating. But they better know what they are doing because if they can't correctly identify a mushroom, they might end up very ill or even dead! Some people think that poisonous mushrooms are brightly colored and the edible ones aren't. That's not always true!

The mushroom shown in Figure 7-11 is poisonous, though it looks very similar to edible mushrooms. It even tastes good! It is called a *death cap* and is the most poisonous mushroom in the world. Eating more than half of one can cause death in three days or less.

E Equipment

- mushrooms (purchased from a grocery store, e.g., button mushrooms)
- magnifying glass
- scalpel

7-11
The appropriately named death cap

PROCEDURE

❶ Examine the edible mushrooms that you have with a magnifying glass.

1. In what group of mushrooms is your edible mushroom—club, sac, zygote, or imperfect?
2. In what group of mushrooms is the death cap?

❷ Slice your mushroom in half with a scalpel or knife.

3. What structure can you observe when you slice the mushroom open that isn't otherwise visible?

ANALYSIS

4. How is your edible mushroom similar to the death cap in Figure 7-11?
5. How is your edible mushroom different from the death cap in Figure 7-11?
6. Why do you think so many people have been killed by the death cap?

CONCLUSION

You could use the differences between your edible mushroom and the death cap to avoid eating it. But there are edible mushrooms that look more similar to the death cap than yours does. Your life could depend on accurate identification!

Mushrooms are a resource that God has given us to use wisely. We can develop our knowledge to eat them safely. But if you ever get lost in the woods and are looking for some food, don't eat the mushrooms!

7B REVIEW QUESTIONS

1. What do we call the filamentous cells of a fungus?
2. Spore-producing structures in fungi are called ______.
3. A mushroom produces spores on its ______.
   a. cap
   b. gills
   c. hyphae
   d. stipe
4. Describe three ways in which fungi can obtain energy.
5. What two organisms make up a lichen? How does each benefit from the relationship?
6. How are fungi classified?
7. What do all imperfect fungi have in common?
8. Name a fungal product that you use.
9. (True or False) Fungi cause many plant diseases, but they are not known to cause any human diseases.

CHAPTER 7

CHAPTER SUMMARY

7A PROTISTS

- Protists are mostly micro scopic, unicellular, eukaryotic organisms that are not animals, plants, or fungi.
- Protist classification is changing as we learn more about their structure and mode of living.
- Protists move using cilia, flagella, or pseudopods.
- Some protists get energy by consuming other organisms or food particles or by making their own food. Some can do both. Some protists are parasites.
- Most protists reproduce asexually through mitosis. Many are also capable of sexual reproduction using gametes. Some can exchange DNA through conjugation.
- Protists play many important roles in the environment, including serving as food for other organisms, providing oxygen through photosynthesis, and acting as agents of disease.
- Even though they are unicellular, protists demonstrate amazing complexity.

7A TERMS

protist	134
cilium	136
pseudopod	136
consumer	137
conjugation	138
plankton	138

7B FUNGI

- Fungi cannot make their own food and have cell walls made of chitin.
- Fungi are classified on the basis of the types of spores and spore-forming structures that they produce.
- Most fungal cells form filamentous hyphae, which can intertwine to form larger structures such as fruiting bodies.
- Most fungi are saprophytes that obtain energy from dead organic matter. Some fungi are parasites and obtain energy from a living host. Some fungi obtain energy through symbiotic relationships with plant roots or algae.
- Perfect fungi can reproduce both sexually and asexually; imperfect fungi reproduce only asexually. Sexual reproduction usually takes place during periods of unfavorable environmental conditions and results in the production of spores.
- The most vital role of fungi is as decomposers that help recycle organic matter. Fungi also help plants obtain nutrients, provide food and medicine for people, and cause disease in plants, animals, and humans.

7B TERMS

spore	142
fungus	144
hypha	145
stipe	145
fruiting body	145
gill	145
cap	145
saprophyte	146
parasite	146
symbiosis	146
lichen	146
perfect fungus	147
imperfect fungus	147
decomposer	148

CHAPTER REVIEW QUESTIONS

Recalling Facts

1. All protists are ______.
   a. consumers
   b. eukaryotes
   c. producers
   d. prokaryotes
2. (True or False) All protists are microscopic organisms.
3. List three ways that protists move.
4. The term *amoeboid movement* describes movement using what structures?
5. Describe how a paramecium captures food.
6. (True or False) Some protists can reproduce both sexually and asexually.
7. Apart from serving as food for other organisms, what vital service do planktonic protists provide for the environment?
8. Name a human disease caused by a protist.
9. What kind of protist causes red tides?
10. How are fungi classified?
11. Name three characteristics that all fungi share.
12. Which fungal structure serves as a building block for larger structures?

Refer to the figure below to answer Questions 13–16.

13. Identify Structure 1.
14. Identify Structure 2.
15. Identify Structure 3.
16. Identify Structure 5.

17. (True or False) Fungi are producers.
18. How do most fungi obtain energy?
19. Why do plants form symbiotic relationships with fungi?

Understanding Concepts

20. Why has classifying protists proved to be such a difficult task?
21. Refer to the figures below. Match the type of movement structure with the number of the correct organism.

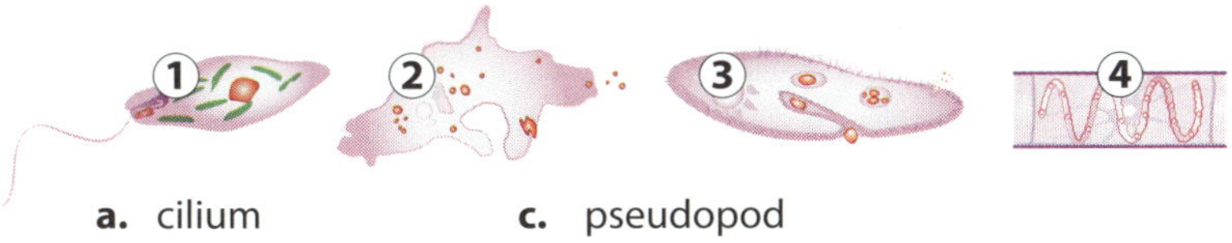

   a. cilium
   b. flagellum
   c. pseudopod
   d. no structure for movement
22. Why is conjugation considered a type of sexual reproduction?
23. Why is it an advantage for a protist to be both a producer *and* a consumer?
24. Why do photosynthetic protists such as diatoms often lack any special structures for movement?
25. The fungus shown on the right produces spores in tiny sac-shaped structures. What kind of fungus is it?

26. A homeowner notices some mushrooms growing in his lawn and removes them. Explain why this will not eliminate his mushroom problem.
27. The puffball shown on the left does not look like a typical mushroom, yet is classified in the same group as the edible white mushroom *Agaricus bisporus*. Why is this so?
28. Under what conditions do fungi usually switch to sexual reproduction?
29. Describe one beneficial aspect of lichens and one harmful one.

Critical Thinking

30. What would happen if protists were removed from the ocean food chain?
31. When observing ciliates such as paramecia under a microscope, scientists often add a few cotton fibers to the liquid medium. Why do you suppose they do this?
32. How do parasitic protists such as *Giardia lamblia* illustrate the effects of the Fall?
33. How do fungi demonstrate the wisdom of God in providing for His creation?

Use the Case Study below to answer Questions 34–36.

34. Why has the classification of slime molds changed so often?
35. Think back to the discussion of biblical kinds in Chapter 5. On the basis of what you learned there, do you think that evolutionists will ever be able to classify slime molds (and other protists) to their satisfaction?
36. Read Genesis 1. When were slime molds most likely created?

CASE STUDY WHENCE THE SLIME MOLD?

As you read in the Chapter opener, slime molds are unusual creatures. Evolutionists have been attempting to piece together their evolutionary history for many years. Slime molds were once considered to be a species of fungus, hence the word *mold* in their name. It's not hard to see why people thought slime molds were fungi. They found them on rotting logs and leaf litter, along with mushrooms, shelf fungi, and such. It was only later that scientists discovered that slime molds were moving about and eating the bacteria growing on the logs and leaves—hardly fungus-like behavior! So slime molds were moved out of kingdom Fungi.

For now, slime molds are considered part of kingdom Protozoa. Evolutionists think that their closest relatives are amoebas. But the question of where to place slime molds on the evolutionary tree of life remains unsettled. The term *slime mold* itself is applied to many organisms that evolutionists do not think are related. A group like this is described as *polyphyletic* (from a Greek word meaning "many races"). Evolutionists believe that the organisms in such a group come from different common ancestors. Nowadays it is common to use DNA analysis to try to establish the presumed evolutionary relationships between organisms. In doing so, evolutionists often discover that some organisms are not as closely related as they once thought.

THE PLANT KINGDOM 8

CORN-FUELED CAR CRAZE

What do the corn in this field and NASCAR® have in common? Believe it or not, the popular racing series owes its existence to the humble corn plant—in a roundabout sort of way. In the American Southeast, corn was long used to make illegal whiskey, called *moonshine*. During Prohibition, *bootleggers* transported the moonshine around the country from distilleries hidden deep in the Appalachian Mountains. They needed cars fast enough to outrun local law enforcement but otherwise plain enough to not draw unwanted attention. It wasn't long before these car owners started racing to see who had the fastest *stock car* (a car not designed as a racing car but made from regular production, or *stock*, parts). Prohibition ended in 1933, but the racing continued.

Today, stock car racing is one of the most popular spectator sports in America. And the link between NASCAR and corn alcohol is still strong—today's cars run on a fuel blend containing 15% ethanol, an alcohol made mostly from corn.

8A Questions

- What is a plant?
- How are plants different from other living things?
- How are plants classified?
- How are vascular and nonvascular plants different?

8A
CLASSIFYING PLANTS

How are plants different from other living things?

Corn is just one of the many thousands of species of living things that we call *plants*. As you'll soon see in this and the following chapter, they serve the creation in a variety of vital ways. God's original good design for plants included that they would provide food for both humans and animals (Gen. 1:29–30). In addition to food, plants and plant products have been used for fuel, shelter, and clothing. Plants provide oxygen and recycle carbon dioxide. Many plants, such as beautiful flowers or majestic trees, are also pleasing to the eye and inspirational to the human spirit. But plants and the ways that we use them have also been affected by the Fall. Just as corn can be used to make a tasty tortilla chip or illegal moonshine, many plants can be used for either good or evil.

8-1

Like other instances in which humans have tried to classify living things, some plants bend the rules. For instance, these plants at left have lost the ability to photosynthesize and live as parasites on fungi.

8.1 WHAT MAKES A PLANT A PLANT?

Plants look like plants, right? Well, most of the time they do. But how about the organism in the image on the right? Is it a plant?

The living stones shown are indeed plants, but they sure don't look like it. So what do we mean exactly when we use the word "plant"? As with other groups of living things that we've already looked at, scientists don't always agree on which organisms are plants. But generally they agree that plants have the characteristics described below.

PLANT FEATURES

Plants are eukaryotic and most are multicellular. They have *plastids*, such as chloroplasts, which are a kind of cellular organelle.

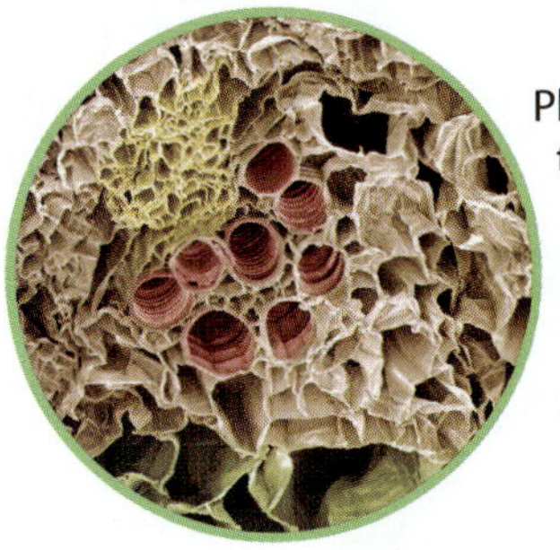

Plants have tissues, especially tissues for transporting materials.

Plants have cell walls made of **cellulose**, which in turn is made of long chains of sugars.

Most plants are producers that get their energy from photosynthesis. To prevent their tissues from drying out in bright sunlight, plants have a waxy coating called a **cuticle** that helps prevent water loss.

Plants have a multi-stage life cycle that consists of both a gamete-producing stage and a spore-producing stage.

8.2 PLANT CLASSIFICATION

Mosses, ferns, pine trees, and tulips are all plants and meet the definition of a plant given in the previous section. But each belongs to a different plant group. These major groups differ in several ways.

One important trait used to classify plants is whether they reproduce using **seeds**. All plants use spores during reproduction, but some like mosses and ferns do not produce seeds in addition to spores. Most of the plants that you are familiar with reproduce with seeds, but they differ in how they make those seeds. Some plants, such as petunias, use flowers and fruits to produce seeds. Other plants, like pine trees, use cones to produce seeds. These structures are described later and in Chapter 9.

Another trait used to classify plants is whether the plant has **vascular tissue**. Vascular tissue conducts water and other materials in some plants. Plants that have vascular tissue are called *vascular plants*. Those that do not have vascular tissue are called *nonvascular plants*.

All nonvascular plants are seedless. Some vascular plants also reproduce without seeds. Other vascular plants produce seeds. Thus, plants can be divided into three major categories: (1) nonvascular plants, (2) seedless vascular plants, and (3) seed-bearing vascular plants (or simply, seed plants).

8.3 NONVASCULAR PLANTS

In nonvascular plants, sometimes called *bryophytes*, water and other materials are transported by diffusion. These plants are fairly short, mainly because diffusion can transport water only so high. They sometimes have parts that look like roots, stems, or leaves, but since they lack vascular tissue, these structures are not true roots, stems, and leaves. What resemble leaves are often only layers of single cells. The lack of vascular tissue usually restricts these plants to living near water or in shady places. All nonvascular plants are seedless.

EXPLORING BRYOPHYTES

Bryophyte Structure. A typical bryophyte consists of two sections. The lower part, the **gametophyte**, is green and leafy with attached, root-like *rhizoids* (RY zoyds). This part of a bryophyte is always present. Rhizoids anchor the plant and absorb water and minerals. The upper part of a bryophyte, the **sporophyte**, consists of the *stalk*, which grows from the leafy part, and the *capsule* on the end of the stalk. These are involved in producing spores and are not always present.

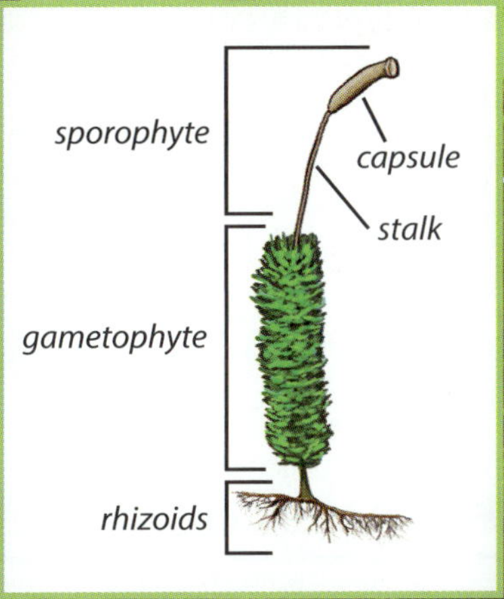

Liverworts. The suffix *-wort* comes from an Old English word that means "plant or herb." Liverworts were used in ancient times as a treatment for liver disease, possibly because the leaves of some resemble the shape of a human liver. The sporophyte of a liverwort looks like a tiny palm tree. The flat, leaflike structure is the gametophyte.

Hornworts. Named for their horn-shaped sporophytes, hornworts are the smallest group of bryophytes. Only a few hundred species are known. Like mosses, they have thin, leafy structures attached by rhizoids and transport water by diffusion. A good place to find both hornworts and liverworts is on rocks along a forest stream. They frequently grow among mosses.

Moss. Thick carpets of moss are often seen on boulders near streams, forest floors, shady stone paths, or any place with sufficient moisture. Most mosses are less than 10 cm tall, but one Australian species can reach 50 cm. One type of moss called *sphagnum*, or peat moss, is particularly useful to humans. Dried peat is used by gardeners and landscapers as a soil conditioner. Peat can also be burned. It is still used as fuel in some parts of the world.

8.4 SEEDLESS VASCULAR PLANTS

Vascular plants can grow taller and live in drier areas than nonvascular plants because they can move water through vascular tissues. But the seedless vascular plants also use water as nonvascular plants do: to transport certain cells involved in sexual reproduction. Although they can survive with less water than nonvascular plants, seedless vascular plants must have a wet period for reproduction to occur.

SEEDLESS VASCULAR PLANTS

Ferns. When we think of ferns, we usually picture them growing in thick masses on forest floors. But some of the world's more than 10,000 species of ferns can also be found in harsh environments such as deserts and swamps. Beautiful ferns are common houseplants. Fern leaves, called *fronds*, come in shapes that look like deer antlers, snake tongues, four-leaf clovers, and fans. Frond blades are often divided into many smaller segments. Small, circular, spore-producing structures called *sori* (s. sorus) grow on the undersides of the fronds.

As fronds emerge upward, they first appear as tight coils. These young fronds are called *fiddleheads* because they resemble the scroll of a violin head. As the fronds continue to grow, they unroll and their blades expand.

In most ferns, fronds are attached to an underground stem called a *rhizome*. Unlike the rhizoids of nonvascular plants, rhizomes are true stems; they contain vascular tissues. Roots grow downward from the rhizome into the soil. Many ferns spread their rhizomes out through the soil and send up new plants to the surface. This is an example of asexual reproduction.

Club Mosses. Named for their club-shaped sporophytes, club mosses (which are not true mosses) are *evergreen* plants (plants that stay green year-round), which add a touch of color to fall and winter forests. They are sometimes used in Christmas decorations, and their spores were once used as flash powder in early photography.

Horsetails. Because horsetails have silica (a hard mineral) in some of their cell walls, they are abrasive. American pioneers used these plants to scrub pots and pans. From this use came one of this plant's common names—scouring rush.

MOSSES AND FERNS UP CLOSE

How are nonvascular and seedless vascular plants similar, and how are they different?

Equipment

- live moss and fern specimens
- hand lens
- squeeze pipette
- probe
- tweezers
- dissecting tray or paper towels

If you've ever taken a hike through a cool, shady forest, you've probably seen mosses and ferns in the wild. In many places, they often grow together. They form part of the forest understory, those plants that are shade tolerant and can live beneath the spreading canopies of large trees. Ferns and mosses also often have very specific habitat needs. Their presence or absence can be an indicator of forest health. How else are mosses and ferns similar? How are they different? Take a look and see!

PROCEDURE

Obtain moss and fern specimens from your instructor. Using a probe and tweezers, carefully tease apart one moss plant from the rest of the sample. Examine the form and structures of both the moss and fern using a hand lens as necessary. Locate the gametophyte, sporophyte (both stalk and spore capsule, if present), and rhizoids of the moss. Locate the frond, sori (if present), rhizome, and roots of the fern. As you observe your specimens, answer the following questions.

1. In what ways are the moss and fern similar? How are they different?
2. Visually speaking, which generation, the gametophyte or sporophyte, is dominant for each plant? How can you tell?
3. Which plant do you think is better adapted to survive a long period without water?
4. Feel the texture of the fern frond. On the basis of what you feel, do you think the frond is adapted to shed water from its upper surface? Explain.
5. Test your hypothesis by using the squeeze pipette to place several drops of water on the frond's upper surface. Describe what happens.
6. Was your hypothesis confirmed? Explain.
7. Why would a moss not need to shed water?

8.5 SEED PLANTS

Seed plants are the most familiar type of plant. These plants do not rely on water for fertilization. They have vascular tissue and can be found in both wet and dry environments. They reproduce by seeds in addition to spores.

8-2

An angiosperm seed, such as this apricot seed, is completely surrounded by fruit.

Seeds are very different from spores. Single-celled spores are in a precarious position with few energy reserves and only a cell wall to protect them from the environment. God designed these plants to produce lots of spores since in nature very few successfully grow into a plant large enough to see. In contrast, seeds are multicellular and contain a young plant called an **embryo**. Seeds usually have abundant energy reserves and have one or more barriers to protect them from the environment. There are two main groups of seed plants—**angiosperms** (AN jee uh SPURMZ) and **gymnosperms** (JIM nuh SPURMZ).

Angiosperms, the largest group of plants in the world, differ from gymnosperms in two main ways. First, angiosperm seeds are fully covered, shielding them from the environment. In fact, the word *angiosperm* means "covered seed." The seeds produced on these plants are completely surrounded by a protective covering called a *fruit*. Fruits develop from flowers; thus, all angiosperms are flowering plants, and all flowering plants are angiosperms. This fact highlights the second difference between angiosperms and gymnosperms—gymnosperms do not produce flowers.

Gymnosperm seeds are not fully covered. The word *gymnosperm* means "bare seed," and gymnosperm seeds are all exposed in some manner. The gymnosperms that you are most likely familiar with, such as pine trees, produce seed-bearing *cones*. The seeds of a mature pinecone can often be found lying on the scales of the cone.

Gymnosperms include some interesting plants called *cycads* and *ginkgoes*. But the most common gymnosperms and the ones most important to us are the *conifers*. Conifers include pines, yews, cypresses, and redwoods. These are used mostly for lumber and paper pulp. Your pencil, the paper in books, and probably much of your home are made from conifers.

8-3

Seeds can be found on the scales of pinecones.

SEED PLANT SAMPLER

ANGIOSPERMS

Angiosperms, the flowering plants, include everything from tiny species such as these 0.3 mm long duckweed …

… to the towering mountain ash of Australia that can soar to heights of nearly 100 m.

People depend on flowering plants for food. Tomato plants, watermelon vines, plum trees—they're all angiosperms.

The tulips in this flower bed are angiosperms. So is the grass!

SEED PLANT SAMPLER

GYMNOSPERMS

Cone-bearing trees such as pines, redwoods, cedars, and spruces are familiar gymnosperms.

Perhaps less familiar are the exotic, tropical cycads …

… and the ginkgo tree with its unusual leaves—even though it is a common ornamental tree.

WORLDVIEW SLEUTHING: MEDICINAL PLANTS

Did you know that a dog could be poisoned by a common ornamental flower? The foxglove flower is beautiful but deadly. But it can also be used to save lives. The foxglove's poison, digoxin, is used to make a drug that treats heart issues. Many drugs are poisons when taken in large doses. And many poisonous plants can be used to make medicines when processed properly.

INTRODUCTION

You are a science writer for a website. Your next assignment is to prepare a short presentation on poisonous plants that are used as medicines.

TASK

Research a plant that produces a poison that can also be used as a medicine. Examples include the opium poppy, deadly nightshade, and rosy periwinkle. Create a PowerPoint® or Prezi™ presentation that is based on your results. Include in your presentation where the plant grows, when its medical use was discovered, and what diseases the medicine treats.

PROCEDURE

1. Do a web search using the keywords "medicines from poisonous plants." Find a plant that you would like to research.
2. Do keyword searches using the name of the plant and the name of the medicine that is made from it.
3. Note how the worldviews of the authors of the websites affect their view of using plants for medicine.
4. Plan your presentation and collect any photos or videos you need. Be careful to give proper credit to your sources.
5. Create your presentation and show it to another person.
6. Deliver your presentation to your class or family.

CONCLUSION

Many poisonous plants have been used throughout the centuries for medicines. Pharmacists and scientists suspect that many more exciting medicines are waiting to be discovered.

8A REVIEW QUESTIONS

1. Which of the following is *not* true of most plants?
   - **a.** They have tissues.
   - **b.** They are eukaryotic.
   - **c.** They have plastids.
   - **d.** They are consumers.
2. Recall some of the characteristics of fungi from Chapter 7 (pp. 142–49). Describe at least two major differences between fungi and plants.
3. You also saw in Chapter 7 that some protists, such as euglenas, can produce their own food through photosynthesis. Why are they not considered plants?
4. What two characteristics are used to classify plants into their three primary groupings?
5. Classify each of the plants in the images on the right as either a nonvascular plant, seedless vascular plant, or seed plant.
6. What are the two major groups of seed plants, and what is the difference between them?

a.

b.

c.

8B Questions

- What are the various parts of plants?
- What do the different parts of plants do?
- How does a plant stand upright?

PLANT PARTS

① *Roots* hold a plant securely in the soil and absorb water and nutrients.

② *Stems* produce and support leaves. They also transport water and nutrients to the leaves and sugars to the roots.

③ *Leaves* absorb energy from the sun and use it to make sugars.

④ *Nodes* are the points along stems where new leaves, branches, or flowers are produced. Roots do not have nodes.

8B PLANT STRUCTURE

What are plants made of?

What comes to mind when you hear the words *lettuce*, *carrots*, and *tomatoes*? Perhaps you thought of a tasty salad. You probably didn't think about how lettuce, carrots, and tomatoes are each different parts of a plant. We eat the leaves of lettuce, the roots of carrots, and the fruits of tomato plants—we even eat the stems of some plants. Leaves, stems, roots, and fruits are not only different parts of plants; they're also structurally different and serve different purposes as well.

Evolutionists suggest that plants first appeared in the world's oceans as single-celled algae and then slowly evolved the structures needed to live on land. But as we've seen in previous chapters, evolution cannot explain how a single-celled alga could produce the many millions of DNA changes that would be needed to start making complex tissues such as roots, stems, and leaves. On the other hand, God's reliable account of the Creation Week in Genesis 1 tells us that plants were brought forth from the ground, not from the oceans. The seemingly endless variety of plants, both in form and in function, is well within the creative abilities of an infinitely powerful and wise Creator. Take a moment to thank God for creating the plant kingdom as we look at what a plant is made of—from the ground up.

8.6 EXTERNAL STRUCTURE

Roots

Have you ever had to pull weeds as a chore? When you pull up a dandelion or clump of crabgrass, you usually bring up some roots, too, perhaps entangled in a clod of dirt. All the roots of a plant are referred to as its *root system*. Roots anchor a plant in the soil and absorb water and dissolved minerals. Absorption is aided by many tiny *root hairs* that increase the surface area of the roots. Some roots also store food, such as the carrot in the earlier salad example.

There are two main types of root systems. A *taproot system* has one or a few main roots that are thicker and longer than the other roots of the plant. The much smaller roots that branch from the main root are called *lateral roots*. Usually, taproot systems reach relatively deep into the ground. Mature dandelions have deep taproots that make them difficult to pull out of your lawn. Mesquites, trees of the arid American Southwest, can have taproots up to 25 m long!

The other type of root system is a *fibrous root system*. Fibrous root systems have clusters of roots that are all about equal in size. These roots may branch several times. Grasses usually have a fibrous root system.

Stems

Plant stems produce leaves, branches, or flowers at points called *nodes*. Depending on the species, each node may produce a single new structure or multiple structures. Stems also carry substances to and from the leaves and roots. The trunk and branches of an oak tree are stems. They are quite different from the soft stems of plants such as tomatoes. *Woody* stems are hard and not very flexible. They have strong, thick cells. Woody stems are capable of supporting a lot of weight.

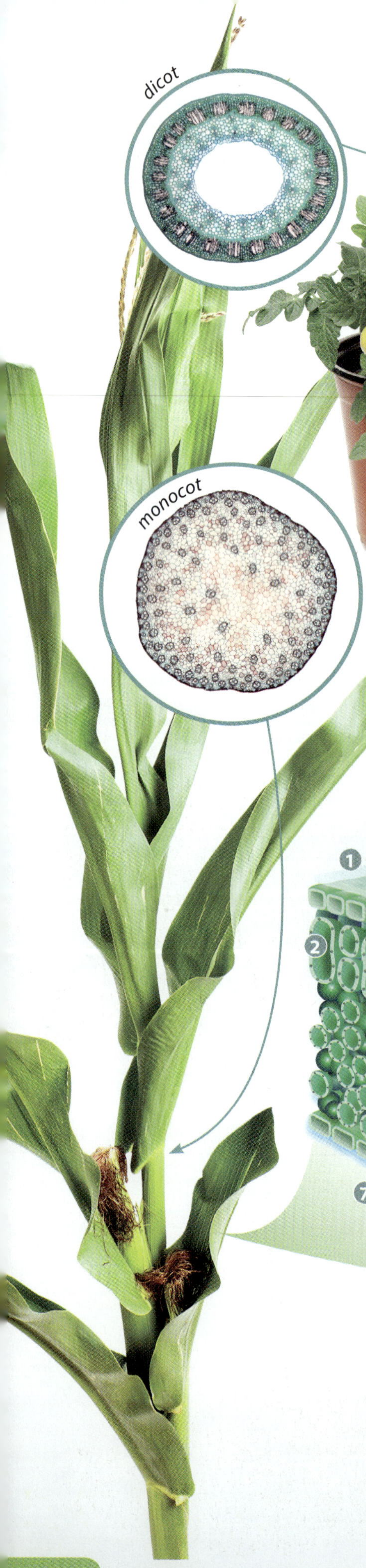

The softer and more flexible type of stem is called a *herbaceous* (hur BAY shuhs) stem. Herbaceous stems have some of the same types of supporting cells that woody stems have, but not as many. New stems in woody plants often begin as herbaceous stems and then become woody as they grow older.

Within a herbaceous stem, vascular tissues are normally clustered together in bundles called (not surprisingly) *vascular bundles*. Plants whose vascular bundles are evenly distributed throughout their stems are called **monocots**. In **dicots**, the vascular bundles are found in a ring around the outer portion of the stem.

Leaves

Leaves have an important job. They make food for the entire plant through photosynthesis. Both the external and internal structures of leaves are designed to help leaves do this task efficiently. Let's take a look at the structure of a leaf, and then we'll see how those structures work together to make food.

The waxy *cuticle* 1 covers the outer surfaces of the leaf and helps prevent water loss through evaporation.

Tightly packed cells of the **palisade layer** 2 make most of the plant's food. The cytoplasm in palisade layer cells circulates chloroplasts so that each gets plenty of light.

3 **Veins** are the vascular tissue of the leaf. A vein is a type of *vascular bundle*.

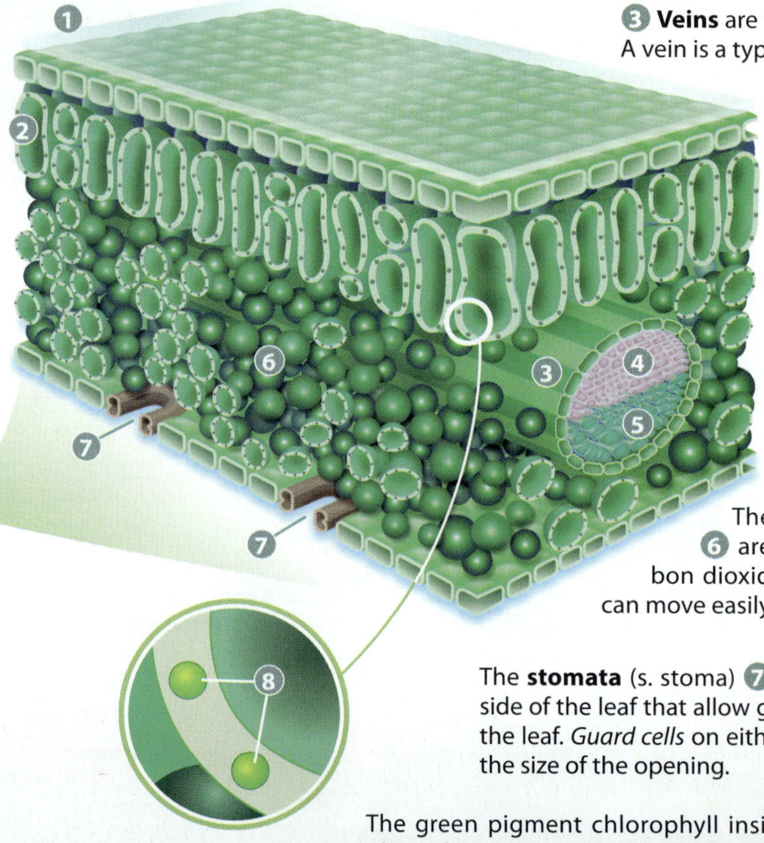

Xylem 4 is a vascular tissue that moves water and nutrients into the leaf.

Phloem 5 is a vascular tissue that moves finished sugars out of the leaf.

The cells in the **spongy layer** 6 are packed loosely so that carbon dioxide, water vapor, and oxygen can move easily to the palisade layer.

The **stomata** (s. stoma) 7 are openings in the underside of the leaf that allow gases to move into and out of the leaf. *Guard cells* on either side of the stoma regulate the size of the opening.

The green pigment chlorophyll inside the *chloroplasts* 8 gives plants their green color. This is where photosynthesis takes place.

Some of the sugars made by the leaf's cells are used for their own energy needs. The rest of the sugar is moved out of the leaf through the phloem portion of the veins. It travels to the other parts of the plant to provide food for cells in the stem and roots. Plants convert some of the sugar into cellulose to build new structures. Any excess sugar may be converted into *starches* or *oils* and stored for later use. The carrot from our salad example is a type of storage structure. Fruits are storage structures too.

8-4

Corn plants store excess energy in the form of either corn starch or corn oil. Check the food labels in your pantry and you'll see these two ingredients listed often.

Bark

As the herbaceous stem of a young woody plant grows, the vascular bundles join together to form a ring of phloem outside a ring of xylem. That thin ring of phloem and the tissue on the outside of the tree trunk form the *bark*. Between the xylem and the phloem is a layer of tissue that makes new xylem and phloem cells. The rings of xylem cells are pushed toward the inside of the stem as they thicken and toughen, forming wood. The phloem cells are pushed toward the outside, forming an inner portion of the bark. The layer of phloem in a tree is not as thick as the layers of xylem, but it is essential to the tree for moving sugars throughout the stem and down to the roots.

Have you ever tried to determine the age of a tree by counting the **annual rings** in an old stump? The rings form because xylem grows differently at different times of the year. During the growing season, when water and sunlight are abundant, xylem cells grow rapidly and have thin cell walls. These cells form the light-colored rings within the tree. The xylem cells of the darker rings are formed when water and sunlight are less available to the tree. During those times, the new xylem cells are smaller and have thicker cell walls.

The outermost layer of the bark is called the *cork*. Cork is made of dead cells with thick cell walls. These cells form a tough waterproof coating that keeps harmful organisms out and moisture in. The **cork cambium**, a layer of living cells just under the dead cork, constantly divides by mitosis, producing new cork cells. The young cork cells grow, form thick cell walls, and then die. As a tree grows larger, the dead cork cells and phloem cells on its trunk and branches do not expand. Instead, they are pushed apart as new cells are formed underneath. This breaking apart causes the rough texture of bark.

8-5

As trees grow, their expanding stems cause the older, outer layers of cork to move apart, creating many furrows. Many trees can be identified by the pattern of furrows within their bark.

8-6
Despite their lower branches being scorched, these pine trees survived a controlled burn thanks to their bark. Controlled burns reduce the amount of leaf litter on the forest floor and return nutrients to the soil.

Bark is important to trees. It protects the delicate vascular tissue beneath the bark and prevents moisture loss. Many trees have bark that produces sticky resin to defend against insects. Other trees have thick, fire-resistant bark that allows them to survive in areas where wildfires are frequent.

8.7 HOW DOES A PLANT STAND UPRIGHT?

Two structures within plant cells are especially important to the survival of plants—cell walls and central vacuoles. Plant cell walls are more than just the outer boundaries of cells. They also support the plant. The cell wall of an individual plant cell is not especially strong, but there are millions of such cells in a plant. Together, the microscopic cell walls can support or bear the weight of the entire plant, even gigantic oak trees. It's similar to how a man, by himself, probably can't lift a piano but together with several of his friends likely could.

The cell walls of plants are made largely of cellulose, which is made of long chains of sugar molecules. Other substances produced by cells bond the cellulose molecules together to make the cell walls rigid. These cellulose molecules contain a lot of energy. When cellulose molecules are burned, their stored energy is given off as heat and light. That's something to be thankful for when sitting in front of a warm campfire on a chilly fall evening.

Plants are also supported by **turgor pressure,** a kind of pressure caused by the presence of water inside a plant's central vacuoles. The more water inside the vacuoles, the greater the pressure exerted within the cells. When many of the cells in a plant part are swollen due to turgor pressure, the plant part is stiff. This stiffness by itself is enough to support most leaves and flower parts and many small plants. If a plant loses too much water through evaporation, the pressure within its central vacuoles drops, and the plant wilts. This is why a houseplant may need watering only once a week, but plants in a hot, sunny garden may need watering several times a day.

In addition to cell walls and turgor pressure, some plant stems contain an abundance of **fibers**. Vascular bundles are often supported by thick-walled fiber cells. The cell walls of fibers are so thick and strong that they alone could support the entire plant. Cornstalks, for example, are supported primarily by the fibers around the vascular bundles of their stems. That's why a dead cornstalk can remain standing even when it's dried out. Fibers are also extremely long, narrow cells. Since they are so long and strong, they can be used to make rope and cloth. The fibers of the flax plant, for instance, are used to make a common type of fabric known as *linen*.

8B REVIEW QUESTIONS

1. List the three main parts of any plant and describe the main functions of each.
2. Which type of stem is soft and flexible?
3. What does xylem transport? What does phloem transport?
4. Describe the two parts of an annual ring.
5. List two functions of bark.
6. Which kind of root system does the plant in the photo on the left have?
7. What would happen if a leaf's palisade layer were on the bottom side of the leaf instead of the top side? Explain.
8. Describe two characteristics of all plant cells that help support the entire plant.

CHAPTER SUMMARY

8A CLASSIFYING PLANTS

- Plants are eukaryotic, multicellular organisms that produce their own food through photosynthesis.
- Plants differ from other organisms in having plastids and cell walls made of cellulose.
- Plants are classified according to whether they produce seeds and spores or seeds alone and whether they contain vascular tissue.
- The three major groupings of plants are nonvascular plants, seedless vascular plants, and seed plants.
- Nonvascular plants have no vascular tissue and reproduce through spores. Seedless vascular plants have vascular tissue and reproduce through spores. Seed plants have vascular tissue and reproduce through seeds.

8A TERMS

8B TERMS

8B PLANT STRUCTURE

- The main parts of plants are roots, stems, and leaves.
- Roots anchor plants in soil and absorb water and nutrients.
- Stems produce and support leaves and transport substances between roots and leaves.
- Leaves make food for plants through photosynthesis.
- Plants with woody stems produce bark that protects the plant and helps it retain moisture.
- Cell walls, turgor pressure, and fibers help plants to stand upright.

CHAPTER REVIEW QUESTIONS

Recalling Facts

1. List the main characteristics of plants.
2. What are the three main groups of nonvascular plants?
3. How is the *rhizome* of a fern different from the *rhizoids* of a liverwort?
4. How is the seed of a gymnosperm different from that of an angiosperm?
5. From which part of a plant do fruits develop?

For questions 6–14, match the following plant structures with their function.

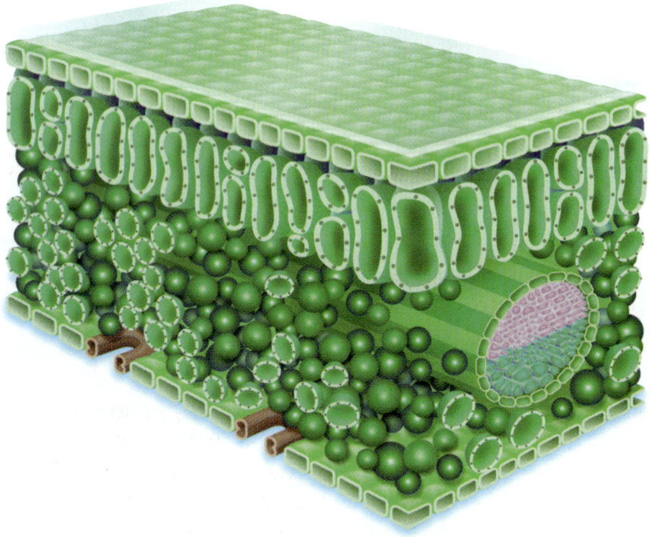

a. cuticle
b. chloroplast
c. vein
d. phloem
e. guard cell
f. palisade layer
g. spongy layer
h. xylem
i. stoma

6. allows gases to move easily within a leaf
7. controls the size of an opening in a leaf
8. transports water and nutrients into a leaf
9. the covering that prevents water loss from a leaf
10. packed cells that make most of a plant's food
11. the organelle where photosynthesis takes place
12. a bundle of vascular tissue
13. the opening into a leaf
14. transports sugars out of a leaf
15. From which part of a stem does a new leaf arise?
16. From which part of a plant stem are fabrics such as linen and burlap made?

Understanding Concepts

17. List two ways that plants are similar to protists. List two ways in which they are different.
18. A botanist discovers a new species of plant that has vascular tissue and reproduces using only spores. To which group of plants does it belong?
19. Why can't bryophytes grow tall?
20. Why are the rhizoids on the moss plant shown at right not considered to be true roots?
21. Like mosses, ferns reproduce through spores. Why are ferns not grouped together with bryophytes?
22. Create a concept map using the following terms: plants, nonvascular plants, seedless vascular plants, seed plants, gymnosperms, angiosperms, club moss, yellow pine, rose, maidenhair fern, and liverwort.
23. Name two plants that have woody stems and two that have herbaceous stems.

24. Seedless vascular plants, such as ferns, have vascular tissue, so why do they still need to live in a moist environment?
25. About how old is the tree whose trunk cross-section is shown at right?
26. What two cellular structures are mainly responsible for allowing a plant to stand upright?

Critical Thinking

27. It may not have occurred to you, but many people in the United States live in homes made from plant stems. Why is this so?
28. Cacti tend to have thicker cuticles than other kinds of plants. Why do you think this is so?
29. The annual rings in a particular tree show a series of a dozen rings spaced very close together. Give a possible explanation for the cause of this pattern.
30. Suppose plant cells had no central vacuoles. How might this affect a plant's ability to stand upright?
31. How is photosynthesis indirectly responsible for helping a plant stand upright?

ETHICS The Wise Use of Plants

Wheat, corn, barley, rice—all of these are very important foodstuffs. Humans have depended on them for centuries. In fact, the first role for plants described in the Bible, all the way back in the first chapter of Genesis, was serving as food. Since then, people have discovered many other good uses for plants. Plants provide us with fuel to keep us warm, fibers to clothe ourselves, and medicines to treat disease.

But mankind's sinful nature has also led people to put plants to less noble uses as well. As you saw in the Chapter opener, corn has been used to make illegal liquor. Just about any kind of plant matter, especially fruits and grains, can be used to create alcohol. How can a Christian determine whether the use of a particular plant is good?

Use the Ethics box above to answer Questions 32–36.

32. Do a web search using the keywords "ethical use of plants" and "ethical consumption of plants." Write a brief summary of what you learn about this issue.
33. What does the Bible say about the issue?
34. On the basis of your answer to Question 33, what are the acceptable and unacceptable options?
35. What may be the consequences of the acceptable options that you identified in Question 34?
36. What view or action should you take regarding the ethical use of plants?

9 PLANT FUNCTIONS

CHRISTMAS BEAUTIES

During the month of December, poinsettias start showing up everywhere. These plants with their bright red and green leaves have become symbols of the Christmas season. But it requires a great deal of work to make these plants look Christmassy. The wild poinsettia is a Mexican shrub over ten feet tall. The plants must be exposed to a certain amount of darkness in order to produce their brilliant red leaves. Is it worth all the time and money spent to produce a plant that serves only as a decoration?

9A

PLANT RESPONSES

How do plants sense and respond to their environment?

9.1 GROWTH CONTROL

In Chapter 8, we learned that plants are useful for food and for medicines. But people are also drawn to beauty because they are made in the image of God. We can use plants for all these purposes.

One way to use plants wisely is to understand how they respond to their environment. Have you ever seen a vine climbing a trellis? How is a plant able to twine around the frame? An oak tree can't do that, and neither can a blade of grass. The vine can because it displays a type of tropism that the oak tree and the grass don't have. A **tropism** occurs when a plant grows toward or away from a *stimulus*, which is anything that causes a response from a plant. Growth toward the stimulus is called a *positive tropism*; growth away from a stimulus is a *negative tropism*. The most common stimuli that produce tropisms in plants are light, gravity, and touch.

9A Questions

- Why do plants bend toward light?
- What controls how plants grow?
- Why do some plants bloom only at certain times of year?
- Should plants be used for purely decorative purposes?

TROPISMS

1 ***Light.*** Stems demonstrate *positive phototropism* by growing toward light. Roots grow away from light, showing *negative phototropism*.

2 ***Gravity.*** Stems grow away from the pull of gravity; this is *negative gravitropism*. Roots grow toward the pull of gravity; this is *positive gravitropism*.

3 ***Touch.*** The stems of some plants have the ability to grow toward an object they touch. This type of tropism is called *thigmotropism*. As the stem grows in response to the touch, it begins to curl around the object. Many climbing vines such as this black-eyed susan vine exhibit thigmotropism.

Hormones

How does a plant part grow toward or away from light? In fact, how does a plant know to grow at all? The answer is hormones. **Hormones** are small chemicals made by some plant tissues that affect how other plant structures grow. They move easily from cell to cell and even from one tissue to another. They can cause plant cells to grow in size, divide, or stop growing. There are many different plant hormones, but four are most important.

HORMONES

1. **Auxins** were among the first plant hormones discovered. They cause shoots and stems to grow by causing their cells to enlarge. They also slow down cell division.

2. **Cytokinins** do the opposite of what auxins do. They cause the cells to divide. Together, auxins and cytokinins help control whether a plant will be tall and thin or short and bushy. Cytokinins also help a plant that doesn't have enough water.

3. If auxins and cytokinins encourage the plant to grow, what keeps plants from growing out of control? **Abscisic acids** do that job by inhibiting several of the other plant hormones. They also prevent seeds from sprouting before they are released from the parent plant.

4. The gas **ethylene** causes fruit to ripen. It also causes trees to close off the xylem and phloem to their leaves before the leaves drop off. This prevents a tree from losing a lot of sap as the leaves fall.

So how do hormones affect plant growth? Let's consider one example. Sometimes a plant receives light from mainly one direction, such as when a plant is indoors near a window. Hormones cause the dark side of the stem to grow faster than the lighted side. As the dark side gets longer, the whole plant grows in the direction of the light. This allows the leaves better access to light for photosynthesis.

9-1
Plants grown near a window often bend toward the light source.

9.2 FLOWERING CONTROL

A chrysanthemum remains without blooms all summer long. Then as the days become shorter and cooler in early fall and many plants have finished flowering for the year, the mum begins to blossom. Early researchers thought that flowering was controlled by the length of daylight, so they called plants such as mums and asters **short-day plants**. These plants actually do not bloom until the *nights* are longer than a certain length of time, called the *critical night length*.

Other plants such as irises and snapdragons bloom only when the nights are shorter than the critical night length. These plants are **long-day plants**.

Some plants are described as **day-neutral plants** because they flower whenever the conditions such as temperature and moisture are right, regardless of the length of the night.

9-2
Cucumbers are day-neutral plants.

Short-day and long-day plants have chemicals that help control when the plant flowers. The chemicals are sensitive to the lengths of daylight and darkness. When the length of darkness is correct, the plant flowers.

A plant's sensitivity to the length of darkness is part of the reason that it takes so much work to produce poinsettias for Christmas. Poinsettias are short-day plants, so growers must give them at least thirteen hours of darkness for six weeks or more starting in early October. But since poinsettias are native to Mexico, they also need warm temperatures, so they must be kept in heated greenhouses. This takes a lot of energy. Is it really worth it to produce a plant that serves no purpose other than decoration?

9-3
Irises and snapdragons are long-day plants, so they bloom during the summer.

The Bible certainly never forbids the use of plants for decorative purposes, and God Himself demonstrated a love of beauty in His design of plants and animals. He created the Garden of Eden to be a paradise, a beautiful place. When He gave Moses instructions for the tabernacle, His design consisted of many beautiful symbols, including plants that served no purpose other than to glorify Himself through their beauty. Some parts of the tabernacle were made from the beautiful and useful products of plants, including woods and dyes. God included recipes for incense that used perfumes from plants. From these facts, we can discern that raising plants for their beauty is consistent with a biblical worldview.

9A REVIEW QUESTIONS

1. Write the definition of tropism in your own words.
2. A vine grows around a trellis. Is the plant displaying positive or negative thigmotropism? Explain.
3. The leaves of a sensitive plant fold up when touched. The effect is the same whether the top, bottom, or side of the leaf is touched. Is this a tropism? Explain.
4. What three things do plant hormones control?
5. Which plant hormone inhibits several of the other hormones?
6. A tomato plant flowers in the spring, summer, and fall as long as the temperature is warm enough. Is it a long-day, short-day, or day-neutral plant?
7. During which season do most long-day plants flower?
8. Does the Bible support the use of plants purely for beauty? Explain.

LONG-LASTING FLOWERS

What makes cut flowers last longer?

Roses are beloved by many people, and florists sell millions of them every year, especially around Valentine's Day. But cut flowers rarely last much longer than a week. Many popular ideas suggest adding substances to the water in the vase to make the flowers stay fresh longer. But do any of these actually work?

Equipment

E

roses of one color with fertilizer packet (12)

vases, 1 L (4)

aspirin tablet, 325 mg

lemon-lime soda, 60 mL

water

PROCEDURE

1. Pour 700 mL of water into each 1 L vase.
2. Pour the fertilizer packet from the bouquet into Vase 2.
3. Drop an aspirin tablet into Vase 3.
4. Pour 60 mL of lemon-lime soda into Vase 4.
5. Place three roses in each of the four vases.

1. Which vase do you think will do the best? The answer to that question is your hypothesis.
2. After one day, do any of the vases of flowers look any worse than the others? If so, which one(s)?
3. After five days, have any of the roses started to fade? If so, which one(s)?
4. After ten days, do any of the vases of roses look better than the others? If so, which one(s)?
5. Rank the vases of roses from those that remained fresh looking the longest to those that faded most quickly.
6. Compare your observations to your hypothesis. Try to explain the differences, if any.
7. What else could you test to lengthen the life of cut flowers?
8. Why is this scientific work valuable from a Christian worldview?

9B

PLANT REPRODUCTION

9B Questions

- How can seedless plants reproduce?
- How does reproduction differ in flowering and nonflowering plants?

How do plants reproduce?

Have you ever had a vegetable garden? If so, you've seen how some plants reproduce. First a plant flowers, and then, as the flowers die, you can see tiny fruit where the flowers used to be. Then the fruit grows larger and eventually ripens until it's ready to eat. You're probably already familiar with some aspects of plant reproduction, such as the roles of flowers and seeds. But what about plants like mosses and liverworts that don't produce seeds? And how about pine trees that don't produce flowers? How do plants like these reproduce? Let's examine that question. As we do, take note of the similarities and differences in how reproduction occurs in the various plant groups.

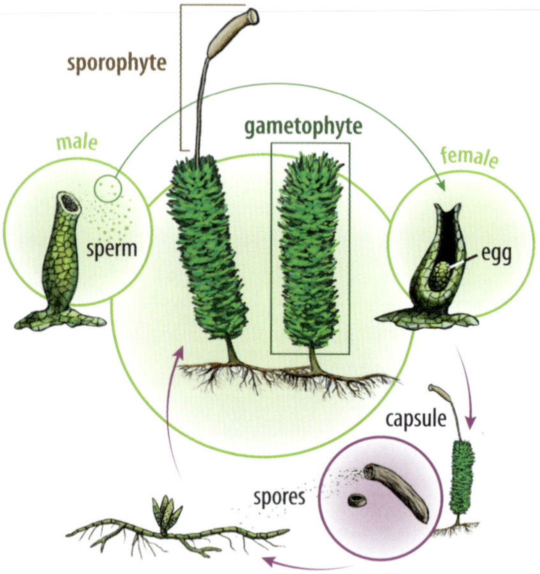

9.3 REPRODUCTION IN NONVASCULAR PLANTS

In mosses, liverworts, and hornworts, the gametes needed for sexual reproduction are produced in different areas on the top of the green leafy part of the plant, the gametophyte. Water is necessary for fertilization because the sperm must swim to the unmoving eggs. After fertilization, zygotes remain attached to their gametophytes and grow into stalks and capsules, which together make up the sporophytes. Thousands of spores are produced in each capsule. When conditions are right, the capsules pop open and spores are released, falling on nearby soil or being transported farther away by the wind. A small percentage of these spores grow into new moss plants. This repeating cycle of gametophyte and sporophyte structures is called **alternation of generations**.

9-4

Both gametophytes and sporophytes are plainly seen in this photo of moss.

9.4 REPRODUCTION IN SEEDLESS VASCULAR PLANTS

Just like other plants, ferns have alternation of generations. On the underside of many mature fronds are sori. You remember from Chapter 8 that these structures produce spores. Thus the dominant fern stage is the sporophyte. Tiny fern spores are scattered by the wind. Those that fall on suitable soil grow into a small, heart-shaped plant that is easily overlooked. This gametophyte plant usually grows no larger than a fingernail. It produces eggs and sperm (gametes) and requires water for the sperm to swim to the eggs and fertilize them. After fertilization, the zygote, which remains attached to the small plant, will grow into a recognizable fern.

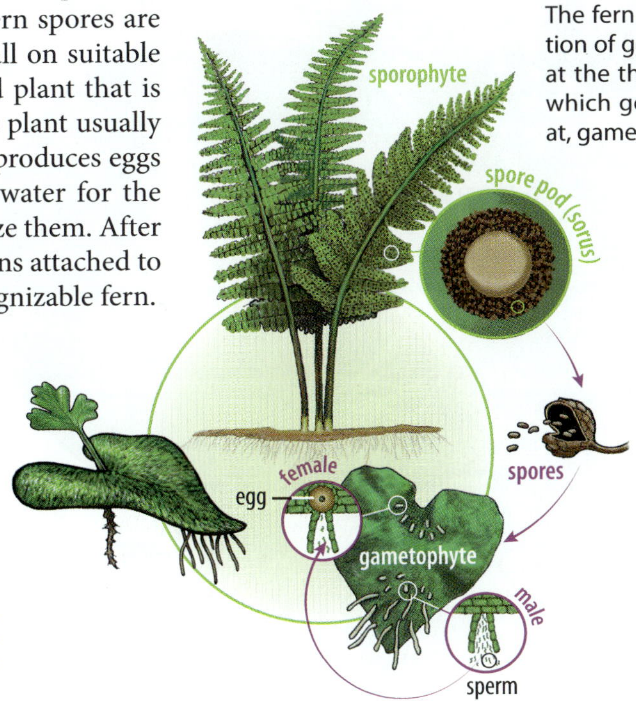

9-5

The fern life cycle exhibits alternation of generations. When you look at the thick carpet of ferns below, which generation are you looking at, gametophyte or sporophyte?

9.5 REPRODUCTION IN GYMNOSPERMS

Like mosses and ferns, gymnosperms such as pine trees do not have flowers. But unlike mosses and ferns, gymnosperms produce seeds. When you look at a pine tree, you are seeing the sporophyte. It produces spores inside the cones. These spores form microscopic male and female gametophyte structures inside the cones. Male cones release pollen containing the male gametes. Wind carries the pollen grains to female cones to fertilize the eggs. As the zygotes develop, they are enclosed in seeds inside the female cones. When the seeds are mature, the cones open, releasing the seeds into the air. A few of these seeds will *germinate* and grow into new trees.

9-6

These male cones are releasing pollen to be carried by the wind to female cones.

9.6 REPRODUCTION IN ANGIOSPERMS

Like gymnosperms, angiosperms encase their zygotes inside seeds. But angiosperm ovaries also swell to form a **fruit** that surrounds and protects the seeds. Angiosperms also have flowers, something that no other type of plant has. Flowers come in many shapes and sizes, but most have four main parts, though some flowers lack one or more of them.

FLOWERS

The **stamens** ❶ are the male part of the flower. The pollen-producing *anthers* grow at the top of stalks called *filaments*.

The **carpel** ❷ is the female part of the flower. Pollen is carried by the wind or animals from the stamens to the sticky *stigma* at the top of the carpel. The stigma sits on top of a tube called the *style*. At the base of the style is the *ovary* that contains one or more *ovules*. Each ovule holds an unfertilized egg.

The often-colorful **petals** ❸ help attract pollinators to many flowers.

The **sepals** ❹ are leaflike structures positioned outside the petals. The sepals protect the flower while it is developing. While they are usually green, some are as colorful as the petals.

When a pollen grain lands on another flower's stigma, the pollen produces a pollen tube that grows down through the style to the ovary. The gametes travel down the pollen tube to fertilize an egg in an ovule. The fertilized egg develops into the embryo that will form the new plant. Each ovule forms a seed that protects the new plant, and the ovary forms the fruit.

Angiosperm fruits have an astounding variety of shapes, sizes, and textures. Some, like the blackberry, are small, round, and soft. Others, such as the Atlantic Giant pumpkin, are enormous and hard.

Some fruits are spread by the wind. Other fruits contain a lot of sugar, which attracts animals that disperse the seeds by eating the fruit.

Once a seed is released from the fruit, the seed needs the right conditions to sprout, including the correct temperature and moisture. If those conditions occur, the seed germinates as the sprout emerges. It can then begin to photosynthesize and grow into a mature plant.

Most angiosperms reproduce sexually using flowers. But a few can also reproduce asexually. Part of the parent plant breaks off and grows into a new plant.

9-7
This bird gains a nutritious snack by eating these fruits. It also spreads the tree's seeds.

9-8
This hen and chicks reproduces asexually. The smaller plants started as part of the parent and then took root as separate plants.

9B REVIEW QUESTIONS

1. Why would seedless plants have a hard time reproducing in a desert?
2. What two main groups of plants produce seeds?
3. What two reproductive structures of angiosperms do gymnosperms lack?
4. What structure carries the male gametes in both gymnosperms and angiosperms?
5. What type of plant produces the structures shown below?

FROM SLAVE TO SCIENTIST: SERVING AS A HORTICULTURALIST

Cotton was king in Alabama at the beginning of the twentieth century. In fact, it had been king for decades. But all the years of growing nothing but cotton had removed most of the nutrients from the soil. Then around 1915 boll weevils appeared in the state and destroyed much of the cotton crop. Farmers and sharecroppers, already struggling due to the poor soil conditions, were facing bankruptcy.

At the same time, George Washington Carver worked as a teacher and horticulturalist at Tuskegee Institute (now Tuskegee University). He realized that planting peanuts instead of cotton would help farmers. Peanuts return nitrogen to the soil, improving farmland. They also are not affected by boll weevils. To convince people to begin planting peanuts, Carver began publishing information on how to grow and use peanuts. He also traveled throughout Alabama to teach farmers the importance of growing more than one crop. As a result, the state's economy improved greatly.

Some horticulturalists, like George Washington Carver, find new and better ways to grow crops. Others experiment with new ways to control crop pests, while some help create better landscaping techniques in urban areas. Still others care for lawns and fields for private landowners or the government.

GEORGE WASHINGTON CARVER'S TESTIMONY

"I was just a mere boy when converted, hardly ten years old. There isn't much of a story to it. God just came into my heart one afternoon while I was alone in the loft of our big barn while I was shelling corn to carry to the mill to be ground into meal.

A dear little white boy, one of our neighbors, about my age came by one Saturday morning, and in talking and playing he told me he was going to Sunday school tomorrow morning. I was eager to know what a Sunday school was. He said they sang hymns and prayed. I asked him what prayer was and what they said. I do not remember what he said; only remember that as soon as he left I climbed up into the loft, knelt down by the barrel of corn, and prayed as best I could. I do not remember what I said. I only recall that I felt so good that I prayed several times before I quit.

My brother and myself were the only colored children in that neighborhood and, of course, we could not go to church or Sunday school, or school of any kind.

That was my simple conversion, and I have tried to keep the faith."

—G. W. Carver; Letter to Isabelle Coleman; July 24, 1931

CHAPTER 9

CHAPTER SUMMARY

9A TERMS	
tropism	177
hormone	178
auxin	178
cytokinin	178
abscisic acid	178
ethylene	178
short-day plant	179
long-day plant	179
day-neutral plant	179

9A PLANT RESPONSES

- A plant exhibits a tropism when it grows toward or away from a stimulus such as light, gravity, or touch.
- Some plant tissues make hormones that affect how other plant structures grow.
- The critical night length determines when long-day and short-day plants bloom. Day-neutral plants bloom whenever the conditions are right.
- The Creation Mandate justifies using plants for beauty.

9B PLANT REPRODUCTION

- All plants exhibit alternation of generations with sporophyte and gametophyte structures.
- Seedless plants have gametophytes, producing gametes that unite to form a zygote that grows directly into the sporophyte structure.
- In plants that produce seeds, the zygote is enclosed inside the protective seed.
- Gymnosperms produce seeds inside cones that are released unprotected into the environment.
- Angiosperms produce seeds that are surrounded by a fruit.

9B TERMS	
alternation of generations	182
fruit	185
stamen	185
carpel	185
petal	185
sepal	185

CHAPTER REVIEW QUESTIONS

Recalling Facts

1. What is a positive tropism?
2. What stimuli from the environment affect the direction in which plants grow?
3. What is a hormone?
4. Which hormones together determine the shape of a plant?
5. What determines when a short-day or long-day plant flowers?
6. What determines when a day-neutral plant flowers?
7. What types of plants are seedless?
8. What is a fern's reproductive structure called?
9. What structures produce pollen in gymnosperms and angiosperms?
10. What is a fruit?
11. What are two differences between cones and fruits?

Understanding Concepts

12. Describe the positive and negative forms of phototropism and gravitropism in stems and roots.
13. Compare the action of auxins and cytokinins.
14. An aster grows throughout the summer and then blooms in late September or even early October. What type of plant is it?
15. Explain why a wet environment is necessary for reproduction in seedless plants.
16. Explain why plants that produce seeds can reproduce even in dry environments.
17. Organize the following terms in a Venn diagram, including circles with the headings "Angiosperms" and "Gymnosperms": *pollen*, *stamen*, *carpel*, *egg*, *fruit*, and *cone*. (See Appendix B for an example of a Venn diagram.)

Critical Thinking

18. What would happen if roots exhibited negative gravitropism?
19. A pear tree's fruit fails to ripen each year. A scientist discovers that the tree does not produce enough of one of the major plant hormones. Which hormone is the tree likely lacking? Explain.
20. A plant in a greenhouse is kept in the dark for fourteen hours a day. When the period of darkness is decreased to thirteen hours a day, the plant flowers. Is this a short-day plant or a long-day plant? Explain.
21. Use biblical teaching to defend using plants purely for beauty.
22. If honeybees became extinct, would it affect gymnosperms or angiosperms more? Explain.

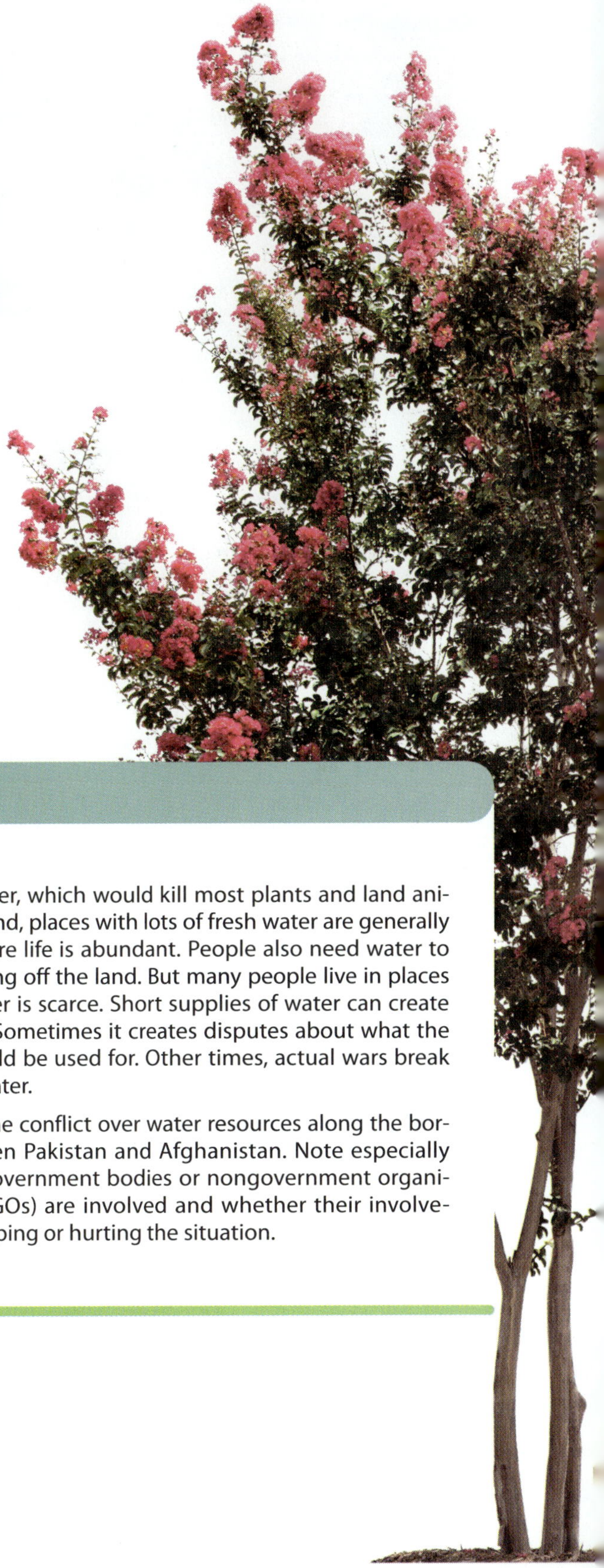

ETHICS — Water Wars

Water—plants need it to survive. In fact, all life needs it to survive, but there is only so much of it to go around. Though 71% of the earth is covered with water, most of it is saltwater, which would kill most plants and land animals. On land, places with lots of fresh water are generally places where life is abundant. People also need water to make a living off the land. But many people live in places where water is scarce. Short supplies of water can create problems. Sometimes it creates disputes about what the water should be used for. Other times, actual wars break out over water.

Research the conflict over water resources along the border between Pakistan and Afghanistan. Note especially whether government bodies or nongovernment organizations (NGOs) are involved and whether their involvement is helping or hurting the situation.

Use the Ethics box above to answer Questions 23–27.

23. What information can I get about this issue?
24. What does the Bible say about this issue?
25. What are the acceptable and unacceptable options?
26. What may be the consequences of the acceptable options?
27. What action should I take?

unit 3 THE ANIMAL KINGDOM

FROM A MOLECULAR GENETICIST TO YOU

I have been passionate about science for what seems like my whole life. Initially I wanted to become a medical doctor, but then I heard a scientist talk about his research studying cells and DNA and I was hooked! As I studied through college and graduate school about God's amazing, complex, intricate design in living things, it was hard for me to believe how anyone could study that and believe that all life came about by random chance over millions of years.

Several years ago at Answers in Genesis, I was given the opportunity to lead a group of scientists to categorize all living and extinct animals (known only from the fossil record) into their distinct "kinds." This was a huge job that needed to be done to estimate the number of animals that were on Noah's Ark. We then used this information in several exhibits at the Ark Encounter in Kentucky. Science confirms God's Word, which states that God created animals according to their kind and that they were to reproduce within those kinds.

Dr. Georgia Purdom

10 ANIMAL CLASSIFICATION

PANDA-MONIUM

It's a bear. No, wait—it's a raccoon. Wait, yes—it's a bear. Scientists are sure of it this time. Well, almost certainly sure, maybe.

The cuddly looking giant panda with its distinctive black-and-white markings is instantly recognizable, but it has puzzled scientists for many years. Why is that? It's because evolutionary scientists assume that pandas evolved from a related kind of animal. Since pandas share certain similarities with both bears and raccoons, evolutionists long debated which of the two groups pandas belonged to. They *think* that DNA studies have settled the issue for good, but have they? Like any scientific model, the classification of animals can change as more data is gathered. And thanks to evolutionary theory, taxonomy (the science of classification) changes all the time.

10A

CHARACTERISTICS OF ANIMALS

How are animals classified?

As you learned back in Chapter 1, scientists classify living things. They do so to see both how living things are alike and how they are different. Classifying animals makes it easier to study them. It also helps scientists share their findings with others in their field. But before we can classify animals, we first have to define what an animal is.

10A Questions

- What is an animal?
- What characteristics are used to classify animals?
- Are there different ways to classify animals?

10.1 WHAT IS AN ANIMAL?

Do you have a pet? If so, then you are probably very familiar with some of the characteristics of animals. Dogs, cats, kangaroos, prairie chickens—it's pretty obvious when you see one of these that it's an animal. But what about this thingy on the right? Is it an animal? How could you tell? Are those tentacles or maybe some kind of weird leaves? Hmmm.

How would a *zoologist* (a scientist who studies animals) decide whether this is an animal? There are certain characteristics that animals have in common.

Animals are made of many eukaryotic cells. They are also consumers. In addition to those two characteristics, most animals have cells that are specialized to do different jobs and are organized into tissues, organs, and systems. Most animals can also move during at least part of their lives, eat their food rather than merely absorb it, and reproduce sexually (though some can reproduce asexually).

The strange organism is eukaryotic and multicellular, has tissues and organs, consumes food (the feathery parts catch particles of food that drift by), can move, and reproduces sexually. So is it an animal? Yes! It's a *crinoid*—a type of animal similar to a sea star.

Bilateral symmetry. This crab has *bilateral symmetry*. An imaginary line drawn down its middle divides the crab into two halves that look very much alike. Notice that there is only one way to divide the crab into mirror-image halves, as indicated by the dashed line. We sometimes refer to this as an animal's *midline*.

10.2 CHARACTERISTICS USED TO CLASSIFY ANIMALS

How is a crinoid different from a crayfish? What makes an earthworm different from an aardvark? Beyond the characteristics listed in Subsection 10.1, which are common to all animals, animals can be sorted and grouped on the basis of the presence or absence of certain other characteristics. What are some of these characteristics?

Symmetry

Symmetry is a mathematical term. It's the property of an object that means that it can be divided by one or more imaginary lines into mirror-image halves.

Radial symmetry. This starfish, though, is a bit different. There's more than one possible imaginary line that can divide it into mirror-image halves. This is an example of *radial symmetry*. Think of it like the spokes on a bicycle wheel that radiate from a central hub.

Asymmetry. Have fun trying to find an imaginary line that will divide this tube sponge into mirror-image halves! There isn't one. This lack of mirror-image halves is called *asymmetry*.

Heads

Most animals have their sensory organs and brain clustered at one end of the body—the head. This arrangement is called *cephalization* (SEF uh la ZAY shun). Having the sense organs together at one end of the body helps an animal determine the direction that sensory cues such as light and smells are coming from.

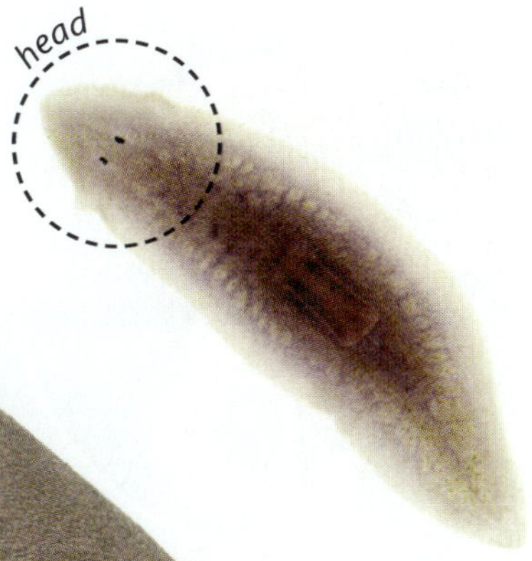

Guts and Body Cavities

The word "guts" usually brings something unpleasant to mind, but it's actually a science term. An animal's *gut* is its digestive tract—the part of its body that takes in food, digests it, and absorbs it. You might be surprised to learn that not all animals have one! The sponge that you just read about is one such animal. The individual cells of its body trap food particles and digest them. It has no specialized digestive structures.

Animals can also be classified on the basis of the presence or absence of a *body cavity*, a hollow within the body that contains most of the animal's vital organs. Most kinds of animals with complete guts have a true body cavity, but some do not.

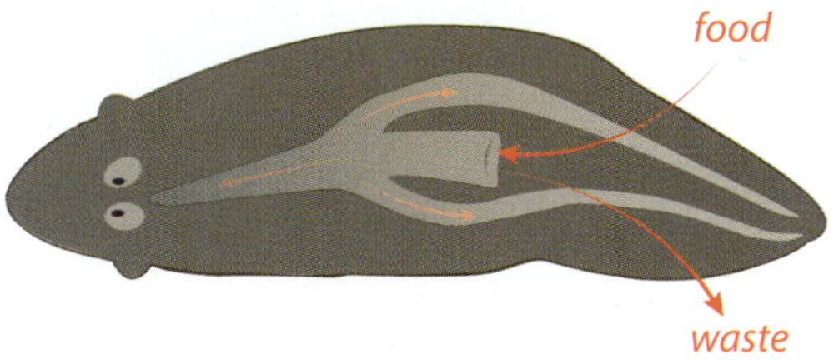

Incomplete guts. Planarians like this one have incomplete guts. A planarian's digestive tract has only one opening. The planarian draws in food through this opening, digests the food, and then expels wastes through the same opening.

Complete guts. Most of the animals you are familiar with, like this aardvark, have complete guts. Their digestive tracts have two openings. Food enters the opening at one end of the tract and is digested. Wastes are then expelled through an opening at the other end of the tract.

Chordates. Yes, this is an animal! It's a sea squirt. The adult sea squirt doesn't have a backbone or even a notochord. But larval sea squirts do have notochords, so sea squirts *are* chordates, but not vertebrates.

Presence of a Backbone

Another characteristic used to classify animals is the presence of a flexible rod called a **notochord** at any time during the animal's life. Animals that have this structure are called *chordates*. In most chordates, the notochord develops into part of the spinal column, also called the *backbone*. Made of individual bones called *vertebrae*, the backbone protects the animal's spinal cord. Such animals are called **vertebrates**. A few chordate animals don't ever develop a spinal column. Some of them keep their notochords for their entire lives, while others lose their notochords as they grow and develop. The great majority of animals, over 95% in fact, never have a notochord at any point in their lives and never have a bony spinal column. These animals are called **invertebrates**.

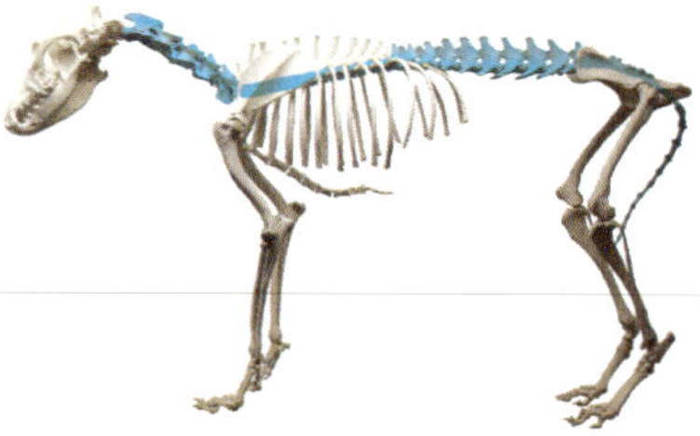

Invertebrates. This mantis has never had a notochord. Instead, it supports its body with a hard outer covering called an *exoskeleton*.

Vertebrates. Like all vertebrates, the Australian dingo has a bony spinal column made of vertebrae. Vertebrates are the most numerous category of chordates.

Temperature Control

Some animals can regulate their internal body temperature, keeping it at the ideal level for their needs. Other animals lack this ability. Their body temperatures are determined by the temperature of their environments. Animals that can regulate their inner temperatures are **endotherms**. Animals that can't are **ectotherms**.

Endotherms. This snow leopard lives in cold places, but it uses its fur and its internal body processes to stay warm.

Ectotherms. This lizard can't regulate its body temperature without help from its environment. A sunny rock helps the lizard warm up.

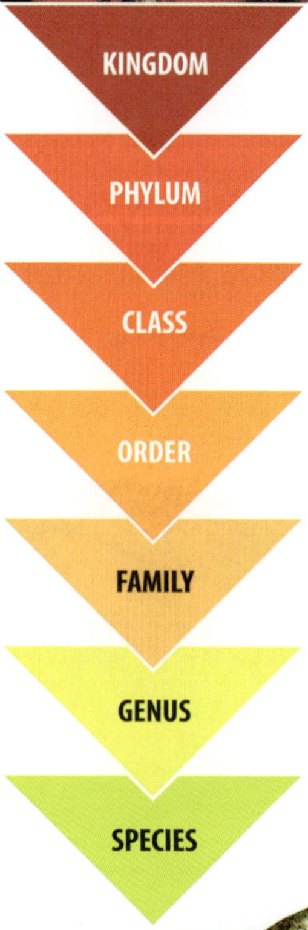

10.3 CLASSIFYING ANIMALS

Our modern classification system has its roots in the work of Carolus Linnaeus, whom you read about in Chapter 1. Linnaeus was a Christian and a creationist. He believed that God had created distinct kinds of plants and animals, just as the Bible says. Linnaeus's classification system reflected this belief. Within that system, he attempted to place all living things into groups "after their kind," to use the biblical language. He even said that animals are classified "by sight," that is, according to their appearance.

The Linnaean system is still used today. But as we've seen throughout this textbook, evolutionists have made many changes to the way in which organisms are classified. This is due mostly to their continuing efforts to try and figure out how plants and animals have evolved from common ancestors. Since their ideas about how different organisms are related change frequently, their classifications of those organisms change frequently too. These ever-changing classification models can be very confusing—even for scientists!

Creationists have suggested a solution for this problem of revolving-door taxonomy. The Bible clearly teaches that God created distinct kinds of animals. If different species of animals can interbreed, then it follows that they must be part of the same biblical kind. But a flounder cannot interbreed with, say, a trout. Though they are both fish, they are definitely two different kinds with a boundary between

10-1

The scientific name of the giraffe haplochromine, *Nimbochromis venustus*, has changed at least three times since it was first described in 1908. Its original genus, *Haplochromis*, has been divided into at least forty-one separate genera by some taxonomists.

them. This is true for other kinds of animals as well. Creationist taxonomists have attempted to define where the boundaries are between the various kinds of animals. They call this effort **baraminology**, from two Hebrew words *bara* and *min* that together mean "created kind." God established the kinds of animals during the Creation Week; baraminologists are trying to discover what those kinds are. In contrast, evolutionary taxonomists are searching for relationships that, according to the Bible, do not exist.

10A REVIEW QUESTIONS

1. What characteristics do *all* animals have in common?
2. Give an example of how a particular characteristic can be used to classify animals.
3. Identify the symmetry of each of the animals below.

frog

jellyfish

sponge

ladybird beetle

4. What is the relationship between chordates and vertebrates?
5. (True or False) Most animals are invertebrates.
6. How is the modern classification system different from the classical Linnaean system?
7. What do baraminologists try to do?

10B

INVERTEBRATES

What are invertebrates?

10B Questions

- What is an invertebrate?
- What are some of the major invertebrate groups?

10.4 INVERTEBRATE CHARACTERISTICS

As we begin our survey of kingdom Animalia, it's good to keep one thing in mind: *all* living things are vastly complex! We believe that God created all living things as distinct kinds during Creation Week. But creationists *can* talk about *relative complexity*. That is, the animals that you'll see later in this chapter are relatively more complex than the animals presented earlier in the chapter. Compared with sponges and worms, for example, mammals have relatively more tissues, more organs, more complex structures, and more varied behaviors. But we should always remember that each kind of animal, whether worm or walrus, bears witness of the infinite creative ability of an all-wise God.

What makes an animal an invertebrate? Though we already saw the answer to that question in Subsection 10.2, the name itself also gives a clue. You may recall that the prefix *in-* means "without," and the word *vertebrate* refers to the vertebrae in a backbone. So invertebrates are simply animals without backbones. Invertebrates are also ectotherms. But beyond these two shared characteristics, invertebrates are extremely varied.

10.5 INVERTEBRATE CLASSIFICATION

Some animals are very different from the organisms you usually think of as animals. You may even have trouble identifying them as animals. But they have all the animal characteristics described in Subsection 10.1. As you look at the groups of invertebrates, consider these questions: What kind of symmetry do these animals have? Do they have a head? Do they have a complete or incomplete gut? This will help you to see how these animals are classified. Let's explore the major groups of invertebrates.

Sponges

Sponges are animals with no symmetry, no cephalization, and no gut. They are quite unlike any other group of animals. Their bodies have many pores through which water flows, bringing food and oxygen in and carrying wastes out. The phylum name Porifera (puh RIF er ruh) means "pore-bearer." All sponges live in water, and most species live in the oceans.

Cnidarians

Jellyfish and anemones belong to phylum Cnidaria. All **cnidarians** (ni DAIR ee enz) have stinging cells called *cnidocytes*, which they use to capture food. Cnidarians have radial symmetry. Most cnidarians are marine animals and many are important members of coral reef communities. In fact, corals are cnidarians too!

Worms

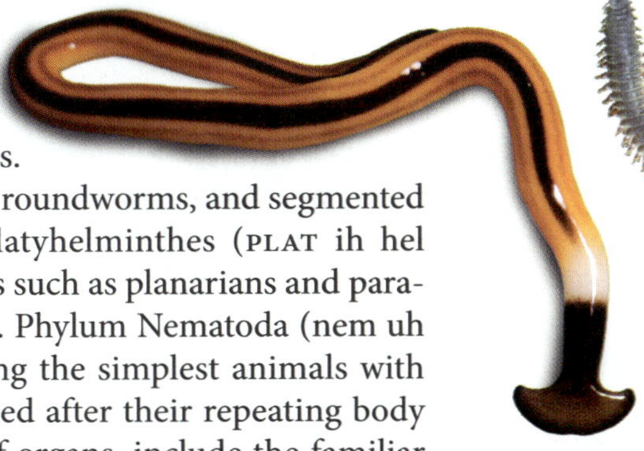

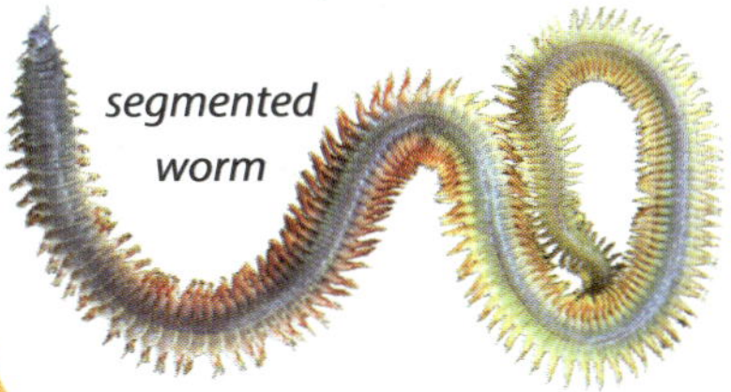

Worms are long, soft-bodied animals with bilateral symmetry. They are classified into many different groups. Three of the major ones are flatworms, roundworms, and segmented worms. The *flatworms* of phylum Platyhelminthes (PLAT ih hel MINTH eez) include free-living species such as planarians and parasitic species like tapeworms and flukes. Phylum Nematoda (nem uh TOE duh), the *roundworms*, are among the simplest animals with complete guts. *Segmented worms*, named after their repeating body segments that contain the same sets of organs, include the familiar earthworm, marine bristle worms, and leeches. These worms are classified in phylum Annelida (uh NELL i duh).

Echinoderms

Sea stars, sea urchins, and sea cucumbers are **echinoderms** (ih KYE nuh DURMZ). Their phylum name, Echinodermata, refers to their spiny skin. Echinoderms live in the world's oceans, from tide pools to the deepest sea floors. Most have radial symmetry.

sea cucumber

Mollusks

Mollusks in phylum Mollusca include three major groups. *Gastropods* include snails and slugs. Octopuses and squids are *cephalopods*. And *bivalves* are animals with two-piece, hinged shells such as clams, oysters, and scallops. The name *mollusk* refers to the soft bodies of these animals. Most rely on one or two hard shells for protection, except for the cephalopods. Many studies have shown octopuses to be among the most intelligent of all invertebrate animals.

nautilus

octopus

giant clam

green lacewing

Arthropods

Phylum Arthropoda (ar THRAH puh duh), the **arthropods**, has the most species of any animal phylum. The phylum name refers to the group's jointed *appendages*, such as legs and antennae. Along with the shelled mollusks, arthropods also have an **exoskeleton**, a support structure on the outside of the animal's body.

hornet

rainbow grasshopper

Insects

Within the arthropods, *insects* are the largest class. They are easy to tell apart from other arthropods—adult insects have three body segments and six legs. There are almost a million identified species of insects, and many more unidentified species probably exist. Insects include the only invertebrates that can fly.

elephant hawk moth

Crustaceans

Another important arthropod group are the *crustaceans*—the shrimps, crabs, and lobsters. Unlike insects, crustaceans have more than six legs—crabs have eight, while shrimps and lobsters have ten.

ghost crab

Coleman shrimp

blue freshwater lobster

jumping spider

tick

scorpion

Chelicerates

Spiders, scorpions, and ticks are *chelicerates* (kuh LISS uh RETS). These arthropods have two body segments and eight legs. The group takes its name from their chelicerae, a kind of mouth part. In spiders, the chelicerae are modified into fangs for injecting venom.

Table 1

Invertebrate Characteristics

	Sponges	Cnidarians	Segmented Worms	Echinoderms	Mollusks	Arthropods

Symmetry	none	radial	bilateral	radial	bilateral	bilateral
Cephalization	no	no	yes	no	yes	yes
Gut	none	incomplete	complete	complete	complete	complete

10B REVIEW QUESTIONS

1. Why are sea anemones, earthworms, and squid all classified as invertebrates?
2. List the six major groups of invertebrates.
3. Which invertebrate group has no symmetry, no cephalization, and no gut?
4. Which two invertebrate animal groups have radial symmetry?
5. (True or False) Because most worms are round in cross-section, they are considered to have radial symmetry.
6. The spiny-skinned invertebrates are ___________.
   - **a.** mollusks
   - **b.** echinoderms
   - **c.** crustaceans
   - **d.** arthropods
7. What characteristic can be used to distinguish different groups of arthropods?
8. You find an animal with an exoskeleton and jointed legs. Which invertebrate group does it belong to?

10C

VERTEBRATES

What are vertebrates?

American kestrel

10.6 VERTEBRATE CHARACTERISTICS

Vertebrates are the most numerous group of chordate animals, making up about 97% of phylum Chordata. They're also the animals that you're probably most familiar with. Even though less than 5% of all animals are vertebrates, there is an amazing diversity among this group. Yet all vertebrates have some traits in common. Remember that a vertebra is one of the bones along the back. All the animals in the vertebrate group have vertebrae and skulls as adults. They also have one other main characteristic in common. What do you think this other characteristic is?

10C Questions

- What are vertebrates?
- What are the characteristics of the different kinds of vertebrate animals?

rainbowfish

red panda

elephant shrew

thorny devil

The other characteristic is that all vertebrates have internal skeletons, or **endoskeletons**, that are alive! Invertebrate skeletons, such as the exoskeletons of insects and the shells of clams, are nonliving secretions. Because the bone or cartilage of vertebrate endoskeletons are living tissues, they are made of cells, require energy, and are capable of growth. Let's look at some different groups within phylum Chordata.

garter snake

10.7 FISH

Fish are aquatic vertebrates, most of which are ectothermic. They have fins, gills, and streamlined bodies that are designed for living in the world's oceans, lakes, and rivers. Fish are the single largest group of vertebrate animals. In fact, there are as many species of fish as in all other vertebrate groups combined. Fish can be classified into one of three basic groups.

Jawless Fish

Jawless fish lack the jaws found in other vertebrates. Unlike most other fish, they also have no scales or paired fins. They are eel-like creatures with skeletons made of *cartilage*, a firm but flexible kind of tissue. Two examples of jawless fish are lampreys and hagfish.

Bony Fish

The *bony fish* group is by far the largest group of fish and thus the most numerous kind of vertebrate animal. This group includes the fish with which you are most familiar, such as catfish, bass, flounder, goldfish, guppies, and trout. These fish have an endoskeleton made of bone. They also have jaws and paired fins, and most have scales.

Cartilaginous Fish

Cartilaginous (KAR tuhl AJ i nus) *fish* have endoskeletons made of cartilage. They also have jaws, scales, and paired fins. Sharks, skates, and rays are members of this group.

10.8 AMPHIBIANS

Amphibians are ectothermic vertebrates that usually live in water when young but can live on land as adults. The name *amphibian* comes from Greek words that mean "double life." This name reflects that an amphibian has two habitats—water and land. Most amphibians go through metamorphosis from an aquatic juvenile form to an adult form that can live on land. They are the only vertebrates that go through this process. Even though adult amphibians can live on land, not all of them have lungs. Instead, most amphibians can exchange gases through their moist skin. Though some amphibians can lay their eggs on land, most return to water to breed and to lay eggs. The two most numerous groups of amphibians are frogs and toads in one group and salamanders in the other.

green and golden bell frog

Frogs and Toads

Frogs and toads are easily recognized by two distinct features: strong hind legs designed for jumping and the lack of a tail. Since 2012, frogs have also included the world's smallest known vertebrate. Found in the forests of New Guinea, adults of this tiny species (at right) are a mere 7 mm long—less than the width of a dime!

Salamanders

Salamanders, which include newts, are lizard-like amphibians with tails. Like frogs and toads, salamanders must keep their skin moist, so they are never far from wet or at least damp environments. Amphibians are sensitive to toxins in their surroundings, so they are sometimes looked at as *indicator species*. Indicator species can give scientists clues about the health of the animal's environment.

California slender salamander

10-2

The foggy coast of California is home to many slender salamanders.

10.9 REPTILES

Reptiles and amphibians are often lumped together in books. But though reptiles may share some of the secretive habits of amphibians, they are really quite different. Some groups of reptiles do move back and forth from land to the water, but all reptiles have lungs and breathe air. Reptiles also do not experience the life-changing metamorphosis of amphibians. Nor do reptiles have the smooth, moist skin found on most amphibians. Instead, most reptiles have dry, scaly skin that protects their bodies from drying out. Most reptiles also lay eggs that are covered with a leathery shell. This permits the eggs to be laid on land far from water without drying out.

Like fish and amphibians, reptiles are ectotherms. They are divided into four groups: snakes and lizards, crocodiles and alligators, turtles, and tuataras (TOO uh TAR uhz). The tuatara group consists of only two species found in New Zealand. Let's consider the first three groups in more detail.

Snakes and Lizards

California red-sided garter snake

Snakes and lizards together are known as *squamates*. After fish, squamates are the second-largest group of vertebrates. It might seem odd to group snakes and lizards together, but they do share several characteristics. Both are covered by hard *scales*. Reptile scales do not grow as the animal grows, so all squamates shed their skin now and then in a process called *molting*. Unlike in other vertebrates, the upper jaws of squamates are not fused to their skulls. A moveable upper jaw helps snakes in particular to swallow large prey. So what makes snakes and lizards different from each other? Snakes have no legs, no ear openings, and immovable eyelids. Most lizards have four legs, ear openings, and movable eyelids.

leopard gecko

Crocodiles and Alligators

Crocodiles and alligators, both part of a group known as *crocodilians*, can be found in and near the shallow waters of warm streams, rivers, swamps, and lakes. Crocodilians are covered by scale-like *scutes*. The outer layers of the scutes are periodically shed, but unlike squamates, crocodilians do not shed their skin all at one time. Crocodilians also have four-chambered hearts, whereas other reptiles have hearts with three chambers. Alligators and crocodiles can be distinguished by the shape of their snouts. Alligators have broad, rounded snouts, while crocodiles have narrow, pointed snouts. Also, in crocodiles some lower teeth stick outside the jaws when the mouth is closed.

crocodile

alligator

green sea turtle

Turtles

Various names have been given to different kinds of turtles. But whether they live in the ocean as sea turtles do, dwell entirely on land like tortoises, or live in or near fresh water like many terrapins, all belong to the same group. Turtles have two hard, bony shells that cover their bodies, one on the top and one on the bottom.

10.10 BIRDS

The last two groups of vertebrates that we'll look at are both endothermic. These warm-blooded animals can maintain a steady internal body temperature no matter the temperature of their environment.

If a living animal has feathers, it's a bird. Only birds have this kind of body covering. Some of a bird's feathers are specially modified for flight. Birds and bats (which are mammals; see below), are the only endotherms capable of true flight. Besides wings, other bird structures are also custom-built to enable flying. Birds also have distinctive *beaks* that lack teeth. There are about 10,000 species of birds.

10.11 MAMMALS

Mammals are vertebrates that have *hair* and produce *milk*. Milk is a liquid substance that mammal mothers feed their newborn offspring. It provides the fat, protein, carbohydrates, vitamins, minerals, and water that an infant mammal needs. Milk is produced in the mammary (MAM uh ree) glands of the mother mammal. Major differences in the way that mammals reproduce are used to classify mammals into three groups.

10-3
All mammals produce milk to nourish their young.

Placental Mammals

The mammals with which you are most familiar are the **placental mammals**. Most mammals, such as dogs, horses, deer, and whales, are placental mammals. While they are growing inside their mothers, mammal embryos receive oxygen and food and have wastes removed through a structure called the *placenta* (pluh SEN tuh). Developing mammals are connected to the placenta by an *umbilical cord*. Placental mammals are born alive. They do not hatch from eggs or develop in pouches.

Egg-Laying Mammals

Though most mammals give birth to live young, two, the duck-billed platypus and the echidna (ih KID nuh), lay soft-shell eggs. These two are considered mammals because their bodies are covered with fur and they have mammary glands that produce milk to nourish their young. These egg-laying mammals are called *monotremes*.

sugar glider

Marsupials

Marsupials (mar SOO pee ulz) are the third group of mammals. Marsupials do not develop true placentas like most mammals do. Instead, they have a pouch in which they raise their young. The marsupials that you are probably most familiar with are kangaroos, koalas, and opossums. Australia and its surrounding regions harbor a wide range of marsupials, including marsupial versions of mice, moles, flying squirrels, and more. This amazing variety of marsupials testifies of the richness of God's creation.

10C REVIEW QUESTIONS

1. What characteristics do all vertebrate animal groups share?
2. Which vertebrate animal groups are ectothermic?
3. What two characteristics are used to distinguish the three main fish groups?
4. The characteristic that is unique to birds is that they
   - a. can fly.
   - b. lay eggs.
   - c. are endothermic.
   - d. have feathers.
5. What are the two unique characteristics of mammals?
6. (True or False) Egg-laying mammals are called marsupials.

LAB EXERCISE

E **Equipment**

Animal Phylum Bingo cards and tokens

ANIMAL PHYLUM BINGO: THINK YOU KNOW YOUR ANIMAL CLASSIFICATION?

PROCEDURE

1. Your teacher will give you an Animal Phylum Bingo card and tokens.
2. Your teacher will call out the name of an animal phylum. Look through your tokens and find an animal that belongs in that phylum. If you get stuck, remember the characteristics that are unique to that phylum, then look for an animal with those characteristics.
3. Place your token on one of the squares labeled with the correct phylum.
4. Win the round by correctly matching five animals to their phyla on five squares in a row either horizontally, vertically, or diagonally.

CHAPTER SUMMARY

10A CHARACTERISTICS OF ANIMALS

- Animals are eukaryotic, multicellular consumers.
- Most animals can move at some point in their lives, eat food rather than absorb it, and reproduce sexually.
- Animals can be classified on the basis of shared characteristics.
- The modern classification system is a modified Linnaean system that attempts to determine evolutionary relationships between organisms.
- Baraminology is a scientific model that is based on biblical teaching and works to discover the boundaries between created kinds of organisms.

10A TERMS	
symmetry	194
notochord	196
vertebrate	196
invertebrate	196
endotherm	197
ectotherm	197
baraminology	199

10B INVERTEBRATES

- Invertebrates are ectothermic animals without backbones.
- Sponges are aquatic animals with no symmetry, cephalization, or gut.
- Cnidarians are aquatic, stinging organisms with radial symmetry and incomplete guts.
- Soft-bodied worms have bilateral symmetry and cephalization.
- Echinoderms are spiny-skinned animals with radial symmetry and complete guts.
- Mollusks are soft-bodied animals, some of which secrete hard outer coverings. Mollusks have bilateral symmetry, cephalization, and complete guts.
- Arthropods are identified by their jointed legs. They have bilateral symmetry, cephalization, and complete guts.

10B TERMS	
sponge	201
cnidarian	201
echinoderm	202
mollusk	202
arthropod	203
exoskeleton	203

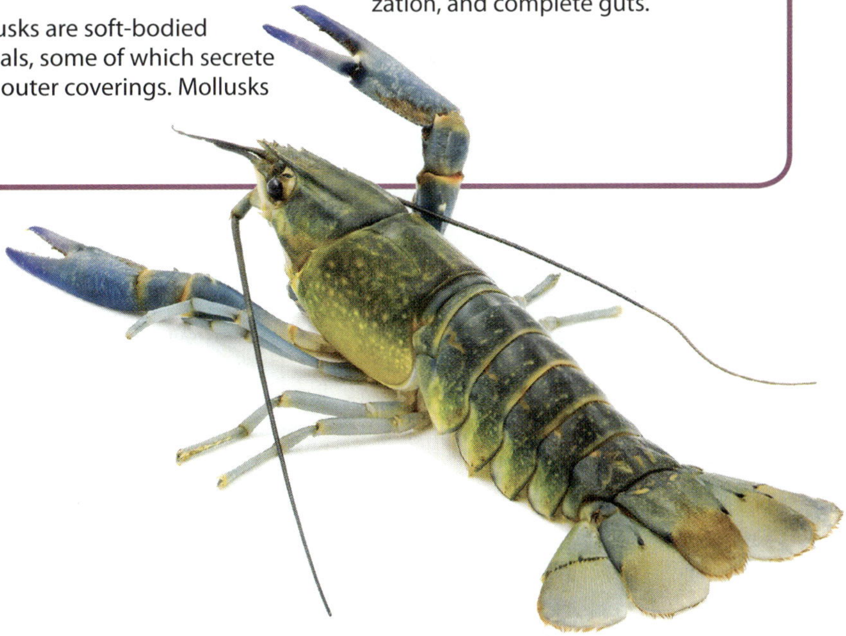

CHAPTER 10

CHAPTER SUMMARY

10C VERTEBRATES

- Vertebrate animals have skulls and backbones made of living bone or cartilage.
- Fish are aquatic, ectothermic vertebrates characterized by the presence of fins and streamlined bodies for swimming and gills for breathing.
- Frogs and toads are tailless amphibians with powerful hind legs for jumping. Salamanders are lizard-like amphibians with tails.
- Reptiles are ectotherms adapted to living and reproducing on land. They have lungs, breathe air, and lay eggs with tough shells.
- The major groups of reptiles are squamates, crocodilians, and turtles.
- Birds are endothermic vertebrates with feathers and beaks, most of whom can fly.
- Mammals are endothermic vertebrates that have hair and produce milk to nourish their young.
- Mammals are further classified by type of reproduction into placental mammals, egg-laying mammals, and marsupials.

10C TERMS

endoskeleton	205
amphibian	207
placental mammal	211
marsupial	212

CHAPTER REVIEW QUESTIONS

Recalling Facts

1. Which of the following is *not* a characteristic shared by all animals?
   - **a.** have specialized cells
   - **b.** can move as adults
   - **c.** are consumers
   - **d.** are eukaryotic
2. All of the following are used to classify major groups of animals except ______.
   - **a.** symmetry
   - **b.** cephalization
   - **c.** color
   - **d.** the presence of a notochord
3. (True or False) The modern classification system attempts to classify animals according to evolutionary relationships.
4. What does the word *invertebrate* literally mean?
5. What characteristic of sponges does the phylum name Porifera refer to?
6. What structures do jellyfish use for stinging and capturing food?
7. Jellyfish and coral belong to phylum ______.
8. A jellyfish has ______ symmetry.
   - **a.** bilateral
   - **b.** radial
   - **c.** asymmetrical
   - **d.** vascular
9. In what phylum do clams, snails, slugs, and octopuses belong?

10. What sea star characteristic does the term *echinoderm* describe?
11. What type of skeleton do insects have?
12. Which is the largest group of invertebrates?
    a. arthropods
    b. echinoderms
    c. mollusks
    d. cnidarians

13. What characteristic distinguishes vertebrates from other animals?
14. What is the largest group of vertebrate animals?
15. What characteristics are used to classify fish into three major groups?
16. Which group of vertebrates is characterized by undergoing metamorphosis from an aquatic to a land-dwelling form?
17. Name two kinds of amphibians.
18. What two groups make up the endothermic vertebrates?
19. Other than the fact that they are covered with feathers, name one characteristic that distinguishes birds from reptiles and one that distinguishes birds from mammals.
20. Which characteristic is unique to mammals?
    a. flight
    b. four legs
    c. lungs
    d. hair
21. What are monotremes? Give two examples.
22. (True or False) Most mammals are placental.

Understanding Concepts

23. What type of symmetry does the organism at right have?
    a. radial
    b. bilateral
    c. invertebral
    d. It has no symmetry.

24. Why is color not a good characteristic to consider when classifying major groups of animals?
25. Why is symmetry used (and not the presence or absence of eukaryotic cells) to classify different kinds of animals?
26. Construct a hierarchy chart of invertebrate classification. Include the following words: *invertebrates*, *cephalopods*, *Porifera*, *Cnidaria*, *gastropods*, *Arthropoda*, *crustaceans*, *Annelida*, *Mollusca*, *chelicerates*, *insects*, *worms*, *Nematoda*, *bivalves*, *Platyhelminthes*, and *Echinodermata*. Also include an example organism for each group in parentheses beside its name. See Appendix B for an example of a hierarchy chart.
27. The animal shown below is a ______.
    a. jawless fish
    b. cartilaginous fish
    c. reptile
    d. bony fish

28. The animal shown above lives on land but requires water in which to lay its eggs. Is it a lizard or a salamander? Defend your answer.

Critical Thinking

29. For a class project, Fred prepares a posterboard with the title "Animals." Fred presents pictures and information about fish, amphibians, reptiles, birds, and mammals. What's wrong with Fred's poster?

30. Create a table that compares the type of symmetry found in the various animal phyla.

31. Read Genesis 1:24–28. Evolutionary taxonomists classify man as part of the mammal group. Although man does share characteristics with the mammals, why is it not proper from a biblical point of view to classify man along with the animals?

ETHICS Managing Animals

INTRODUCTION

There are five species of salmon that return to spawn in streams of the Pacific Northwest. The annual homeward-bound journeys of Chinook, coho, pink, chum, and sockeye salmon are epic tales of survival. Some of these awe-inspiring fish are considered to be endangered. Fisheries management scientists are tasked with putting protections in place that will ensure the long-term survival of salmon. There are laws that govern how many fish can be caught each year. Other laws dictate what kinds of activities can be done in and around the streams that salmon need in order to successfully reproduce.

At first glance, protecting salmon does not seem like it should be a terribly difficult task. But then taxonomy comes into play. As scientists studied these fish, they realized that there were not just differences *between* species of salmon, but also *within* each species. Chinook that spawn in a stream in California, for example, may be quite different from Chinook that spawn far away in Alaska. They may look different, they may behave differently, and they may even have markedly different DNA. These different populations of fish are called *stocks*. Now, instead of managing just five species of salmon, scientists must consider how to protect *over two hundred* separate salmon stocks. Suddenly the task doesn't seem quite so simple!

TASK

You and your team are going to work through the process of analyzing how to ethically manage salmon. You've been doing this on your own up to this point, but now you will work together with a teammate to gather information and share your opinions. You'll still be working through the five-question strategy that you've been using. Starting in Chapter 12, you will get to work through this strategy on your own without the questions present to prompt you.

PROCEDURE

1. Do a keyword search using the words "manage salmon stock."
2. Look for people, industries, and governments that are influenced by salmon-stock management such as farmers, fisherman, etc. Since salmon need rivers and streams to spawn in, especially look for how water use may be affected.
3. Discuss with your team what the Bible says about managing animals like salmon. Consider both biblical principles and biblical outcomes.
4. Discuss with your team the acceptable and unacceptable options in managing salmon as well as the consequences of these options.
5. Each person on the team should plan and write an essay to answer the question, "How should we manage salmon populations?" Write a three-paragraph response on the issues related to managing Pacific salmon. Consider especially any problems that may result from treating a single salmon species as many individual stocks. Be sure to discuss principles from Scripture that may guide Christians in this area. Suggest possible avenues of response for the believer and the possible outcomes of those responses. Lastly, give your opinion on how a believer can make wise decisions about how to use the fisheries resources that God has given us. (*Hint*: Review Subsection 1.3 on pages 6–8 to refresh your memory about Christian ethics.)
6. Share and discuss your essay with other members on your team. Maybe you can persuade someone to agree with you!

CONCLUSION

Salmon are just one group of animals. There are lots of animals in the world that affect people in different ways. Now you have developed a skill to apply when thinking about the ethics of managing animals from the perspective of biblical principles and biblical outcomes.

11 ANIMAL STRUCTURE AND FUNCTION

NGATIDYE!

Hearing someone say "Let's eat!" in the Shona language of Zimbabwe might give an American reason to run the other way. You see, the Shona often make a meal of fried caterpillars, called *mopane worms* (shown at left), with sides of boiled greens and cooked cornmeal. Most Americans think eating bugs is gross! But actually, America is one of the few places in the world where people *don't* regularly eat insects. Around the planet, lots of people dine on ants, crickets, beetle grubs, and many other kinds of edible insects. But before anyone can bite a beetle or gobble a grasshopper, the insects must first grow to an edible size. To do so they must be able to support their bodies, move around, obtain energy, and eliminate wastes. In this chapter, we'll look at some of the ways that animals—including insects—accomplish those tasks.

11A

ENERGY

How do animals get and use energy?

11.1 DIGESTION

Like all living things, animals need energy to live, grow, and reproduce. Animals are consumers. They cannot make their own food as plants do, so they must get their energy from an outside source. They do this by eating other organisms. They have a **digestive system** that breaks down food and absorbs nutrients. Such systems often begin with a mouth for taking in food. An animal's mouth may have special structures, such as jaws and teeth, for crushing and grinding food into small particles. This is called **mechanical digestion**. Animals without teeth, such as earthworms and birds, may have a muscular organ called a *gizzard* that performs the grinding and crushing. During **chemical digestion**, the stomach and intestines use chemicals to break down food molecules into nutrients. The nutrients are then absorbed into the bloodstream by the *intestines*. The leftover matter that can't be digested is expelled through an *anus*. The complete path from mouth to anus is often called a *complete digestive system*. An *incomplete digestive system* has only a single opening for both taking in food and expelling wastes.

As you can see in the diagrams, not all animals have the same parts and pieces to their digestive system. The digestive tracts of vertebrates have additional organs whose secretions further aid digestion. A few animals even have no digestive organs at all. Tapeworms are a kind of parasite that live inside the digestive tracts of other animals. Since they are surrounded by digested food, they simply absorb nutrients through their skin.

11A Questions

- How do animals get energy from their food?
- What kinds of foods do animals eat?
- How do animals maintain homeostasis?
- How do animals eliminate wastes?

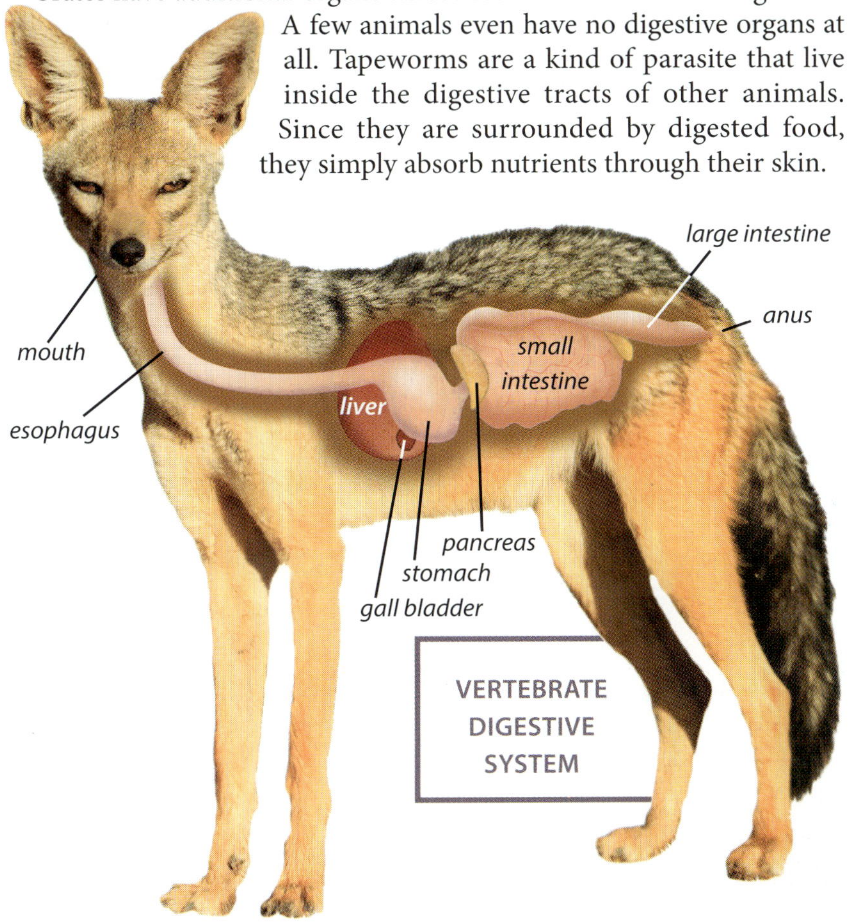

VERTEBRATE DIGESTIVE SYSTEM

INVERTEBRATE DIGESTIVE SYSTEM

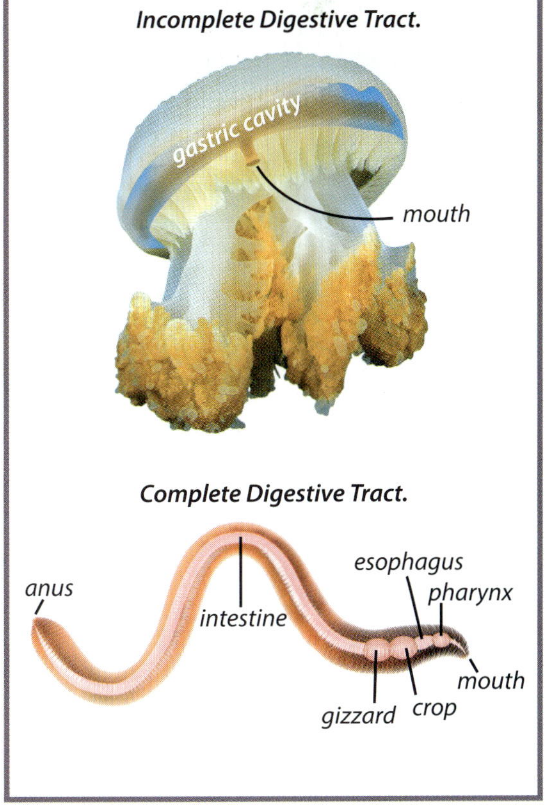

11.2 FEEDING STRATEGIES

As we just saw, God has customized the digestive systems of animals for eating certain kinds of food. Animals can be grouped together on the basis of how they get that food.

Herbivory

Animals that feed on plants are **herbivores**. Many animals feed on only one part of a plant, such as the leaves, fruit, seeds, or nectar. God has provided herbivores with a wide variety of means for obtaining energy from plant sources.

GETTING ENERGY FROM PLANTS

Eating leaves and stems poses special problems for animals. These plant parts are hard to digest and don't contain lots of energy. Many herbivores have mouthparts made for cutting, chewing, or scraping leaves and twigs. Slugs and snails have a file-like structure called a *radula*, kind of like a cheese grater, for rasping away at the surfaces of leaves. Insects use scissors-like *mandibles*. Leaf-eating mammals have blade-shaped *incisor* teeth for cutting and flattened teeth called *molars* that grind up tough plant tissues. Plant-eating animals usually have long digestive tracts and digest their food slowly. This helps them extract as much energy as possible from their leafy dinners.

fruit bat

Fruits have lots of energy-rich sugars in them, so many animals will eat them, even if fruit is not the main part of their diet. Fruit tends to be seasonal, so relatively few animals are adapted for living on fruit alone. Animals that rely on fruit for much of their energy needs include toucans, orangutans, and fruit bats.

sulphur-crested cockatoo

Seeds are another energy-rich, plant-based food source. But many seeds have hard protective shells. Seed-eaters have structures for cracking, gnawing, or boring into these shells to get at the softer seeds inside.

Some animals feed on the fluids of a plant, whether sap from the stems or nectar from the flowers. Aphid mouthparts are modified for piercing a plant's stem to suck out its energy-rich sap. Butterflies, moths, and hummingbirds have long mouthparts or beaks for reaching deep into flowers to get at the sugar-laden nectar inside. Some nectar drinkers, like bees, also feed on protein-rich pollen.

Some animals can even get their energy from wood! Most have bacteria or protists living in their guts that break down the wood, releasing nutrients that the wood-eater can absorb. Most wood-eating animals are insects, such as these termites.

termite mound

Many carnivores, like this jellyfish, use venom to subdue their prey. Venomous animals have different ways of delivering their poisons, such as stingers in wasps and the fangs of rattlesnakes.

Carnivory

Carnivores are animals that get most or all of their energy and nutrients from eating other animals. Most carnivores have specialized body parts and behaviors that enable them to catch their **prey**, the animals that they eat. Let's look at some examples of the tools and techniques that carnivores use to do the job.

Carnivores that prey on mollusks face a hard challenge—the shells of their prey. This hungry starfish used its many tube feet to latch onto a clam's shell and pry it open.

Look closely! Not every carnivore relies on chasing down prey and using sharp teeth for its next meal. Ambush predators often lie disguised, waiting for their food to come to them. This carefully hidden stonefish can suck in an unwary fish or shrimp in less than a tenth of a second!

Most predators have one or more highly developed senses for detecting prey. This mantis relies on keen eyesight. Its powerful forelimbs and rapid movement make it a danger to any passing insect—it can even catch small birds!

Spiders like this orb spider spin strong, sticky traps made of silk. Only the capture silk is sticky, allowing the spider to safely walk on the nonsticky parts of its web.

Big cats like this cheetah are often what we think of when we hear the word *carnivore*—a perfect combination of speed, strength, excellent vision, sharp teeth, and claws. Such tools allow them to tackle even large and potentially dangerous kinds of prey.

Other Methods

There are other ways that animals can get energy besides grazing on grass or ambushing an antelope. Let's take a quick look at some of them.

Filter Feeding

Filter feeding is a good way to get energy that often doesn't require much energy in return. **Filter feeders** have structures that are designed to strain tiny algae, protists, or other food items out of the water. Many filter feeders, such as sponges, stay in one place and let water currents bring their food to them. Other filter feeders, such as the world's largest fish, the whale shark, are more active in their pursuit of food.

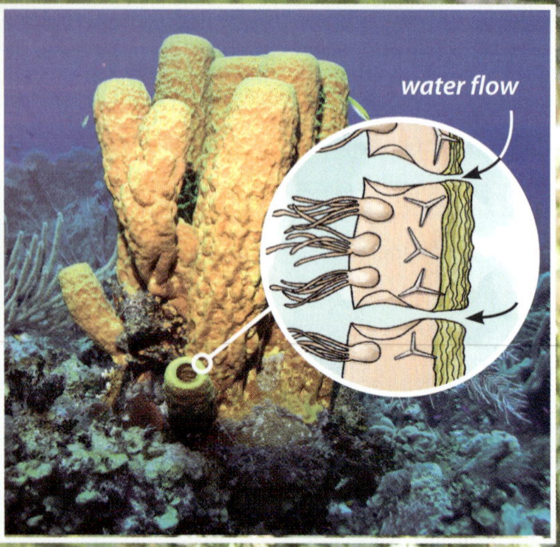

11-1

Collar cells absorb food particles from the water flowing through a sponge (left). Barnacles (right) sweep bits of food out of passing currents with feathery, modified legs.

Omnivory

An **omnivore** will eat both plant and animal matter, though it may have a strong preference for one or the other. American black bears are one example. About 85% of a black bear's diet is plant matter, including grasses, tree shoots, tree bark, fruit, nuts, and roots. But these bears will also eat insects, honey, fish, dead animals, and the young of deer, elk, and moose. In fact, it might be easier to list what they *won't* eat!

Scavenging

Scavengers eat dead animals. Most carnivores will help themselves to the energy available in such seemingly unpleasant meals, but few of them rely solely on them. Vultures, though, are scavenging specialists. Their broad wings are adapted to soaring for many hours while searching for food. Some, such as turkey vultures, have a very good sense of smell for detecting the odor of decaying meat at great distances. They can even safely digest contaminated food that might make other animals sick.

You shouldn't think of all the groups that you've just read about as absolute categories. There's a certain amount of crossover between them. Many herbivores will get some of the protein they need from animal sources. Hummingbirds, for instance, feed on small flying insects such as midges to supplement their normal diet of nectar. Likewise, many carnivores will not turn up their nose at an occasional helping of plant matter, such as grasses or fruit. Getting enough energy and nutrients to survive can be a tough job, and many animals will eat things that we don't usually think of as their normal food.

11-2
Vultures aren't strong enough to hunt and kill their own food and depend on other organisms to do the job for them.

11.3 MAINTAINING HOMEOSTASIS

As we learned in Chapter 1, organisms need to maintain homeostasis—keeping their inner processes stable. The word means "staying the same." Keeping this internal balance is vital because most animals have body systems and processes that work well only under certain ideal conditions.

An animal's external environment affects its internal environment. And not all animals can respond equally well to changes in their environment. Corals, for example, live in tropical seas where long-term conditions usually change very little. They respond poorly even to small changes in their environment. Water that is too warm or too cold may cause corals to expel the algae that normally live inside their bodies. This is called *coral bleaching*. Corals rely on their algae for most of their food needs, so bleaching may lead to death if conditions don't quickly return to normal. The common carp, on the other hand, can survive in water that's anywhere from just above freezing all the way up to 40 °C.

Though they differ in their ability to survive water temperature changes, corals and carp do have one thing in common—they're both ectotherms. Their body temperatures remain the same as that of their surroundings. Since adult corals can't move, they're stuck with whatever temperature the water happens to be. But carp and many other ectothermic animals can respond to a change in the temperature of their environment by moving. If their environment is too

cold or too hot, they can move to a place that is warmer or cooler. Reptiles use exactly this kind of behavior to control their internal temperature. Snakes and lizards can often be seen sunning themselves in early morning sunshine. Later, as the day gets hotter, they will seek out shade to cool off.

Endothermic vertebrates take this temperature control up a notch. Birds and mammals convert much of the stored energy in their food to heat energy, keeping their body systems at or near an ideal operating temperature. They also have ways to cool their bodies if they get too warm and warm them if they get too cool. After a long game of fetch, your dog will release excess heat energy by panting. If your dog's body is too cool after a bath, his muscles will shiver to produce more heat energy. Humans have mechanisms like these too. We also shiver when our bodies get too cool, and we sweat when our bodies get too warm. As the sweat evaporates, heat energy is removed.

Of course, an animal's internal conditions consist of more than just temperature. The cells of its body depend on having just the right amount of water, minerals, oxygen, and nutrients. They need to have wastes and carbon dioxide removed too. All animals have both internal means and external behaviors to make sure that all these things remain in proper balance. But these ways of maintaining homeostasis have limits. If an animal's internal environment gets too far out of balance, its behaviors and internal mechanisms may not be able to correct the problem. In such an instance, the animal will die.

11-3

Homeostasis in action: a lizard raises its body temperature by sunning. Jackals lower their body temperatures by panting.

11-4

Animal cells need water. Without water to maintain homeostasis, an animal will soon die.

FEEDING OUR ANIMAL FRIENDS: SERVING AS AN ANIMAL NUTRITIONIST

What do you like to eat? If you're a middle-school-aged human, it's a safe guess that you probably like pizza and ice cream! But what do meerkats like? Or gemsboks? Or linsangs? Humans keep thousands of kinds of animals as livestock, pets, or in zoos. And all of those different animals need different kinds of food. A person who specializes in understanding the food needs of animals is an *animal nutritionist*.

Animal nutritionists fill many different roles. Some of them study the best diets for livestock like dairy cattle. Cows need proper diets to stay healthy and produce milk. Other nutritionists try to find the best way to feed pets like your family dog or cat. Even though dogs and cats are both carnivores, their food needs are different.

Some nutritionists have to figure out what to feed the huge variety of animals kept in zoos and aquariums. Many of those animals are very fussy about what they will eat. Knowing exactly what to feed them is especially important for animals that are being kept as part of recovery programs for endangered wildlife.

Most animal nutritionists have at least a bachelor's degree with a major in an agricultural or biological science. Many have advanced degrees too. That's a lot of education! But it might be just the ticket for someone who likes working with animals. Captive or domesticated animals depend on their humans to meet their needs. As an animal nutritionist, you can serve God by helping to take proper care of the animal resources that He has given us.

11.4 EXCRETION

Cellular respiration and other cell actions produce wastes that animals must eliminate. These compounds are toxic and could prove fatal to an animal if they are not removed in a timely manner. Ridding the body of these cell waste products is called *excretion*. How an animal excretes wastes depends in large part on what kind of animal it is and whether it lives on land or in water. Some aquatic animals can simply diffuse wastes into the surrounding water. Most animals, though, have special organs that rid their bodies of cell wastes. These may work together as part of an **excretory system**.

Invertebrate excretory organs tend to be tube-like structures that collect cell wastes inside the animal's body and expel them. In insects, these structures empty wastes into the digestive tract where they are expelled along with digestive waste. The excretory systems of vertebrates are more complex. In vertebrates, blood travels through a **kidney** where wastes are filtered out. The filtered wastes are then expelled directly from the body in either a liquid or semisolid form depending on the kind of vertebrate. Other vertebrate organs can excrete some cell wastes as well, including lungs, gills, and even skin.

11A REVIEW QUESTIONS

1. Explain the differences between mechanical digestion and chemical digestion.
2. List the structures found in a typical animal digestive tract in the order that food passes through them.
3. What kind of digestive system does an animal have if it takes in food and expels undigested waste through the same opening?
4. (True or False) Some animals have no digestive system.
5. Identify the typical food source for filter feeders, herbivores, carnivores, omnivores, and scavengers.
6. Explain why the statement, "Herbivores eat only plants," is incorrect.
7. Give an example of a behavior that an animal uses to maintain its body temperature.
8. What is the primary function of the excretory system?
9. What organ filters wastes from blood in vertebrate animals?

11B Questions

- How do animals move substances to and from cells?
- How do animals exchange gases?

11-5
Like other cnidarians, siphonophores have no circulatory system.

11B

TRANSPORT

How do animals circulate nutrients and breathe?

11.5 CIRCULATION

Every cell in an animal's body needs certain things to survive, grow, and reproduce. Cells need nutrients and oxygen. They need to get rid of wastes and carbon dioxide. Animals need a way to transport these things throughout their bodies. This process is called *circulation*. If multiple organs work together to do this job, then the animal has a **circulatory system**. Some animals, such as sponges, cnidarians, and many worms, have no circulatory system. They can absorb nutrients and oxygen through their thin skins directly from their environments into individual cells.

A circulatory system needs at least two things: a fluid, such as blood, for transporting substances, and a way to circulate the fluid. Circulatory systems can be one of two kinds, either *open* or *closed*. Mollusks, echinoderms, arthropods, and other invertebrates have open circulatory systems. The insides of their bodies are bathed in a blood-like fluid called *hemolymph*. The hemolymph is usually circulated by the animal's body movements.

11-6

Some of a glass frog's closed blood vessels are easily seen through its translucent skin.

Closed circulatory systems are more complex than open systems. They include a muscular **heart** for pumping blood through the system. A heart has one or two *atria* (s. atrium) for collecting blood entering the heart. Blood is pumped out of the heart by one or two powerful *ventricles*. The blood circulates through the animal's body inside closed **vessels**. *Arteries* carry blood away from the heart, while *veins* carry blood toward the heart. Tiny *capillaries* connect arteries and veins. Oxygen, nutrients, and other substances diffuse to and from cells through the walls of the capillaries. All vertebrates, some worms (such as earthworms), and cephalopods (squids and octopuses) have closed circulatory systems.

There are differences in the circulatory systems of the many vertebrate groups. Fish have two-chambered hearts and a *single-cycle* pathway for blood. Other vertebrates have *double circulatory systems* with three- or four-chambered hearts and separate pathways for blood traveling to the lungs and blood heading to the rest of the body. Amphibians and most reptiles have three-chambered hearts. Alligators, crocodiles, birds, and mammals have four-chambered hearts. These differences are shown on the right.

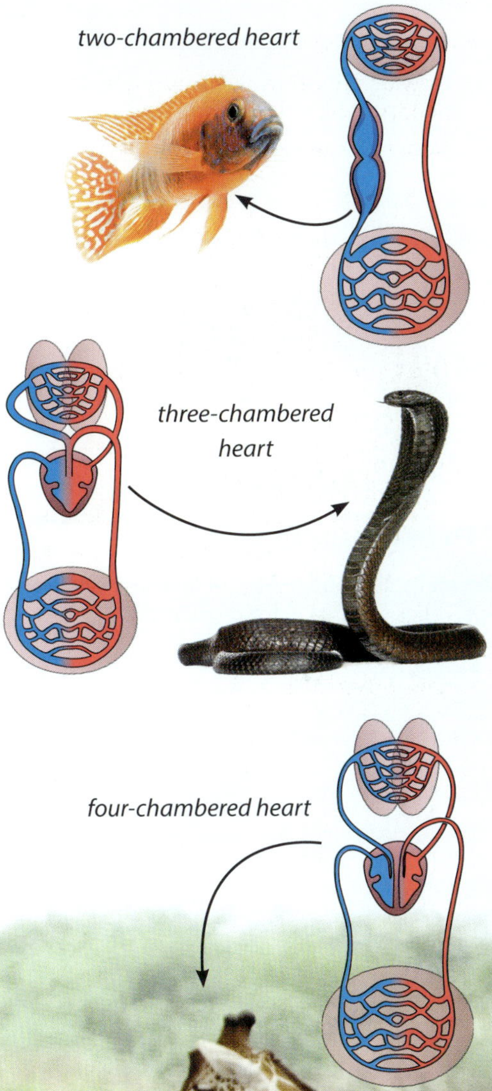

11.6 RESPIRATION

Before an animal's circulatory system can move oxygen and carbon dioxide to and from cells, those gases have to first enter the animal's bloodstream. Gas exchange, the taking in of oxygen and removal of carbon dioxide, is called *respiration*. Respiration is the job of an animal's **respiratory system**. Let's take a look at how different kinds of animals respire.

11-7
The seal salamander, like all lungless salamanders, relies on its skin for respiration.

Skin

Many animals, even some large vertebrates, do at least some of their respiration through their skin. Skin respiration is most efficient in moist environments. That explains why amphibians are the only vertebrate group that does most of their gas exchange through their skin. Compared with other vertebrates, amphibian lungs are weak and inefficient. Some salamanders even have no lungs at all and respire entirely through their skin.

Spiracles

Most insects have tiny openings along the sides of their bodies called **spiracles**. These lead into a branching network of tubes called *tracheae* (s. trachea) that allow air to reach all of the insect's tissues. Insects can control the openings of their spiracles to prevent water loss.

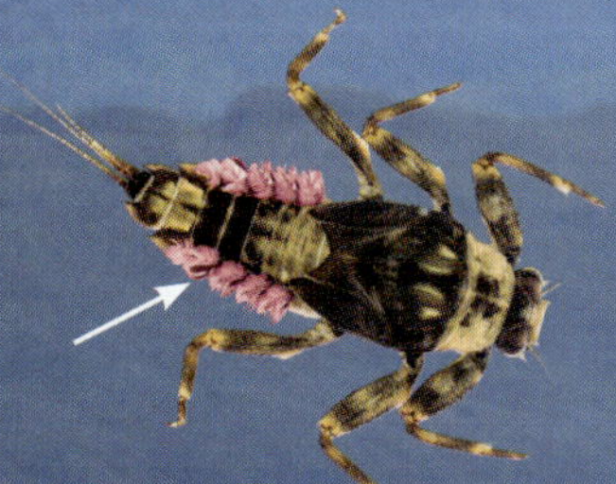

11-8
The feather-like structures along the sides of this mayfly larva are its gills.

Gills

Many kinds of animals spend some or all of their lives in water and use **gills** to exchange gases with their surroundings. Fish, crustaceans, mollusks, and larval amphibians and insects all use gills for respiration. Every kind of gill does two things. First, it provides a large surface area for gas exchange. Second, its outer layer of cells is thin enough to permit gases to diffuse in and out of the gill structure. Gills usually consist of many threadlike *filaments* that increase the area of the gills. As water flows over the gills, gases are exchanged between the passing water and tracheae (in insects) or blood inside the filaments. The blood inside fish gills gives them a red color.

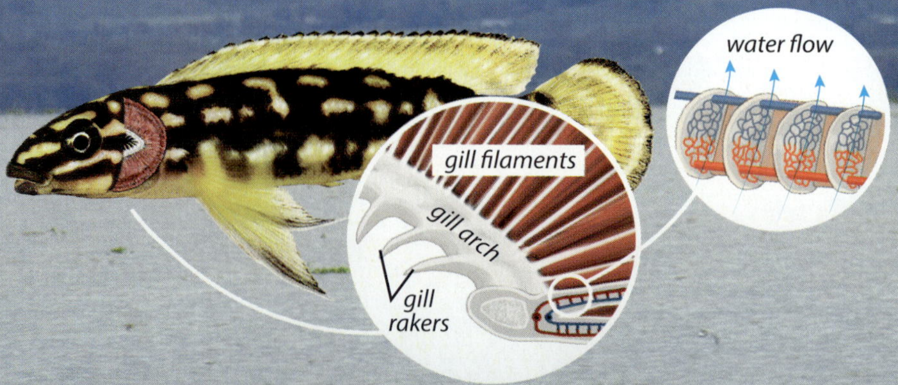

Lungs

Apart from fish, most vertebrate animals respire using **lungs**. Lungs contain spongy tissue that increases the surface area for gas exchange. Air enters the nose or mouth of an animal and passes to the lungs through the *trachea*. Air then enters all parts of a lung through many small passageways in the lung tissue. These passageways are surrounded by capillaries where gas exchange takes place. The respiratory system of each vertebrate group differs slightly in structure. The groups also differ in how they get air into and out of the lungs. You can see some of the differences between birds and mammals in the images on this page.

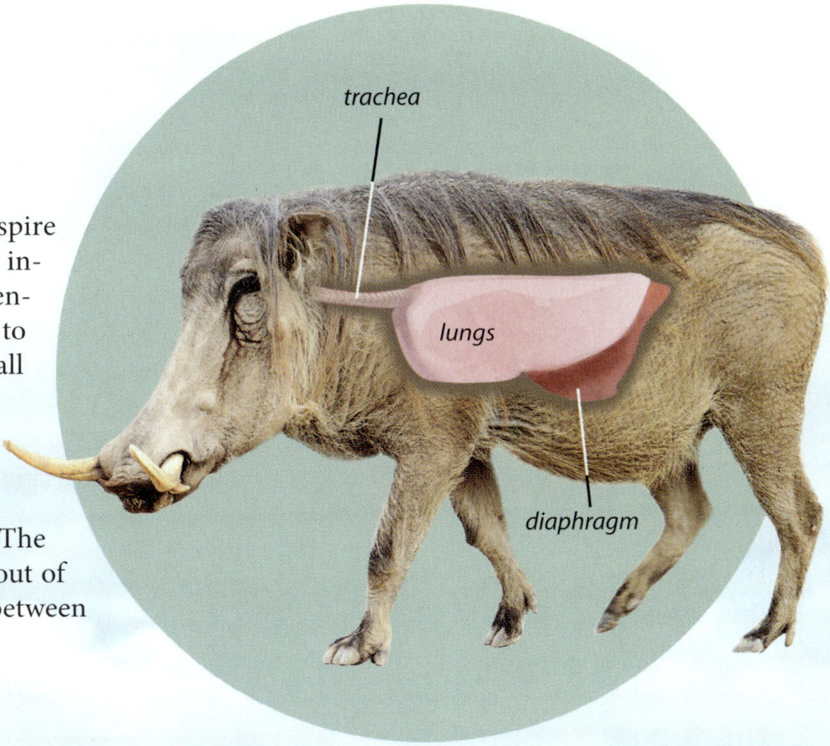

11-9
A bird's respiratory system includes passageways and air sacs that allow fresh air to enter its lungs when it breathes in and when it breathes out. Mammals have a muscle, the *diaphragm*, whose motions draw air into the lungs and force it back out again.

11B REVIEW QUESTIONS

1. What is the main difference between an open circulatory system and a closed one?
2. Which vertebrate group has a circulatory system consisting of only a single pathway for blood?
3. In a double circulatory system, where do the two pathways from the heart lead to?
4. Which vertebrate group is the only one that respires mainly through the skin?
5. What two characteristics of gills help them increase the amount of gas that they can exchange?
6. Compare oxygen transport in insects and mammals.

11C SUPPORT, MOVEMENT, AND CONTROL

How do animals support, move, and control their bodies?

11C Questions
- How do animals support their bodies?
- How do animals move?
- How do animals control their bodies?

11.7 SUPPORT

Have you ever tried to make a tower out of soup? It's a crazy question, right? But grocers make towers out of soup all the time. How do they do it? Their secret is that their towers are made out of soup *in cans*. The rigid metal cans provide support for the soup inside. Animals need a means of support for their bodies as well. Without support, they'd look pretty odd. You would too, for that matter!

For many animals, a support system takes the form of a **skeleton**. Let's take a look at some different kinds of skeletons—not all are made of bones!

Hydrostatic Skeletons

The prefix *hydro-* means "water." *Hydrostatic skeletons* rely on fluid pressure to support an animal. The fluid fills a central body cavity and is surrounded by bands of muscle. The force of the fluid pushing against the bands of muscle gives the animal its shape. Many soft-bodied invertebrates, including worms and cnidarians, have hydrostatic skeletons.

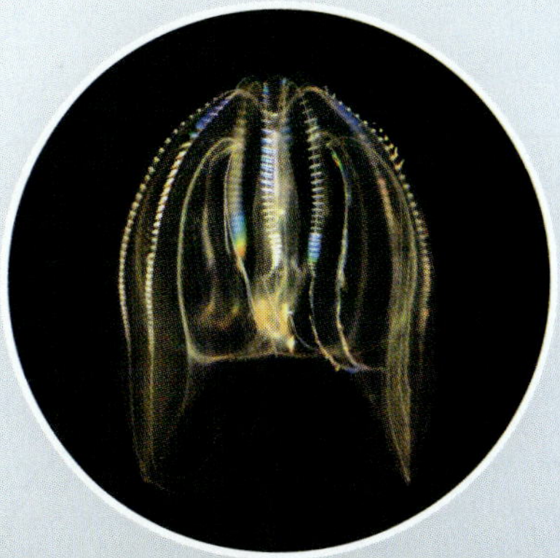

11-10
Otherworldly comb jellies are supported by a hydrostatic skeleton.

Exoskeletons

An *exoskeleton* (*exo* means "outside") is a rigid covering that supports an animal from outside its body. Insects, spiders, crustaceans, and many mollusks have this kind of skeleton. Since the exoskeleton is rigid, it needs places where separate plates meet and allow movement, such as in the legs. These are called *joints*.

Endoskeletons

In contrast to an exoskeleton, an *endoskeleton* is located on the inside of an animal's body. All vertebrates have endoskeletons that are made of either cartilage alone or a combination of bone and cartilage. Endoskeletons made of bone are hardened by the presence of minerals, mainly calcium. Like exoskeletons, endoskeletons also have joints. The bones are connected together by tough bands of tissue called *ligaments*. In addition to supporting an animal's body, endoskeletons act as a reserve of important minerals like calcium. Animals can resorb ("to absorb something again") minerals from their bones if levels in their bodies fall too low. That's part of maintaining homeostasis.

11-11
Can you identify this animal from its endoskeleton alone?

Endoskeletons can support more weight than exoskeletons can. That's why insects and crustaceans are mostly small animals. The largest land animals that ever lived, the sauropod dinosaurs, were vertebrates with endoskeletons.

11-12
Arthropods like these painted ghost crabs are supported by exoskeletons.

11.8 MOVEMENT

Running, crawling, swimming, hopping, gliding, flying—there are almost as many ways for an animal to move as there are animals! When we speak of animal movement, we usually mean *locomotion*—how an animal moves from one place to another. Some animals, such as sponges, are *sessile*, meaning that they remain in one place throughout most of their lives. But most animals need to move in order to obtain food, find shelter, seek mates, or escape predators.

Some animals, like the Portuguese man o' war, get around by *passive locomotion* (see Figure 11-13). They allow their environment to move them. Most animals, though, actively work to get around, and this work is the job of the animal's **muscular system**. In order to locomote, an animal needs to push or pull against something in its environment. An earthworm, for example, has tiny bristles along the sides of its body (left). Using its muscles, it pushes these bristles against the sides of its burrow as it alternates between stretching and pulling its body forward. Octopuses and squids (left) use jet propulsion to get around. Their muscular siphons squeeze out water with great force. As the water squirts in one direction, the cephalopod moves in the opposite direction.

11-13
The float on this Portuguese man o' war acts as a sail, allowing the wind to blow the animal to new places.

Can you guess how animals with exoskeletons and endoskeletons move? Their muscles work together with their skeletons to accomplish movement. Each *skeletal muscle* is attached to a different part of the skeleton. By pulling on that part of the skeleton, the animal's muscles move that part of its body. Skeletal muscles can only *pull* on part of the skeleton, never push. That means that each part of a skeleton that moves needs two muscles. One muscle moves the bone or plate in one direction; its partner moves it in the opposite direction.

Flight is a means of movement that has intrigued humans for centuries. The wings of insects, birds, and other animals that can fly are designed to produce *lift* as they move. Lift is a kind of force that pushes the animal upward. The fins of some sharks produce the same effect in water as a shark swims forward.

Wherever we look in the animal world, we see marvels of God's design. Each animal is equipped with precisely the structures it needs to move, respire, get energy, remove wastes, and do all the other things an animal must do to survive. The endless variety of ways that animals possess to do these things is a vibrant testimony of the infinite creative abilities of Earth's loving Designer.

11-14

A pair of muscles moves the lower leg (tibia) of a grasshopper where it connects to the upper leg (femur). The *flexor* bends the tibia toward the grasshopper's body. The *extensor* bends the tibia away from the grasshopper's body. Which muscle do you think powers the grasshopper's powerful jumps?

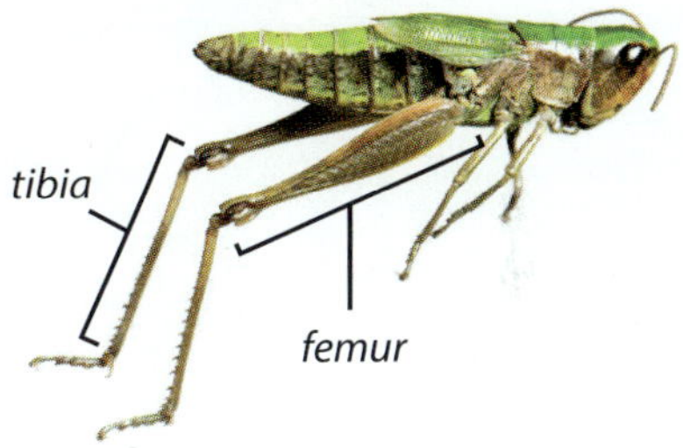

11-15

Head, shoulders, knees, and toes—a vertebrate animal has many individual muscles because it has many body parts that it needs to move.

11.9 CONTROL

Whether it's finding food, running from a predator, or migrating for thousands of miles, animals can do many amazing things! To do them, all of an animal's many systems must work together as one. Ensuring such harmony is the job of the animal's **nervous system**. The nervous system gathers information about the animal's environment and monitors all of the animal's other body systems. This information is sent to other parts of the nervous system in the form of electric signals called *nerve impulses*. When a response is required, the nervous system sends out impulses that control the animal's actions.

Sponges have no nervous system, and cnidarians and most echinoderms have simple *nerve nets* with no brains. Nerve nets allow animals to respond in limited ways to their environment. Animals with bilateral symmetry, though, have more complex systems that allow for a greater range of sensing and control.

CONTROL AND RESPONSE

1 *Getting Information.* As a big cat stalks its prey, special cells called **neurons** collect information. Neurons in the big cat's eyes detect the type and size of the prey. Neurons in its nose detect the prey's smell. The neurons of most animals are bundled together in larger structures called **nerves**.

2 *Processing Information.* Nerves send impulses to and from the animal's brain by way of a *nerve cord*. In vertebrates, the nerve cord is known as the *spinal cord*. The brain, itself a collection of neurons, processes the received impulses and decides how to respond.

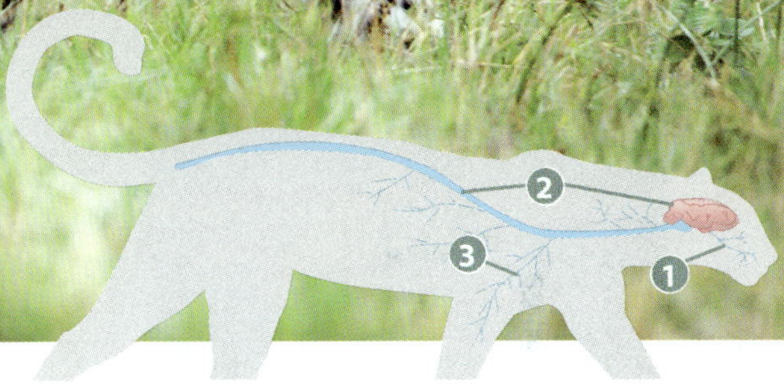

3 *Response.* Once the decision for action is made, the brain sends its own impulses via neurons to the muscles. These signals control which muscles contract and with how much force. At the same time, additional impulses can trigger changes in other systems, such as increasing the animal's heart and breathing rates.

11C REVIEW QUESTIONS

1. Identify the type of skeleton possessed by each animal below.

   a.  b.  c. 

2. (True or False) All vertebrates have endoskeletons made of bone.
3. What additional function does an endoskeleton serve other than providing support?
4. (True or False) Animals need something to push or pull against in order to move.
5. Why do skeletal muscles always work in pairs?
6. What kind of animal movement requires no muscle?
7. Animals such as jellyfish and octopuses shoot water to propel themselves. What do their muscles act on?
8. For animals such as birds and mammals that jump and fly, what do their muscles act on?
9. Which type of nervous system allows an animal a greater range of responses to change in its environment—a nerve net or a system with a brain and nerve cord?
10. What type of cell is found in both an animal's nerves and in its brain?

BIOMIMICRY: IMITATING NATURE

Equipment
art or craft supplies

Animal kingdom solutions for engineering problems

What do kingfishers and bullet trains have in common? Japanese bullet trains can travel at speeds up to 320 kph. Their high speed created a problem. As the trains left tunnels, their sonic boom-like noise woke local residents and disturbed wildlife. The problem was the sudden change in pressure as trains left the dense air inside the tunnels and entered the less-dense air outside.

Engineers turned to nature for possible solutions. They saw that kingfishers faced the same issue when diving from air into water. As kingfishers enter water, their long, tapered beaks neatly push the water aside. Their dives create very little splash. The design of the kingfisher's beak was the answer that the engineers needed. They designed a train whose front mimicked the beak's shape. The trains could then quietly transition from the tunnels to the outside air.

Now it's your turn to think like an engineer and develop an application for a design from the animal kingdom.

PROCEDURE

1. Think back over the many animal structures discussed in this chapter, such as a mammal's grinding molars, a spider's strong silk, or a bird's super-efficient respiratory system. Think of ways that these structures and systems could be applied to solving an everyday problem that humans face.
2. Narrow down your ideas to one problem and one animal-inspired solution. Develop this idea further.
3. Now model your idea! Your model could be a diagram, a drawing, or even an actual 3D model.
4. Present your idea to your class or family. Be sure to describe the problem, the inspiring animal structure or system, and how your solution will fix the problem.

CONCLUSION

Ever hear the old saying, "Don't reinvent the wheel"? Engineers realize that animals and other organisms face many of the same difficulties in life that humans do. By copying nature's structures, engineers help improve the lives of others. As believers, we can give thanks to God for His many clever designs and for our ability to recognize and use them for our own well-being.

GOING FURTHER

If this activity has sparked your interest, do an Internet search using the keyword "biomimicry." You might be amazed by how engineers are putting God's designs to work.

CHAPTER SUMMARY

11A ENERGY

- Animals are consumers and get their energy through digestion.
- Digestion consists of both mechanical and chemical methods.
- A digestive tract usually consists of a mouth, stomach or intestine (or both), and anus, and may include other organs that aid digestion.
- Animals employ a variety of feeding strategies, including filter feeding, eating plants or parts of plants, eating other animals, eating both plant and animal matter, or scavenging.
- Animals have both behaviors and internal mechanisms to help maintain homeostasis.
- If an animal is unable to maintain its internal environment, it will die.

11A TERMS

digestive system	219
mechanical digestion	219
chemical digestion	219
herbivore	220
carnivore	222
prey	222
filter feeder	224
omnivore	224
scavenger	225
excretory system	228
kidney	229

11B TRANSPORT

- An animal's circulatory system moves oxygen and nutrients to its cells and carries carbon dioxide and wastes away from them.
- A circulatory system can be either open or closed, the difference being that a closed system transports blood inside vessels.
- A closed circulatory system consists of a heart for pumping blood and a series of vessels.
- An animal's respiratory system is responsible for exchanging oxygen and carbon dioxide with its environment.
- Animals breathe in a variety of ways, including through their skin or by using spiracles, gills, or lungs.

11B TERMS

circulatory system	230
heart	231
vessel	231
respiratory system	232
spiracle	232
gill	232
lung	233

CHAPTER SUMMARY

11C SUPPORT, MOVEMENT, AND CONTROL

- Most animals rely on a form of skeleton for support, either a hydrostatic skeleton, an exo-skeleton, or an endoskeleton.
- Animals' muscles work in pairs in ways that allow them to move from place to place through their environment.
- An animal's nervous system gathers information about its environment and controls the responses of the animal's body.

11C TERMS	
skeleton	235
muscular system	236
nervous system	238
neuron	239
nerve	239

CHAPTER REVIEW QUESTIONS

Recalling Facts

1. What is the main function of the digestive system?
2. What do the teeth of a cow and the gizzard of a pigeon have in common?
3. Where does most chemical digestion take place?
4. (True or False) All animals can respond equally well to changes in their environments.
5. List three things other than body temperature that an animal needs to regulate to maintain homeostasis.
6. Vertebrate animals filter wastes from their blood through a(n) _____.
7. Excluding blood, what are the two main structures in a vertebrate circulatory system? What are their functions?
8. What fluid is circulated by an open circulatory system? by a closed system?
9. List four different structures that can be used for gas exchange. Give an example of an animal that respires using each structure.
10. What kind of animals typically rely on support from a hydrostatic skeleton?
11. What structures within a skeleton allow different body parts to be moved?
12. What term describes an animal that does not move from place to place?
13. What part(s) of its body do the muscles of a vertebrate animal pull on?

Understanding Concepts

14. Place the following structures in the correct order through which digestion progresses: anus, intestine, mouth, stomach.
15. Explain the difference between complete and incomplete digestive tracts.
16. Give an example of an animal that survives without a digestive tract and explain how it is able to do so.
17. Identify which type of animal (herbivore, carnivore, or scavenger) is most likely to use each food item shown on the left as a main food source.
18. Identify one herbivorous animal and one carnivorous animal and describe at least one body structure that it has for using that feeding strategy.
19. Explain why many animals are neither strictly herbivores nor carnivores.
20. How do ectothermic animals regulate their body temperatures?

a.

b.

c.

21. Explain the differences in excretion between an insect and a mammal.
22. Identify whether each of the animals on the right has an open or closed circulatory system.
23. What are two differences between a fish's circulatory system and that of any other vertebrate?
24. Compare how the surface area for gas exchange is increased in gills and lungs.
25. Explain why a bird's respiratory system can function more efficiently than that of other vertebrates.
26. The hinged shell of a clam and the many armor-like plates of a crab are structurally very different. Why are they both considered a form of exoskeleton?
27. How do the wings of flying animals make flight possible?
28. Why are animals with nerve nets not capable of making complex responses to changes in their environment?
29. Place the following structures in the correct order that a nerve impulse would pass through as a leopard flexes its paw: brain, muscle, nerve, spinal cord.

Critical Thinking

30. The rakers on the gill arches of a fish act as a strainer to prevent objects from passing through and damaging the gills. Which gill arch in the image on the right do you think belongs to a fish that filter feeds? Explain.
31. Create a hierarchy chart that relates the following terms: open, closed, circulatory system, single-loop circulatory system, double-loop circulatory system, heart with three chambers, heart with four chambers (see Appendix B).
32. If a mammal is unable to get enough calcium in its diet, it may need to resorb calcium from its bones. What do you think may be an unwanted side effect of this process?
33. A horse's muscles don't push or pull directly on the ground, so how does a horse move?

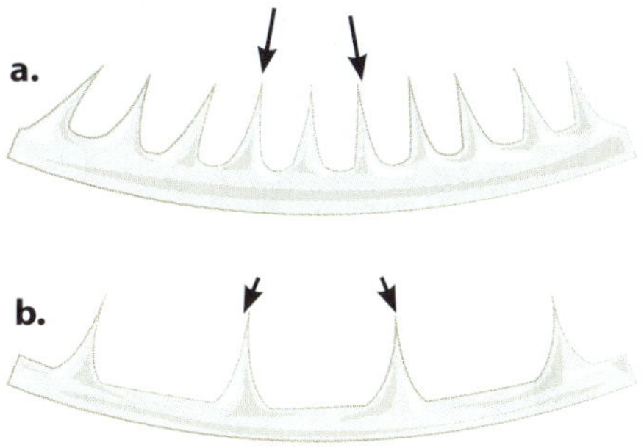

ETHICS — Eating Insects

We saw in the Chapter opener that many people around the world eat insects. Despite this, most Europeans and Americans still cringe at the thought of munching on bugs. But does eating insects pose an ethical problem for Bible-believing Christians? Does the Bible give any guidelines on the eating of insects? Earth's insects are a huge potential source of protein for a growing population, so how can Christians wisely use them?

Use the Ethics box above to answer Question 34.

34. Write a three-paragraph response on the issues related to using insects as food for humans. Be sure to discuss principles from Scripture that may guide Christians in this area. Suggest possible avenues of response for the believer and the possible outcomes of those responses. Lastly, give your opinion on how a believer can make wise decisions about how to use the insect resources that God has given us. (*Hint*: Review Subsection 1.3 on pages 6–8 to refresh your memory about Christian ethics.)

12 ANIMAL REPRODUCTION AND BEHAVIOR

HOG WILD

A herd of pigs tears through a field next to a fence. This fence wasn't built to keep the pigs in—it's to keep them out. These pigs are feral, or wild. They are descended from pigs that long ago escaped from farms and ranches. Since they reproduce very rapidly, they have spread throughout the United States, and they now roam at least thirty-nine states. They destroy crops and root up valuable pasture land. They also eat food that both domesticated and wild animals need. Today they are one of the most destructive invasive large-animal species in the country. How do invasive species become such a problem? How can we control them?

12A

ANIMAL REPRODUCTION

How do animals reproduce after their own kind?

12A Questions

- What organs and structures do animals use to reproduce?
- How are animal eggs fertilized?
- How do animal embryos develop?

12.1 FERTILIZATION

One of the reasons that invasive species such as feral hogs are such a nuisance is that they reproduce very quickly. Reproduction adds new individuals to a population. That's usually a good thing, since God created animals to reproduce after their kind (Gen. 1:24–26). If the number of births in a population is greater than the number of deaths, the population will grow. But if deaths outnumber births, the population will decrease and may even die out.

Most animals reproduce by sexual reproduction. In this type of reproduction, a gamete called a *sperm* from the male's **testes** (s. testis) unites with a gamete called an *egg* from the female's **ovaries** in the process of *fertilization*. Since both the sperm and egg are haploid, they form a diploid zygote.

TYPES OF FERTILIZATION

External Fertilization

Many animals, such as these horseshoe crabs, reproduce using **external fertilization** in which the sperm and egg unite outside the parents' bodies. Typically the male fertilizes the eggs as the female lays them. Because the sperm must be able to swim to the eggs, fertilization occurs in the water.

The sperm must be able to enter the egg, so the eggs do not have shells. Instead, they are usually coated in a jellylike substance like what you see on the fish egg at the top of the next page. Most amphibians, fish, echinoderms, and some mollusks use external fertilization.

Internal Fertilization

Many other animals, such as these cuttlefish, reproduce using **internal fertilization**. The male's sperm and the female's egg unite inside the female's body. Reptiles, birds, mammals, insects, spiders, some fish, and many other animals practice internal fertilization. Because sperm and egg unite inside the female's body, fertilization can take place in water or on land.

12.2 DEVELOPMENT

After sperm and egg unite, the resulting new zygote has a lot of development to do before it is ready to enter the world. Different species develop in different ways, but they can generally be divided into those that develop outside the mother's body and those that develop inside.

External Development

Many animals that fertilize their eggs externally, such as salmon, often leave them without any parental protection. As a result, many of the eggs and offspring are eaten by predators. But species that fertilize their eggs externally often produce hundreds or even thousands of eggs. This makes it likely that some will survive.

Some animals that fertilize their eggs internally eventually release the developing offspring as eggs. Examples include invertebrates, birds, and some fish and reptiles. These eggs typically have a shell that protects them and keeps them from drying out. The offspring are nourished by the nutrient-rich yolk inside the egg.

Internal Development

Other animals reproduce using internal fertilization but do not lay eggs. Instead, the offspring develop inside the mother's body. This process is called *internal development*. In most animals using internal development, including the Pacific rattlesnake, the mothers hold the eggs inside their bodies until they hatch. The young are then born.

However, mammals, such as cows, hold their embryos in a structure called the **uterus**, or womb. The baby is connected to the mother by the *placenta,* which brings it nourishment.

Metamorphosis

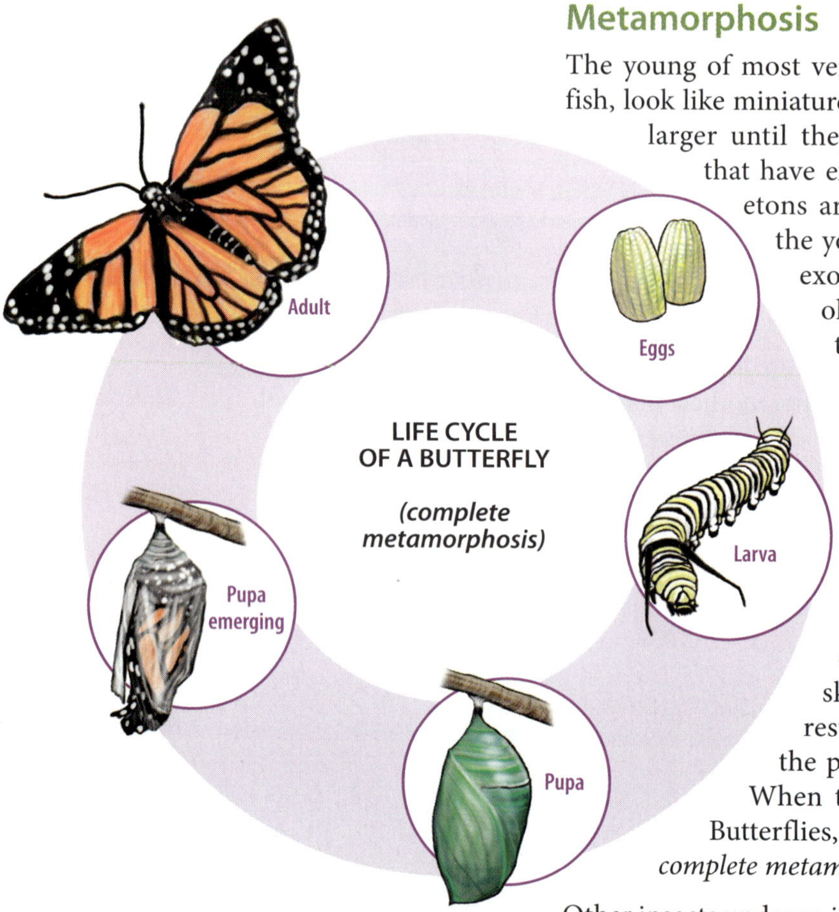

The young of most vertebrates, other than amphibians and some fish, look like miniature versions of their parents. They slowly grow larger until they reach their adult size. But invertebrates that have exoskeletons have a problem. Their exoskeletons are nonliving and cannot grow larger. So as the young invertebrate grows, it forms a new soft exoskeleton and then splits open and exits the old one. The new exoskeleton then hardens to protect the invertebrate. Each time it does this, it grows a little larger until it reaches its adult size. This type of growth that involves a change to the body is called **metamorphosis**. Most insects and some other invertebrates grow this way. Some insects undergo an extreme version of metamorphosis. The offspring hatch from eggs as *larvae* that look very little like their parents. The larvae grow and shed their exoskeleton several times. Then they enter a resting period called the *pupa* stage. During the pupa stage, their bodies radically transform. When they leave the pupa stage, they are adults. Butterflies, moths, and beetles all undergo this type of *complete metamorphosis.*

Other insects undergo *incomplete metamorphosis*, which does not include a pupa stage. Instead, the offspring, called *nymphs*, often look like smaller versions of their parents. Each time they grow a new exoskeleton, they become larger until they eventually reach adult size.

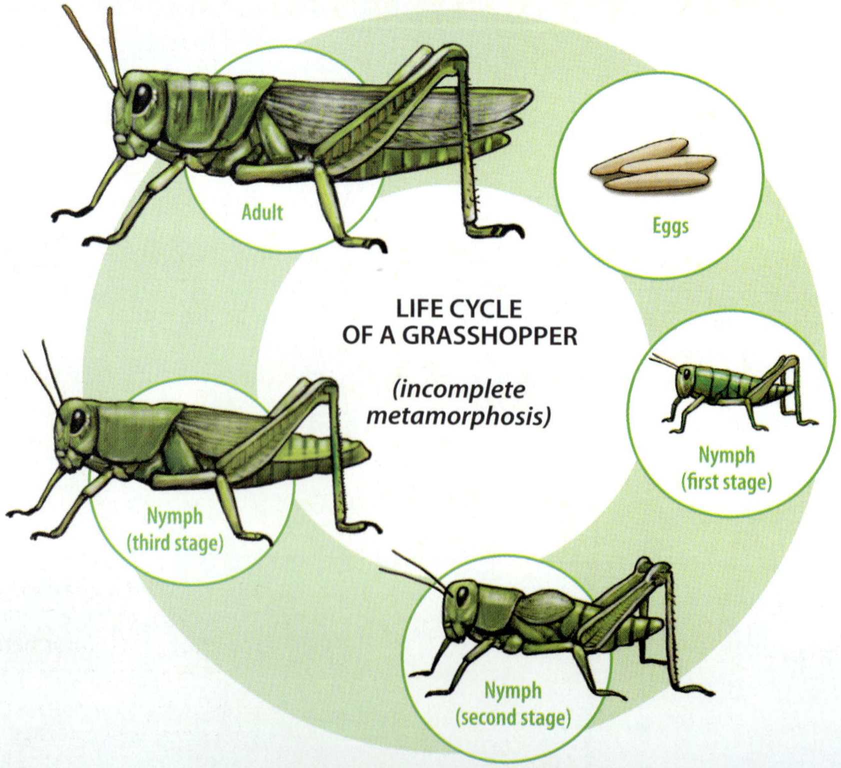

Amphibians also undergo metamorphosis.

TAKING CARE OF THE KIDS

Animals have many different ways of caring for their offspring. Birds, such as this hoopoe, guard and **incubate** their offspring and then feed them until they reach maturity. Mammals produce milk to feed their young.

Some species care for their young in unusual ways. The male yellow-head jawfish, for instance, holds the eggs in his mouth until the young hatch.

Some animals provide no care for their young. The young must be able to care for themselves immediately after they are born or hatch from eggs. This sea turtle is returning to the ocean after laying her eggs in the sand. The few that survive after hatching will grow up without ever seeing their parents.

WORLDVIEW SLEUTHING: DEALING WITH OWLS

Conservation is about managing a habitat to benefit the plants and animals that are supposed to be there while eliminating those that aren't. One of the species that needs preserving is the northern spotted owl. These owls prefer forests with old trees in the northwestern United States. But the same forests are also valued by timber companies for lumber. In the mid-1990s, the federal government tried to discourage timbering in areas where the spotted owls live. It was hoped that the number of owls would increase. But in recent years, the owls have faced a new threat: barred owls. The barred owl is a species that historically has not inhabited this area.

Barred owls are from the eastern United States. But they have recently expanded their range to include the spotted owl's territory. And since barred owls are larger and more aggressive, they often drive spotted owls away from nesting and hunting sites. Federal agencies are now considering shooting barred owls in the hope that the spotted owl population will recover.

INTRODUCTION

You are an advisor to the Fish and Wildlife Service (FWS). Your assignment is to evaluate the plan to shoot the invasive barred owls. You must then advise the FWS on whether they should implement this plan.

TASK

Research the plan and write a two-paragraph report on it. In the first paragraph, evaluate the plan, including where shooting barred owls has been tried before and whether it has allowed spotted owls to return. In the second paragraph, give your recommendations for actions that should be taken. Be as detailed as your research allows.

PROCEDURE

1. Do a keyword search using the words "shooting barred owls to save spotted owls."
2. Note how the worldviews of the authors of the websites affect their view of the plan.
3. Plan and write your report. Read it and make appropriate revisions to confirm that you have completed the task as outlined above. Have one person proofread your report.
4. Share your report with your class or family.

spotted owl *barred owl*

12A REVIEW QUESTIONS

1. What are the two main reproductive organs in animals?
2. Which haploid gamete is produced by the male and which is produced by the female?
3. Why must externally fertilized eggs be laid in water?
4. Are egg layers internal or external fertilizers?
5. Corn snakes (left) lay eggs with a leathery shell. Are the eggs fertilized internally or externally? Explain.
6. What are the two main sources of nourishment for embryos of animals that have internal development?
7. Compare complete and incomplete metamorphosis.

PURPLE FISH, GREEN FISH

Do fish show color preference?

Many people enjoy keeping fish as pets. Today you are going to test whether fish prefer one color over another.

Equipment
small aquarium fish (6–12)
aquarium, 10 gal
cellophane, purple and green
water

E

PROCEDURE

❶ Cover the left half of the aquarium with purple cellophane.

❷ Cover the right half of the aquarium with green cellophane.

❸ Fill the tank with room-temperature water.

❹ Place the fish in the middle of the tank.

❺ Cover the left half of the top of the aquarium with purple cellophane and the right half of the top with green cellophane.

❻ Allow the fish to acclimate for five minutes.

1. Hypothesize whether the fish will prefer the green half or the purple half of the aquarium.

❼ Count and record in a table the number of fish in each half of the aquarium.

❽ Repeat Step ❼ nine times with two minutes between each observation.

❾ Calculate the average number of fish on each side.

2. What was the average number of fish on each side of the aquarium?

3. Did your data confirm your hypothesis?

12B Questions

- What kinds of behaviors do animals display?
- How do animals communicate with one another?

12B

ANIMAL BEHAVIOR

How do animals interact with each other?

12.3 TYPES OF BEHAVIOR

Innate Behaviors

Have you ever seen a bird carrying grass and twigs on its way to build its nest? Birds fill the days of spring with singing, nest building, courtship, and egg laying. Each of these activities is a **behavior**, which is how an animal interacts with other animals or with its environment.

Did you know that each species of bird generally has a distinctive type of nest? Mourning doves build a flimsy platform of twigs, while robins build a cup of mud and grass. How did these birds learn to build their type of nest? Most birds leave their parents as soon as they reach maturity, so they have never seen a nest built before they build their first one. The type of behavior that an animal is born with is called **innate behavior**.

Some innate behaviors are fairly simple, immediate responses called **reflexes.** A dog pants when it gets too hot. An armadillo jumps into the air when it is startled. The animals don't have to think about these actions. God has given animals these behaviors to help them survive.

12-1

The armadillo's reflexive jumping startles a potential predator and gives the armadillo a chance to escape.

Other innate behaviors are quite complex and occur over a longer period of time than a reflex does. These types of innate behaviors are called **instincts**. Birds' nest building is an instinct. So is the hibernation of a groundhog and the nursing instinct of a baby mammal. One of the most extraordinary examples of instinct occurs when animals make seasonal long-distance trips called **migrations**. Although birds have some of the most famous and most spectacular migrations, insects like the monarch and globe skimmer butterflies also have long migrations, especially given their size. Leatherback sea turtles sometimes migrate more than 16,000 kilometers between the beaches where they lay their eggs and the waters where they like to snack on jellyfish.

12-2

The Baltimore oriole builds a distinctive hanging basket nest.

Learned Behaviors

Animals do not have to practice innate behavior. It happens naturally. But they do need to work to learn other behaviors. These **learned behaviors** are actions developed by observation and training. Some animals such as earthworms learn very little. Their actions are mostly innate. But some vertebrates are great learners. The ability of squirrels and raccoons to get into containers for birdseed or table scraps is legendary.

Sometimes this learning comes from the offspring mimicking its parents. This occurs through the process of **imprinting**. The young animal assumes that the nearest moving object is its parent and begins to follow it around and imitate it. Normally this works very well because the closest moving object *is* the parent. But if the parent has died or the young have been removed, then the young might imprint on another animal, a person, or even an object. Scientists have even used this capability to teach orphan geese how to fly by getting them to imprint on a small aircraft. When the geese are ready to learn to fly, they follow the aircraft into the skies.

Many times an animal is born with an innate behavior that it must sharpen over time. Turkey chicks crouch down and freeze by instinct as soon as they sense movement over their heads. This protects them from hawks and other birds of prey. But the chicks also crouch when harmless birds and even falling leaves pass over their heads. Over time, the chicks learn to distinguish a dangerous bird like a hawk from a harmless one like a duck.

12-3
These geese have imprinted on an ultralight. Now the pilot is teaching them to fly.

12.4 ANIMAL COMMUNICATION

Animals interact with one another all the time. Obviously they don't talk as humans do, but they can communicate on a basic level. Animals send and receive communication using several means. Let's look at some common ones.

COMMUNICATION FORMS

Sound. Wolves howl to communicate with pack mates and other packs. Many animals communicate using sound. Male frogs croak in the spring to attract mates. Rattlesnakes rattle their tails to warn other animals to stay away or be bitten. Birds, of course, do a lot of sound communication, especially in the spring.

Sight. The rings of this blue-ringed octopus indicate that it is preparing to attack with its deadly venom. Other animals communicate using body language. Positions of the head, mouth, body, and tail communicate with other individuals. Other animals that live in the dark or deep in the ocean communicate visually by producing light called **bioluminescence**. Some jellyfish produce this light as do fireflies and glowworms.

Chemicals. The male cecropia moth's feathery antennae allow it to detect a female from over a mile away. Female moths release chemicals called **pheromones** that cause other individuals to behave in specific ways. Female moths use their pheromones to attract a mate. A queen bee uses pheromones to alert the other bees in the hive to her presence. Her pheromones are so powerful that they keep the female worker bees from laying eggs!

Touch. Many animals communicate using touch. In chicken flocks, the strongest chicken will peck other chickens to gain access to food and nesting material. The second strongest chicken will do the same to all the other chickens except the most dominant one. This continues on down the line with all the chickens dominating the weakest chicken in the *pecking order*.

Communication Messages

Animals use communication for many purposes. But there are several uses that are common across many animals. Many animals communicate **aggression**, the willingness to attack another individual, over food resources, potential mates, or areas to raise their young. Often an animal will establish a **territory**, an area that the animal claims as its own. Some animals have very large territories covering several square miles. Other animals have very small territories. Many times the individual defending its territory has only to act aggressively to get the intruder to leave. But sometimes such a dispute leads to fighting, though these fights rarely lead to death.

12-4
Each of these northern gannets has a very small territory.

In animal groups, some individuals are more powerful than others. Individuals at the top maintain their *dominance* by acting aggressively toward other individuals of the group. Normally the other individual expresses *submission*. The pecking order you read about earlier is an example of dominance and submission.

12-5
Can you tell which wolf is dominant over the other?

A lot of animals communicate while seeking a mate. This type of behavior, called **courtship**, is different from one species to another. Sometimes the male will perform specific actions, such as singing or dancing, to try to impress a female. In other species, the female attracts the males; sometimes they both take part in courtship behavior.

12-6
This male lesser bird of paradise is performing a courtship display for a female.

Animal communication is an amazing example of the creativity of God, who made all things very good. Human communication is even more spectacular. Human speech can be used to discuss ideas such as art, science, and current events. Written language is uniquely human. And not only is it a result of our special creation by God in His image, it is also the means by which God has revealed Himself to us.

12B REVIEW QUESTIONS

1. What is an innate behavior?
2. An octopus was able to open a jar only after watching another octopus open one. Is this an innate or learned behavior?
3. Why is the monarch butterfly's migration considered an innate behavior?
4. Name four common forms of communication in animals.
5. What is courtship?
6. A male fiddler crab attracts a mate by moving side to side and waving its claws. What form of communication is the crab using for its courtship?

CASE STUDY INVASIVE SPECIES

Invasive species often thrive in new areas because they reproduce rapidly and because they have few predators in their new habitat. When invasive species take over an area, they become a potential threat to native species. Can this be prevented? For example, how can we control the invasion of feral hogs? Several states allow and even encourage the hunting of feral pigs in the hope that the number of hogs will be reduced. If successful, over time this may reduce the amount of damage that they do.

On a different note, why should we even care that millions of hogs are roaming the United States destroying ecosystems? Or for that matter, if the hog can reproduce more quickly than the native wildlife, why should we try to control the hogs?

The answers to these questions lie in the Creation Mandate. God's command to people is to rule the earth and have dominion over all the animals. This includes two aspects. One is that we should try to manage natural resources in order to maximize their usefulness. Invasive species often make an area's resources less useful because they reduce the number of other species and sometimes destroy areas that humans need. The other aspect of the Creation Mandate concerns the care for God's creatures. Dominion calls for humans not to destroy the earth's resources, but to preserve them. Invasive species often threaten to destroy other species, so part of care for God's creatures includes controlling invasive species.

Use the Case Study above to answer Question 7.

7. An activist argues that feral hogs should be left alone, that their existence is just an example of survival of the fittest. From a biblical worldview, what is wrong with this argument?

CHAPTER SUMMARY

12A ANIMAL REPRODUCTION

- Most animals reproduce sexually when a haploid sperm unites with a haploid egg to form a diploid zygote.
- External fertilization occurs when the egg and sperm unite outside the female's body.
- Internal fertilization occurs when the egg and sperm unite inside the female's body.
- External development occurs when eggs are fertilized externally or when internally fertilized eggs are laid.
- Internal development occurs when a parent, usually the mother, holds its young inside its body while it develops.
- Some animals go through a change in form later in life called metamorphosis.

12A TERMS	
testis	245
ovary	245
external fertilization	245
internal fertilization	245
uterus	247
metamorphosis	248
incubate	249

12B TERMS	
behavior	252
innate behavior	252
reflex	252
instinct	252
migration	252
learned behavior	253
imprinting	253
bioluminescence	254
pheromone	254
aggression	255
territory	255
courtship	255

12B ANIMAL BEHAVIOR

- Animal behaviors may be classified as innate or learned.
- Innate behaviors are those that the animal is born with.
- Learned behaviors are those that the animal acquires by watching another individual or through practice and trial and error.
- Common forms of animal communication are sound, sight, chemicals, and touch.
- Animals often communicate with other members of their group about territory and courtship.

CHAPTER REVIEW QUESTIONS

Recalling Facts

1. Which organ produces sperm?
2. Which gamete does the female animal produce?
3. What is the difference between external and internal fertilization?
4. Do most mammals have internal or external development? Explain.
5. What is a learned behavior?
6. Is nest building by birds an innate or learned behavior? Explain.

Understanding Concepts

For Questions 7–10, match each term with its proper location in the process map.

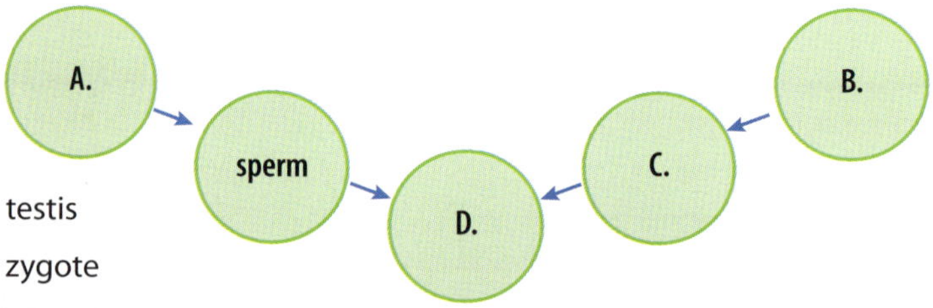

7. testis
8. zygote
9. egg
10. ovary

For Questions 11–13, indicate the type of fertilization and development described.

11. An insect lays eggs on land.
12. The mother provides nourishment to developing offspring.
13. The father releases sperm into the water and dies soon afterward.

Critical Thinking

14. You are studying an animal that produces only one offspring every three or four years. Is it likely to use external or internal fertilization? Explain.
15. Birds are born with the ability to sing a song similar to those sung by their parents. Birds raised separately from other birds of their species usually lack the communication skills of others not separated. Is bird singing an example of an innate behavior, a learned behavior, or both? Explain.
16. What would happen to robins if they lacked the innate ability to build nests and raise young?
17. In the image below, the chameleon on the left is showing dominance. How is the one on the right showing submission?

18. Dogs often lower the front half of their bodies before engaging in play, an action sometimes called a "play bow." Since dogs' play behaviors often mimic aggressive behaviors, what purpose does the play bow serve?

19. The olm, a type of blind cave salamander, does not communicate using sound. How is it likely to communicate? Explain.

20. Conflicts between members of the same species rarely result in one animal killing another. Many male venomous snakes wrestle rather than bite one another. How is this beneficial to the survival of the species?

Read the Ethics box below and then work on Questions 21–22.

ETHICS Animal Testing

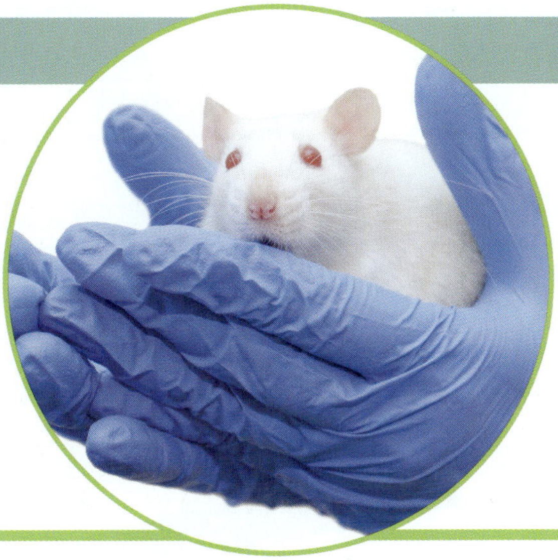

Every year, amazing new medicines and other products are released. But many of these products must go through a long testing process to ensure that they are safe to be used by people. Some testing is done on animals. Often, medicines are tested by creating animals with diseases such as cancer or cystic fibrosis. These animals suffer as a result of these diseases or from the results of the products used on them. Some people view all animal testing as cruelty. Others defend the practice by arguing that many helpful and even life-saving products would not exist without it.

21. Research, plan, and outline a three-paragraph response on the ethics of animal testing. In the first paragraph, discuss principles from Scripture that may guide Christians in this area. In the second paragraph, discuss the possible outcomes of heeding or ignoring the guidelines of Scripture. In the third paragraph, give your opinion on how a believer can make wise decisions about the issue of animal testing.

22. Now, using your outline as a guide, write your response. You can use the rubric below to help you determine whether you have completed all the necessary tasks for writing your response.

	Task completed	Task partially completed	Task not completed
I gathered and used additional information for my response.	two or more sources	just one source	none
I reviewed Scripture for relevant principles on animal testing.	two or more passages	just one passage	none
I stated both the acceptable and unacceptable options.	both	only one	neither
I stated the consequences of the acceptable options.	yes		no
I considered the biblical outcomes of human flourishing, a thriving creation, and God's glory.	all three	only one or two	none
I stated what action should be taken.	yes		no

unit 4 THE HUMAN BODY

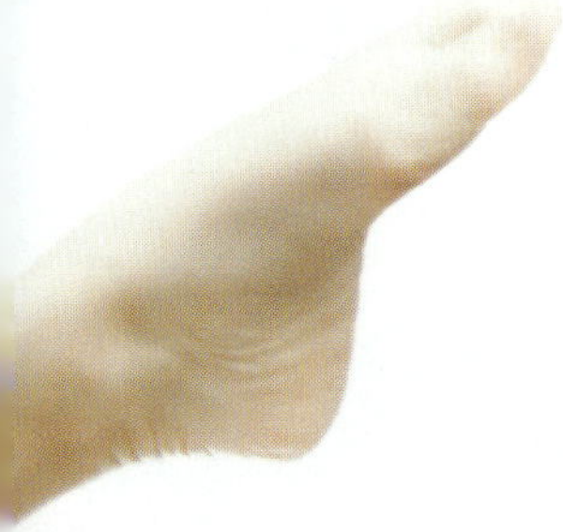

FROM A MEDICAL DOCTOR TO YOU

As a semi-retired emergency room doctor, I now have the privilege of teaching cadaver-based anatomy and physiology at Bob Jones University. Every time I teach about the amazing human body I am stunned at its complexity, which in itself defies evolution. King David summed it up well in Psalm 139:14: "I will praise thee; for I am fearfully and wonderfully made: marvellous are thy works; and that my soul knoweth right well."

Dr. Marc A. Chetta

13 SUPPORT AND MOVEMENT

SHADES OF BROWN

People around the world look different. It is easy to identify with people that look like us and have a similar culture. That's what an ethnic group is: a group of people that share similar physical and cultural characteristics.

But because of the fall of mankind into sin, differences in other people are not always appreciated. People sometimes look down on others from different ethnic groups. That's called *racism*, and racism isn't right.

God made one race, the human race (Gen. 3:20; Acts 17:26). The differences between people around the world are not really significant and are mostly external. We all have skin with varying amounts of brown pigment. We are all descended from one couple, Adam and Eve. Every one of us enjoys the richness of community and the ability to contribute to culture. We are all made in God's image. And we all need the redemption that Jesus, the perfect human, freely offers.

13A

INTRODUCTION TO ANATOMY & PHYSIOLOGY

How does the human body work?

13A Questions

- How are we different from animals?
- What are the four types of tissues?
- How is the human body organized?

13.1 WHAT IS A HUMAN?

Before you study how the human body works, let's take a moment to consider what a human is. Physically, humans share many characteristics with mammals. We have most of the same body systems. We have the same physical needs, including food, water, and oxygen.

But humans are also different from all other creatures because we alone are made in the image of God and have immortal souls. This spiritual nature is what gives us a conscience. Animals can be trained to avoid certain actions but they cannot understand the difference between right and wrong. When the first man, Adam, sinned, all humanity became sinful (Rom. 5:12), and the image of God in man, though still present, was marred. People, as God's image bearers, no longer represented God accurately because of their sin and separation from God. Still, the Bible does not leave us without hope. God has provided a way of salvation from sin by which our relationship with God can be restored. Ultimately, all believers will be restored as proper image bearers when they receive resurrected bodies (1 Cor. 15:42–44).

13.2 TISSUES AND ORGANS

How would you define a bicycle? You could describe it by what it looks like. A bicycle is a metal frame set on two wheels and equipped with pedals and a seat. Or you could describe it by what it does. You can use it to travel or have fun by pedaling it. These two definitions are similar to the two subjects that you will study in this unit. You will study the different organs and structures of your body. This is called **anatomy**. You will also study how those organs and structures function. This is called **physiology**.

Your body is made of trillions of cells. But those cells do not work independently of each other. As you learned in Chapter 2, similar groups of cells are organized into tissues to perform specific functions. Tissues make up organs and organs make up body systems. Let's take a closer look at tissues, organs, and body systems.

TISSUES

ORGANS

Just as cells are organized into tissues, tissues also work together as organs, such as your heart.

The *pericardium* (1), which encloses the heart, is made of connective tissue.

Nerves connect to structures called *nodes* that coordinate the heartbeat.

The *endocardium* (2), which lines the inside of the heart, is made of epithelial tissue.

The majority of the heart is composed of muscle tissue that contracts, producing a heartbeat that pumps blood throughout the body.

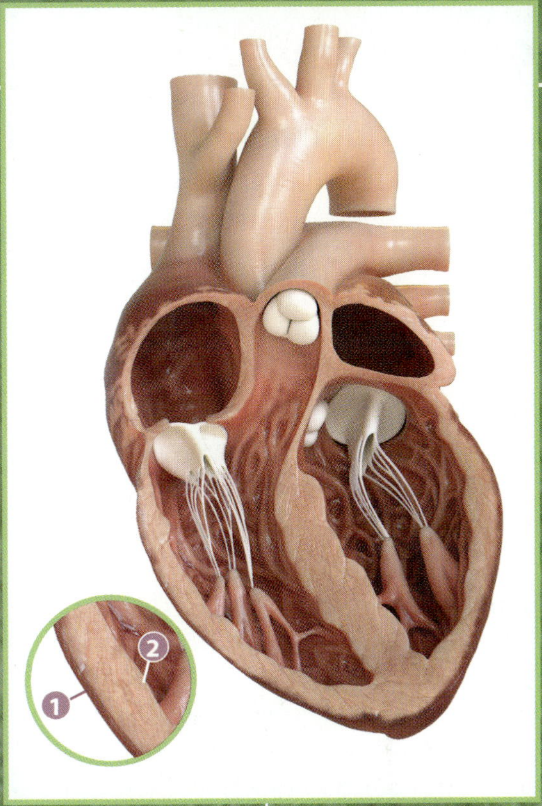

ORGAN SYSTEMS

Organs also work together in groups called *organ systems*. Sometimes an organ can be part of more than one organ system. For instance, the lungs are the main organs of the respiratory system but are also part of the excretory system. This occurs because God designed your organs to work together. Let's look at some of the main organ systems that you will study in this unit.

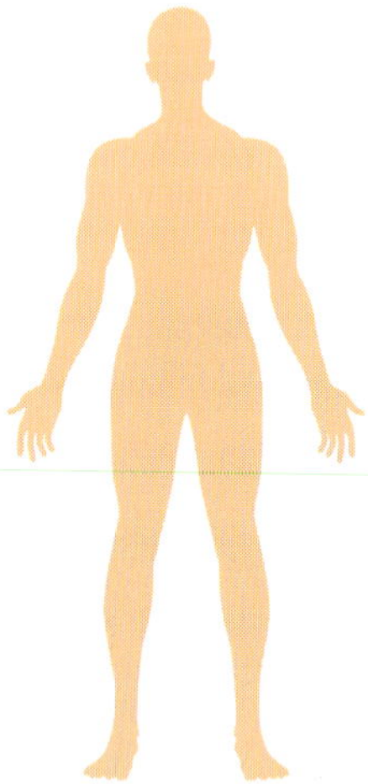

The *integumentary system* provides protection and includes the skin.

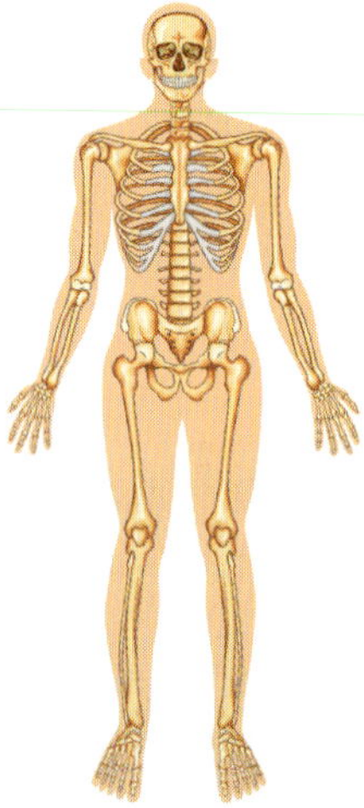

The *skeletal system* protects the internal organs and supports the body by providing a rigid anchoring point for muscles.

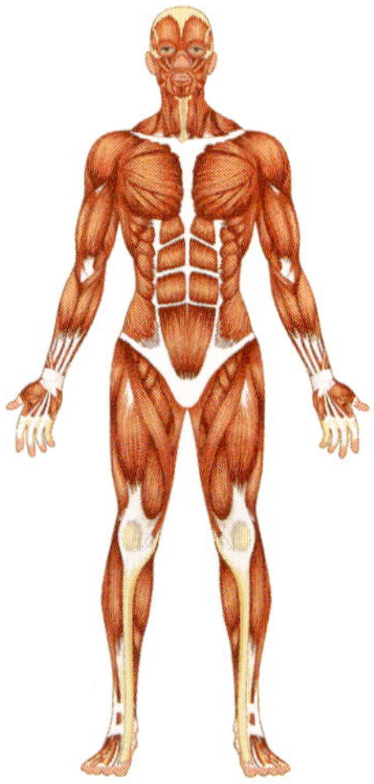

The *muscular system* provides movement by pulling on the bones, causing them to move. Muscles are also involved in digestion and circulation.

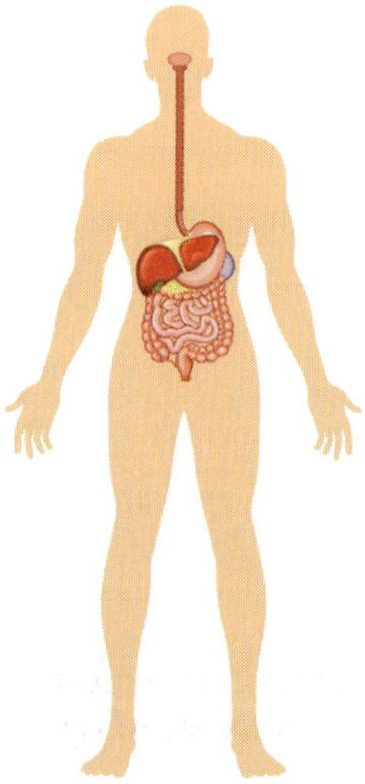

The *digestive system* supplies the body with energy by absorbing food. It also eliminates wastes.

The *reproductive system* creates and protects new human life.

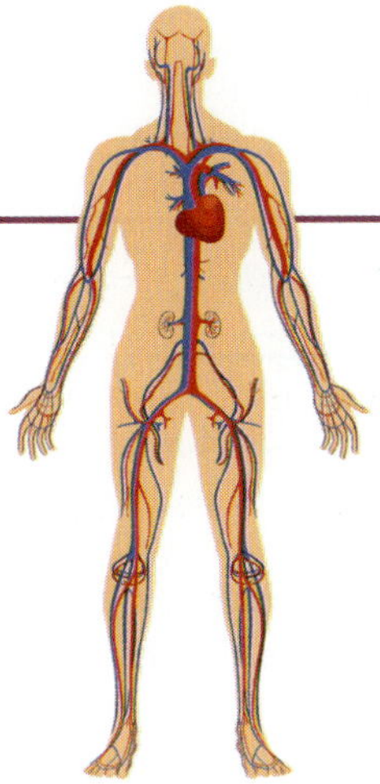

The *circulatory system* transports blood, which carries oxygen, food, and other substances throughout the body.

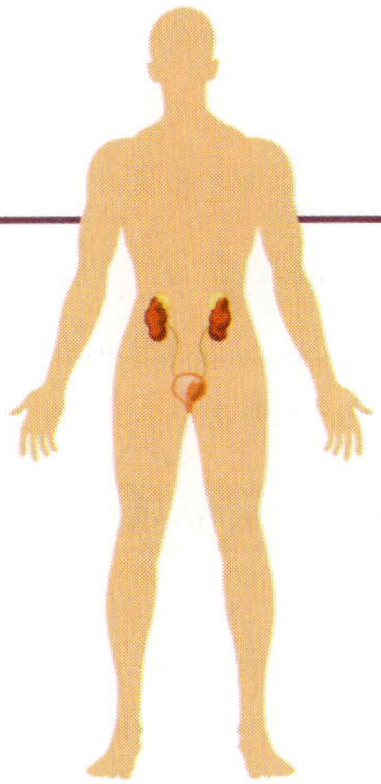

The *urinary system* eliminates certain types of wastes and maintains the right amount of water in the body.

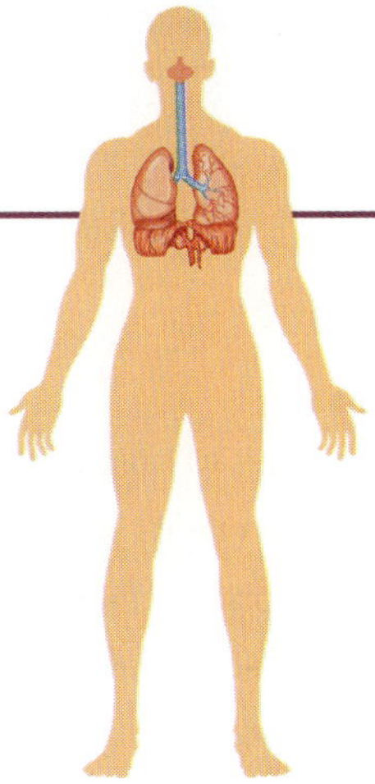

The *respiratory system* exchanges oxygen from the atmosphere with carbon dioxide in the blood stream.

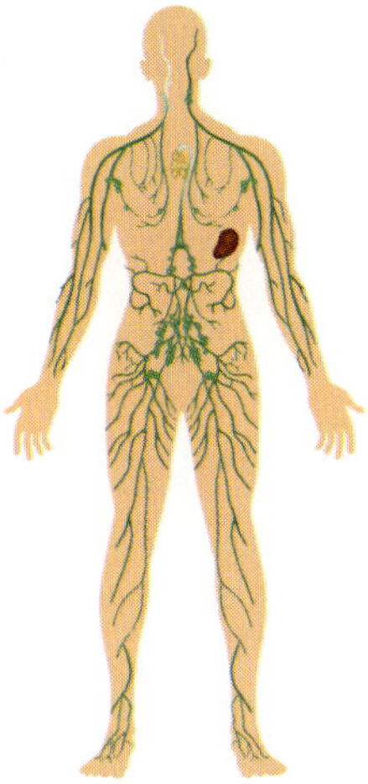

The *lymphatic system* helps protect the body from threats such as bacteria and viruses. It is part of the larger *immune system*.

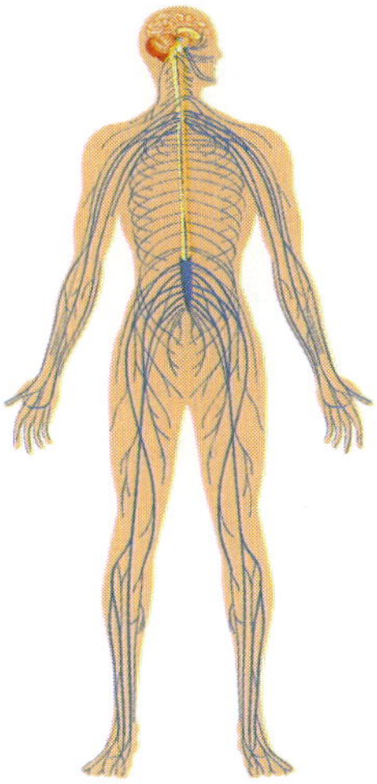

The *nervous system* coordinates and controls movement and processes information from the sensory organs, such as the eyes and ears.

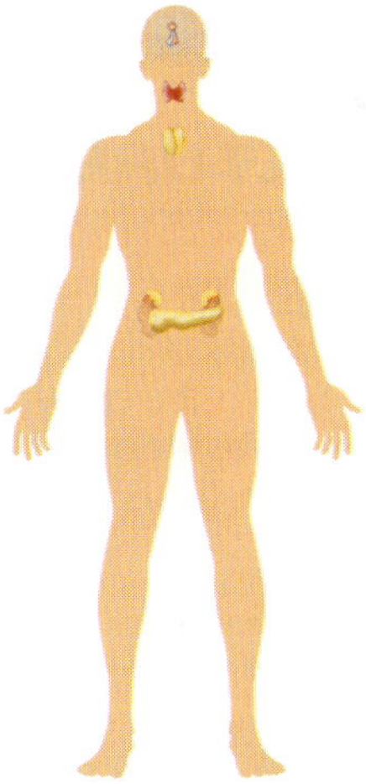

The *endocrine system* secretes chemicals that regulate body functions.

The human body is a marvel of beauty and engineering. As you study anatomy and physiology, praise the Lord of creation for how He has marvelously designed you! As Dr. Chetta pointed out in the Unit opener, you are "fearfully and wonderfully made"!

13A REVIEW QUESTIONS

1. How did God distinguish humans from animals?
2. What is the only hope for sinful man?
3. Describe the function of two types of tissue.
4. Give an example of a cell, a tissue, an organ, and an organ system that involves the heart.
5. Which system is the muscular system anchored to?

13B
SKIN

13B Questions
- What are the layers of human skin?
- How does the skin do its job?
- How are skin colors produced?
- How does the Bible explain the origin of ethnic groups?

Why are there so many different skin colors?

13.3 SKIN STRUCTURE

Have you ever cut your finger? Even a small cut can be quite painful. Yet you can cut your hair without any pain at all. Why is that? It's because your skin is living while your hair is dead.

Your skin, hair, and nails make up your **integumentary system**. Your skin is one of your body's major defenses. Unless it is cut or damaged, few disease-causing organisms can penetrate it. Although we often think of skin as thin and easily damaged, it is actually quite strong. It functions well in keeping your insides in and other things out. Notice the various parts of your skin.

SKIN STRUCTURES

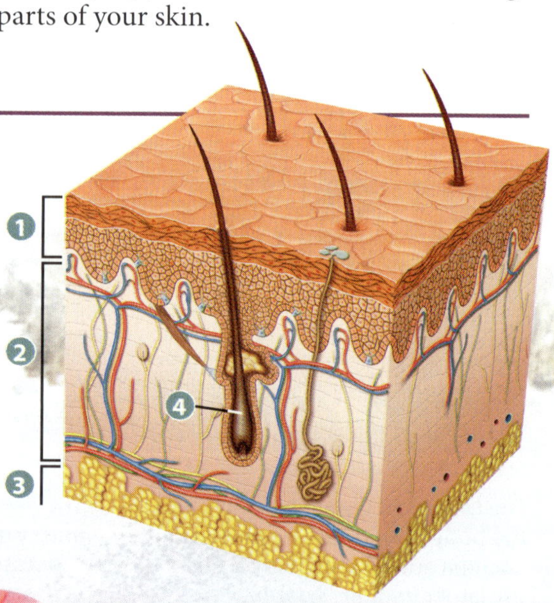

1 The **epidermis** is the outer layer of your skin and is made of epithelial tissue. Its surface consists of dead cells filled with a protein called *keratin*. Living cells deeper in the epidermis eventually fill with keratin and die as they rise to replace the dead cells that are being shed. At the end of your fingers and toes, the epidermis produces nails made of keratin.

2 The **dermis** is the second layer of skin. It is thicker than the epidermis and consists of connective tissue. It contains blood vessels, nerves, follicles, and glands that produce sweat, oil, and wax. Special cells in the dermis make fibers that weave around and between these structures to connect them.

3 The *subcutaneous layer* below the dermis is not a part of the skin. It consists of loosely arranged fat cells and fibers. The fat cells help cushion and insulate your body. The fibers attach the skin to the muscles under the skin. Blood vessels are also found throughout the subcutaneous layer.

4 Hair grows as cells in **hair follicles** divide. By the time the cells reach the epidermis, they have died and filled with keratin. This forms the visible part of the hair. Each hair follicle has an oil gland next to it that secretes oil to keep the skin and hair soft, flexible, and water resistant. **Pores** in the epidermis release the oil onto the skin surface.

So now you know what's in your skin. But how does it work to protect your body and help it maintain balance?

SKIN FUNCTIONS

Sensation. Your skin contains lots of nerve endings that respond to cold, heat, touch, pressure, and pain. Some of the nerves that detect the touch sensation loop around hair follicles. Whenever a hair is bent, the nerve is stimulated and you sense that the hair has been touched.

Protection. In addition to protecting you from germs, your skin also protects you from friction. When skin is exposed to intense friction or heat, the epidermis separates from the dermis. Fluid collects in the space, and a *blister* forms. But if the amount of friction isn't too great, the skin responds by producing more cells in the area experiencing friction. This thickened *callus* protects deeper skin layers.

Temperature Regulation. Your skin also helps keep your body at the correct temperature. When your body gets too warm, more blood begins flowing through the blood vessels in the dermis, releasing heat into the epidermis and the air. At the same time, as sweat from sweat glands evaporates, it transfers even more heat. When you are cold, less blood flows through the skin's blood vessels, and the sweat glands close. As you become even colder, small muscles in your skin contract, forming goose bumps.

Making Vitamin D. We get most of the vitamins that we need from our food. But vitamin D is different. When exposed to a moderate amount of sunlight, our skin makes vitamin D. But too much sunlight can cause a sunburn or even skin cancer.

13.4 SKIN COLOR

If there is only one race, the human race, why are there so many skin colors? The truth is that all humans have the same skin color. We are all different shades of brown. Human skin color is determined by the brown pigment **melanin**. Special cells in the epidermis produce melanin and transport it to other cells.

The genes that code for the production of melanin have two different alleles. One allele codes for a lot of melanin, and the other codes for less. A person with many alleles that code for more melanin will have dark skin. If a person has many alleles that code for less melanin, he will have light skin, often with a slightly reddish look because of the skin's blood vessels. A person with a mix of alleles will have medium-brown skin.

A person's skin shade is the result of genetics and environment. Everyone is born with genes that influence the amount of melanin that his skin produces, so the child of dark-skinned parents will usually also have dark skin. But the skin also produces more melanin when it is exposed to intense sunlight, resulting in a suntan. The skin forms extra melanin to protect itself from the sun's damaging rays. So people become darker when frequently exposed to the sun.

We know from the Bible that all people are descended from Adam and Eve. They probably had a mix of alleles that gave them medium-brown skin. Since everyone alive today is descended from Noah's

three sons and their wives, the same is probably true of them as well. After God confused the language of people at the Tower of Babel, people began to move throughout the world. While we do not know for sure how different groups of people came to have different skin colors, we have some hypotheses that are consistent with the Bible. People with more alleles for melanin were likely better able to survive in parts of the world near the Equator where the sun shines most intensely. People with fewer alleles for melanin were probably better able to survive in areas near the poles. Other people groups continued to have a mixture of alleles resulting in medium-brown skin.

Over the years, sinful human beings began to use skin color as a way of viewing themselves as better than others. This far-too-common racism continues today. A biblical worldview reminds us that God has made all people from one man. All people are made in God's image and have an equal, inherent dignity. A biblical worldview also reminds us that every person is a sinner in need of Christ's salvation. So no person can claim to be superior to someone from a different ethnic group.

All Descended from Adam

"You come of the Lord Adam and the Lady Eve. . . . And that is both honor enough to erect the head of the poorest beggar, and shame enough to bow the shoulders of the greatest emperor on earth."

—*Prince Caspian* by C. S. Lewis

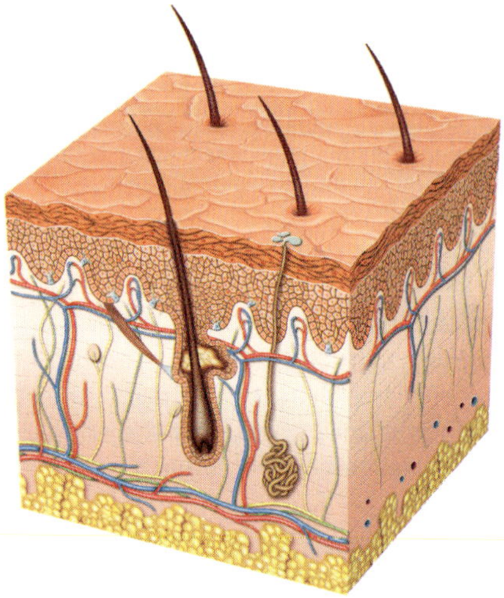

13B REVIEW QUESTIONS

1. What is the name of the top layer of human skin?
2. Which layer of the skin contains nerves, blood vessels, and sweat glands?
3. What types of tissue primarily make up the two layers of skin?
4. What are the four main functions of the skin?
5. How does a callus protect the skin?
6. What pigment gives human skin its color?
7. Explain from Scripture why there is only one human race.

ETHICS

Medicine and Human Dignity

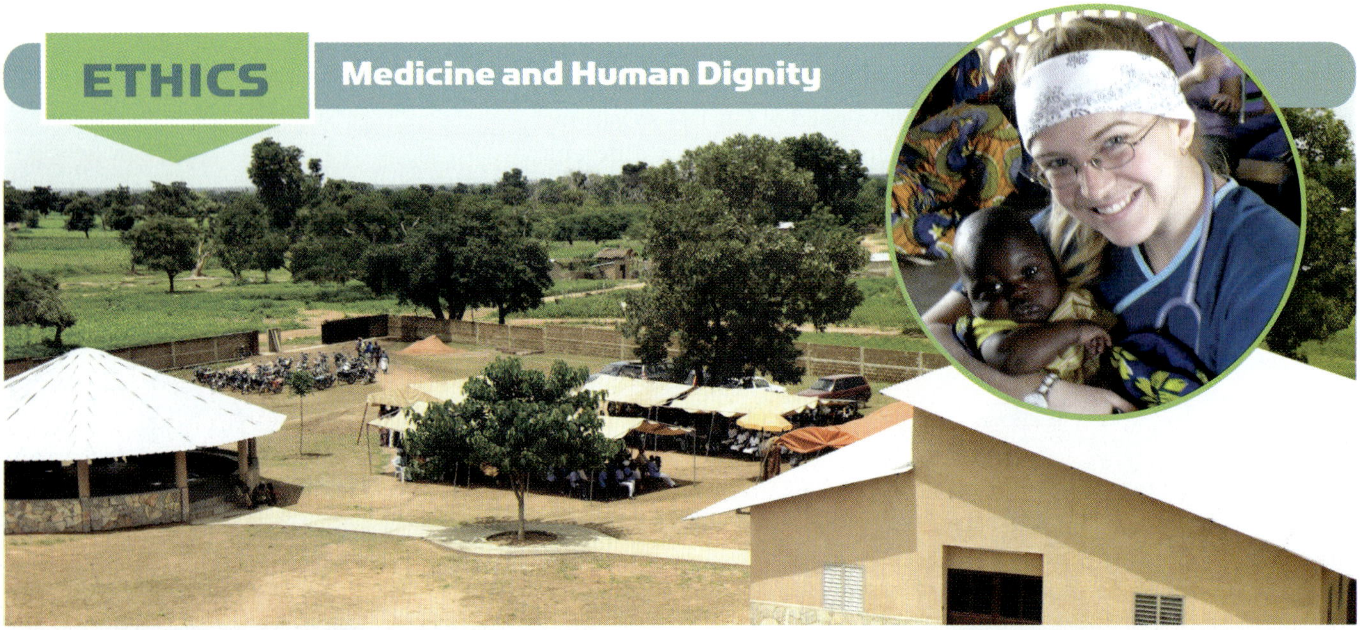

The northern part of Benin is a semiarid land on the edge of the Sahara Desert. It is a very poor area of the country. People there have little access to medical care, and more than one out of every ten children in the country dies before the age of five. Most have not heard the gospel.

One of the people groups in the area is the Fulani tribe. This seminomadic people group rarely seeks medical treatment because they are often treated poorly by the staff of the local medical facilities. This kind of ethnic discrimination is still common in much of the world. People are treated with contempt, given substandard care, or denied medical treatment entirely because of their ethnicity. But in some areas this problem is being addressed. A group of medical missionaries opened Evangelical Baptist Health Center of Boiffo, Benin, in 2017. Many Fulani have begun to come for medical care. When the Christian doctors, nurses, and dentists show them the respect due to them as God's image bearers, they respond by welcoming the opportunity to receive medical care. In the process, local Christians have the opportunity of sharing the gospel with those who come for treatment.

Use the Ethics box above to answer Questions 8–10.

8. Write the five steps of the strategy for considering ethical issues.
9. Research, plan, and outline a three-paragraph essay explaining the scriptural reasons why all people deserve medical care, regardless of ethnicity. Discuss the image of God in man and Christ's command to love our neighbors. Suggest a way that a believer can help provide adequate medical care for those who lack access or are denied access to it.
10. Using your outline as a guide, write your response.

13C

BONES

How does a bone grow and heal?

13C Questions

- What are the body's major bones?
- How is a bone constructed?
- How do bones grow?
- What types of joints does the body have?

13.5 BONE STRUCTURES

Bones are very strong and tough, but occasionally one does break. As many as 6 million people in the United States break a bone each year. But the really amazing part is that bones do not break more often. They endure a lot of stress but usually remain strong and whole. In fact, bones are constantly renewing and strengthening themselves.

Together the bones form the **skeletal system**. Individual bones are made of connective tissue with large amounts of calcium to make them hard and rigid.

SKELETAL SYSTEM STRUCTURE

Most bones are made of several different materials. Each serves a different purpose.

Bones are covered with a tough white tissue called the ❶ **periosteum**. As you grow, the periosteum forms new bone tissue. It also connects to ligaments and tendons.

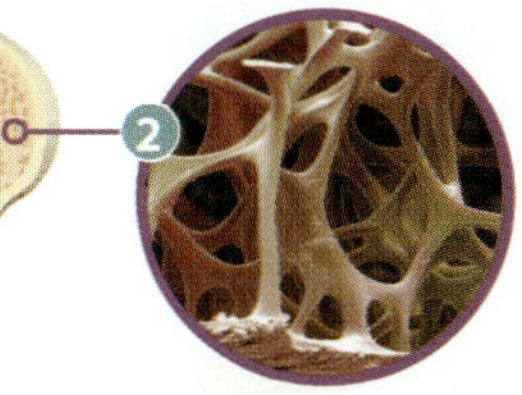

The ends and the center of many bones are made of ❷ **spongy bone**. Though its name suggests otherwise, spongy bone is very hard and rigid, but it is less strong than compact bone because of all the empty spaces in it.

Mature bones are made mostly of hard ❸ **compact bone**. The circular layers of bone, called *osteons*, are created by cells living between the layers. Small channels between the layers carry oxygen and nutrients to the cells.

Many bones have a hollow region at the very center called the marrow cavity. The marrow cavity and the spaces in the spongy bone are filled with **bone marrow**. In children, all bone marrow is *red bone marrow*, which makes blood cells. In adults, most bone marrow has become *yellow bone marrow*, which is primarily made of fat.

A few parts of your skeletal system are made of **cartilage**, a type of connective tissue that is softer and more flexible than bone because it contains much less calcium. Cartilage forms your outer ears and the tip of your nose. It also covers the ends of most of your bones to help them slide smoothly against one another.

Your bones have areas called **growth plates**, which are made of cartilage. Cartilage cells on one side of the plate divide to produce more cartilage cells. Cartilage on the other side changes to bone. Once you have stopped growing, the entire growth plate will become bone.

Major Bones

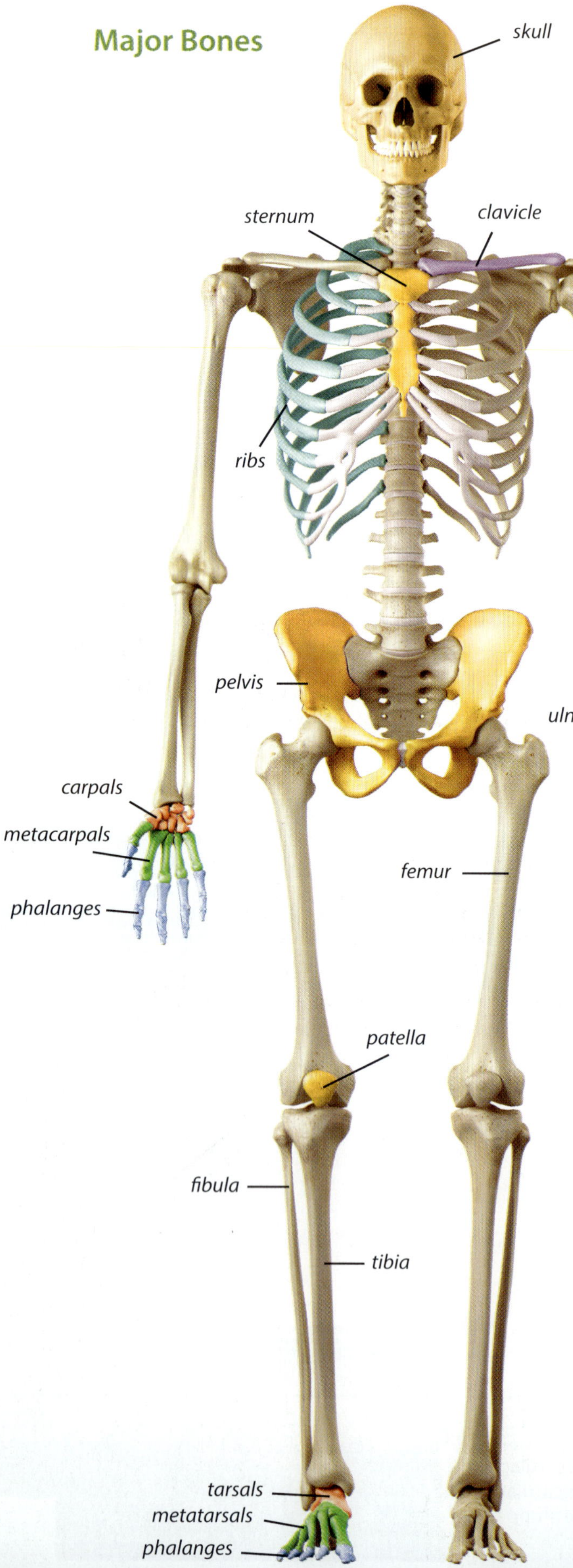

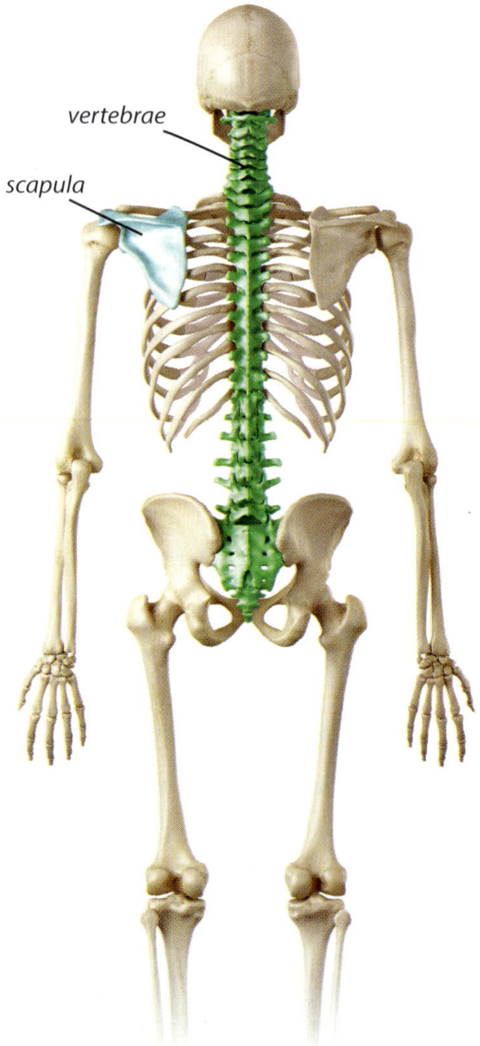

Joints

Two or more bones meet at a **joint**. The bones of most joints are held together by tough, flexible bands of connective tissue called **ligaments** that attach to the periosteum of the bones.

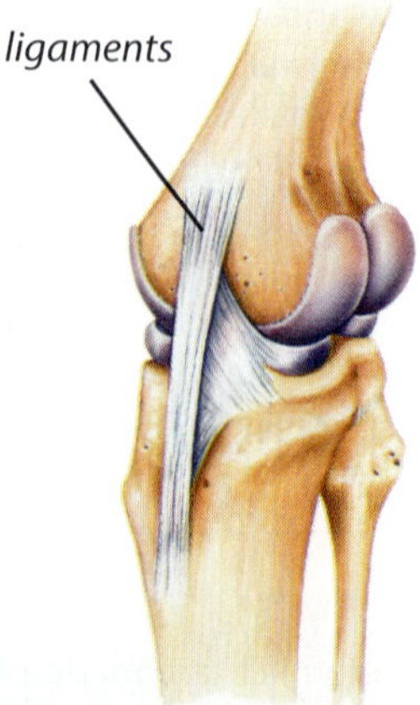

Most joints are *movable joints*. They allow movement in one or more directions. Movable joints include the shoulder, the knee, and the wrist. Other joints prevent any movement because the bones have become permanently fused together. These are called *immovable joints*. The skull has several immovable joints that fuse to form a rigid structure.

13.6 SKELETAL SYSTEM FUNCTION

God designed your skeleton to help your body function and maintain homeostasis. Let's take a closer look at the functions of the skeletal system.

SKELETAL FUNCTIONS

Support and Movement. Your skeleton provides the framework to support your body, but it also does much more. Though a rigid skeleton would support you, it wouldn't allow you to move. Because your skeleton has joints, you can move freely. And because your bones are rigid, they provide firm attachment points for your muscles.

Protection. Your skeleton also protects your body's organs. For example, your skull protects your brain, eyes, and inner ears, your rib cage protects your lungs and heart, and your vertebrae protect your spinal cord.

Mineral Storage. Calcium gives bones their hardness and rigidity. But calcium is also used in other parts of the body. If the body needs calcium, it can remove some from the bones. And if it has extra calcium, it can store it in the skeletal system.

Blood Cell Production. As you learned in the last subsection, your bones have bone marrow inside them. Red bone marrow makes blood cells to replace ones that get old and worn-out.

13C REVIEW QUESTIONS

1. Describe compact bone.
2. What is the name of the hard covering on the surface of bones?
3. What part of a bone grows?
4. Name two bones of the arm, excluding the wrist and hand.
5. What type of joint is the elbow?
6. Name three functions of the skeletal system.

CASE STUDY JOINTS ARE LEVERS

Did you know that you are carrying lots of levers around with you? Each one of your joints is a lever. As you may have learned in the past, a lever is a rigid bar that turns on a point called a *fulcrum*. Levers come in three different forms.

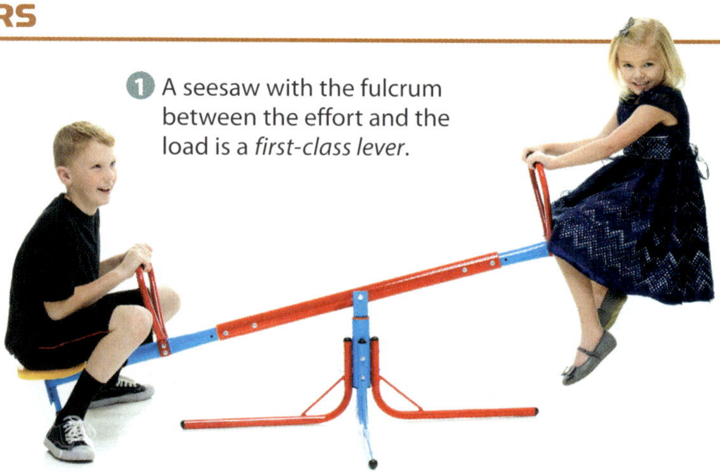

1 A seesaw with the fulcrum between the effort and the load is a *first-class lever*.

2 A wheelbarrow with the load between the effort and the fulcrum is a *second-class lever*.

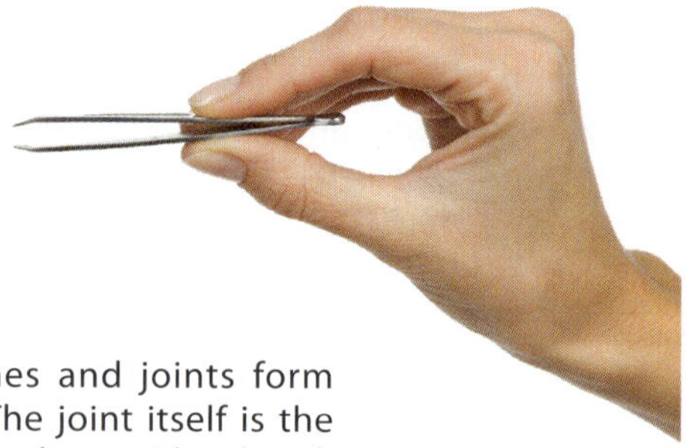

3 A pair of tweezers, illustrating a situation where the effort is between the fulcrum and the load, is a *third-class lever*.

Many of your bones and joints form third-class levers. The joint itself is the fulcrum and the muscle provides the effort. The load is typically on the end of the bone opposite the joint.

Use the Case Study above to answer Questions 7–9.

7. When you lift a barbell, you are using your bicep muscle, which is attached to your radius, to exert an effort. The fulcrum is your elbow and the load is the barbell. What kind of lever is your elbow?
8. When you kick a soccer ball, your tibialis anterior (muscle in lower leg) exerts an effort. Your ankle is the fulcrum, and your foot contacts the ball, the load. What kind of lever is your ankle?
9. Why do you think that most body joints are examples of third-class levers?

13D

MUSCLES

How can I move my arm just by thinking about it?

13.7 TYPES OF MUSCLES

God gave you a skeleton as a solid framework, but it wouldn't be much use without a way to move your bones. That is the role of your **muscular system**. Your body moves when your muscles become shorter and thicker, or *contract*. Since your muscles move your body by contracting, they can only pull. They never push.

You can control some muscles at will. For instance, you can close and open your hand when you wish. Muscles that you can control are called **voluntary muscles**. They are usually attached to bones by bands of connective tissue called **tendons**, so they are sometimes called **skeletal muscles**. Under a microscope, skeletal muscles show a pattern of dark and light bands called *striations*. These bands occur where layers of different kinds of protein meet. Because of these striations, skeletal muscles are also called **striated muscle**.

13D Questions

- Where are voluntary muscles?
- Where are involuntary muscles?
- What is the difference between a ligament and a tendon?
- What is the difference between skeletal, smooth, and cardiac muscle?
- How do muscles work?

COMMON SKELETAL MUSCLES

Muscles are named according to various characteristics. Some of the most common are listed below.

Some muscles have names that indicate shape.

deltoid—triangular-shaped muscle

trapezius—trapezoid-shaped muscle

oblique—muscle arranged in a slanting direction

biceps—muscle with two connection points on one end

triceps—muscle with three connection points on one end

Some muscles have names that come from attached bones.

tibialis anterior—muscle attached to the tibia

quadriceps femoris—muscle with four connection points on one end and attached to the femur

Some muscles have names that indicate size.

major—large muscle

minor—small muscle

maximus—largest muscle in that region of the body

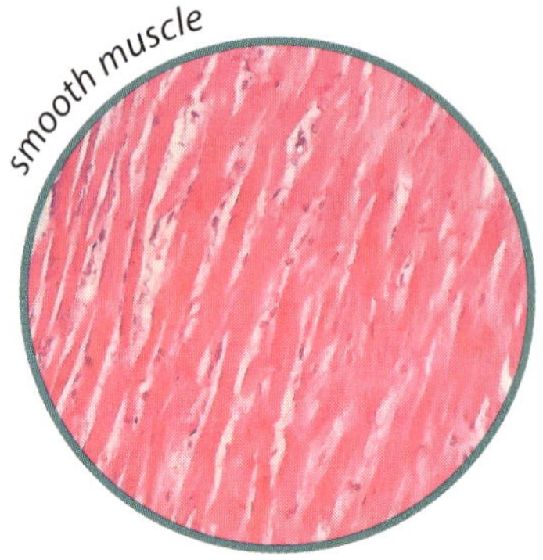

You also have a number of muscle tissues that you cannot control at will. These muscles are called **involuntary muscles**. Your nervous system automatically controls involuntary muscles without you having to think about it. Since most involuntary muscles do not have striations, they are called **smooth muscles**. They are found in the stomach, intestines, blood vessels, and other internal organs. But the involuntary muscle of the heart does have striations and is called **cardiac muscle**.

Involuntary muscles generally contract more slowly than voluntary muscles, but they can function longer before tiring. In fact, your heart beats over 40 million times a year without stopping.

13.8 MUSCLE FUNCTION

Muscles Working Together

Muscles contract to move your body or to move substances inside your body. As you read in Chapter 11, skeletal muscles usually work in pairs. When the bicep muscle of your arm contracts, your arm bends at the elbow. At the same time the tricep muscle in the back of your arm stretches as it relaxes. When the triceps contracts, the biceps stretches as it relaxes and the arm straightens.

Often the ends of a skeletal muscle will be attached to two different bones located on each side of a joint. When the muscle contracts, one of the bones moves while the other remains stationary. For example, the biceps connects the scapula and the radius. When it contracts, the radius moves but the scapula does not.

A few skeletal muscles do not extend across joints. Instead they cause movement of the skin or other muscles. Examples include the muscles that change the size of the eyes and mouth.

Smooth muscles usually occur in the wall of many organs. You rely on smooth muscles when you eat. As food passes through the esophagus on its way to the stomach, the smooth muscles in the part of the esophagus wall behind the food contract, pushing the food forward. At the same time, the muscle in the esophagus wall just ahead of the food relaxes, making it easier for the food to move forward. This makes it possible to swallow food even when you are upside down. And because smooth muscle is involuntary muscle, this all happens without you ever having to think about it.

Cardiac muscle contracts together to pump blood throughout the circulatory system. When the muscle in the heart contracts, blood is forced from the heart into the blood vessels. The heart's contractions are strong enough to force blood through all the vessels throughout the body until it reaches the heart again.

Muscles and Energy

Muscles need a constant supply of energy to function properly. Stored energy comes to muscles in the form of glucose carried by the blood. Muscle cells use cellular respiration to break down glucose to produce ATP (see page 39). The muscle cells then use the ATP to make the muscle cells contract.

CELLULAR RESPIRATION

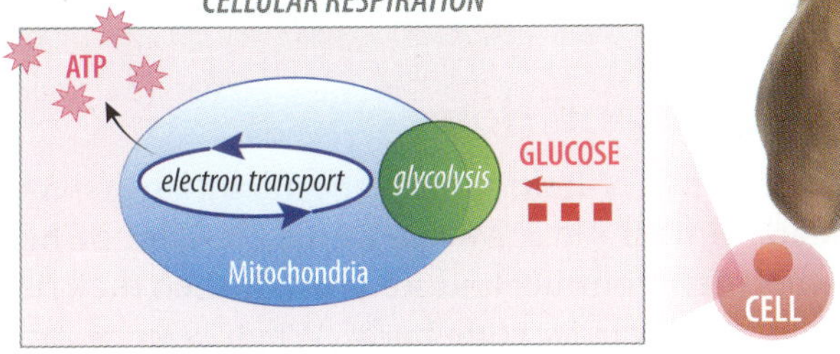

Cellular respiration to make ATP requires oxygen. Muscle cells use oxygen carried by the blood from your lungs. When your muscles are doing a lot of work, you begin to breathe rapidly and deeply to increase the amount of oxygen available to your muscles. But sometimes muscles work so hard that they cannot get enough oxygen. When this happens, they begin to use lactic acid fermentation to break down more glucose (see page 40). Lactic acid fermentation does not produce as much ATP as cellular respiration would. But it does allow muscle cells to continue to function for a short time until the lungs are able to provide enough oxygen to the muscles.

LACTIC ACID FERMENTATION

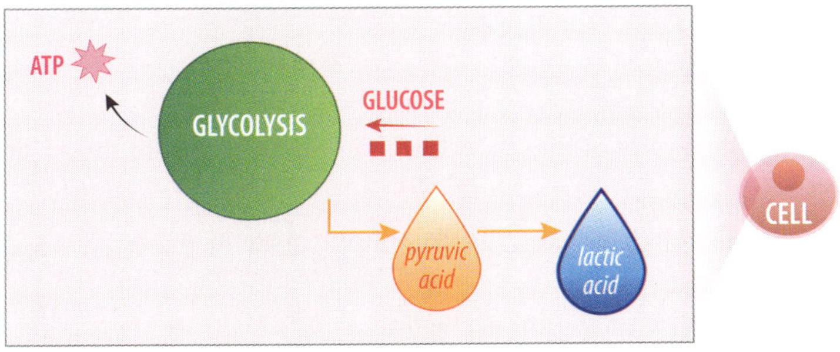

When your muscles have completed this kind of hard work, the lactic acid is converted back to glucose to be reused. As the amount of oxygen in your body returns to normal, your breathing slows to your normal rate.

Muscles and Homeostasis

Your muscles help your body maintain its constant temperature. Muscles produce a lot of heat when they contract and some even when they rest. This heat warms your body. When your body is cold, you may begin to shiver, and goose bumps may appear. These occur because muscles are contracting to produce heat and raise your body's temperature.

13D REVIEW QUESTIONS

1. Are the muscles that control your intestines voluntary or involuntary? Explain.
2. Copy the table below and complete with the words *skeletal muscle*, *smooth muscle*, or *cardiac muscle*.

	Striations	**No Striations**
Voluntary		
Involuntary		

3. Indicate which type of muscle is found in the liver, triceps, heart, and biceps.
4. The gastrocnemius and the tibialis anterior work as a pair. When the tibialis anterior muscle contracts, what will the gastrocnemius do?
5. If your muscles begin producing lactic acid, what is in short supply?

CHAPTER SUMMARY

13A INTRODUCTION TO ANATOMY AND PHYSIOLOGY

- Humans are special creations made in God's image.
- The human body is composed of connective, epithelial, muscle, and nerve tissue.
- The human body can be divided into eleven organ systems.

13A TERMS	
anatomy	264
physiology	264
nerve tissue	264
connective tissue	264
muscle tissue	264
epithelial tissue	264

13B SKIN

- Skin is made of two layers—the epidermis and the dermis.
- Skin senses the body's environment, protects it from harmful substances, regulates its temperature, and manufactures vitamin D.
- Skin color is caused by the pigment melanin.
- Since all people are descended from Adam and Eve, all people are part of one race—the human race.

13B TERMS	
integumentary system	268
epidermis	268
dermis	268
hair follicle	268
pore	268
melanin	270

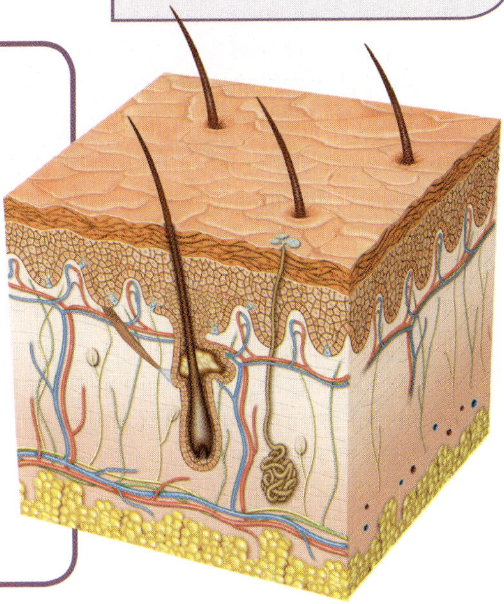

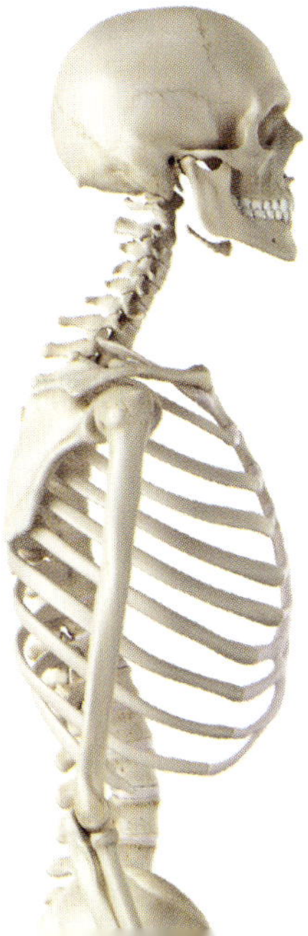

13C BONES

- Bones are made of spongy bone and compact bone.
- Bones grow longer as cartilage cells in the growth plates divide and become bone.
- Joints are either movable or immovable.

13C TERMS	
skeletal system	273
periosteum	273
spongy bone	273
compact bone	273
bone marrow	273
cartilage	273
growth plate	273
joint	274
ligament	274

CHAPTER 13

CHAPTER SUMMARY

13D MUSCLES

- You can control voluntary muscles but not involuntary muscles.
- Skeletal muscles are voluntary muscles; smooth and cardiac muscles are involuntary muscles.
- Skeletal muscles often function in pairs.
- Ligaments attach bone to bone. Tendons attach muscle to bone.
- Muscles use lactic acid fermentation to produce ATP when they do not have enough oxygen.

13D TERMS	
muscular system	277
voluntary muscle	277
tendon	277
skeletal muscle	277
striated muscle	277
involuntary muscle	280
smooth muscle	280
cardiac muscle	280

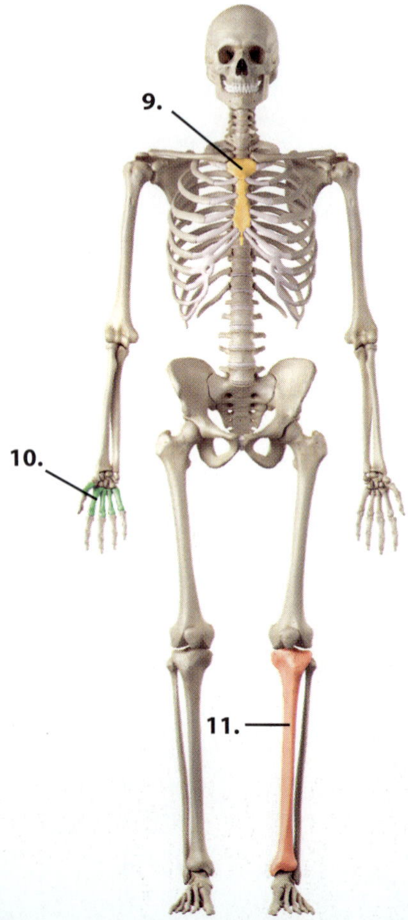

CHAPTER REVIEW QUESTIONS

Recalling Facts

1. What makes humans different from all other creatures?
2. What are the four tissue types found in the human body?
3. What are the three levels of organization above the cell level?
4. What protein fills dead skin cells on the surface of the epidermis?
5. What fluid does the skin produce to cool the body down?
6. What two factors influence the shade of a person's skin?
7. Different ethnic groups with different skin shades may have begun to appear after what event?
8. Describe spongy bone.

For Questions 9–11, write the name of the bone indicated in the image on the left.

9.
10.
11.
12. Name the three bones of the arm.

Understanding Concepts

13. What is the difference between a human choosing not to steal and a dog being trained to avoid a road?
14. What organ and organ system does a periosteum cell belong to?
15. Which organ system is the biceps part of?
16. Which organ system is the only one with striated involuntary muscle?

17. Do biblical principles allow room for denying medical care to a person on the basis of his ethnicity?

18. Curl and uncurl your fingers. What bones are moving?

19. What is the difference between ligaments and tendons?

Critical Thinking

20. You see a sample of the skeletal structure of the end of the human nose on a microscope slide. The sample is almost entirely cartilage. Did it come from an adult or a child?

21. You are a doctor seeing a patient. A test reveals that his muscles are producing a high amount of lactic acid. The patient does not exercise a lot. What organ system should you test to determine whether it is working properly?

Use the Punnett square and information below to answer Questions 22–27.

	ABC	ABc	AbC	aBC	Abc	aBc	abC	abc
ABC	AABBCC	AABBCc	AABbCC	AaBBCC	AABbCc	AaBBCc	AaBbCC	AaBbCc
ABc	AABBCc	AABBcc	AABbCc	AaBBCc	AABbcc	AaBBcc	AaBbCc	AaBbcc
AbC	AABbCC	AABbCc	AAbbCC	AaBbCC	AAbbCc	AaBbCc	AabbCC	AabbCc
aBC	AaBBCC	AaBBCc	AaBbCC	aaBBCC	AaBbCc	aaBBCc	aaBbCC	aaBbCc
Abc	AABbCc	AABbcc	AAbbCc	AaBbCc	AAbbcc	AaBbcc	AabbCc	Aabbcc
aBc	AaBBCc	AaBBcc	AaBbCc	aaBBCc	AaBbcc	aaBBcc	aaBbCc	aaBbcc
abC	AaBbCC	AaBbCc	AabbCC	aaBbCC	AabbCc	aaBbCc	aabbCC	aabbCc
abc	AaBbCc	AaBbcc	AabbCc	aaBbCc	Aabbcc	aaBbcc	aabbCc	aabbcc

Skin color is determined by over 300 different genes. Let's consider just three of the most important. In the Punnett square above, an uppercase letter represents the dominant allele that codes for lots of melanin. A lowercase letter represents the recessive allele that codes for less melanin. The Punnett square represents the possible skin shades of children of two people with medium-brown skin. Both parents have the genotype AaBbCc.

22. Assume that having five or more dominant alleles produces dark brown skin. Is it possible for this couple to have children with dark brown skin?

23. Out of the sixty-four possible squares in the Punnett square, how many contain the genes to produce dark brown skin?

24. Assume that anyone with at least five recessive alleles has light-toned skin. Is it possible for this couple to have children with light skin?

25. Out of the sixty-four possible squares in the Punnett square, how many contain the genes to produce light skin?

26. What would be the skin shade of most of this couple's children?

27. Is it possible that this couple could have a child with dark skin and another with light skin?

14 ENERGY

EATING AND EMOTIONS

Food affects people's emotions. Do you get angry when you're hungry? When you're stressed, do you stuff a chocolate bar in your mouth or grab a bag of potato chips? People often look to food to get relief from the stresses of life. This can become a problem if a person develops habits that lead to an unbalanced diet and an unhealthy weight.

On top of these pressures is the one that many teen girls in Western cultures especially feel—the need to have a perfect body. So how can people reconcile the desire to eat what they want with the desire to look good? Some give up on their appearance, becoming overweight or even obese. Some develop emotional problems, refusing food. This is called *anorexia*. Some will overeat, feel guilty, and then purge themselves in a condition known as *bulimia*. How can we get energy from food and enjoy it in the way that God intended?

14A

FOOD AND NUTRITION

How do we get energy from food?

14.1 NUTRIENTS IN THE BODY

Obesity, anorexia, and bulimia are things that many Americans struggle with. But true pleasure doesn't come from eating, and true value doesn't come from having a perfect-looking body. Those come only from having a personal relationship with Jesus, the Bread of Life. Only He is able to meet all our needs. We all need the truth found in God's Word—food for the soul.

But God also provides nutritionally for our bodies. And He wants us to enjoy the food that He gives us. We need to use food in a way that shows that we are exercising dominion not only over God's world but also over our own bodies. Understanding what is in different kinds of food helps us to develop a diet that cares for our bodies, God's creation. Let's find out what's in food.

14A Questions

- How many nutrients are there?
- What do nutrients do in the body?
- How can I know whether my favorite food is good for me?

NUTRIENTS IN YOUR FOOD

Breads, cereals, desserts, and starchy vegetables are typical sources of **carbohydrates** (or *carbs*). *Simple carbohydrates* include smaller molecules, such as glucose. *Complex carbohydrates* include large molecules, like starch and cellulose. Your body converts most other carbs into glucose to fuel cellular respiration.

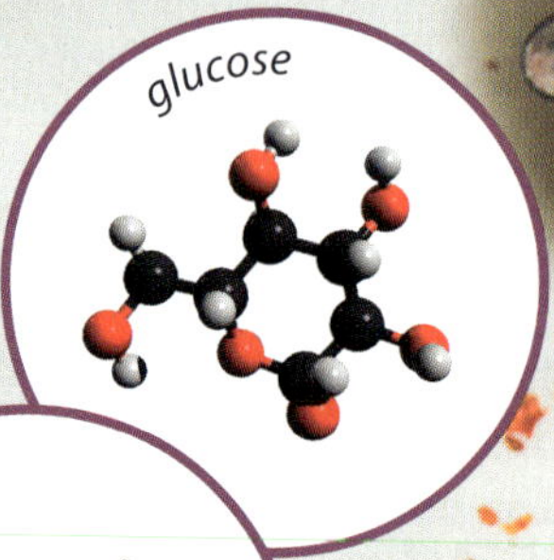

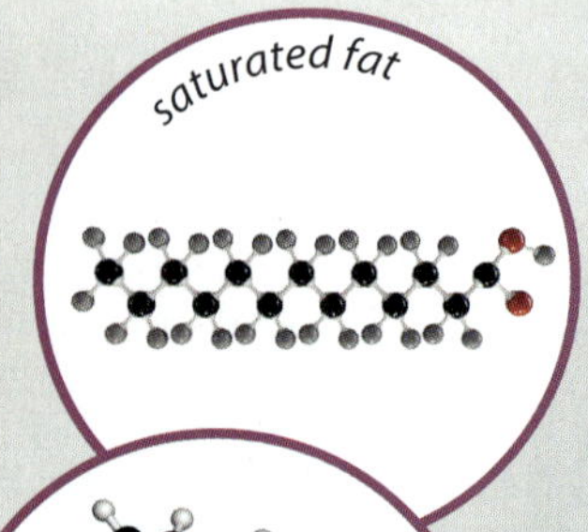

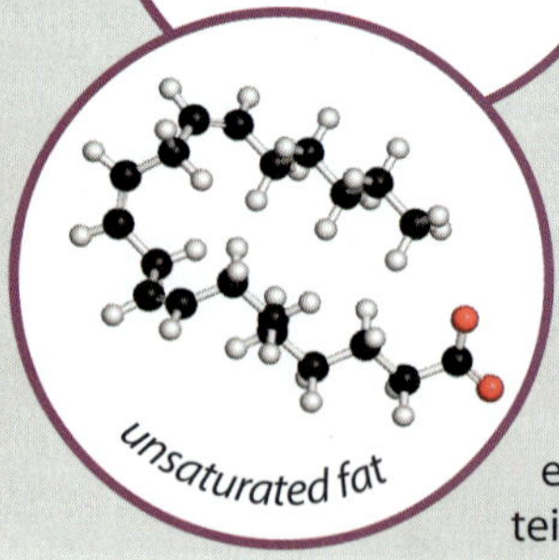

Fats commonly occur in meat, dairy products, and eggs. *Saturated fats* tend to be solids at room temperature, while *unsaturated fats* are usually liquids. The body uses fats to protect internal structures and nerves and to form cellular membranes. Because the body also uses fats to store extra energy, eating too many carbs, fats, or proteins can cause the body to store too much fat.

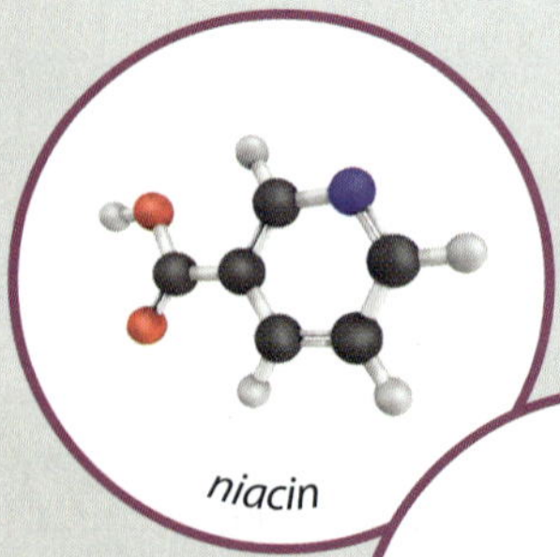

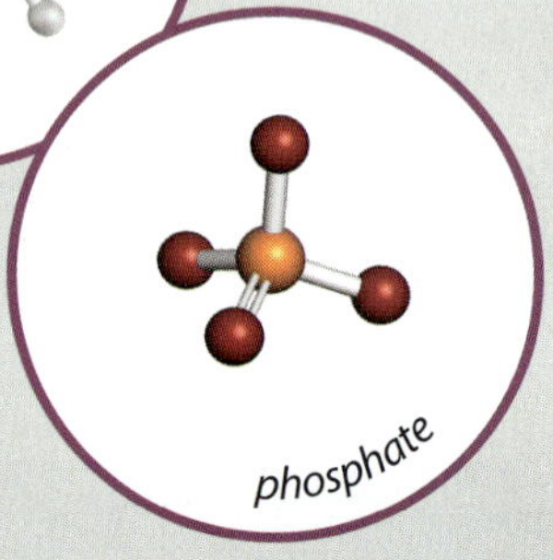

Minerals are *inorganic substances* (molecules without carbon), such as calcium and phosphate, that the body uses to control processes and build certain structures, like bones.

Vitamins are often found in fruits, vegetables, nuts, and dairy products. Your body needs a small amount of each vitamin to function properly. But the body cannot make vitamins, so you must obtain them from food.

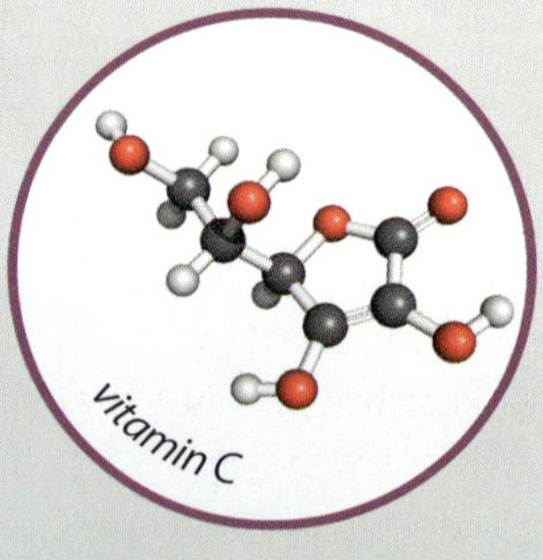

Proteins are found in meat, eggs, nuts, and beans. As you learned in Chapter 3, proteins are made of amino acids. Your body can make about half of the twenty amino acids that it needs; these are called *nonessential amino acids*. You must receive the other half of the amino acids from your diet—the *essential amino acids*. Your body uses proteins to build structures and control reactions.

Water is essential for life. Most of the chemical reactions in your body take place in water. Water also dissolves foods, maintains the fluid balance in the blood, and helps regulate body temperature. You need about a half gallon of water a day.

Of the six classes of nutrients, your body can use carbohydrates, fats, and proteins to fuel cellular respiration, although it typically uses carbs first. It cannot use vitamins, minerals, or water to power cellular respiration, but that doesn't mean that you can do without them. Your body needs each class of nutrient in order to properly maintain homeostasis.

14.2 NUTRITION IN YOUR DIET

Your car can't get very far without fuel. And neither can you. Your body needs energy to operate. The energy that your body takes in is usually measured in **Calories**. It's important to get enough energy from your diet, but many people in the United States have the opposite problem: they are getting too much energy. This results in obesity and related health problems.

But healthy eating is more than just getting the proper number of Calories. And though a healthy diet for one person may look different from another person's healthy diet, there are certain principles that everyone should follow.

HEALTHY EATING

Liquids. You need to drink plenty of liquids each day. This can include things like juice and tea, but you should also drink several glasses of water a day.

Fruits & Vegetables. We all need lots of fresh fruits and vegetables. You don't have to become a vegetarian, but everyone should eat these nutritious foods. Many of the vitamins and minerals you need come from fruits and vegetables.

It's important to form good eating habits. It may take some research or require talking to a doctor or trained nutritionist. The good news is that it's actually possible to eat healthy food and still enjoy your meals.

Balanced Nutrients. Your body needs all the nutrients. That includes carbs, fats, and proteins. Completely cutting out any one of them or eating just one of them does not make for a healthy diet. A healthy diet includes all three along with lots of water and plenty of foods containing vitamins and minerals.

Steamed Broccoli

Nutrition Facts	
Serving Size 3/4 cup (85g)	
Amount Per Serving	
Calories 30	Calories from Fat 0
	% Daily Value*
Total Fat 0g	**0%**
Saturated Fat 0g	**0%**
Trans Fat 0g	
Cholesterol 0mg	**0%**
Sodium 20mg	**1%**
Total Carbohydrate 4g	**1%**
Dietary Fiber 2g	**8%**
Sugars 2g	
Protein 1g	
Vitamin A	2%
Calcium	2%
Vitamin C	50%
Iron	0%

*Percent Daily Values (PDV) are based on a 2,000 calorie diet.

WHAT'S IN MY FOOD?

This meal looks delicious, right? But how healthy is it? Today you are going to find out.

Food packages have a panel called a *nutrition facts label*. This label lists the amounts of carbs, fats, and proteins, as well as sodium, cholesterol, and certain vitamins and minerals that the food contains. These are most helpful if the food is eaten without mixing it with other ingredients. While food is not usually eaten this way, the nutrition facts label can at least give you an idea of how healthy your diet is.

Today you are going to use the nutrition facts labels for the components of this meal to estimate how healthy it is.

PROCEDURE

1. Read the serving size at the top of each label.

1. Do the serving sizes seem reasonable?

A meal can be healthy and taste great too.

❷ On each nutrition facts label, locate the number of Calories per serving.

2. Assume that you eat two servings of mashed potatoes and one serving of the steak and broccoli. How many Calories would you consume in this meal?

The number of Calories that a person needs each day varies depending on gender, age, and physical activity. Most seventh graders need around 2000 Calories a day.

3. Assume that this is your largest meal today and should account for about 36% of your Calories for the day. How many Calories should you have at this meal?

❸ On each nutrition facts label, locate the number of grams of carbohydrates, fats, and proteins per serving.

❹ On a piece of paper, copy the table below. In the *Grams* row, enter the amounts of carbs, fats, and proteins in this meal. Don't forget to double the mass from the potatoes since you're eating two servings of those.

❺ A gram of fat yields about 9 Calories. A gram of protein or carbohydrate yields about 4 Calories. In the *Calories* row of your table, enter the number of Calories for each group in this meal.

❻ For each nutrient, divide the number of Calories for that nutrient by the total number of Calories in this meal. In the *% Calories* row of your table, enter this percentage.

	Carbohydrates	**Fats**	**Proteins**
Grams			
Calories			
% Calories			
Suggested Daily %	45%–65%	10%–30%	25%–35%

❼ Compare the percentage of each nutrient in this meal with the suggested daily percentages.

4. Is this a balanced meal according to the USDA guidelines?

5. Assuming that this is your evening meal, what kind of meal might you have eaten for lunch to make up for any imbalance in this meal?

6. After all that calculating and determining that you have a few more Calories to spare, you decide to finish dinner with a piece of apple pie and a scoop of ice cream, adding another 260 Calories to this meal. How many Calories would this meal be now?

7. How can learning how to analyze the nutrition of food help build a right view of nutrition?

Mashed Potatoes

Nutrition Facts
Serving Size 2/3 cup (147g)

Amount Per Serving	
Calories 170 Calories from Fat 60	
	% Daily Value*
Total Fat 7g	**11%**
Saturated Fat 4.5g	**23%**
Trans Fat 0g	
Cholesterol 20mg	**7%**
Sodium 430mg	**18%**
Total Carbohydrate 26g	**9%**
Dietary Fiber 2g	**8%**
Sugars 2g	
Protein 3g	
Vitamin A	4%
Calcium	2%
Vitamin C	15%
Iron	2%

*Percent Daily Values (PDV) are based on a 2,000 calorie diet.

The suggested daily percentages of each nutrient come from the 2010 dietary guidelines prepared by the US Department of Agriculture.

Pan-Fried Sirloin Steak

Nutrition Facts
Serving Size 8 oz (224g)
Servings 1

Amount Per Serving	
Calories 280 Calories from Fat 70	
	% Daily Value*
Total Fat 8g	**6%**
Saturated Fat 3g	**8%**
Cholesterol 130mg	**22%**
Sodium 100mg	**2%**
Total Carbohydrate 0g	**0%**
Protein 50g	**49%**
Iron	8%
Riboflavin	8%
Vitamin B6	25%
Phosphorus	20%
Selenium	45%
Thiamin	0%
Niacin	35%
Vitamin B12	20%
Zinc	30%

Not a significant source of dietary fiber, sugars, vitamin A, vitamin C or calcium.

*Percent Daily Values (PDV) are based on a 2,000 calorie diet.

14A REVIEW QUESTIONS

1. Name three of the classes of nutrients.
2. Identify each of the following as one of the six classes of nutrients.
   a. essential amino acids
   b. salt (NaCl)
   c. unsaturated fats
   d. sucrose (sugar)
   e. riboflavin (Vitamin B12)
3. Which nutrients does the body get energy from?
4. Which class of nutrients is common in bread?
5. Which class of nutrients does the body use to store energy?
6. Why would powdered donuts and ice cream not be a healthy meal?

14B

DIGESTION

Could the order of the digestive organs be changed?

14B Questions

- What organs are part of the digestive system?
- In what order does food go through the digestive system?
- What is the difference between mechanical and chemical digestion?
- How does chemical digestion work?

14.3 DIGESTIVE PROCESSES

Imagine eating your favorite food. (Hopefully, you're not reading this while you're hungry.) Your body can use it as fuel, but your cells can't run cellular respiration on bits of beef or make cell membranes from tiny pieces of lettuce. Your cells need to break down food into molecules in order to use it. This process is called **digestion**.

Digestion occurs in two forms. *Mechanical digestion* occurs when food is crushed or ground into smaller, more easily digested pieces. This often includes mixing the ground pieces with digestive enzymes or acid. *Chemical digestion* breaks the food particles into molecules using enzymes or acids. Carbohydrates are broken into simple sugars, and proteins are broken into amino acids. Fat molecules are separated from one another so that the body can use them more easily.

Once mechanical and chemical digestion have broken down your food into small enough molecules, your body begins to take in the nutrients, a process called *absorption*. After your body has absorbed the nutrients, it uses them to power cellular respiration and to make new cells and other structures in a process called *assimilation*.

14.4 DIGESTIVE ORGANS

Digestion takes place in a 9 m tube called the **alimentary canal**. This tube is composed of several different digestive organs and is aided by several *accessory organs*. Let's take a closer look at these.

DIGESTIVE TRACT

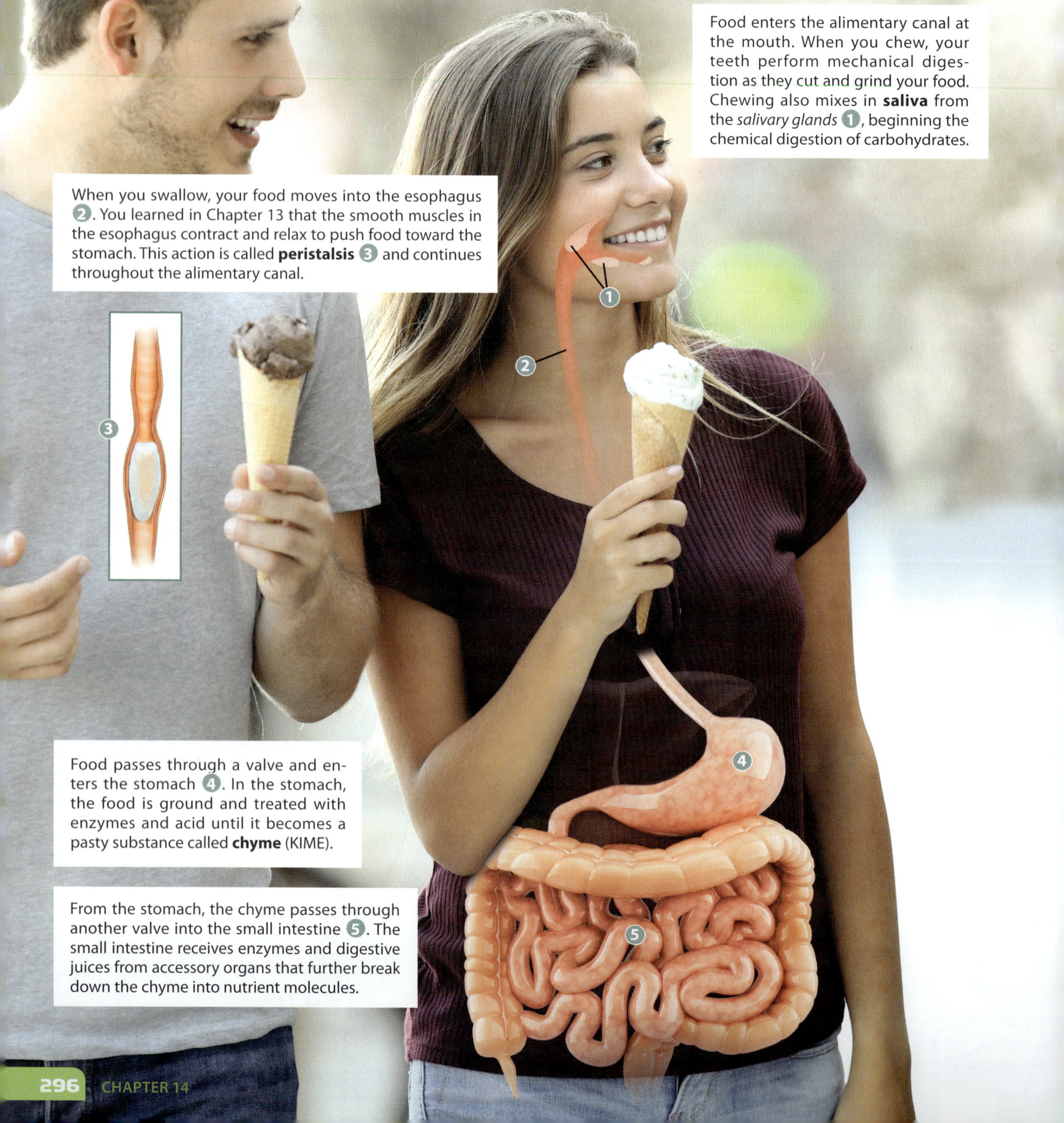

Food enters the alimentary canal at the mouth. When you chew, your teeth perform mechanical digestion as they cut and grind your food. Chewing also mixes in **saliva** from the *salivary glands* 1, beginning the chemical digestion of carbohydrates.

When you swallow, your food moves into the esophagus 2. You learned in Chapter 13 that the smooth muscles in the esophagus contract and relax to push food toward the stomach. This action is called **peristalsis** 3 and continues throughout the alimentary canal.

Food passes through a valve and enters the stomach 4. In the stomach, the food is ground and treated with enzymes and acid until it becomes a pasty substance called **chyme** (KIME).

From the stomach, the chyme passes through another valve into the small intestine 5. The small intestine receives enzymes and digestive juices from accessory organs that further break down the chyme into nutrient molecules.

The *pancreas* 6 produces more digestive enzymes than any other organ of the digestive system. It releases enzymes into the small intestine that aid in the chemical digestion of sugars, proteins, and fats. It also releases a fluid that neutralizes the acid in the chyme.

The liver 7 performs over a hundred functions in the body. One of them is to produce *bile*, which is stored in the *gallbladder* 8 until it is needed. The gallbladder releases the bile into the small intestine, where it helps digest fats.

Most absorption takes place in the small intestine. Its walls have many microscopic, finger-like structures called **villi** (s. villus) 9. The villi provide a large internal surface area for absorbing nutrients from digested food.

The remaining undigested food enters the large intestine, or colon 10. Its primary function is to absorb water and minerals. The material that is left moves into the final structure of the large intestine, the **rectum** 11, and is then eliminated as waste.

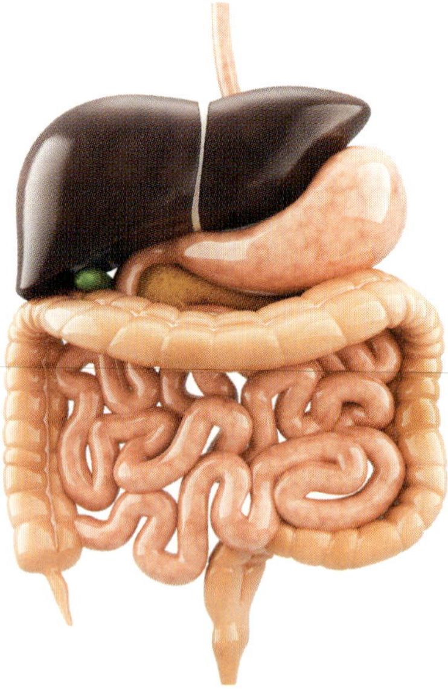

Your digestive system works to extract the nutrients that you need from your food. But of course that means that your food has to contain all the nutrients. This is why eating a balanced diet is important. Taking care of the body God has made for each of us is one way of giving Him glory.

14B REVIEW QUESTIONS

1. Define *chemical digestion*.
2. Does the stomach perform mechanical digestion, chemical digestion, or both? Explain.
3. Which organs produce enzymes or digestive juices for chemical digestion?
4. List the following organs of the alimentary canal in the order that food passes through them: esophagus, large intestine, mouth, and small intestine.
5. Describe villi.
6. What nutrients does the large intestine absorb?

14C Questions

- What are the types of excretion?
- What organs make up the urinary system?
- How do kidneys remove wastes from blood?
- What is the path of urine through the urinary system?
- Why are performance-enhancing drugs dangerous?

14C THE URINARY SYSTEM

How does the body remove wastes?

14.5 EXCRETORY SYSTEM

The primary function of the *excretory system* is to remove wastes from the body. In doing so, it helps the body maintain homeostasis. The excretory system includes organs that are part of other systems as well. Your lungs expel carbon dioxide from your body, so they are part of both the respiratory and excretory systems. Sweat glands also release some wastes, but their main role is to help control the body's temperature, so they are more commonly studied as part of the skin. You learned about sweat glands in Chapter 13, and you will study the lungs in Chapter 15. In this section, you will study the urinary system.

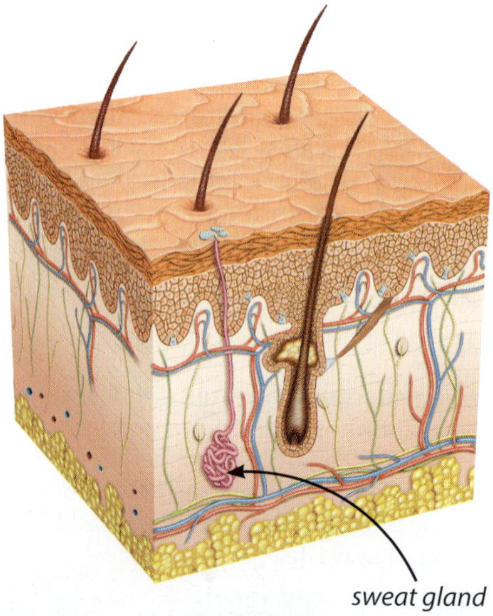

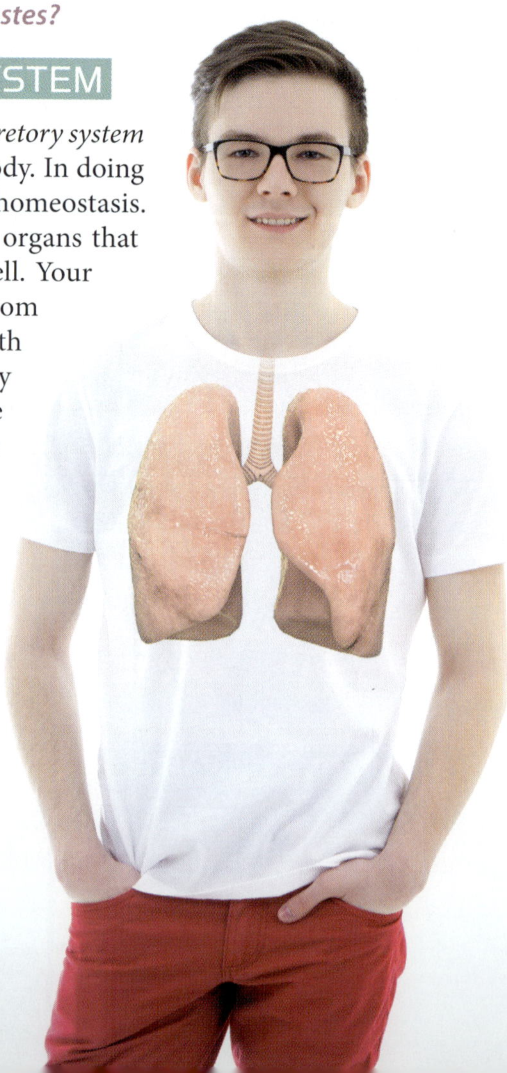

14.6 KIDNEY FUNCTION

As the body uses proteins, it produces a toxic waste molecule called *ammonia*. Your liver transforms the ammonia to less toxic urea. Blood then transports the urea to your *kidneys*, which filter out the urea. Your kidneys are located in your lower back, one on each side of the backbone. And they really do look like large kidney beans, which is why the beans were called kidney beans in the first place.

REMOVING UREA

Each kidney ① has about one million tiny structures called **nephrons** that filter the blood.

Filtering takes place in two steps. First, most of the water from the plasma leaves the blood vessels and enters the nephrons. This water contains many dissolved substances, including urea. The water then flows along the nephron *tubule*.

In the second step, most of the water, nutrients, and substances such as salts and calcium return to the blood by passing through the wall of the tubule back into the blood vessels.

Waste products, substances like drugs, and some water remain in the nephrons as a fluid called **urine**.

The urine from the nephrons collects in the kidneys and then passes into the **ureters** ②, a pair of very thin tubes that use peristalsis to move the urine to a muscular storage organ called the **bladder** ③. The urine is held in the bladder until it can be removed from the body through another tiny muscular tube called the **urethra** ④.

In addition to removing urea from your blood, your kidneys also help keep the right amount of water in your body. Having too little water can be very dangerous, but so can having too much. When your body has more water than it needs, your kidneys release the excess in urine. When your body needs more water, your kidneys release more water back into your blood vessels. In this way, your urinary system helps maintain your body's homeostasis by helping to keep your body properly hydrated.

14C REVIEW QUESTIONS

1. What two structures are part of the excretory system but not part of the urinary system?
2. What four organs are part of the urinary system?
3. What structure does water enter to be filtered as it leaves the blood vessels?
4. What would happen if the second step of filtering didn't occur?
5. Put the following events (a–e) in order:
   a. urine passes through ureter
   b. water and dissolved substances leave blood vessels
   c. urine passes through the urethra
   d. urine collects in the bladder
   e. water and salts return to the blood
6. What organ connects a kidney to the bladder?

ETHICS Doping in Sports

The pitcher prepares to throw the ball. The windup, the pitch, then the batter swings. The ball flies over the heads of the outfielders and into the stands as the crowd goes wild. Home run!

Baseball has been called America's pastime. But in recent decades, the sport has been rocked by the discovery that some of its most prominent players have used performance-enhancing drugs. Major League Baseball, like most sports organizations, forbids athletes from taking such drugs. Besides being cheating, the use of these drugs does a great deal of harm to athletes' bodies. Long-term health problems often result for those who take them.

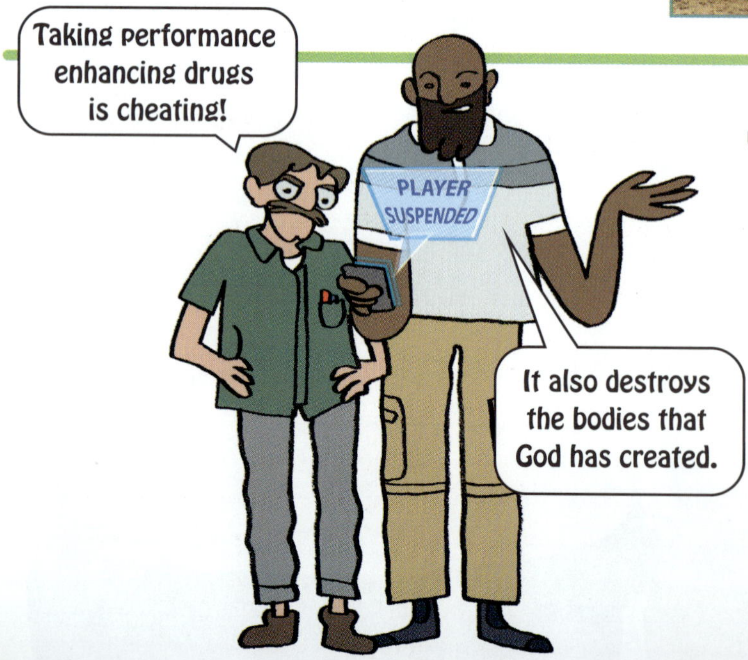

Use the Ethics box above to answer Questions 7–8.

7. Research, plan, and outline a three-paragraph essay explaining the Scriptural reasons why athletes should avoid using drugs to enhance their performance. Discuss how doping violates fair play and God's command to respect the body that He has created. Explain the difference between taking performance-enhancing drugs and working out or practicing to become a better player. Suggest a way that a believer can discourage the use of drugs in sports.
8. Using your outline as a guide, write your response.

CHAPTER SUMMARY

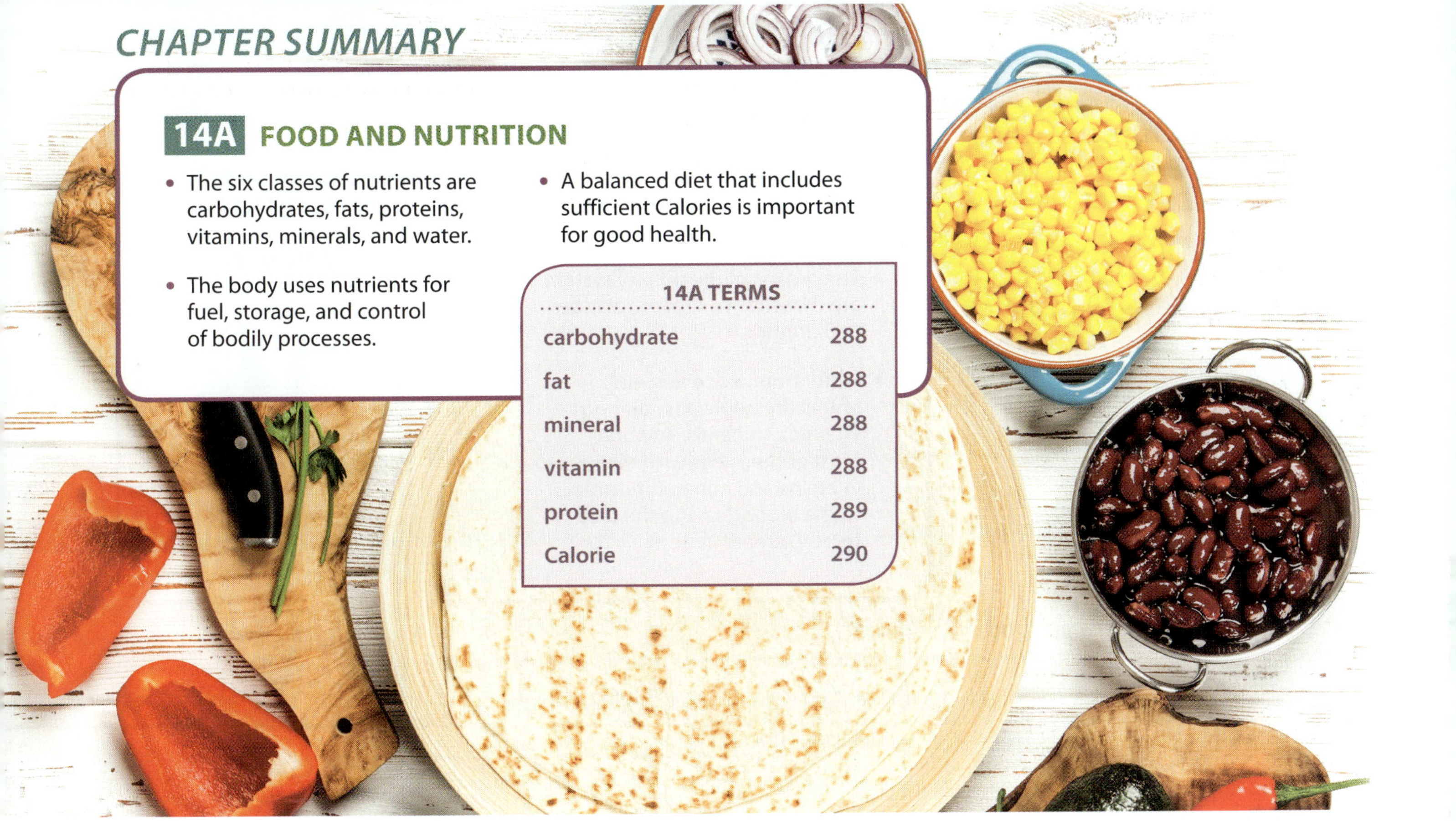

14A FOOD AND NUTRITION

- The six classes of nutrients are carbohydrates, fats, proteins, vitamins, minerals, and water.
- The body uses nutrients for fuel, storage, and control of bodily processes.
- A balanced diet that includes sufficient Calories is important for good health.

14A TERMS	
carbohydrate	288
fat	288
mineral	288
vitamin	288
protein	289
Calorie	290

14B DIGESTION

- Mechanical digestion occurs when food is crushed or ground into small pieces.
- Chemical digestion occurs when food is treated with acids or enzymes. It releases nutrients from food to be absorbed by the body.
- The digestive organs consist of the alimentary canal and the accessory organs.
- Food enters the body at the mouth, then travels through the esophagus, stomach, and small and large intestines.

14B TERMS	
digestion	295
alimentary canal	296
saliva	296
peristalsis	296
chyme	296
villus	297
rectum	297

CHAPTER SUMMARY

14C TERMS	
nephron	299
urine	299
ureter	299
bladder	299
urethra	299

14C THE URINARY SYSTEM

- Excretion includes exhaling carbon dioxide, sweating, and producing urine.
- The organs of the urinary system are the kidneys, ureters, bladder, and urethra.
- Water from blood vessels enters the nephrons carrying vitamins, minerals, and urea. Most of the water with the vitamins and minerals returns to the blood, leaving the urea to be excreted in urine.
- Urine travels in the ureters from the kidneys to the bladder, where it is stored until it is expelled from the body through the urethra.

CHAPTER REVIEW QUESTIONS

Recalling Facts

1. List the six classes of nutrients.
2. Which is the primary nutrient used to fuel cellular respiration?
3. Which class of nutrient includes inorganic substances that the body uses for building structures and controlling processes?
4. What is the role of water in the body?
5. Which organs perform both mechanical and chemical digestion?
6. How does digestion affect carbohydrates?
7. How does mechanical digestion aid chemical digestion?
8. In which organ is water absorbed?
9. Which organ of the urinary system filters urea from the blood?
10. What is a nephron?

Understanding Concepts

11. Which two nutrients would be most abundant in a Thanksgiving turkey?
12. Contrast mechanical digestion and chemical digestion.
13. Where does chemical digestion begin?

Use the diagram on the right to answer Questions 14–15.

14. Is organ e primarily involved in mechanical or chemical digestion?

15. Indicate which organs are accessory organs.

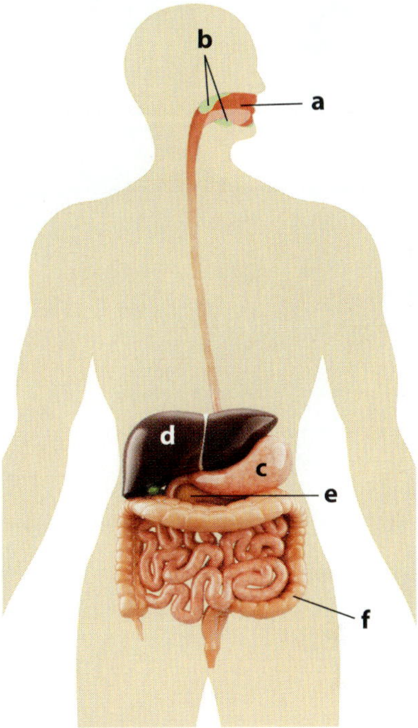

16. What two substances are removed by the excretory system?

17. List the following organs of the urinary system in the order that urine passes through them: bladder, kidney, ureter, urethra.

Critical Thinking

18. How would eating a starchy vegetable, such as a baked potato, along with a hamburger make it a more balanced meal?

19. If the pancreas stopped making digestive enzymes, how would the small intestine be affected?

20. What would happen if the first step of filtering in the nephron did not occur?

21. Some people suffer from kidney stones. These occur when some of the salts or urea in the urine form particles in the kidneys. Why would this be a problem?

CASE STUDY IN LOVE WITH SUGAR

Look at this picture. Notice the chocolate syrup running down the ice cream and pooling around the base of a luscious brownie. Don't you want to eat it? But wait—is it healthy?

Many people love desserts. But desserts are often extremely high in simple carbohydrates or sugars. In fact, people in the United States eat way too much sugar. It is recommended that no more than 10% of your daily Calories should come from sugars, but for the average American that figure is around 14%.

We tend to consume large amounts of sugar because we love to eat desserts and drink soda and other sugary drinks. They taste great, but they usually contain few vitamins, minerals, protein, or anything other than sugar.

Use the Case Study above to answer Questions 22–24.

22. Is it necessary to avoid all dessert and sweetened drinks in order to be healthy? Explain.

23. What should be a Christian's primary motivation for eating a healthy diet?

24. Evaluate your own typical diet for imbalances that may be affecting your health now or may lead to health problems in the future. What changes should you make now?

15 TRANSPORT

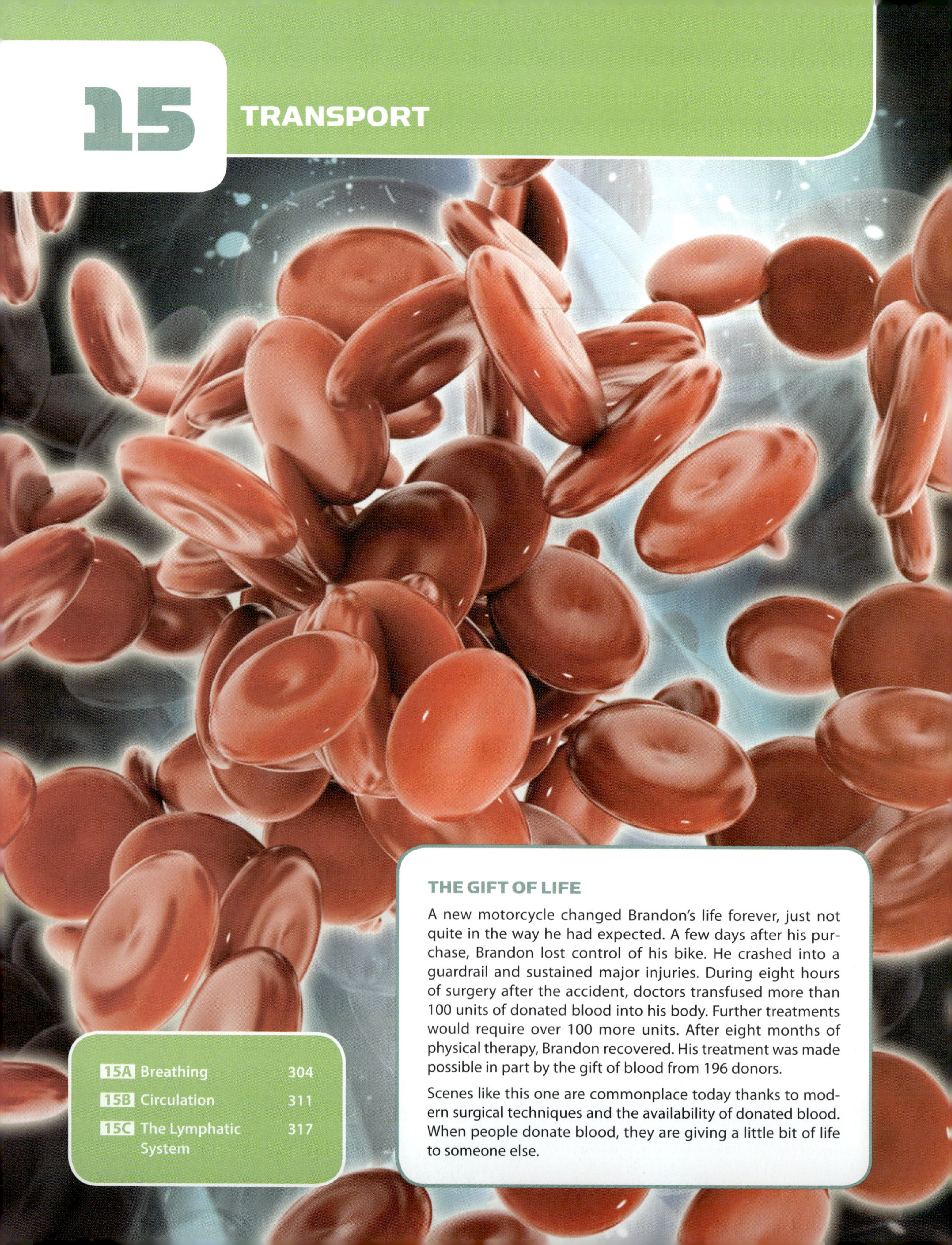

THE GIFT OF LIFE

A new motorcycle changed Brandon's life forever, just not quite in the way he had expected. A few days after his purchase, Brandon lost control of his bike. He crashed into a guardrail and sustained major injuries. During eight hours of surgery after the accident, doctors transfused more than 100 units of donated blood into his body. Further treatments would require over 100 more units. After eight months of physical therapy, Brandon recovered. His treatment was made possible in part by the gift of blood from 196 donors.

Scenes like this one are commonplace today thanks to modern surgical techniques and the availability of donated blood. When people donate blood, they are giving a little bit of life to someone else.

15A

BREATHING

How does breathing keep me alive?

15.1 THE NEED FOR AIR

To stay alive, your body is constantly transporting material from place to place. These materials include blood and air. We don't normally think too much about breathing since it doesn't require any conscious effort. But your respiratory system supplies a vital substance for your body—oxygen. As we learned in Chapter 2, cellular respiration requires oxygen. Without it, your cells can't produce the energy that your body needs.

It's your respiratory system's job to keep up a fresh supply of oxygen for your body. The lungs are the main organs for doing this. They deliver oxygen to your blood while removing carbon dioxide. The air you inhale (breathe in) is the source of oxygen. The air you exhale (breathe out) contains the carbon dioxide produced during cellular respiration.

A tough hike can leave you gasping for air. But what really makes a person breathe faster? You might think that it's because the body needs more oxygen to get energy from cellular respiration. While it's true that increased activity does raise your body's need for oxygen, it's actually the need to get rid of more carbon dioxide that increases your breathing rate. A certain amount of carbon dioxide helps regulate the chemistry of your blood, but above that level it quickly becomes toxic.

15A Questions

- Why do we breathe?
- How does breathing help maintain homeostasis?
- What are the parts of the respiratory system?
- How does breathing work?

15-1

More energy means more carbon dioxide to get rid of.

15.2 THE RESPIRATORY SYSTEM

Take a deep breath and read on as we examine the structures that air passes through on its way to delivering its load of oxygen to your body.

As air enters your nose, your **nasal cavity** 1 warms and moistens it. Otherwise, your respiratory system could easily dry out.

Air next enters the **pharynx** 2, which connects the passages of your nose and mouth to the passage-way that leads to your stomach, your *esophagus*.

The multi-purpose **larynx** 3 is made of muscle and cartilage and has several important jobs. One is to help prevent food from entering your trachea. It also contains your *vocal cords*, which enable you to speak.

As air leaves your larynx, it passes through your **trachea** 4, or windpipe. This branches into two smaller tubes called **bronchi** 5 (s. bronchus), one leading to each lung.

The spongy tissue of each lung 6 is made of millions of tiny sacs called **alveoli** 7 (s. alveolus). This is where gas exchange takes place. Each alveolus is surrounded by a network of tiny blood vessels. The walls of the alveoli and blood vessels are very thin, allowing gas to easily diffuse between the two.

As we've seen, no organ system in the body can do its job alone. Your respiratory system works together with several other systems to fulfill its mission of delivering oxygen to cells. Muscles of the muscular system are responsible for pumping air into and out of your lungs. The rate at which this is done is governed by your nervous system. Blood transports the oxygen to cells through the vessels of the circulatory system. Your lungs even play a role in your excretory system since they rid your body of waste carbon dioxide.

15.3 BREATHING

Breathing is the process of inhaling and exhaling air. The air is not actually moved by the lungs but by muscles surrounding the lungs.

You inhale by making your chest cavity larger. This works much like sipping a beverage through a straw. To drink through a straw, you create a vacuum in your mouth by making the back of your mouth cavity larger. Since the straw then forms the only opening into your mouth, whatever is at the end of the straw rushes in to fill the space. Let's examine how this works in the human body.

At the start of a breath, your lungs have a certain *resting volume*. To increase this volume, you make your chest larger by contracting your *diaphragm*, a dome-shaped muscle located beneath your lungs. It forms a flexible wall that divides your body into upper and lower cavities. You don't have to think about contracting your diaphragm—it's a kind of involuntary muscle. When the diaphragm contracts, the dome flattens, pushing the organs in your stomach area down and making your chest larger. This increases the volume of your lungs and creates a vacuum, just like in the straw example. Air then rushes through the trachea into the lungs. Muscles that raise and expand your rib cage also help make your chest larger.

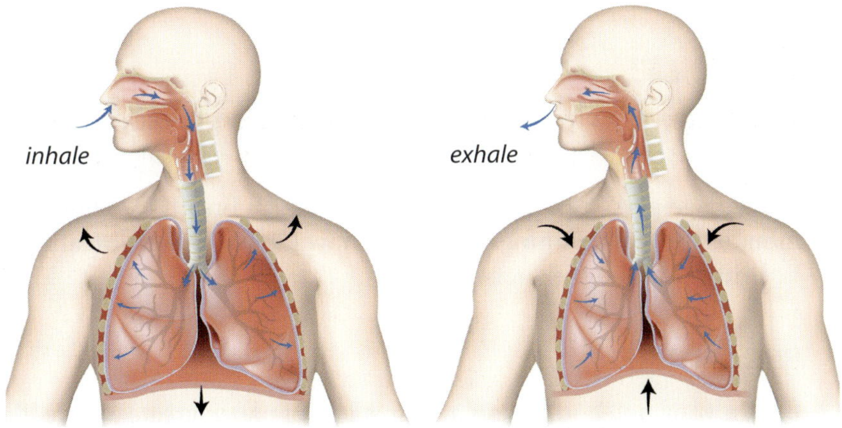

15-2

Breathing involves movement of both the diaphragm and the ribs.

You exhale by making your chest smaller. Perhaps you're tempted to think that your diaphragm pushes upward to force the air out, but it doesn't. Your diaphragm and rib muscles simply relax. As they do, your chest returns to its resting volume, forcing air out.

When resting, a person usually inhales and then exhales about a half liter of air with each breath. By taking a deep breath, you can inhale several times that amount. A person uses this extra lung capacity when breathing heavily during exercise, taking a deep breath to swim underwater, and so forth. If the alveoli become diseased and do not work properly, a person will become short of breath more quickly than normal.

THUNDERSTORM ASTHMA

CAN THUNDERSTORMS TRIGGER ASTHMA ATTACKS?

Thunderstorms are nothing new in Melbourne, Australia's second-largest city. But on November 21, 2016, a thunderstorm struck the city of nearly five million with a new kind of deadly fury. And the deaths were not due to lightning or high winds, as you might expect. In the storm's wake, some 8500 people swamped the emergency rooms of the city. All of them had had severe asthma attacks, and nine died.

Asthma is a common illness that occurs when a person's lungs become inflamed, often in response to tiny particles in the air such as pollen. During an asthma attack, a person suffers bouts of coughing, wheezing, and shortness of breath. In developed countries like Australia, deaths from asthma are rare. That's what made the Melbourne cases so shocking. But perhaps more surprising, this wasn't the first time such an event had happened. Scientists first became aware of what is called *thunderstorm asthma* after three outbreaks in the 1980s. A current theory is that the storms break pollen grains, especially rye grass pollen, into hundreds of smaller pieces that can be inhaled deep into a person's lungs. Many of the thunderstorm asthma victims had never had an asthma attack before, but they did have a history of hay fever, an allergic response to pollen.

Are allergy sufferers who live near rye grass lawns at risk of death due to thunderstorm asthma? Is rye grass pollen even the real culprit? Although the evidence suggests that the answer is "yes," neither of the questions has yet been scientifically tested. In the meantime, Melbourne residents will remain uneasy when they see thunderclouds build up during hay fever season.

LAB EXERCISE

HOLD IT!

How Does Exercise Affect Holding Your Breath?

Equipment
timer or stopwatch
lab notebook or paper
pencil or pen

Student Name	Time #1	Time #2

An average person in good health can hold his breath for about two minutes. But lots of things can affect how long a person's breath can be held. How about exercise?

PROCEDURE

❶ In your notebook or on a clean sheet of paper, create a table like the one shown on the left. Answer the questions in your notebook as well.

1. What do you think? After jogging, will you be able to hold your breath for a longer period of time or a shorter period of time?

❷ Take as big a breath as you can, and hold it for as long as you can. Keep your mouth closed and your nose pinched closed. Have a partner time how long you can hold this breath. Record your time in the *Time #1* column of your table.

❸ Jog in place rapidly for one minute.

❹ Again, take as big a breath as you can and hold it for as long as you can. Keep your mouth closed and your nose pinched closed. Have a partner time how long you can hold this breath. Record your time in your table in the *Time #2* column.

❺ Also record the results of each member of your group.

ANALYSIS

2. How does your ability to hold your breath after jogging compare with your ability after resting? Was there a difference?
3. How can you explain the results obtained by your group?
4. What do you think are some other factors that might affect how long a person can hold his breath?

When you exercise in a way that raises your heart rate and breathing rate, it's called *cardiovascular exercise*. Jogging is an example. It's important to exercise hard enough to get your heart and lungs moving but not so much that you overexert yourself. Training coaches suggest exercising hard enough so that you can't sing but can talk while exercising.

5. How is this recommendation related to lung capacity during exercise?

CONCLUSION

It's important for you to be able to listen to your body, especially when it comes to breathing. Doing so helps you to take care of the body that God has given you. Many processes rely on your ability to breathe, so it's important to exercise those lungs!

15A REVIEW QUESTIONS

1. What is the main organ of the respiratory system?
2. (True or False) Your breathing rate increases in response to cells needing more oxygen.
3. Place the following parts of the respiratory system in the correct order that air from the outside passes through them: bronchi, larynx, lungs, trachea.
4. Where in a lung does gas exchange actually take place?
5. What muscle is most responsible for moving air in and out of the lungs?
6. During which of the two parts of a breath are muscles contracting to do work?

15B CIRCULATION

How does my body move stuff around inside?

15B Questions

- Why do we need blood?
- How does blood flow help maintain homeostasis?
- What are the parts of the circulatory system?
- What path does blood take through the body?

15.4 THE BODY'S DELIVERY SYSTEM

Think of the muscles that allow your eye to scan the sentences on this page. They need oxygen, but they're not near your lungs. How does an alveolus in your lung get oxygen to those muscles in your eyes? For that matter, how does any cell in one part of your body get the things that it needs from other parts of your body? It turns out that your body has a built-in delivery service for moving substances around your body. It's called your *circulatory system.*

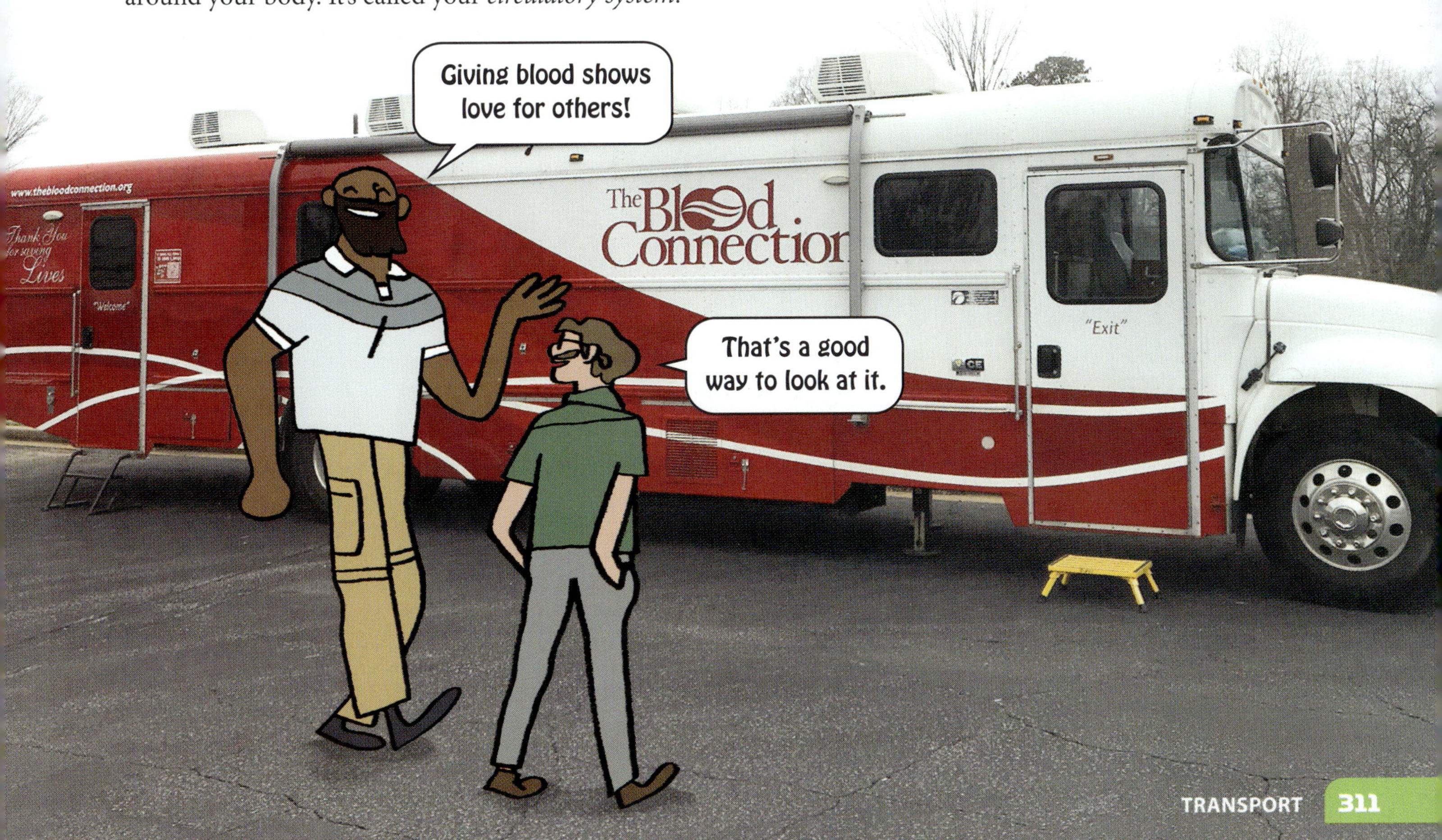

Your circulatory system plays a vital role in helping your body maintain homeostasis. If nutrients obtained from your food by the digestive system are not transported to cells, those cells will run out of energy and shut down. If wastes like carbon dioxide are not carried away to parts of the excretory system, the cells' environment quickly becomes toxic. The system is kept hard at work by a powerful pump—your heart—controlled by your nervous system. Day after day, your heart keeps things moving. How does this system work? Let's take a look!

THE CIRCULATORY SYSTEM

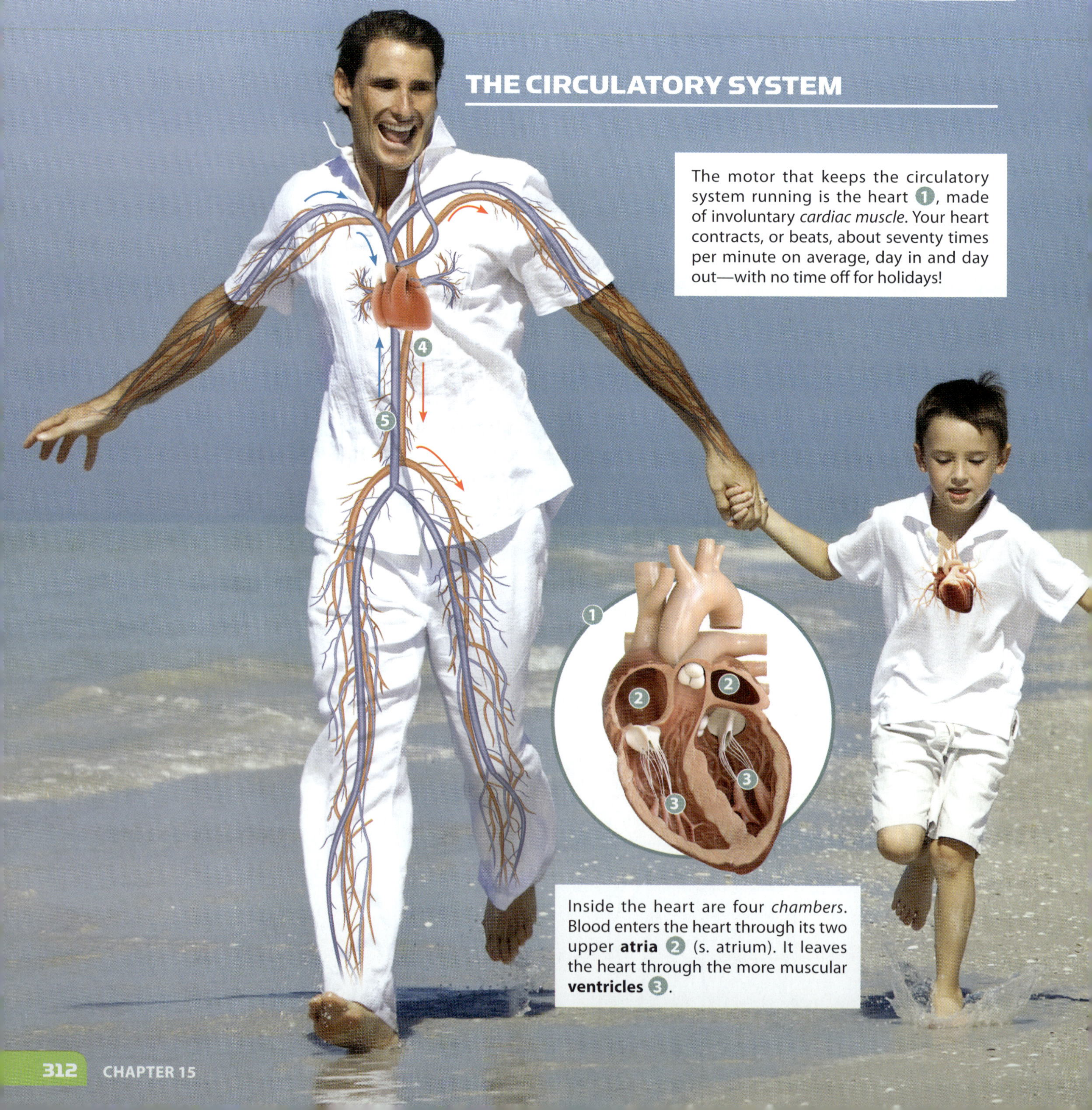

The motor that keeps the circulatory system running is the heart (1), made of involuntary *cardiac muscle*. Your heart contracts, or beats, about seventy times per minute on average, day in and day out—with no time off for holidays!

Inside the heart are four *chambers*. Blood enters the heart through its two upper **atria** (2) (s. atrium). It leaves the heart through the more muscular **ventricles** (3).

Blood travels through your body in a closed system of *vessels*. Blood flows *away* from the heart through **arteries** (4) and *returns* through **veins** (5).

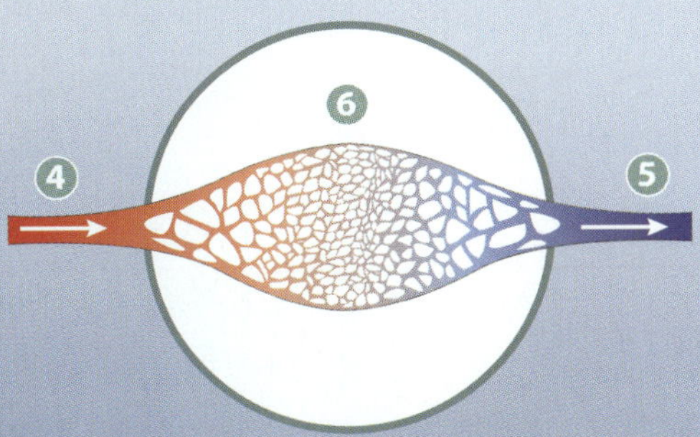

Arteries are connected to veins by tiny **capillaries** (6), the smallest vessels in your body. Some are so small that only a single red blood cell can pass through at a time!

The interior walls of both arteries and veins are lined with involuntary muscle. The walls of arteries are thicker, and the muscle contracts regularly to prevent the backflow of blood. Veins have thinner walls; they have additional help in preventing backflow in the form of **valves** (7)—flaps that allow blood to flow only in one direction. The flow of blood through your heart is also controlled by valves.

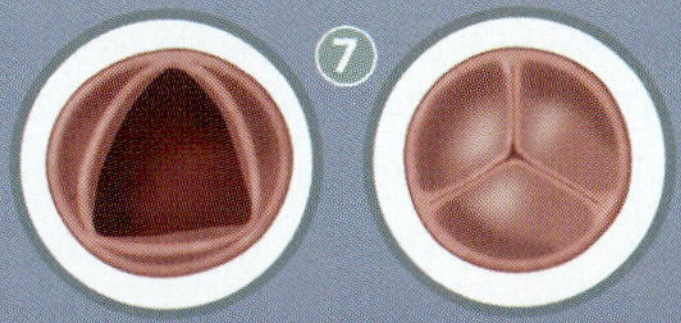

15.5 BLOOD

If you think of your circulatory system as a network of roads and highways, then blood cells are the vehicles traveling on them. Let's take a closer look at the components of blood.

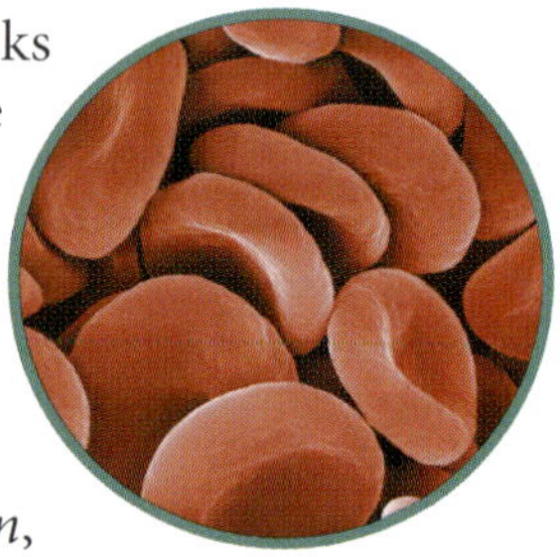

Red blood cells are the delivery trucks traveling through your body. Unlike your other cells, red blood cells don't have a nucleus, so they can't make copies of themselves. Instead, they are created by the red bone marrow inside your bones. Red blood cells contain large amounts of *hemoglobin*, the molecule that binds to and carries oxygen. Hemoglobin is what makes your blood appear red. Red blood cells carrying oxygen fresh from your lungs make blood appear bright red. Once most of this oxygen has been delivered to cells, your blood changes color to dark red. It's not really blue, as some people think—it just looks that way through your skin. Red blood cells wear out quickly, lasting only three or four months before needing replacement. Your bone marrow replaces about *two million* red blood cells every second!

White blood cells (which are also made by bone marrow) are the police patrol cars of your circulatory highway. They're much less common than red blood cells, being outnumbered about 600 to 1. White blood cells are constantly on the lookout for foreign invaders in the body, such as viruses or bacteria. How they deal with those invaders is covered in Chapter 16.

Platelets are your circulatory system's on-call emergency construction workers. Wherever there is a break in the system, such as a cut, platelets begin to clump together, forming a *clot*. This is the first step in your body's efforts to repair itself.

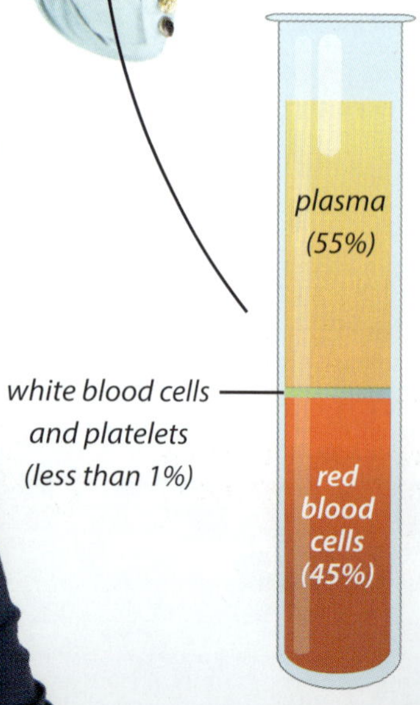

Slightly more than half the volume of your blood isn't made up of cells at all. Instead, it's plasma. **Plasma** is mostly water. In addition to carrying blood cells, plasma transports nutrients, minerals, proteins, hormones, and other substances. It takes about one minute for your plasma to make a complete trip through your circulatory system, carrying vital supplies to and from cells. In fact, plasma flows so efficiently through the system that red blood cells are made with no other means of moving themselves.

15.6 BLOOD FLOW

Speaking of flowing through the system, how does blood get from one part of your body to another? Let's examine the pathway that a red blood cell takes during its sixty-second jaunt through your body. We'll start at the beginning—at your heart!

As blood moves through your heart, it passes through a series of four chambers. Valves ① prevent the backflow of blood into the chambers so that blood will travel in only one direction. The *lub-dub* sound of a heartbeat is created when different heart valves snap shut.

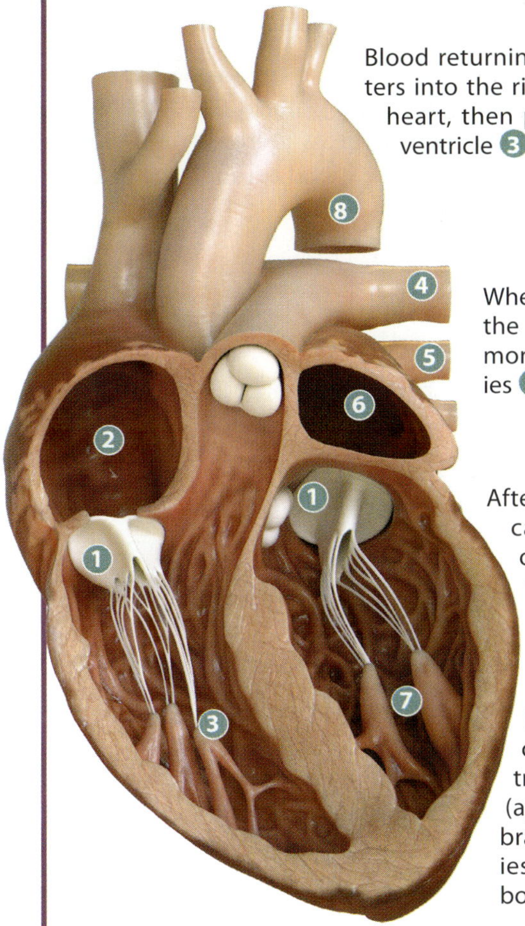

Blood returning from the body enters into the right atrium ② of the heart, then passes into the right ventricle ③.

When the right ventricle contracts, the blood moves through the pulmonary (PULL muh NEHR ee) arteries ④ and on to the lungs.

After the blood has dropped off its carbon dioxide and picked up oxygen at the lungs, it returns through the pulmonary veins ⑤ and enters the left atrium ⑥.

Blood next enters the left ventricle ⑦. As the left ventricle contracts, blood rushes into the aorta (ay OR tuh) ⑧. This large vessel branches into many smaller arteries to supply blood to the entire body.

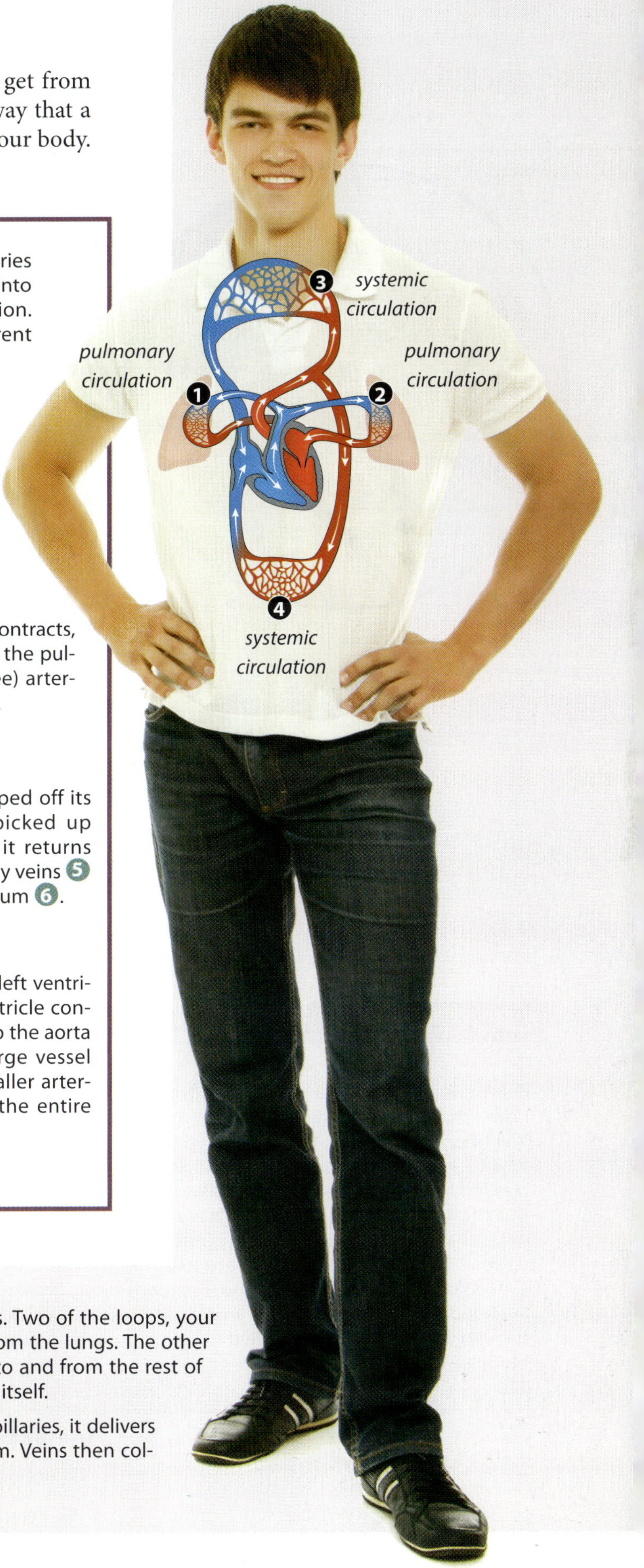

Your circulatory system consists of a number of loops. Two of the loops, your *pulmonary circulation* (❶ & ❷), carry blood to and from the lungs. The other two, your *systemic circulation* (❸ & ❹), carry blood to and from the rest of your body—including the muscle tissues of your heart itself.

As the blood in the systemic loops passes through capillaries, it delivers oxygen to cells and absorbs carbon dioxide from them. Veins then collect the oxygen-poor blood and return it to the heart.

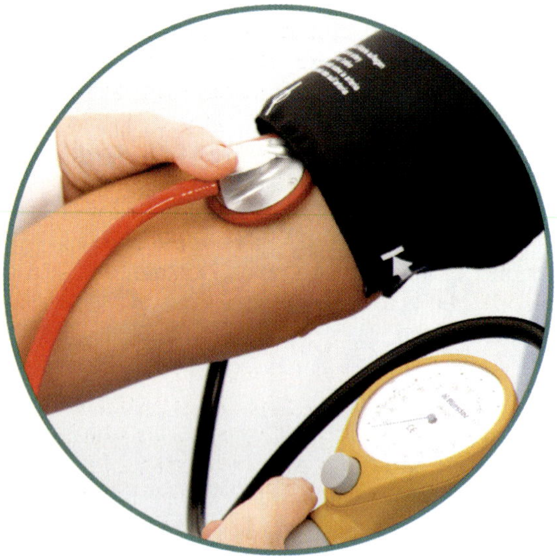

In the previous section, we saw how the circulatory system works together with the respiratory system to exchange carbon dioxide for oxygen. In this section, we see that the circulatory system works with several other systems as well. Blood cells are made inside your bones, which are part of your skeletal system. White blood cells fight disease as part of your immune system. The nutrients in plasma are the products of processes that take place in your digestive system. God has designed many marvelous systems that work together to do all the things necessary to keep your body alive and humming.

As the heart pumps, the blood passes into the arteries with such force that they swell slightly. You feel this force when you take your **pulse**. By pressing an artery against a bone, you can feel the force of blood pushing against the pressure caused by your finger. Your pulse rate tells you how fast your heart is beating.

The pressure of circulating blood against the walls of the arteries is called **blood pressure**. Even between heartbeats there is some pressure in the arteries. This pressure forces the blood through the arteries and on into the capillaries of the body.

15B REVIEW QUESTIONS

1. What is the purpose of the circulatory system?
2. Name, in order, the kinds of blood vessels that your blood passes through after leaving your heart.
3. What structures prevent the backflow of blood in the heart and veins?
4. List the four main components of blood.
5. Describe one way that your circulatory system works together with another system in your body.
6. After entering the heart, why is blood next sent to the lungs?
7. What causes blood pressure?

CASE STUDY SEGREGATED BLOOD?

Our ability to give and store blood for later use is somewhat new—the first blood banks were not established until World War I. On December 7, 1941, the United States entered World War II. As part of ramping up to go to war, the country needed to begin large-scale stockpiling of donated blood. But the United States Army and Navy informed the Red Cross that they wanted blood only from white donors. The nation's prevailing ideas about race extended beyond just skin color. Ironically, the man most responsible for making large blood banks possible was an African American, Dr. Charles Richard Drew. The American Red Cross continued to segregate blood until 1950.

Use the Case Study above to answer Questions 8–9.

8. Why did people think that blood should be segregated?
9. Read Acts 17:26. What light does this verse shed on the question of race?

NEEDLEWORK: SERVING AS A PHLEBOTOMIST

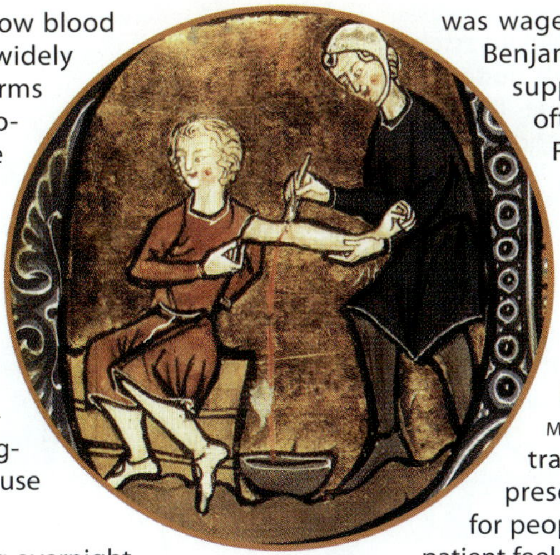

People haven't always understood how blood circulates in the body. It was once widely believed that blood pools in one's arms and legs and becomes stagnant. People thought that this could cause fevers, headaches, and even acne! Doctors in those days treated these conditions by removing some of the patient's blood, a practice called *bloodletting*. Today we don't normally use this treatment for disease, but people do still need to sometimes have their blood drawn. Sometimes blood is collected to help diagnose a disease. Blood is also drawn to use in transfusions.

Doctors didn't give up on bloodletting overnight. In the early nineteenth century, a kind of "doctors' war" was waged over the practice. Some, like Dr. Benjamin Rush (1746–1813), were strong supporters of bloodletting and used it often on their patients. Later studies by France's Dr. Pierre Louis (1787–1872) showed that bloodletting was not as effective as earlier physicians had thought. Despite this, some doctors continued to let blood even into the early twentieth century.

A person who is trained to draw blood is a *phlebotomist* (fleh BOT uh MIST). By collecting blood for testing or transfusions, a phlebotomist can help preserve and improve the quality of life for people. A good phlebotomist can help a patient feel better about what is often a dreaded but necessary procedure.

15C THE LYMPHATIC SYSTEM

What does the lymphatic system do?

15C Questions

- What is lymph?
- How does lymph help maintain homeostasis?
- What are the parts of the lymphatic system?
- How does lymph flow through the body?

15.7 LYMPH

Your circulatory system is not without its issues. Most importantly, its capillaries leak. This isn't a design flaw—there's actually a good reason for it. In order for gases to be exchanged between your blood and your body cells, your capillary walls are only one cell layer thick, and the cells don't fit together tightly. Red blood cells are too large to slip out through the gaps—but plasma can and does. In fact, about *twenty liters* of plasma leaks out every day! Happily, about seventeen of those liters get reabsorbed right back into your bloodstream. But what about the other three liters? That's where your *lymphatic system* comes in. One of its jobs is to collect some of the leaked plasma, called **lymph** while it is in the lymphatic system, and return it to your heart. Along the way, the lymphatic system cleanses the lymph of possible pathogens. Because of this, the lymphatic system plays an important role in your body's defense against disease. You'll read more about that defense in the next chapter.

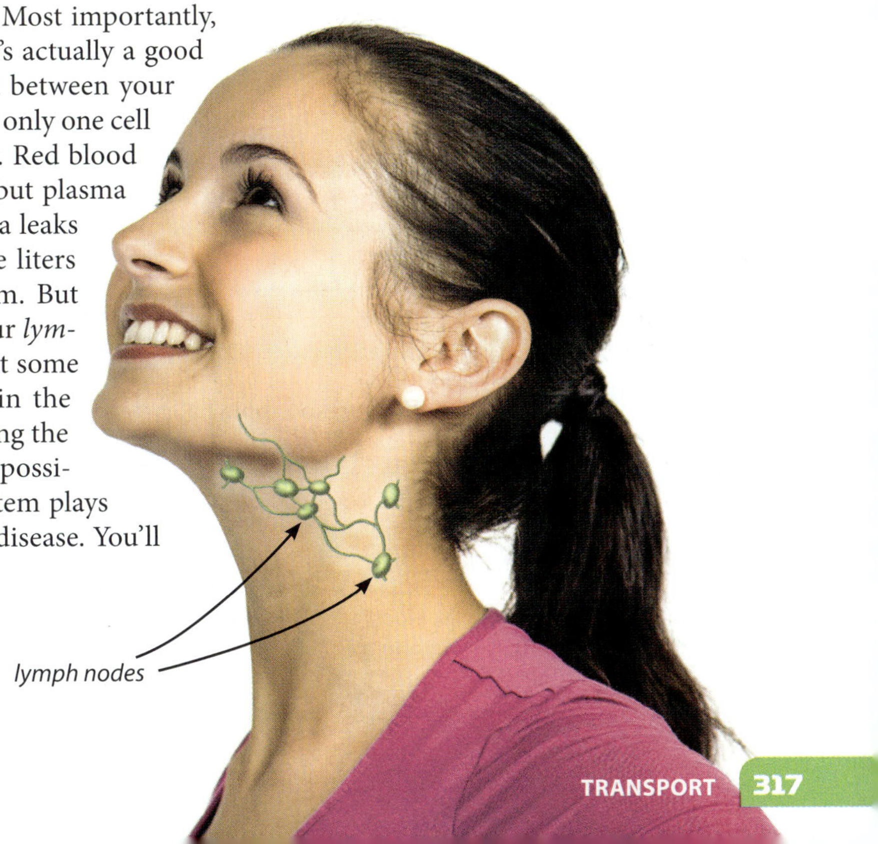

Let's explore how the structure of the lymphatic system defends your health.

THE LYMPHATIC SYSTEM

Lymph flows through *lymph vessels* 1 and passes through **lymph nodes** 2 along the way. The lymph nodes filter dead cells, bacteria, viruses, and other particles out of your lymph.

Your *spleen* 3 recycles worn-out red blood cells and produces and stores certain white blood cells.

Other white blood cells mature in your *thymus* 4. This organ is most active during your childhood and greatly reduces in size during your early teen years.

Your lymph eventually travels to one of two large veins on either side of the body near your collarbones. At this point, lymph rejoins your blood stream.

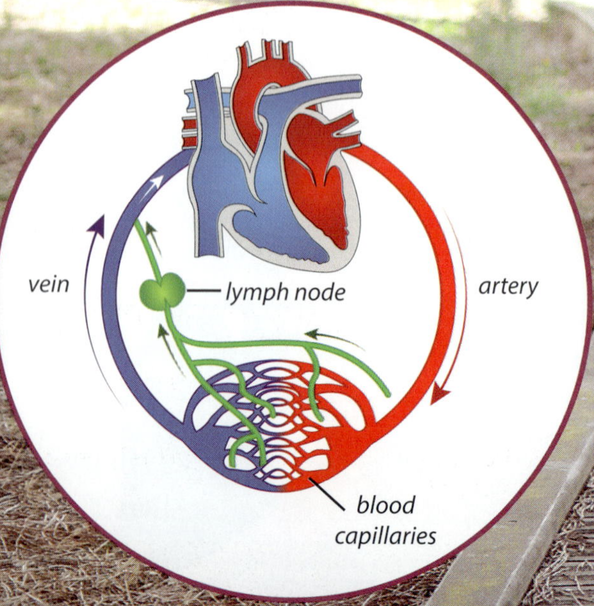

15.8 LYMPH FLOW

Your lymphatic system doesn't have a built-in pump. There's no "heart" that keeps lymph flowing. So how does it get from one part of your body to another? Answer: your muscles! The vessels that carry lymph through your body pass through the muscles of your muscular system. As your muscles contract and relax, the lymph vessels within them get squeezed. This squeezing forces the lymph through the lymph vessels. It's yet another wonderful example of how God has designed the systems of your body to work together to maintain homeostasis.

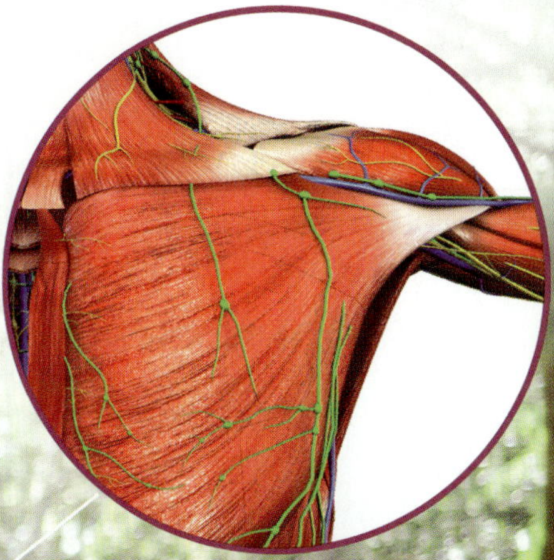

15C REVIEW QUESTIONS

1. What is the origin of lymph?
2. How does the lymphatic system work with the circulatory system?
3. Where in the lymphatic system are particles filtered out of the lymph?
4. Which lymphatic system organ is much smaller in adults than in children?
5. How is lymph circulated through the lymphatic system?

CHAPTER SUMMARY

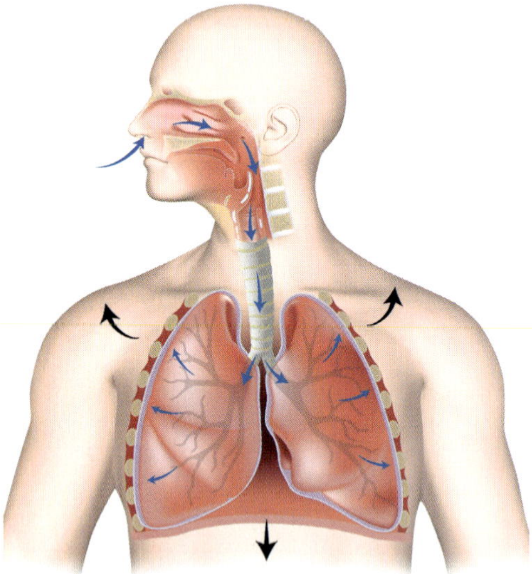

15A BREATHING

- The respiratory system obtains the oxygen needed for cellular respiration and rids the body of carbon dioxide.
- Gas exchange between the body and its environment takes place in the lungs.
- Air enters the body through the nasal cavity, then passes through the pharynx, larynx, trachea, and bronchi before entering the lungs.
- The work of breathing is done by the actions of the diaphragm.

15A TERMS	
nasal cavity	306
pharynx	306
larynx	306
trachea	307
bronchus	307
alveolus	307

15B CIRCULATION

- The circulatory system moves nutrients, oxygen, wastes, and other substances from one part of the body to another.
- The circulatory system consists of a pump, the heart, that moves blood through a system of arteries, capillaries, and veins.
- Red blood cells pick up oxygen in the lungs and transport it to cells. They pick up carbon dioxide at the cells and return it to the lungs.

15B TERMS	
atrium	312
ventricle	312
artery	313
vein	313
capillary	313
valve	313
red blood cell	314
white blood cell	314
platelet	314
plasma	314
pulse	316
blood pressure	316

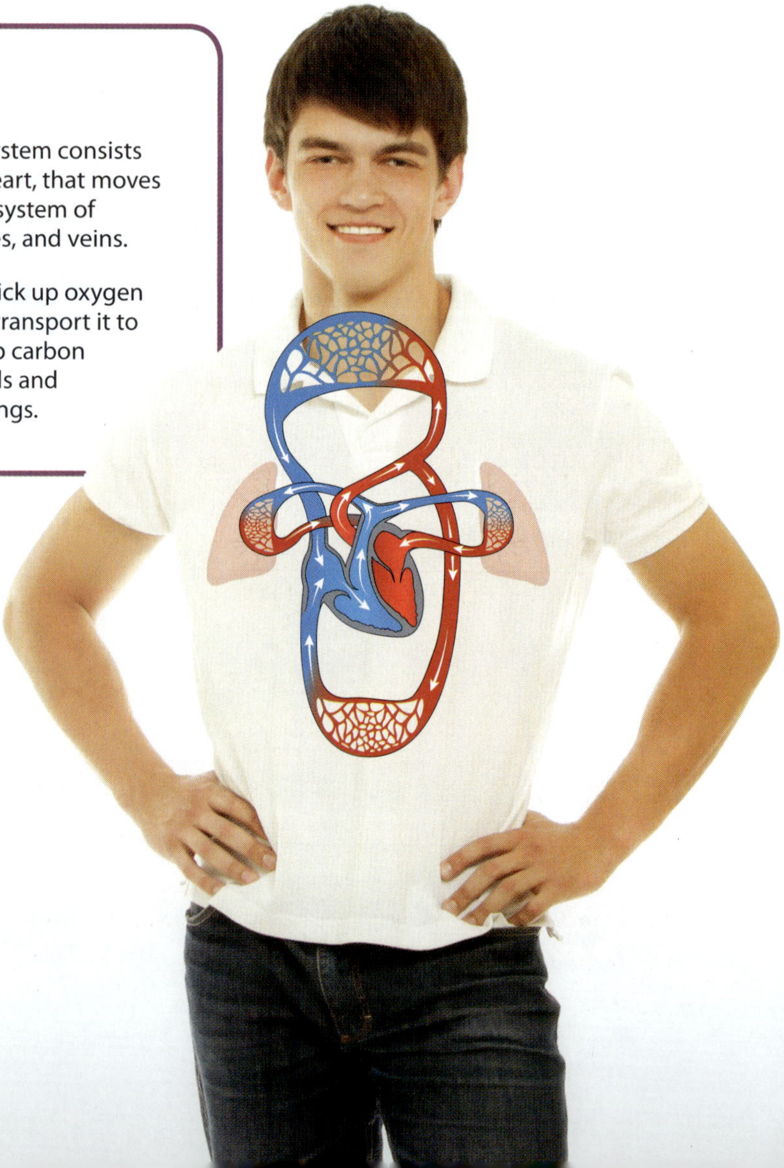

CHAPTER SUMMARY

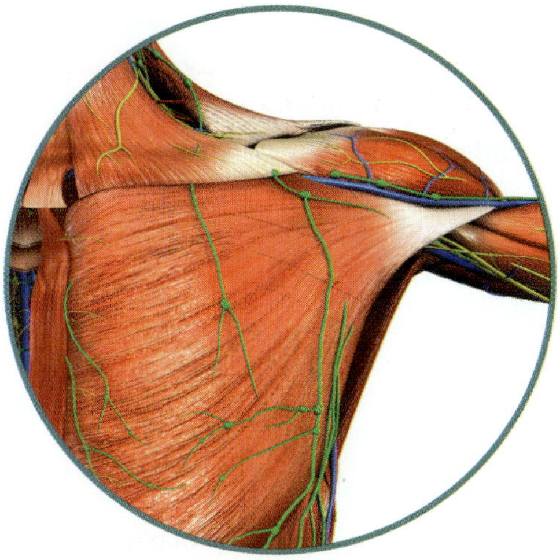

15C THE LYMPHATIC SYSTEM

- The lymphatic system consists of lymph vessels, lymph nodes, and several accessory organs.
- The lymphatic system returns plasma that has leaked from the circulatory system back to that system.
- Lymph nodes filter dead cells, viruses, bacteria, and other particles from lymph.
- Lymph is moved through lymphatic vessels by actions of the muscular system.

15C TERMS	
lymph	317
lymph node	318

CHAPTER REVIEW QUESTIONS

Recalling Facts

1. What is the primary function of the respiratory system?
2. What happens to air as it passes through the nasal cavity?
3. What two organ systems join at the pharynx?
4. The trachea branches into two smaller passageways called ______.
5. How does oxygen in the alveoli enter into the capillaries of the circulatory system?
6. (True or False) A person's lung capacity is the amount of air that he can breathe in during a normal breath.
7. List the components of blood in order from the greatest to least percentage.

Match each blood component in Questions 8–11 with its function.

- **a.** plasma
- **b.** red blood cells
- **c.** white blood cells
- **d.** platelets

8. carries oxygen and carbon dioxide
9. functions in clotting
10. removes foreign substances, such as viruses
11. transports blood cells, nutrients, hormones, and other substances
12. What organs do pulmonary arteries and pulmonary veins connect?
13. Blood enters into the heart's ______ and exits from its ______.
14. What causes blood pressure?
15. From which blood component is lymph derived?
16. To which organ system is lymph returned?
17. What is the chief function of lymph nodes?

The current record for free diving (holding one's breath and diving without scuba gear) is four minutes and twenty-four seconds.

18. What is one function of the spleen that is *not* also a function of the thymus?

Understanding Concepts

19. Draw a sketch of the respiratory system. Label the nasal cavity, pharynx, larynx, trachea, bronchi, and lungs.

20. Place the following events in the order in which they occur using the numbers provided.

    1. the diaphragm relaxes
    2. air enters the lungs
    3. the diaphragm contracts
    4. air exits the lungs
    5. oxygen is exchanged for carbon dioxide

21. How does a person breathe even if he is not awake?

22. Compare the ways that arteries and veins maintain the flow of blood in one direction.

23. Why is the pulmonary vein called a vein even though it carries oxygenated blood?

Use the diagram to the right to answer Questions 24–25.

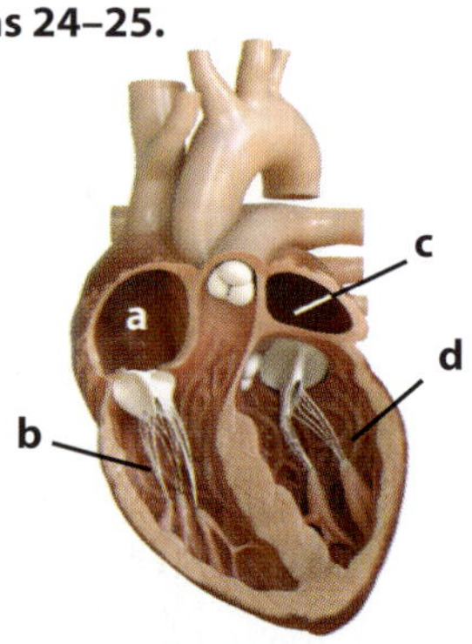

24. Which chambers are atria?

25. From which chamber does blood leave the heart to travel through the body?

26. Describe how the lymphatic system works closely with two other organ systems.

Critical Thinking

27. Why does a person's breathing rate decrease while he is asleep?

28. Why is blood considered a tissue even though it is a liquid?

29. What might happen if a person has weak or defective heart valves? What would be the effect of this condition?

30. Why is the human circulatory system described as a double circulatory system?

31. Why do people often experience swollen lymph nodes when they are sick?

32. Create a T-chart that shows the similarities and differences between the circulatory and lymphatic systems.

ETHICS Organ Donation

The year 1954 gave us the first successful kidney transplant. Since then, thousands of people have received the gift of a new kidney. As new advances in medicine are made, the list of human organs and tissues that can be transplanted continues to grow. Today, doctors can transplant skin, bone, hearts, lungs, and many other body parts—even entire hands and faces. But there are problems. The need for donations often far exceeds the availability of donors.

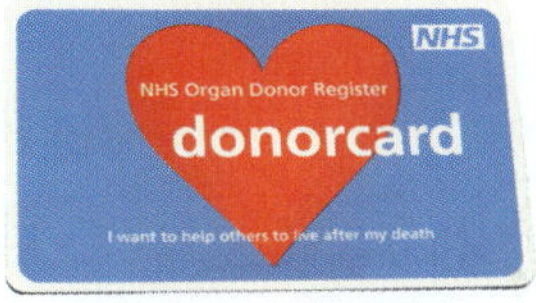

Most donors are people who have only just been declared dead, limiting the availability of tissues and organs. On top of this is the question of defining when a person has truly died. Modern life support systems can keep a person "alive" long after he has no measurable brain activity. Where's the balance between holding out hope for a person who may recover from a nonresponsive state and withholding hope from a person in dire need of a transplant?

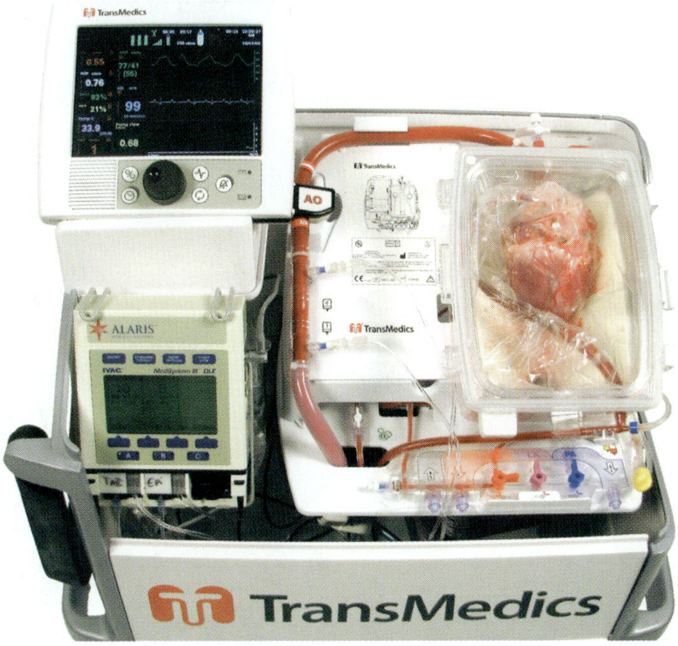

Use the Ethics box above to answer Question 33.

33. Write a three-paragraph response on the ethical use of donated organs. In the first paragraph, discuss principles from Scripture that may guide Christians in this area. In the second paragraph, discuss the possible outcomes of heeding or ignoring the guidelines of Scripture. In the third paragraph, give your opinion on how organ donations should be used.

16 CONTROL

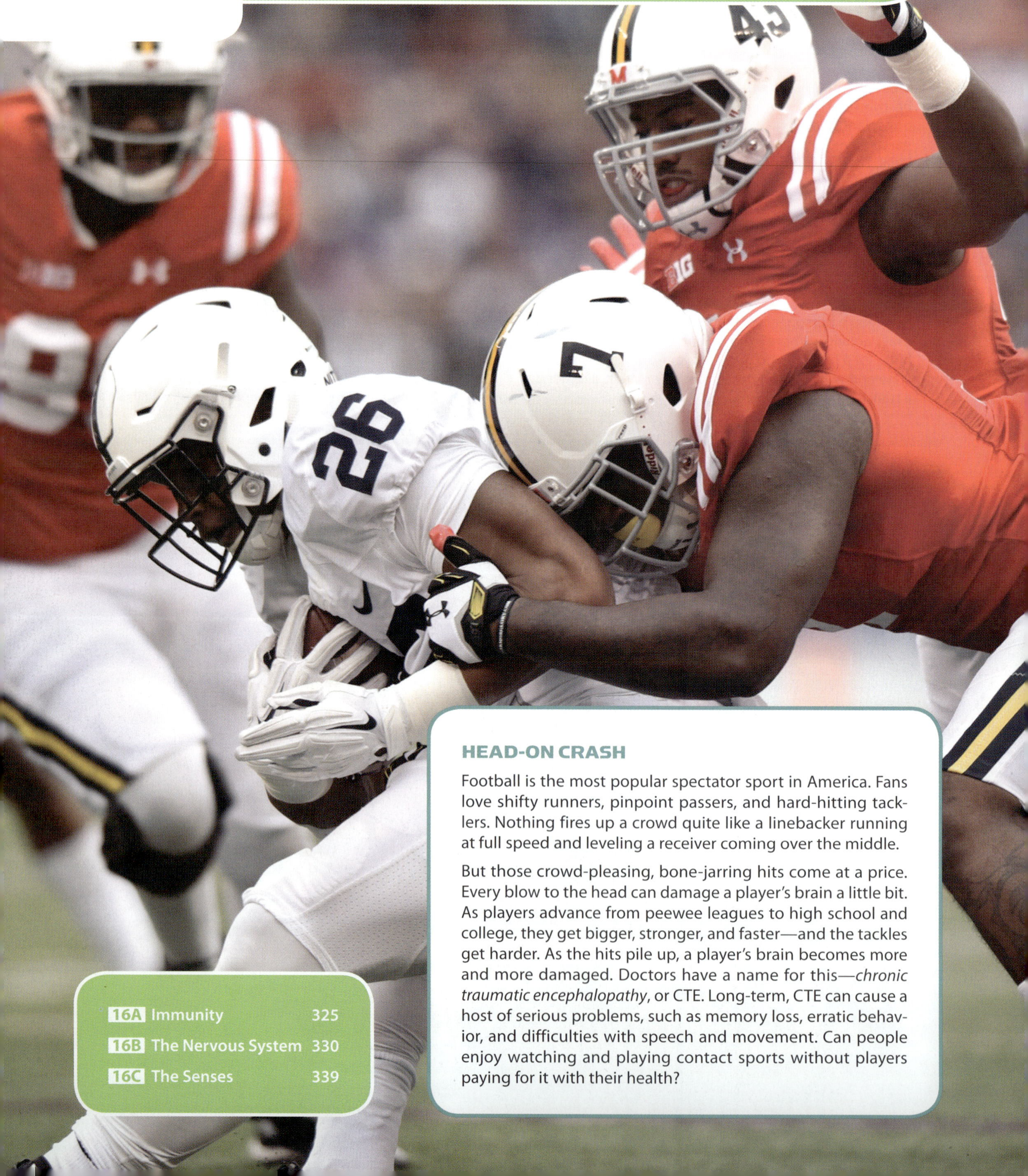

HEAD-ON CRASH

Football is the most popular spectator sport in America. Fans love shifty runners, pinpoint passers, and hard-hitting tacklers. Nothing fires up a crowd quite like a linebacker running at full speed and leveling a receiver coming over the middle.

But those crowd-pleasing, bone-jarring hits come at a price. Every blow to the head can damage a player's brain a little bit. As players advance from peewee leagues to high school and college, they get bigger, stronger, and faster—and the tackles get harder. As the hits pile up, a player's brain becomes more and more damaged. Doctors have a name for this—*chronic traumatic encephalopathy*, or CTE. Long-term, CTE can cause a host of serious problems, such as memory loss, erratic behavior, and difficulties with speech and movement. Can people enjoy watching and playing contact sports without players paying for it with their health?

16A
IMMUNITY

How does my body fight disease?

16.1 THE BODY'S DEFENSES

Contact sports like football can cause severe health problems, but a person can always opt not to play. In our fallen world, though, humans face real health threats that they can't just walk away from. We need to understand the body's limits and its defenses so that we can help protect God's image bearers.

Viruses, bacteria, and microorganisms are lurking all around us. Each one has the potential to invade a human body and wreak havoc. But God didn't leave His image bearers without any shield against disease. He has built into our bodies a superb network of defenses designed to detect, isolate, and destroy harmful substances. This defensive network is known as the *immune system*. The long-term resistance to disease that this system provides is called *immunity*.

16A Questions

- What are my body's lines of defense?
- What is an inflammatory response?
- What is the difference between active and passive immunity?

Your immune system consists of two kinds of defense. The first kind is designed to keep out any sort of disease-causing agent, or *pathogen*. Since this first type of defense isn't picky about what pathogens it will defend against, we call it a *nonspecific defense*. We call the other kind of defense a *specific defense* since it consists of cells that are trained to find and attack only certain kinds of pathogens. These two kinds of defensive systems combine to make a potent one-two punch for knocking out threats. Let's take a look at some of the key pieces in your body's lines of defense.

NONSPECIFIC DEFENSES

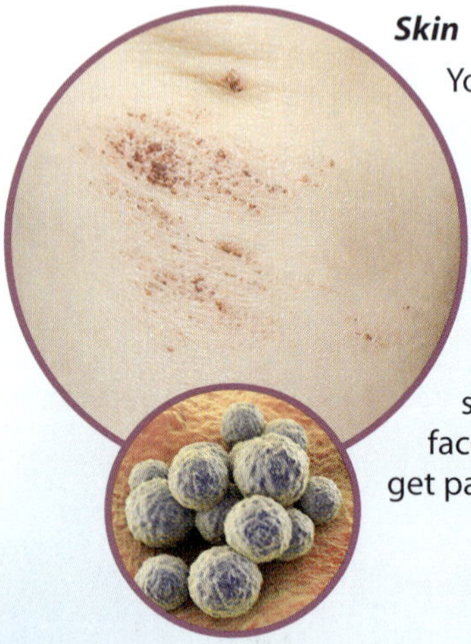

Skin

Your body's first line of defense against invaders is its skin. As we saw in Chapter 13, your integumentary system provides a protective covering for the outside of your body. Although the skin is thin, the connections between its cells are very strong. That makes it tough for germs to get through. The skin also makes some chemicals that kill germs on its surface. Usually the only way an organism can get past your skin is through a cut or scrape.

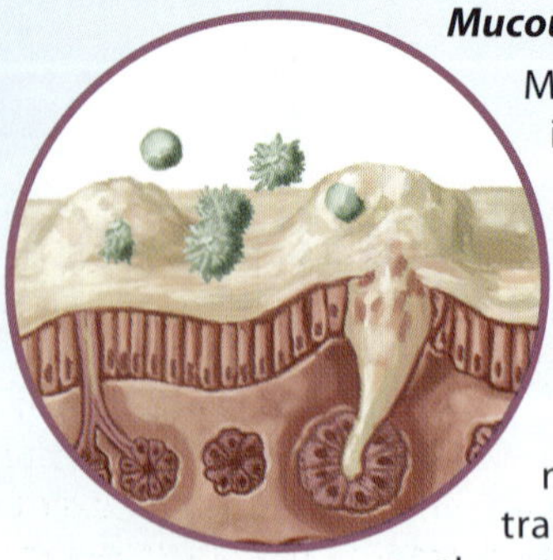

Mucous Membranes

Mucous membranes line the insides of body cavities such as your mouth and nose. These membranes produce sticky *mucus* that traps foreign matter. Some mucous membranes have cilia that move the mucus and trapped materials to the throat where they are swallowed. Stomach acid then kills any germs trapped in the mucus. Like the skin, this barrier can be breached only through small tears or scrapes.

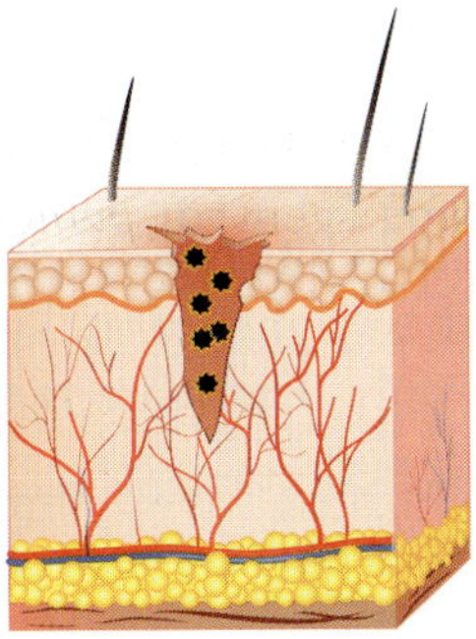

Inflammation

If a pathogen gets through the skin or mucous membranes, the next line of defense springs into action—inflammation. **Inflammation** is a local response to invaders. When cells are damaged, they release special chemicals that cause changes in nearby blood vessels. Inflammation deals with invaders that affect only a small area.

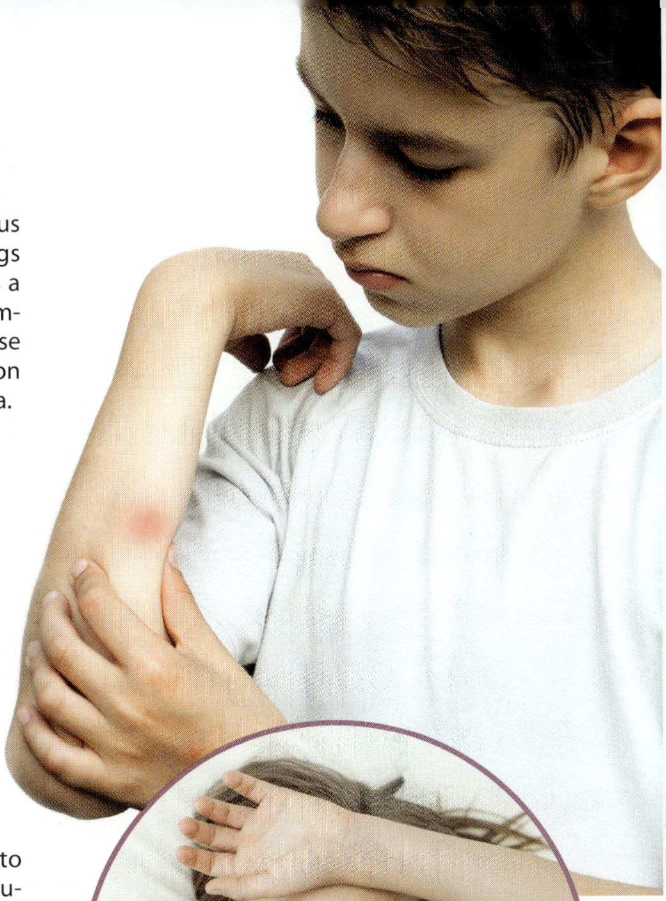

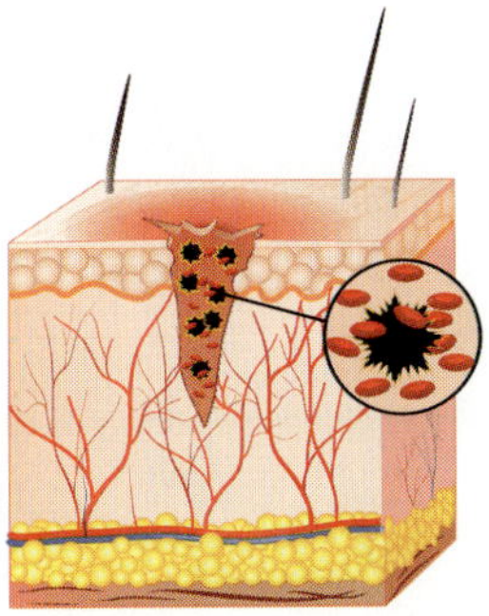

First, the blood vessels in the area increase in size to supply more blood to the area.

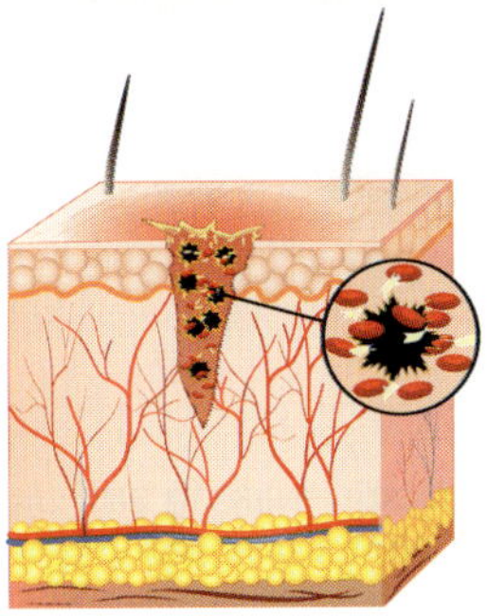

Second, the capillaries allow more plasma to reach the affected site. The plasma brings nutrients and other substances to fight the pathogens. If the inflammation site is near the surface of the skin, the area may be swollen, red, painful, and slightly warmer than the surrounding skin.

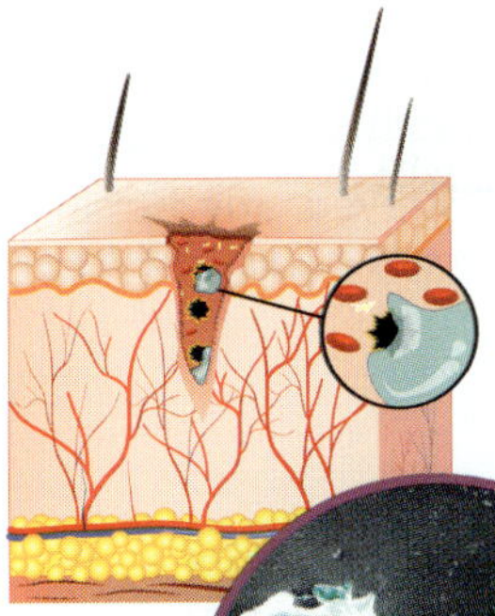

Third, special white blood cells called macrophages slip through the capillary walls. The macrophages attack the foreign organisms and kill them. How do the macrophages know what to attack? All organisms contain chemicals called **antigens** on their cell membranes. Your body recognizes antigens from other organisms as "foreign." Macrophages key in on those foreign antigens and attack them.

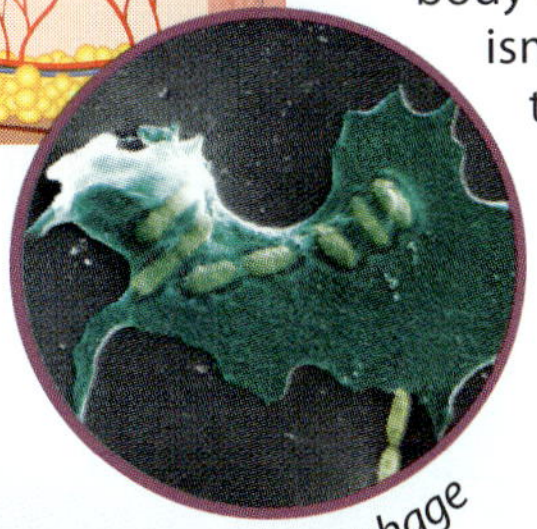

macrophage

Fever

If the invader multiplies and travels to affect the whole body, the body responds by causing a *fever*. A fever is an increase in the body's normal temperature. You might think that a fever is bad. But the temperature increase often slows the rate at which the invaders can reproduce. Bacteria, for example, have very narrow temperature ranges in which they grow best. Of course, if the temperature goes too high or is raised for too long, the body's tissues may be damaged.

16.2 SPECIFIC DEFENSES

The body's final line of defense is a group of special white blood cells called *lymphocytes* (LIM fuh SITES). They pick their fights carefully, responding to and attacking specific pathogens only. There are two types of lymphocytes—T cells and B cells.

IMMUNE RESPONSE

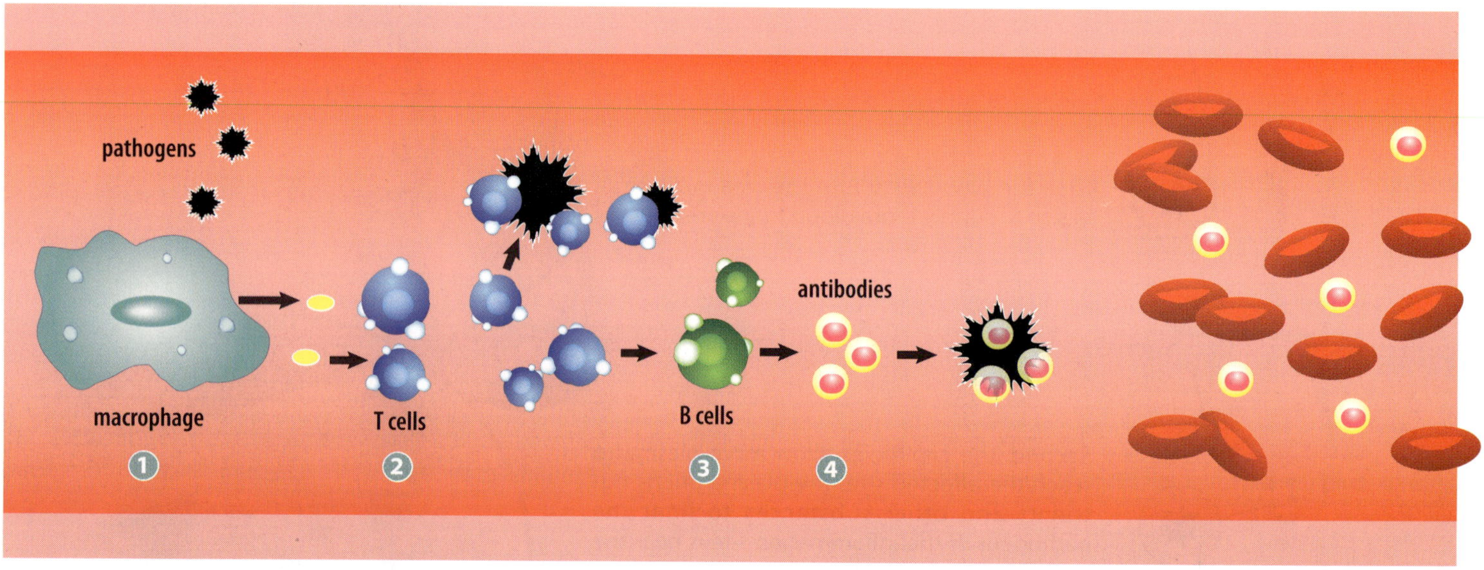

Millions of **T cells** exist in the blood. Once a macrophage ❶ identifies a foreign antigen on an invading pathogen, the macrophage releases a chemical that causes T cells to spring into action.

First, T cells ❷ quickly multiply. Some of the T cells attack the pathogen. Others attack infected body cells. Still other T cells alert **B cells** ❸ that a foreign antigen has been found.

Once activated by the T cells, B cells begin making antibodies ❹. **Antibodies** are proteins that react against specific antigens. The antibodies destroy the pathogen or disable it so that other cells can destroy it. Antibodies are so specific that an antibody that destroys one pathogen may have no effect on another.

Once the pathogen has been destroyed, any surplus antibodies remain in the bloodstream in case they are needed again.

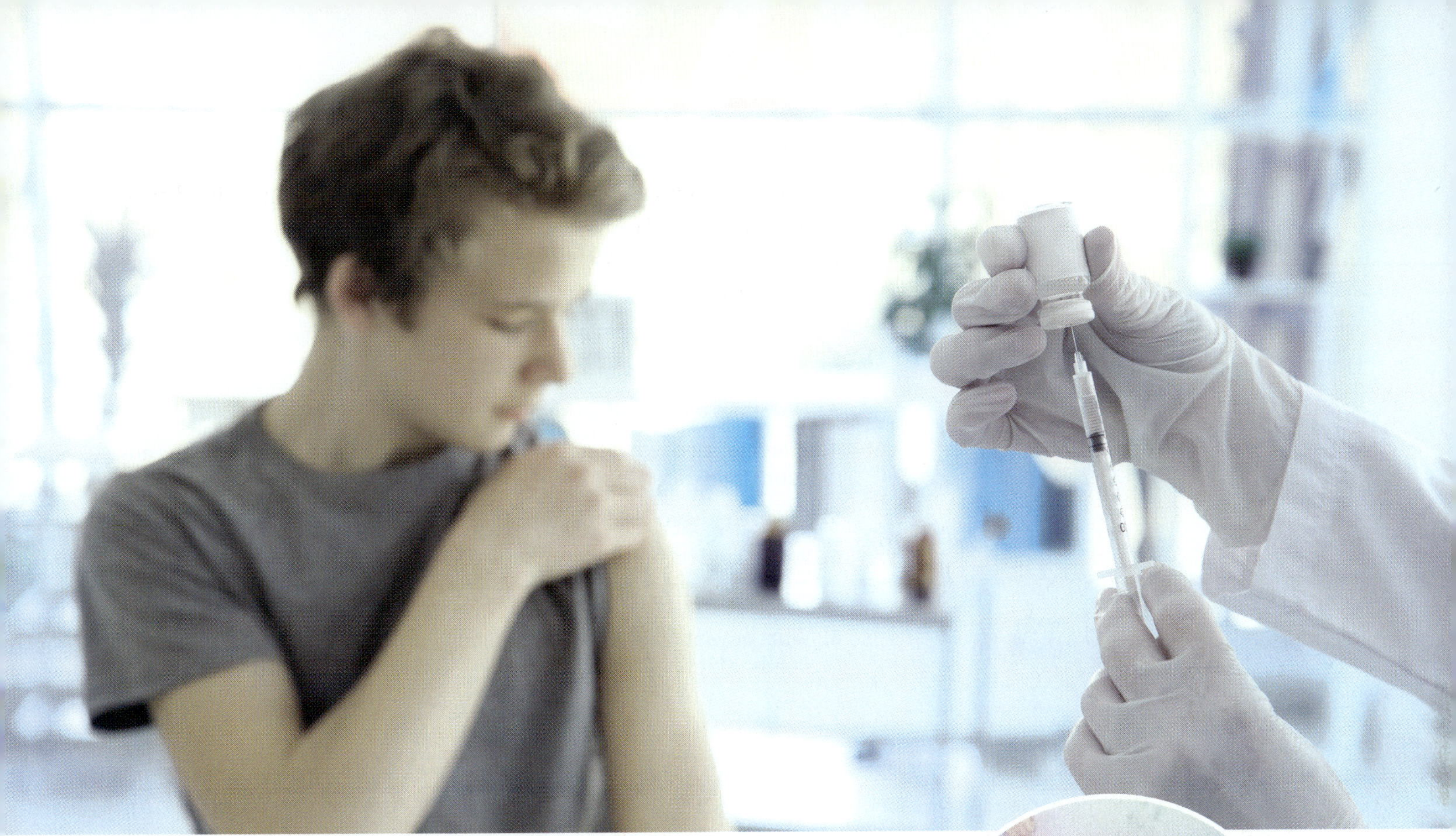

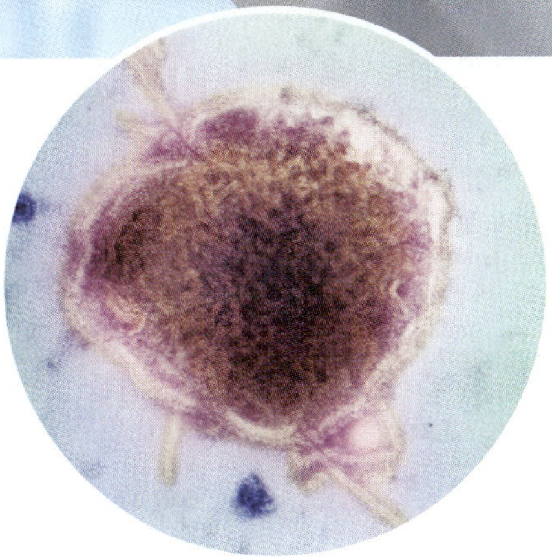

16.3 ACTIVE AND PASSIVE IMMUNITY

When the fight against a disease-causing agent is over, your body doesn't just lower its guard. It remains alert in case the same pathogen comes around again. Besides the antibodies created during the initial infection, some surplus T cells and B cells are also produced. These *memory cells* can "remember" the pathogen that caused the disease. Instead of having to start a new immune response from scratch, antibodies and *memory cells* can respond quickly and stop a pathogen cold. This fast-acting defense built up as a response to a prior infection is your body's **active immunity**.

Another way to develop active immunity is to receive a *vaccine*. Vaccines contain dead or weakened pathogens that still contain antigens. Your body recognizes the antigens as foreign and kicks your immune system into gear. Since the vaccine contains a weakened form of the pathogen, you don't get a full-blown case of the disease. If you meet that pathogen in the future, your body has the weapons to fight it off quickly before you become sick.

A second kind of immunity occurs when you receive ready-made antibodies from an outside source. Since your body receives the antibodies, but doesn't make them, we call this **passive immunity**. This is the type of immunity that newborn babies receive when antibodies are passed to them from their mothers. These antibodies don't last long, so infants often receive vaccines to strengthen their immune systems against common childhood diseases. People can also be treated with antibodies from humans or animals. Since these antibodies also do not last long, booster shots are needed to maintain the immunity they provide.

16A REVIEW QUESTIONS

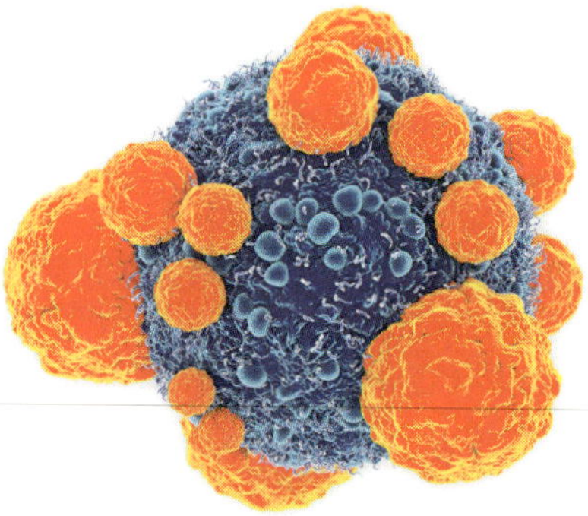

1. Why is your skin considered a nonspecific form of defense?
2. Which of the following are examples of an inflammatory response?
   - **a.** a swollen cut
   - **b.** swollen eyes from pink eye
   - **c.** swollen eyes from allergies
3. How does a fever aid an immune response?
4. When inhaled, what lines of defense would an airborne cold virus encounter (in order) in the human body?
5. Compare the roles of macrophages, T cells, and B cells.
6. Which kind of immunity is produced as a result of a prior infection?
7. Which kind of immunity is produced as a result of a vaccination?
8. What advantage does active immunity provide compared with passive immunity?

16B Questions

- What are the parts of the nervous system?
- How do nerves relay messages?
- How does my brain control my body?
- How does head trauma affect an athlete's health?

16B THE NERVOUS SYSTEM

How does my body control itself?

16.4 COMMAND AND CONTROL

Your body is made of a large number of parts that do very different jobs. Your stomach digests food, your muscles move you around, your lungs exchange gases, and so on. Some of these jobs—like walking from one store to the next at the mall—you have to think about in order to get them done. Others get done whether you think about them or not, like breathing. Directing all these tasks and making sure that they get done efficiently and on time requires a vast network of command and control. That network is your *nervous system*. This system not only controls your bodily movements and internal functions—it's also responsible for your awareness, emotions, and will.

If you think about it, what your body is really doing is gathering, processing, and responding to information from its environment. Each piece of information that your body can react to is called a *stimulus* (p. stimuli). How your body responds to a stimulus is called a *response* (makes sense, doesn't it?).

The basic building blocks of your nervous system are several different kinds of neurons, or nerve cells. *Neurons* are cells that can create or transfer *nerve impulses*, the electrical signals that control your body. Although they have different shapes and sizes, all neurons have the same basic parts.

The largest part of a neuron is the *cell body* (1), which contains cytoplasm, organelles, and a nucleus (2).

Branching off the cell body are short extensions called *dendrites* (3). These receive impulses and carry them toward the cell body.

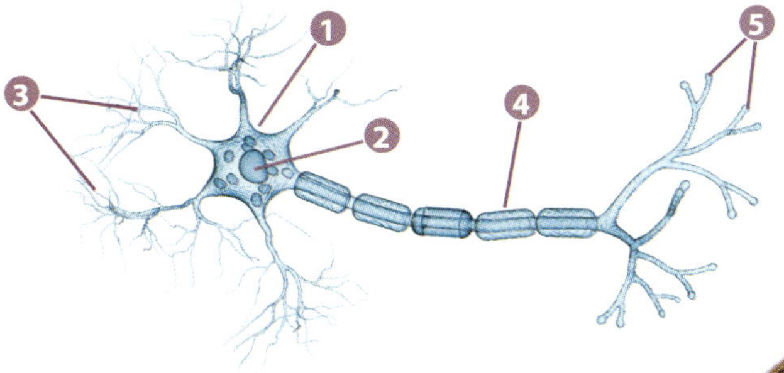

Axons (4) are usually much longer than dendrites and carry impulses away from the cell body. Most neurons have several dendrites but usually only one axon.

At the far end of the axon are branched extensions (5) that can transfer the impulse to other cells, including the dendrites of other neurons.

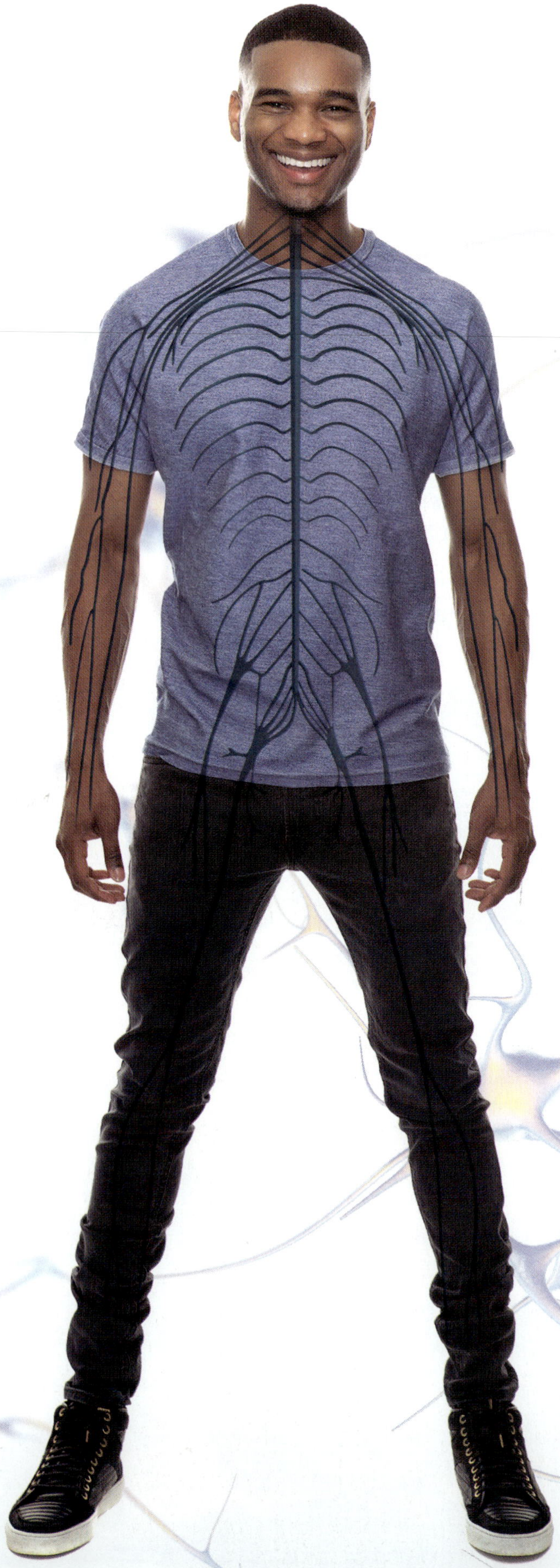

Your body has three kinds of neurons. *Sensory neurons* pick up information about your environment and transmit it to your brain or spinal cord. *Motor neurons* carry signals from your brain or spinal cord to the parts of your body. *Interneurons* connect sensory and motor neurons.

It usually takes more than one neuron to transmit an impulse to or from your brain. But neurons don't connect directly to one another, so how does the impulse get from one neuron to the next? A nerve impulse travels along the cell membrane of a neuron. When it reaches an axon tip, it runs into a small gap between that axon and the dendrites of the next neuron. That gap is called a *synapse*. The neuron carrying the impulse emits special chemicals that bridge the synapse. When the chemicals reach the next neuron, that neuron "fires," sending along an impulse of its own. Motor neurons use these chemical messengers to trigger muscle contractions.

Neurons combine to form the larger structures of the nervous system, such as your brain and spinal cord. Those two organs together make up the **central nervous system**. The remainder of the nervous system is called the **peripheral** (peh RIF er al) **nervous system**. Neurons that branch off the central nervous system are bundled together in protective coverings. These long, thin structures are nerves. They lead to sensory organs, muscles, skin, and many other structures of the body.

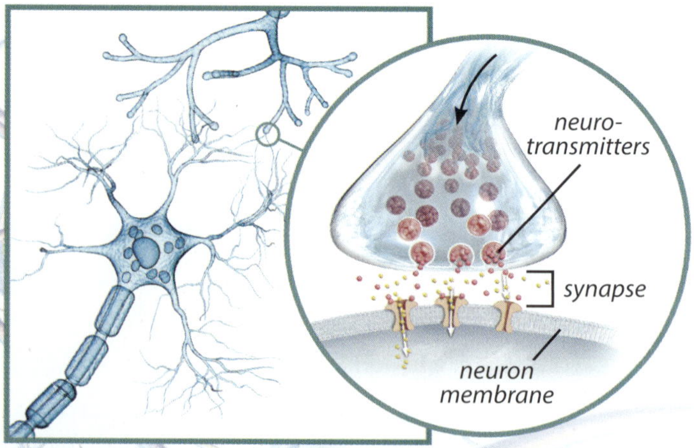

16.5 THE CENTRAL NERVOUS SYSTEM

The Brain

The main control hub for your nervous system is your brain. An adult human brain weighs only about 1.4 kg, but it has over sixteen *billion* neurons. Twelve pairs of *cranial nerves* branch from the brain to various structures of the head, neck, and internal organs. These nerves carry impulses to control the actions of those parts of the body. The brain is divided into three main parts: the cerebrum (suh REE brum), the cerebellum (SEHR uh BEL um), and the brain stem.

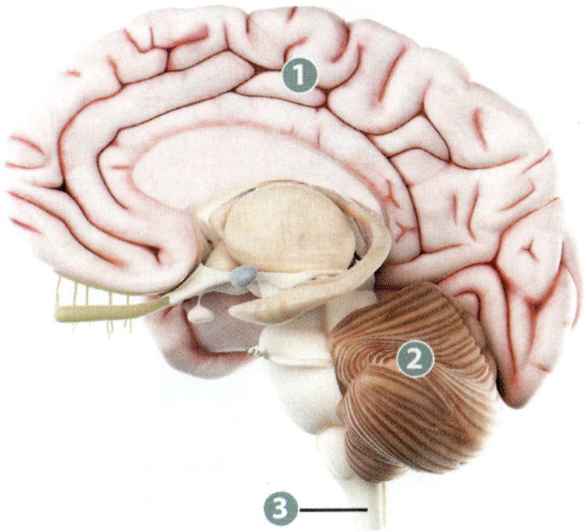

16-1
The cerebrum 1, cerebellum 2, and brain stem 3 of a human brain

The Cerebrum

The **cerebrum** is the largest portion of the brain. It consists of two dome-shaped halves called *hemispheres*. A layer of white connective tissue joins the two halves together. The natural color of the outer cerebrum is a grayish red. For this reason it's sometimes called *gray matter*. The gray comes from the color of the neurons, and the red from the color of the blood that nourishes the brain. The cerebrum also has many wrinkles and folds. These wrinkles increase the surface area of the brain. Since most of your conscious thinking takes place in the neurons near the surface of the cerebrum, the increased surface area means a greater capacity for thinking.

The cerebrum is responsible for controlling conscious activities, such as walking or riding a bike. Different parts of the cerebrum also receive and process the signals coming from your sensory organs like eyes and ears. The cerebrum also stores your memories. So reading this book and recalling what you read for a quiz are both tasks that are handled by your cerebrum. In fact, the ability to read, speak, and understand language itself is another function of the cerebrum.

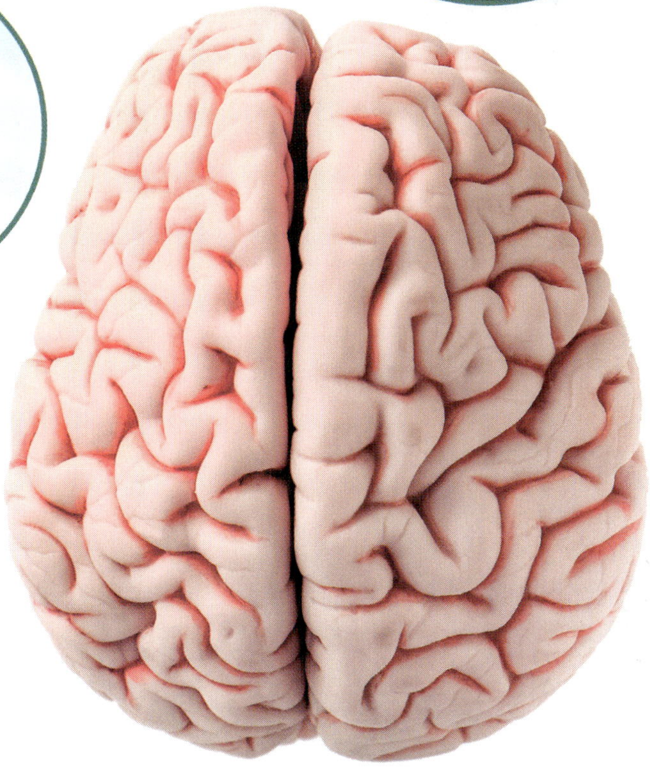

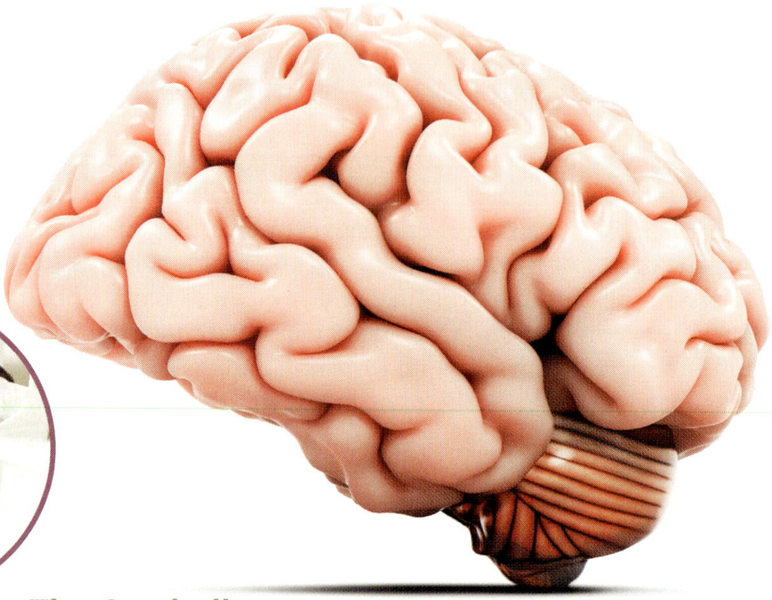

The Cerebellum

The nerves that connect your cerebrum to your body pass through the **cerebellum.** It is the second-largest part of the brain. The cerebellum helps with your *motor control*, the coordination of all the muscles that work together to complete a motion. This is totally involuntary. For example, suppose you need to write with a pen. Your hand contains over thirty muscles. If you had to think about controlling each one, you'd never get any writing done! But once your cerebrum makes the decision to write, it sends this information to the cerebellum. The cerebellum then works out which muscles will need to be stimulated and how much each muscle needs to contract. The result is that you can move your pen in the smooth and precise manner needed for writing. In like fashion, your cerebellum also helps maintain your posture and balance. It even sends impulses to your cerebrum regarding your body's position. This is why you know what your hands are doing even if you aren't looking at them.

The Brain Stem

The **brain stem** is below the cerebrum and in front of the cerebellum. It helps control breathing, heartbeat, blood pressure, coughing, sneezing, and swallowing. It also controls your cycles of waking and sleeping and helps coordinate your posture and balance.

The Spinal Cord

The **spinal cord** is continuous with the brain stem and extends down your back. It relays impulses between your brain and the rest of your body. In an adult the spinal cord is about 42–45 cm long. It is protected by the vertebral column (backbone), which forms a bony tube that surrounds the spinal cord. Between the vertebrae are *spinal nerves* that branch off the spinal cord. These nerves go to the various parts of the body.

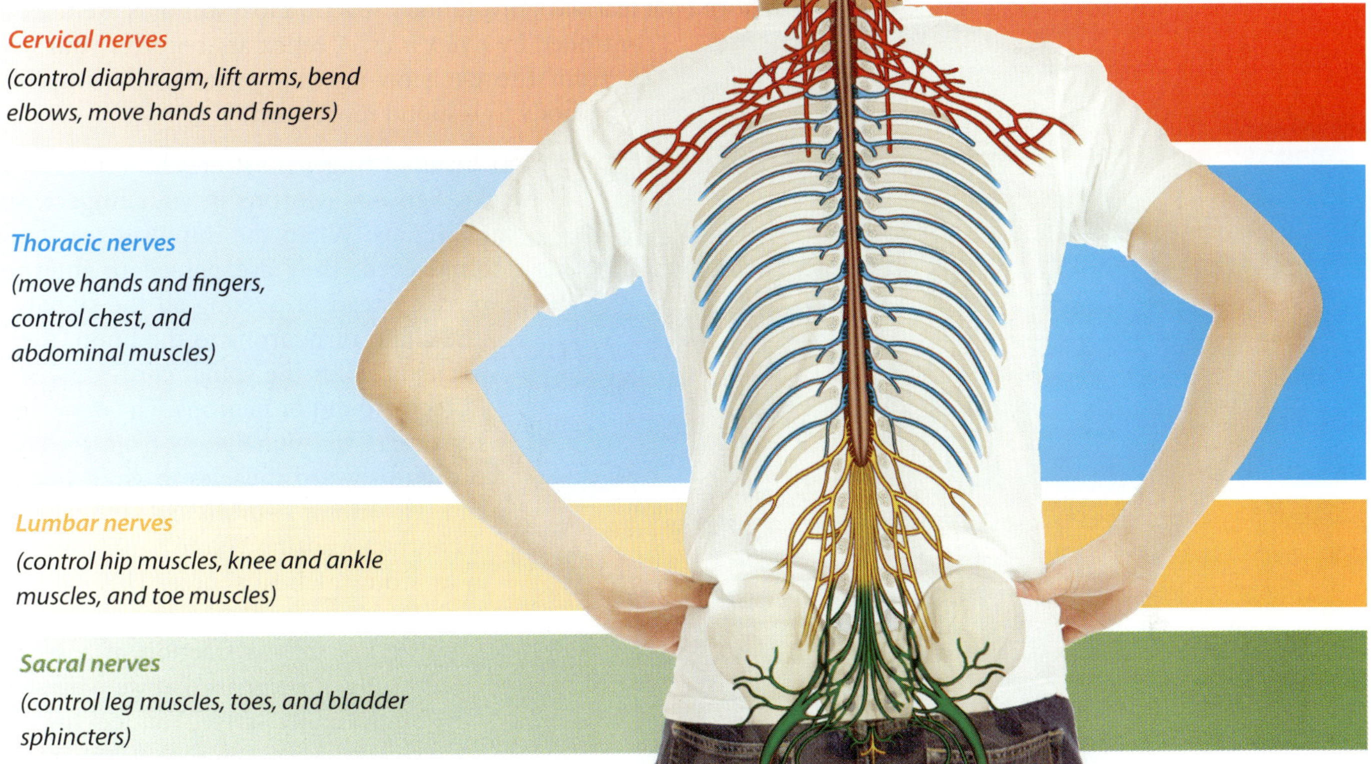

The intricate relationship between all the parts of the nervous system clearly points to a designer. From the Bible, we know that that Designer is Almighty God. Our ability to make decisions in the cerebrum wouldn't help much if the cerebellum weren't there to coordinate the actions. Coordinating the actions would do no good if the spinal cord were not there to transmit the impulses to appropriate muscles.

The actions of those muscles bring many more parts into play. Cellular respiration must provide ATP energy for the muscles to function. That requires food from the digestive system and oxygen from the respiratory system. But having oxygen in the lungs doesn't help if it can't get to the body cells. So the circulatory system is also vital. And we aren't done yet. There is much more to the amazing control system that God has designed for the human body—your sense organs.

16.6 REFLEXES

Different kinds of nerve impulses travel at different speeds, anywhere from less than 1 m/s to as much as 119 m/s. That sounds fast, but there are instances when even that speed isn't fast enough. If you touch a very hot object, for example, the burning cells in your finger are getting damaged *right now*. Even though your brain is only three feet away, your finger has no time to send a pain impulse and wait for your brain to make a decision on how to respond. But your body has a way to deal with this problem. It's called a reflex—an immediate, inborn (not learned), involuntary reaction to a stimulus. Reflexes are controlled by *reflex arcs*. A reflex arc removes your brain from the decision-making process, allowing your body to respond more quickly.

Your burning finger illustrates how a reflex arc works. Sensory neurons in your finger create pain signals. When the impulses reach your spinal cord, they are passed to interneurons. Instead of passing all the signals up the chain of neurons to the brain, the interneurons in the spinal cord route some directly to motor neurons. The motor neurons carry the signal away from the spinal cord and toward muscles in your arm and hand. The motor neurons signal the muscles to contract, causing you to yank your hand away from the heat source. This entire action, from sensing the heat to moving your hand, takes only a fraction of a second. Meanwhile, the signals that were passed to the brain finally arrive. You feel the pain, but the reflex arc has already moved your hand out of immediate danger.

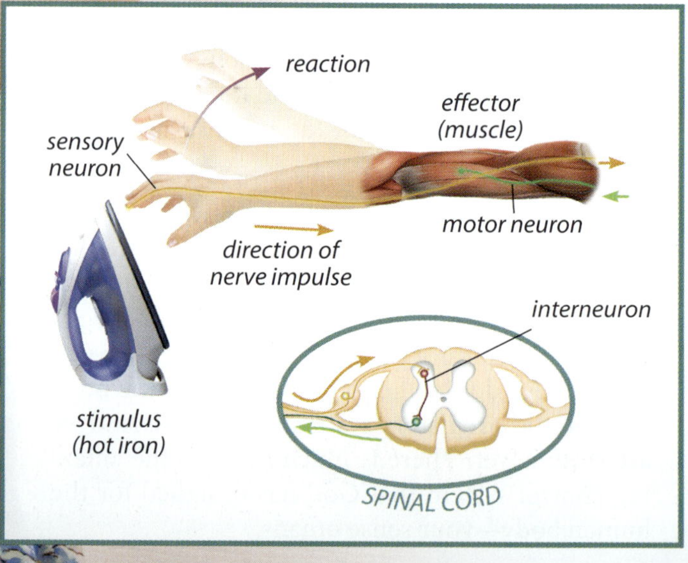

Without reflexes, we could be injured much more easily. The intricate design of reflexes is one way that God has provided for our well-being. Your body has many different reflexes. Sneezing, coughing, and even breathing can all be reflexes. When a doctor taps you on the knee to make your leg jerk, he is testing a reflex. When he flashes a light into your eye to see how quickly your eye responds, he is testing another reflex. If you are in an accident, a physician may test your reflexes to see whether your nervous system has been damaged. By knowing where the neurons of certain reflex arcs are located, a physician can find the damage even though it may not be visible on an x-ray.

WORLDVIEW SLEUTHING: SCREEN ADDICTION

Do you know someone who can't go five minutes without checking social media on his phone? How about someone who plays video games for hours every day? Maybe you've even jokingly told such a person that he's addicted to his phone or game console. But can a person become *actually* addicted to his electronic devices?

INTRODUCTION

You are a school nurse. It seems to you that students have been spending a lot more time on their smartphones—even when the students are together in groups. Teachers have also remarked to you that their students seem less interested in school these days. They don't seem to be as curious about the world as they used to be or as creative when it comes to problem-solving. Are these observations connected somehow?

TASK

Research the phenomenon of screen addiction. On the basis of your findings, create a PowerPoint or Prezi presentation that you will use to inform teachers at your school site. Include in your presentation the causes and symptoms of screen addiction and how it can be prevented.

PROCEDURE

1. Do a keyword search using the words "screen addiction." Especially look for information provided by people who have done actual research on the issue.
2. Note how the worldview of the authors of the websites affects their view of screen addiction. Also note whether there is broad agreement on the issue.
3. Plan your presentation and collect any photos or videos you need. Be careful to give proper credit to your sources.
4. Create your presentation and show it to another person.
5. Deliver your presentation to your class or family.

CONCLUSION

Screens have become a part of our daily lives. It's important to find and maintain a healthy balance with regard to their usage.

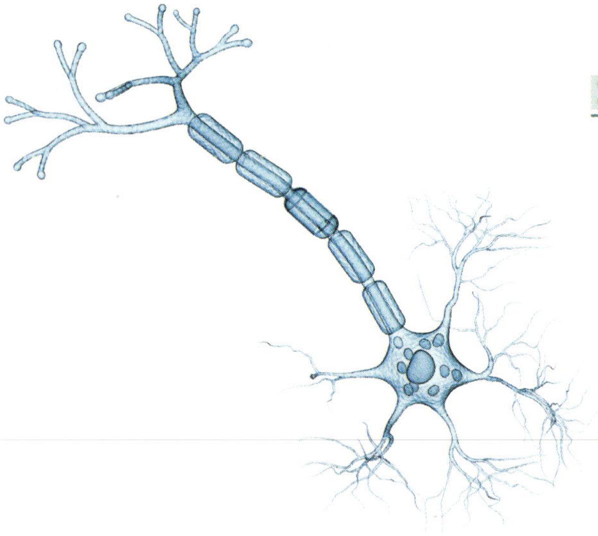

16B REVIEW QUESTIONS

1. Sketch a neuron and label its three main parts.
2. Which kind of neuron carries impulses from the brain to parts of the body?
3. What two organs make up the central nervous system?
4. Which part of your brain are you using to interpret the words in this question?
5. Through what structure(s) are impulses passed from the spinal cord to the rest of the body?
6. Place the following terms in the correct order to describe the path of a reflex arc: motor neuron, response, sensory neuron, spinal cord, and stimulus.

ETHICS Sports Safety

Few moments in sports can quiet a crowd quite like seeing an athlete lying motionless on the field after taking a blow to the head. Even if the player hasn't suffered a spinal cord injury, it's still possible that he has been concussed. What is a concussion? The German word for it describes it well—it translates as "brain shaking." Depending on the severity of the concussion, it may take weeks or even months for an athlete to fully recover.

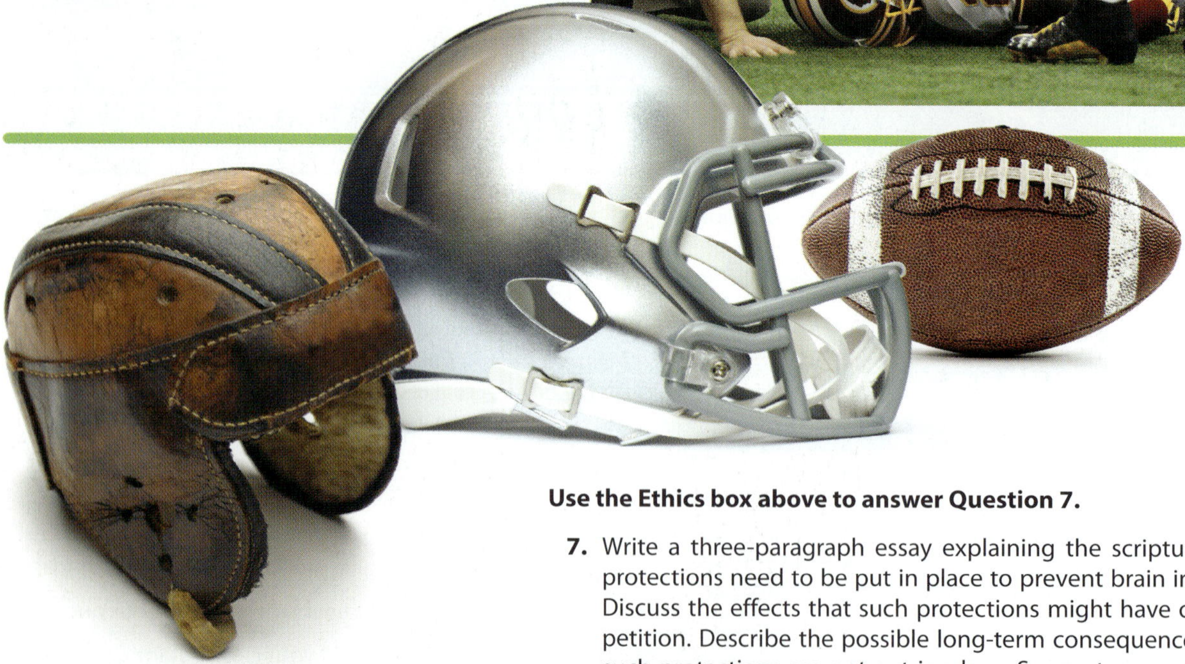

Use the Ethics box above to answer Question 7.

7. Write a three-paragraph essay explaining the scriptural reasons that protections need to be put in place to prevent brain injuries in sports. Discuss the effects that such protections might have on athletic competition. Describe the possible long-term consequences for athletes if such protections are *not* put in place. Suggest a way that people can both continue to enjoy sports *and* look after the well-being of athletes.

16C

THE SENSES

How do I get information about my surroundings?

16C Questions

- What are my body's sensory organs?
- How do sensory organs receive information about the body's environment?
- How do sensory organs help maintain homeostasis?

16.7 SENSE ORGANS

So now we've seen the structure of your nervous system and learned how it reacts to a stimulus. But how does your nervous system gather information about your surroundings? God designed your body to receive many different kinds of stimuli. Scientists aren't even sure just how many senses the human body has. The usual list—sight, smell, taste, touch, and hearing—is far too short. You can also sense temperature, balance, motion, position, and pain, to name a few of the more familiar sensations. To pick up on all these cues, your body has *sense organs* with special neurons called *receptors* that are sensitive to certain stimuli. Think of freshly baked chocolate chip cookies. Your eyes see the cookies, your nose smells them, your fingers feel them, and your tongue tastes them. The receptors in each of these sense organs send impulses to your central nervous system. You become aware of the stimuli when the impulses reach your cerebrum.

The mechanisms that allow you to sense various stimuli, interpret them, and respond to them are very complicated. In this course we will only scratch the surface of how complex they really are. But even that much information reveals some of the creativity, wisdom, and power of the one who created them. It is very easy to take our senses for granted. After all, we use them every day without really thinking about it. As you take time to purposefully learn about the senses that God created for you, also take time to praise Him for His wonderful works (Ps. 107:8).

16.8 THE EYE

The eye is marvelously designed to receive and focus light, sending exact and continuous impulses to the brain. To do this in a space about the size of a golf ball, the parts of the eye must work together precisely.

STRUCTURE OF THE EYE

The eyeball has three main layers. The tough outside tissue that forms the white of the eye helps maintain the shape of the eyeball. At the front of this layer is the clear *cornea* 1 through which light passes into the eye.

The middle layer of the eye contains lots of blood vessels for nourishing the eye. The portion of this layer near the front of the eye forms the colored *iris* 2.

The black spot in the center of the iris is the **pupil** 3. The pupil is actually an opening in the iris that allows light to enter into the eye. The size of the pupil opening is controlled by muscles in the iris.

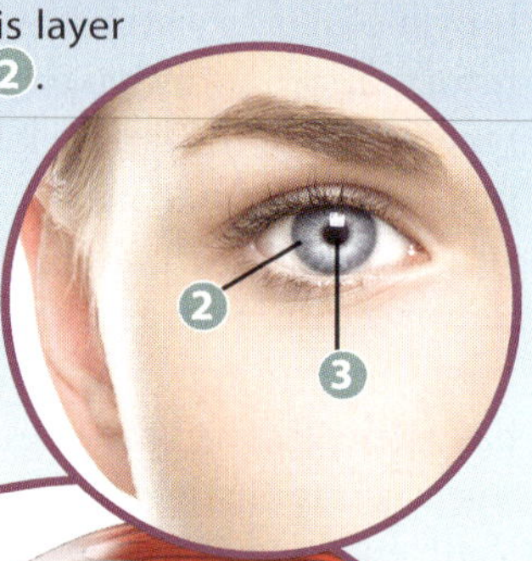

Light passing through the pupil enters the *lens* 4. The lens bends the light to focus it on the retina.

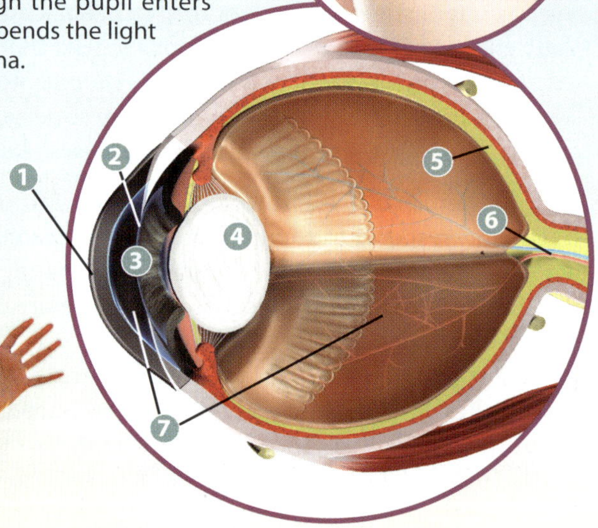

The **retina** 5 is the inner layer of the eye that begins along the sides of the lens and continues around to the back of the eye. In the retina are receptors that are sensitive to light.

Nerve impulses are sent from the retina's receptors to the brain via the *optic nerve* 6.

There are two chambers 7 in the eye filled with a clear fluid. The fluid in front of the lens provides energy for the cells of the cornea. This fluid is constantly being made and then absorbed by the blood.

The chamber behind the lens is filled with a clear, jellylike fluid that helps keep the eyeball round. The fluid in this chamber forms in the eye before you are born and normally remains the same all your life.

How the Eye Works

Tracing the path of light through the eye is a simple way to understand how the eye works. Light reflecting off an object first passes through the cornea. As the light reaches the pupil, the eye makes quick adjustments. If you are in bright light, reflexes cause the iris to make the pupil smaller because the sensitive retina may be damaged by too much light. In dim light the pupil opens wide to allow as much light as possible to enter the eye.

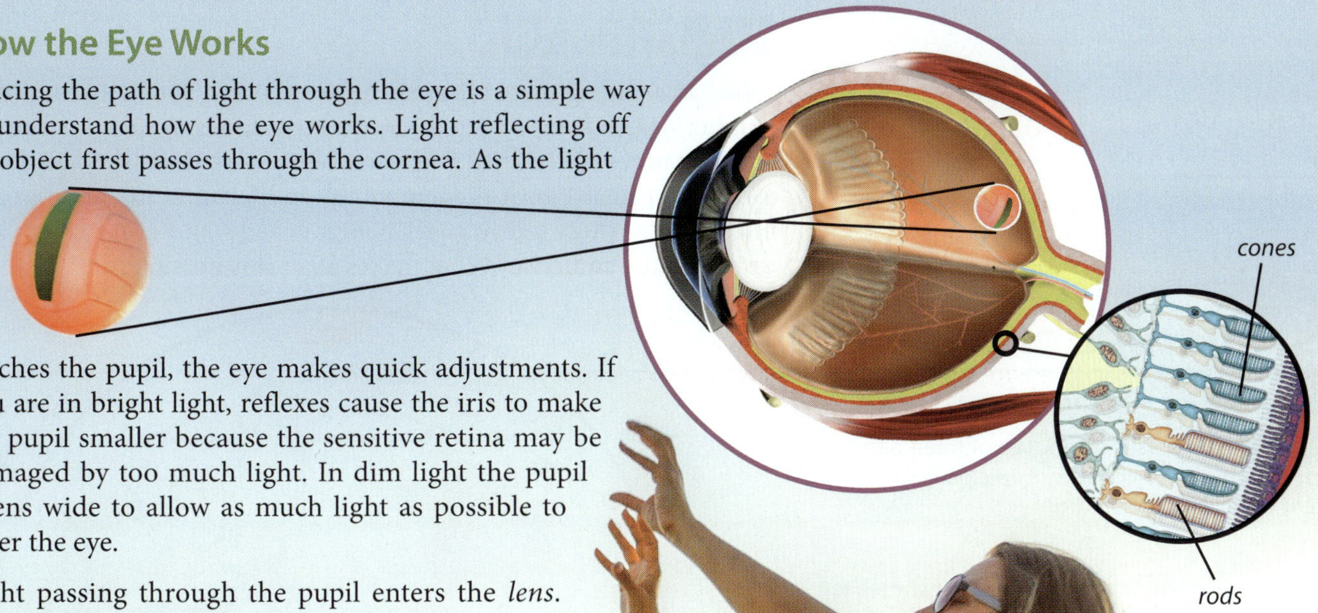

Light passing through the pupil enters the *lens*. The lens bends the light and focuses it as an image on the retina. Surprisingly, the focused image is upside down! Your brain interprets it as a right-side-up image.

The retina contains millions of receptors that are sensitive to light. They are of two types: rods and cones. The *rods* are responsible for vision in dim light. They are stimulated by visible light, regardless of its color. So when you are looking at something in dim light, it appears only black and white. The *cones* are sensitive to the various colors of visible light, but it must be much brighter to stimulate a cone than to stimulate a rod. That is why we see colors most clearly in well-lit areas.

Once receptors in the retina are stimulated, they transmit impulses along the *optic nerve* to the brain. The vision centers of the brain receive the impulses and interpret them as vision. Your brain then decides what to do with the information. All these parts must work correctly, in the right order, for you to see.

16.9 THE EAR

What an amazing thing is human hearing! When you listen to a string quartet, you can easily detect the different tones of each instrument and the notes of the different parts being played. At the very same time, you can also hear the hum of the auditorium's air conditioning, the rustle of sheet music pages being turned—and every cough and sniffle in the audience! What makes all of this possible?

STRUCTURE OF THE EAR

Each of your ears is composed of three basic regions. The *outer ear* ① consists of an external ear flap, the ear canal, and the eardrum.

The eardrum ② is a membrane that stretches over the opening to the middle ear.

The *middle ear* ③ is an air-filled space with three tiny ear bones named for their shapes. These bones are the hammer ④, the anvil ⑤, and the stirrup ⑥.

The *inner ear* ⑦ consists of the cochlea ⑧ and the semicircular canals ⑨.

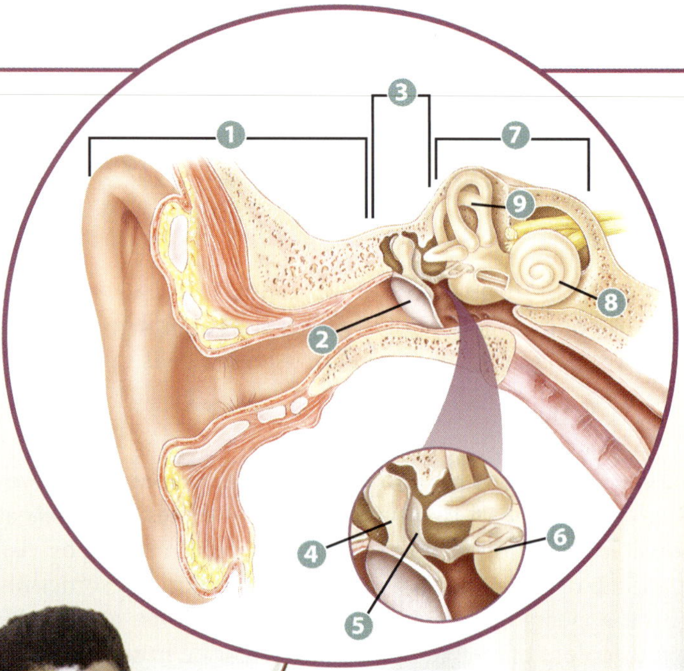

How Do We Hear Sound?

Sound is a form of energy that travels in waves. Your external ear directs these waves into your ear canal. At the end of the canal, sound waves hit the **eardrum**, causing it to vibrate. The hammer is attached to the inner surface of the eardrum. As the eardrum vibrates, the hammer moves with it. The hammer moves the anvil, and the anvil moves the stirrup. In this way, the ear bones pass the vibrations of the eardrum to the inner ear.

The **cochlea** (KOHK lee uh) is a coiled tubular structure that resembles a snail shell. The inside of the cochlea contains fluid. At its base is a small, membrane-covered opening for the stirrup. When vibrations from the eardrum pass through the ear bones, the stirrup vibrates the membrane-covered opening. The membrane then pushes against the fluid in the cochlea. These vibrations form waves in the cochlear fluid that cause hairlike neurons inside the cochlea to bend. When these neurons bend, they start nerve impulses that travel to the brain. The brain interprets the impulses as sound.

Your ears also play a critical role in maintaining balance (see Figure 16-2). Fluid inside the *semicircular canals* moves as your head moves, sending nerve impulses to the brain. The brain interprets the impulses and helps you know what position your head is in, even when your eyes are closed.

16-2
How do you think your semicircular canals are affected by rapid spinning?

16.10 SMELL, TASTE, AND TOUCH

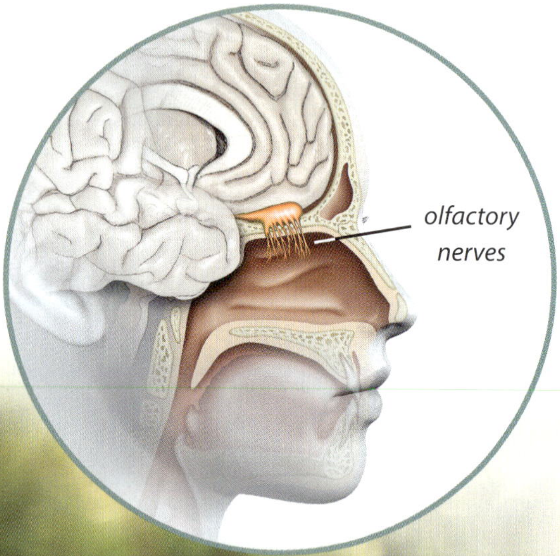

Your nose allows you to sense odors in the air. When you inhale molecules floating in the air, they land on the mucous membrane that lines the inner surface of the nose. There are sensory nerve endings in the mucous membrane. When the airborne molecules contact the mucous membrane, these neurons are stimulated, sending impulses to the brain.

In addition to being used for chewing, speaking, and swallowing, your tongue has many microscopic *taste buds*. Each taste bud is a cluster of cells. Most of these cells serve to protect the neurons, which are located inside a tiny hole. When a fluid containing a dissolved substance enters the hole and stimulates the neurons, you taste something. Since only fluids can enter the hole of the taste bud, solid foods must be chewed and mixed with saliva before you can taste them.

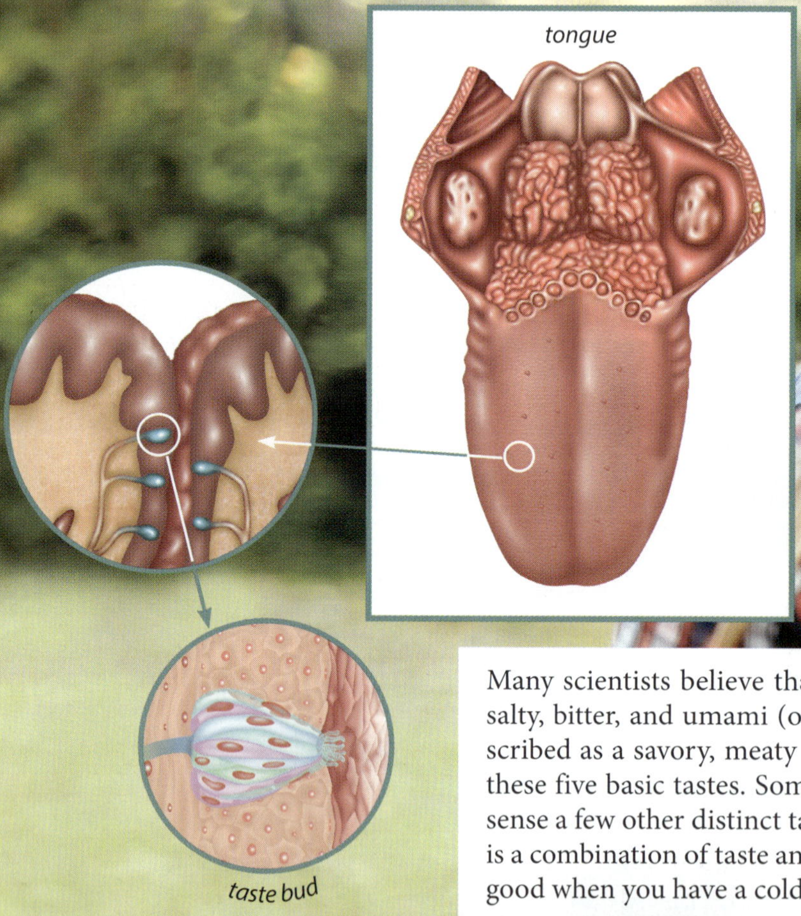

Many scientists believe that there are five basic tastes: sweet, sour, salty, bitter, and umami (oo MAH mee). (Umami is sometimes described as a savory, meaty taste.) Most flavors are a combination of these five basic tastes. Some scientists believe that humans can also sense a few other distinct tastes. Most of the flavor of the food we eat is a combination of taste and smell. This is why food does not taste as good when you have a cold and cannot smell it.

Human skin has several different types of sensory neurons, each able to respond to a different stimulus. Some neurons respond to cold, while others respond to heat. Near the surface of the skin are neurons for touch. Deep in the skin are neurons that sense pressure. Similar pressure-sensing neurons are located elsewhere in the body, such as in the joints and the stomach area. Neurons that sense pain are located not only in the skin but also in almost every other tissue of the body. Not all areas of skin have the same number of sensory receptors. In more sensitive areas, such as your fingertips, there may be many. In other areas, such as the skin of your back, there are fewer.

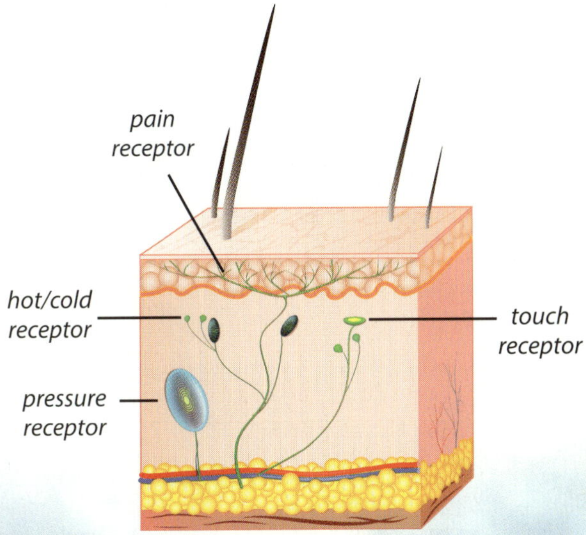

Our senses allow us to enjoy the world that God created for us to live in. Thanks to His gift of sensory organs, we can delight in the colors of a painting, the sound of a favorite song, and the smell of a delicious meal. But our senses are also vital for maintaining homeostasis. Backing away from a hot campfire, regaining your balance after stumbling, stepping back onto the curb when you hear an onrushing car—these responses to potential dangers all begin when sensory neurons are stimulated. Take a moment to thank God for the marvelously designed sensory organs He created for you!

16C REVIEW QUESTIONS

1. Sketch a diagram of the human eye and label the cornea, pupil, lens, and retina.
2. Your brain cannot detect light, so how does it "see" your environment?
3. Place the following structures in the correct order in which a sound wave would encounter them as it is converted into a nerve impulse: anvil, cochlea, ear canal, eardrum, hammer, stirrup.
4. Where are sound receptors located within the ear's structure?
5. What sense would be affected if a person had no semicircular canals?
6. What are the different kinds of receptors that are stimulated when you walk barefoot across a cold tile floor?
7. You go outside on a cold day and your ears start to get cold. How does your nervous system help your body maintain homeostasis?

CHAPTER SUMMARY

16A IMMUNITY

- The immune system consists of nonspecific defenses that defend against any kind of pathogen and specific defenses that work only against specific pathogens.
- When damaged cells trigger an inflammatory response, blood flow is increased to the affected site, bringing additional macrophages to fight off possible infection.
- Active immunity is the body's ability to rapidly respond to a specific pathogen. It happens because a person has had a previous encounter with that pathogen.
- People develop passive immunity when they get antibodies from an outside source.

16A TERMS

inflammation	327
antigen	327
T cell	328
B cell	328
antibody	328
active immunity	329
passive immunity	329

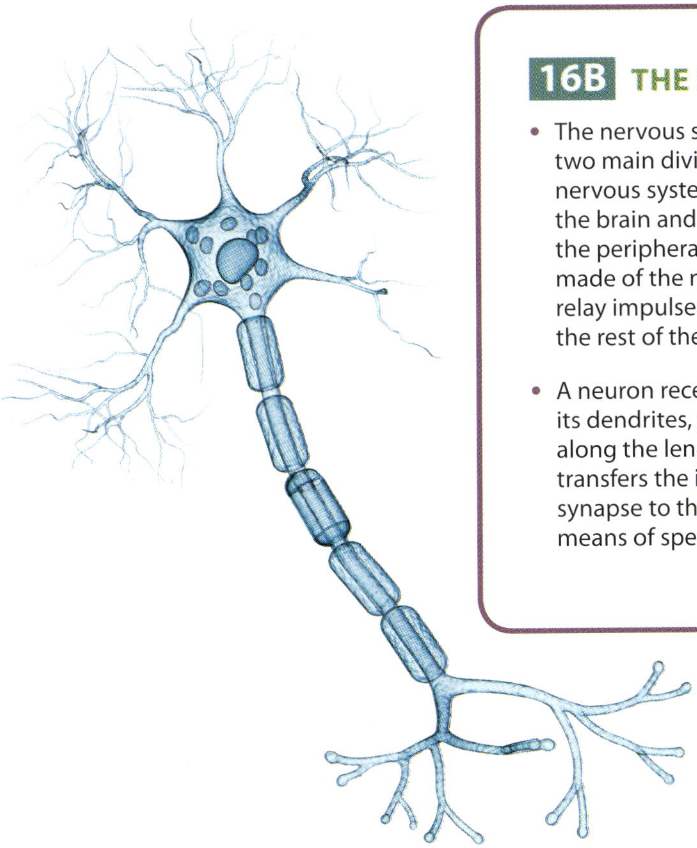

16B THE NERVOUS SYSTEM

- The nervous system consists of two main divisions: the central nervous system, composed of the brain and spinal cord, and the peripheral nervous system, made of the nerves that relay impulses to and from the rest of the body.
- A neuron receives an impulse at its dendrites, passes the impulse along the length of its axon, and transfers the impulse across a synapse to the next neuron by means of special chemicals.
- The brain controls and regulates the body's activities. Conscious activity is controlled by the cerebrum; involuntary activity is controlled by the cerebellum and brain stem.

16B TERMS

central nervous system	332
peripheral nervous system	332
cerebrum	333
cerebellum	334
brain stem	334
spinal cord	335

CHAPTER SUMMARY

16C THE SENSES

- Sensory organs, such as eyes and ears, contain sensory neurons that collect information about the body's environment.
- A sensory neuron triggers an impulse in response to a specific kind of stimulus, such as light, sound, or odor.
- Responding to stimuli from the environment helps a body maintain homeostasis.

16C TERMS	
pupil	340
retina	340
eardrum	343
cochlea	343

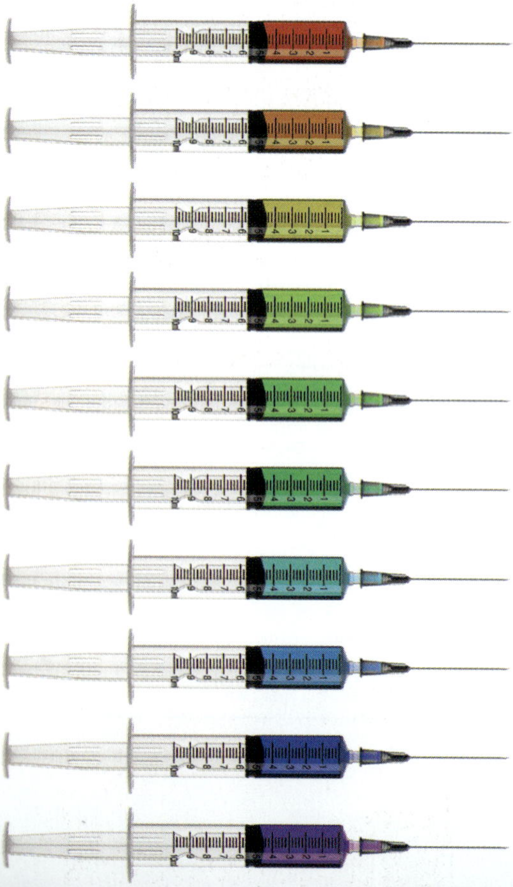

CHAPTER REVIEW QUESTIONS

Recalling Facts

1. List three forms of nonspecific defense that your body uses to fight pathogens.
2. Describe the three parts of an inflammatory response.
3. What type of white blood cell is involved in inflammation?
4. Tell whether each of the following statements is true of T cells, B cells, both, or neither.
   a. It is a type of lymphocyte.
   b. It produces antibodies.
   c. It is part of a specific defense.
   d. It is activated by macrophages.
   e. It is a type of red blood cell.
5. Describe two ways that a person can acquire immunity to a disease.
6. Which part of a neuron receives an incoming nerve impulse? Which part passes the impulse to another neuron?
7. Do neurons carry nerve impulses toward the brain, away from the brain, or both?
8. Why do neurons need chemicals to pass a nerve impulse from one neuron to another?

9. Which portion of the brain helps with motor control?
10. Through what portion of the central nervous system do nerve impulses from your body pass before arriving at your brain?
11. Why does a reflex arc not include the brain as part of its stimulus-response sequence?
12. Why is it incorrect to refer to "the five senses"?
13. What are the two types of sensory neurons in the human eye? What does each detect?
14. Identify the labeled structures in the ear diagram below.

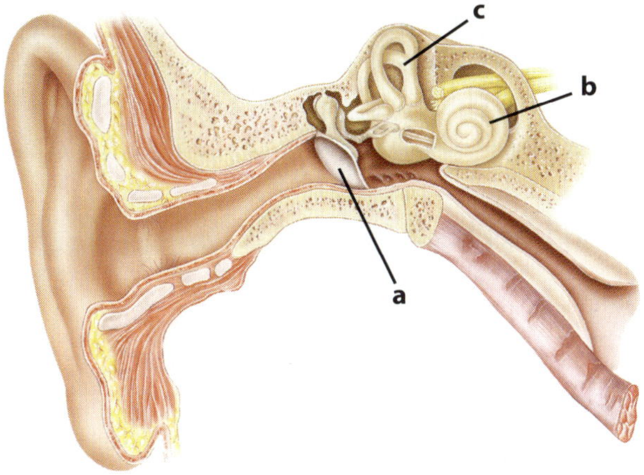

15. Which structures in the human ear are part of the skeletal system?
16. Why must you chew solid foods before you can taste them?

Understanding Concepts

17. Explain what a pathogen is and give an example of one.
18. How does an inflammatory response help to fight an infection?
19. Explain why a person's response to a bee sting is a good example of the inflammatory response.
20. How does a fever help to fight an infection?
21. If you receive an injection of a vaccine, does it provide you with active immunity or passive immunity? Explain.
22. For each action, name the part of the brain that controls it.
    a. deciding to go outside
    b. breathing
    c. balancing on one leg
23. Imagine a lemon. Why is it not necessary for you to actually have a lemon in your presence to know how it smells and tastes?
24. Does a reflex arc bypass the central nervous system? Explain.
25. Why would a parent tell his child to hold his nose when taking a foul-tasting medicine?

Critical Thinking

26. How is a fever similar to an inflammatory response? How is it different?
27. What would be the long-term effect if B cells did not produce antibodies during an immune response?

28. Explain how having an ear on each side of the human head helps to determine from which direction a sound comes.

29. From your own experiences, indicate what kinds of sensory neurons can be found in the tongue other than taste receptors.

30. How might a sense of taste contribute to maintaining homeostasis?

31. Some children are very sensitive to sounds, light, or touch, a condition called *sensory processing disorder (SPD)*. Is SPD more likely to be the result of a problem with the brain or with the sense organs themselves? Explain.

The limit of what sounds people can hear is called the *threshold of hearing*. This is different for men and women, and it changes as people age. The graph below shows the frequency and decibel levels of certain sounds that men and women of a variety of ages can hear. Use the graph to answer Questions 32–35.

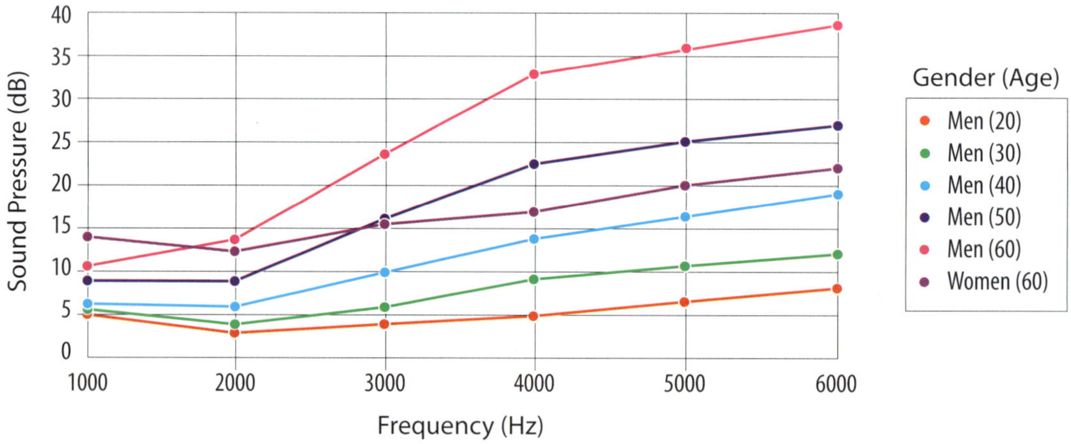

32. How do the thresholds of hearing for men and women at age 60 compare?

33. How do the thresholds of hearing for men at different ages for sounds of similar frequency compare?

34. Sound pressure is the loudness of a sound. Which sounds can people hear better, lower or higher frequency sounds?

35. Suggest an explanation for why certain people hear better than others.

REPRODUCTION, GROWTH, AND DEVELOPMENT

17

GENDER CONFUSION

In your school there are two restroom doors, one for girls and one for guys. Most people aren't confused about which one to use. But there is a growing voice among people who say that the idea of only two genders is outdated. They think of gender as a range of possibilities between male and female. On Facebook, for example, people can pick from dozens of gender options! And if none of those fit, they can create their own.

There is a gender crisis in America and around the world. Are there people who are trapped in a body that doesn't fit who they really are? And how do they know who they really are?

17A Questions

- What does the Bible say about gender?
- What are the parts of the endocrine system?
- How does the endocrine system work?
- How does the endocrine system help balance my body's processes?
- How does a person's body change during puberty?

Male Karyotype

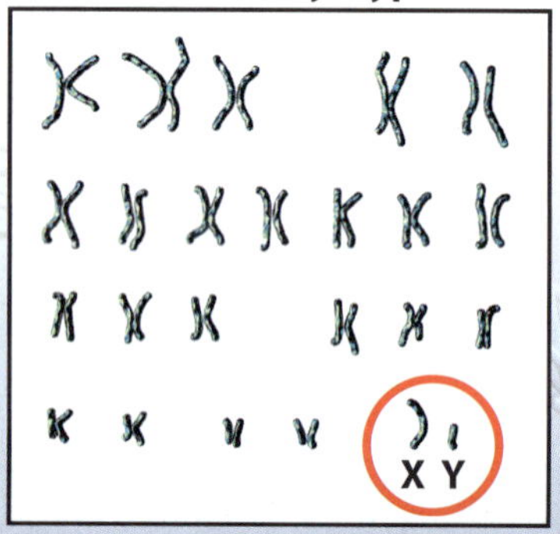

Female Karyotype

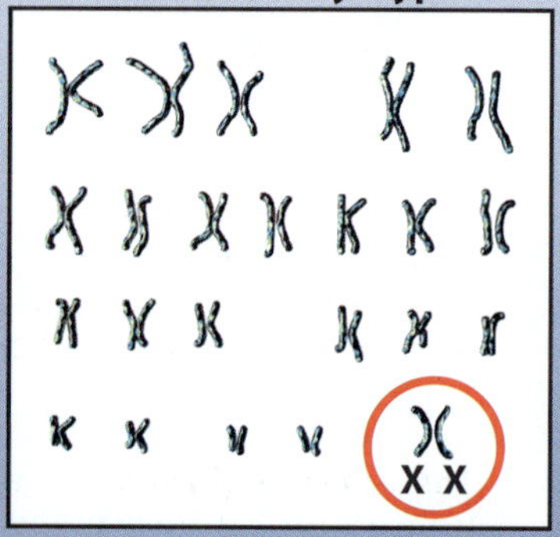

17A THE ENDOCRINE SYSTEM

How do the cells in my body communicate even though they aren't connected?

17.1 STRUCTURES OF THE ENDOCRINE SYSTEM

In every one of your cells, with the exception of gametes, there are twenty-three pairs of chromosomes, which carry your DNA. The last chromosome pair is either an XX chromosome pair (female) or an XY chromosome pair (male). This means that God has created every person as either male or female down to each individual cell (Gen. 1:27). He knew you before you were even born (Ps. 139:13–16).

One of the body systems that God has given you is the *endocrine system*. He has customized it to monitor your body's processes. These processes are the same for both guys and girls, but God has also customized the endocrine system to serve both genders differently.

The nervous system produces quick, short-lasting responses. In contrast, the endocrine system produces slower, longer-lasting responses. While the nervous system uses nerve impulses to relay messages to the body, the endocrine system makes use of hormones.

Hormones are chemicals that act as messages in the body. They affect both the way that your body responds to stimuli and how it develops over time. Girls and guys develop differently in response to the different hormones that they produce. The organs of the endocrine system are all glands that produce and release hormones to influence the body.

THE ENDOCRINE SYSTEM

The **hypothalamus** 1 is part of the brain, so it links the nervous system to the endocrine system. It makes hormones that work like switches to turn glands in the endocrine system on and off.

When the hypothalamus turns on the **pituitary gland** 2, it produces a variety of hormones that help control your body's processes. One of these is *growth hormone*. It does just what you might think—it makes you grow!

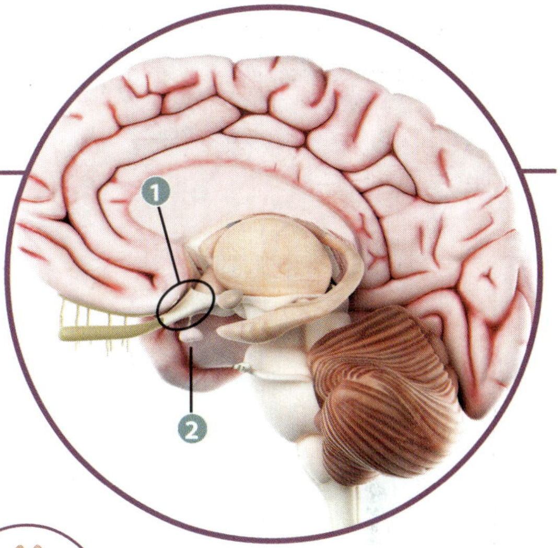

The **thyroid** 3 produces hormones that control how fast your body burns food to produce energy. It can also affect growth.

The pancreas 4 helps the digestive system by producing and secreting *insulin* and *glucagon*. These two hormones regulate the sugar glucose in the bloodstream.

The **adrenal glands** 5 secrete hormones that help you out in stressful situations, like getting an adrenaline rush while riding a roller coaster! One of the hormones that does this is adrenaline, or *epinephrine*. This hormone works with the nervous and digestive systems to put your body in emergency standby mode.

The *ovaries* 6, found only in females, release a hormone called *estrogen*, which works over time to help females reach physical maturity.

The *testes* 7 (s. testis), found only in males, release a hormone called *testosterone*, which works over time to help males reach physical maturity.

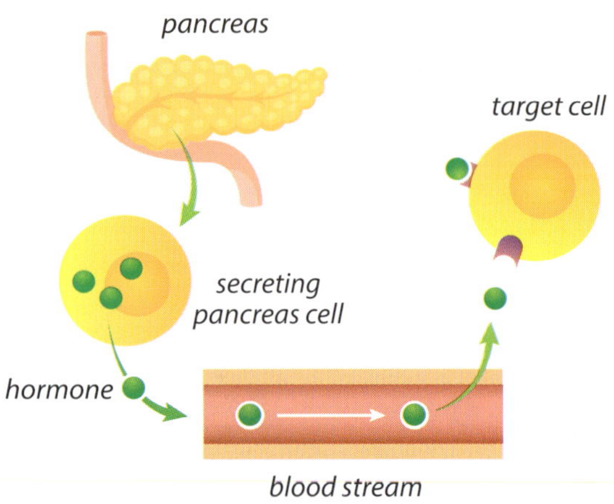

17.2 FUNCTION OF THE ENDOCRINE SYSTEM

So you've learned about the parts of the endocrine system. But how do they work? The glands of the endocrine system are different from other glands in the body. They don't have *ducts* (tubes) to carry the hormones that they produce. Instead they release their hormones into the circulatory system. The blood then carries the hormones throughout the body.

Let's trace the path of a hormone. Your pancreas monitors glucose levels in your blood. When you eat, your glucose levels rise. The pancreas gets a green light to start making insulin. Insulin works like a key with a specific shape that fits a specific "door." It lets your cells absorb glucose so that they can use the energy. After a pancreas cell makes insulin, it releases it into the bloodstream. There it finds a target cell to "unlock" so that the glucose can provide energy. When blood glucose levels drop, the pancreas gets a red light. It stops producing insulin. When glucose levels get too low, the pancreas produces glucagon to break down molecules into glucose in order to restore homeostasis.

NEGATIVE FEEDBACK SYSTEM

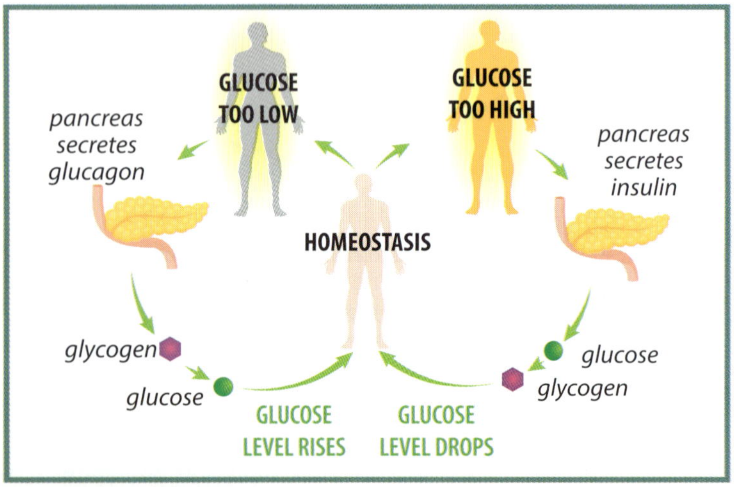

Since the pancreas turns itself off after glucose levels reach a certain condition, this process is called a *negative feedback system*. The result of the hormone's work turns off production of that same hormone after the body reaches homeostasis. Many of the endocrine glands work this same way. People with diabetes have a problem somewhere in this process, so they have a hard time properly regulating the glucose in their blood.

Insulin production is an example of how the endocrine system serves guys and girls the same way. But how does it work in different ways for guys and girls?

As guys and girls approach ages ten to fourteen, their hypothalamus signals their bodies to start producing sex hormones. A female's ovaries start producing estrogen. A male's testes start producing testosterone. These hormones start to cause big changes in the body as the transition to physical maturity begins. This phase of life is called **puberty**. For an example, a young person in puberty may notice that his voice is getting deeper (true for both genders, though much more noticeably in males).

One of the most significant changes for males is the development and nourishment of *sperm*, a gamete. For females, the development of the **menstrual cycle** is also a major change. This cycle involves a female's ovaries producing estrogen in greater and greater amounts until the ovaries produce a single egg, or **ovum**, also a gamete. This process is called **ovulation**. The hormones involved in ovulation don't turn themselves off as in a negative feedback process. Instead, the action of the hormone results in more hormone being produced. This is called *positive feedback*. This positive feedback does *not* lead to homeostasis; there is a bigger goal: producing and protecting new life if pregnancy occurs.

Your body is changing quickly, and that can affect the way that you feel about yourself. Big hormone swings are part of this. Things seem confusing that never seemed confusing before. How should you deal with these changes and feelings?

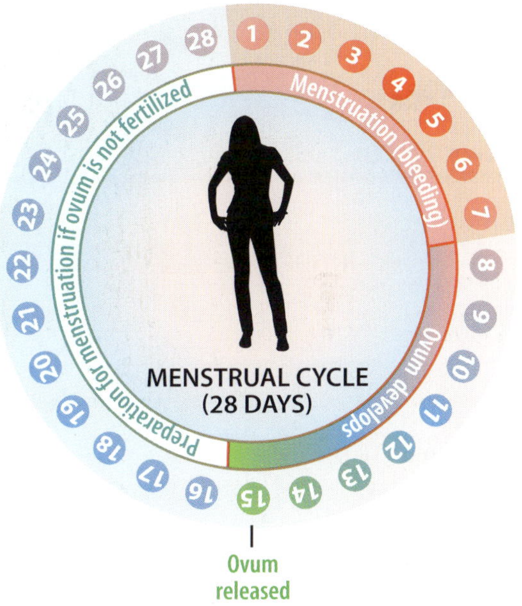

God doesn't leave you to figure life out on your own. He has given you parents, and He has given you His Word. In a world full of darkness and confusion, God gives light and clarity through His Word. The Creator made His world to work a certain way, and He tells us how to best live in that world. Don't let your body's changes make you question the truth of God's Word. You can bear God's image with joy and fulfillment when you live out His plan for you.

17A REVIEW QUESTIONS

1. What do we call the chemicals that act as messages to the human body?
2. Use a table to contrast the endocrine and nervous systems. Contrast their method of communication, the paths through which their messages travel, and their response times. (See Appendix B for an example of a table graphic organizer.)
3. List the hormone(s) that each of the following endocrine system organs produce.
   - **a.** pituitary gland
   - **b.** pancreas
   - **c.** adrenal glands
   - **d.** ovaries
   - **e.** testes
4. How is the hypothalamus different from the other structures of the endocrine system listed on page 353?
5. Order the following events in a gland of the endocrine system chronologically.
   - **a.** A hormone enters the bloodstream.
   - **b.** A hormone reaches a target cell.
   - **c.** A hormone is produced by an endocrine cell.
6. List one change to the body of a person of your gender mentioned in the text that is the result of puberty.
7. What is the difference between a positive and a negative feedback system?

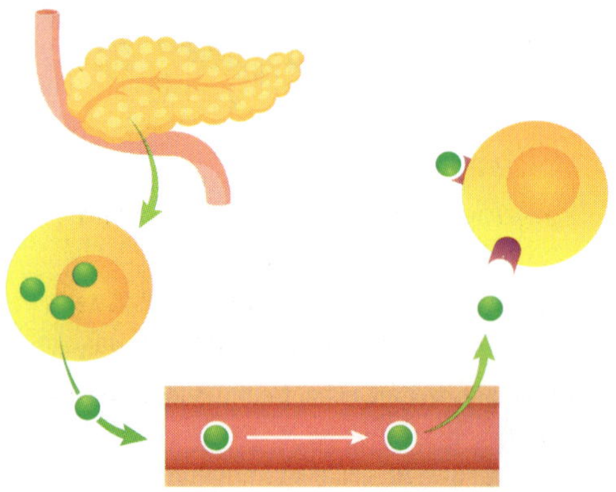

CASE STUDY WHICH DOOR?

In 2013, the state of California passed a very unusual law. It was the first state to allow students from kindergarten to twelfth grade to use whichever bathroom they want. Students can pick either door on the basis of which gender they choose to identify with. They can also likewise choose whether to play on a girls or boys sports team.

Use the Case Study above to answer Questions 8–11.

8. Why do you think California legislators would approve such a law?
9. What kinds of problems could this law produce in schools and society?
10. How would you respond to a person who told you that they think it's old-fashioned to believe that there are only two genders?
11. How should a Christian help a person who is struggling with the gender that God has given him?

FROM A MEDICAL DOCTOR TO YOU

Sex. How do you act or feel when you hear this word? Giggles? Confusion? Nausea? Embarrassment? I remember how I felt as a middle-schooler when this word came up. But know this—sex is a good gift that God created for marriage (Heb. 13:4). It all started in Genesis 2 when God recognized that it wasn't good for man to be alone. He created Eve to share in His good plan.

Even though you aren't ready for marriage right now, you can begin to understand who God created you to be. You can begin to understand why God saves sex for marriage. Sadly, as a doctor I have seen what can happen when people sneak over the fence into dangerous territory. Many in our culture think about sex in ways that are different than the way that God does. It always hurts people when they think about sex in a way that strays from His revelation.

As you learn about reproduction in this section, your teacher and parents want your questions to get answers. Remember that God's plan for human reproduction is a good gift!

Dr. Jeremy Byrd, MD

17B

THE REPRODUCTIVE SYSTEM

What is God's design for human reproduction?

17B Questions

- What does the Bible say about human reproduction?
- How is the reproductive system different from all the other systems in the body?
- How does the human body produce new life?

17.3 REPRODUCTION AND A BIBLICAL WORLDVIEW

We've seen how the body's systems work together to maintain balance, or homeostasis. However, there is one body system that we haven't considered. The goal of the *reproductive system* is not to protect the body's balance; it is to create and protect new life. God created the reproductive system so that we could fulfill an important part of the Creation Mandate—procreation. God made reproduction something that a husband and wife can enjoy for His glory. Children are the beautiful result of that obedience. And it's good for society when we follow God's plan for the family.

We live in a fallen world where people are immoral and experience the problems of sin and disease. People twist God's gift of human reproduction because of their sin nature. That makes it harder to live out God's plan for the family. But that doesn't mean that families should stop having children. In fact, it was through the conception and birth of one Child that we have hope. Through Jesus' redemption we can experience the true freedom that comes from obeying God's commands. Christian families that follow God's pattern can show hope and freedom to a world that desperately needs it.

The Virgin Birth

Christ's birth, growth, and development followed those of any normal child. His conception, however, didn't. When we say that Christ was born of a virgin, we mean that God miraculously allowed Jesus to grow in Mary's womb without procreation. This wasn't a normal function of the reproductive system; it was a miracle.

17.4 STRUCTURES OF THE REPRODUCTIVE SYSTEM

The main organs of the reproductive system are the **gonads**, the organs that produce gametes. In females, the ovaries usually release one ovum every month. In males, the testes produce sperm continuously.

THE REPRODUCTIVE SYSTEM

1. ***Testes.*** Males have two testes that produce sperm. Other structures produce *semen*, a fluid that nourishes and protects sperm. Since males carry the XY chromosome pair, they will pass on either the X or the Y to each of their children.

17.5 FUNCTION OF THE REPRODUCTIVE SYSTEM

The function of the reproductive system is simple: to produce and protect new life. We've already studied the female menstrual cycle. The walls of the uterus become thicker over the course of about a month to prepare for pregnancy. If pregnancy does not occur, the inside layer is shed during menstruation.

But what happens if the ovum *is* fertilized? And how does fertilization work?

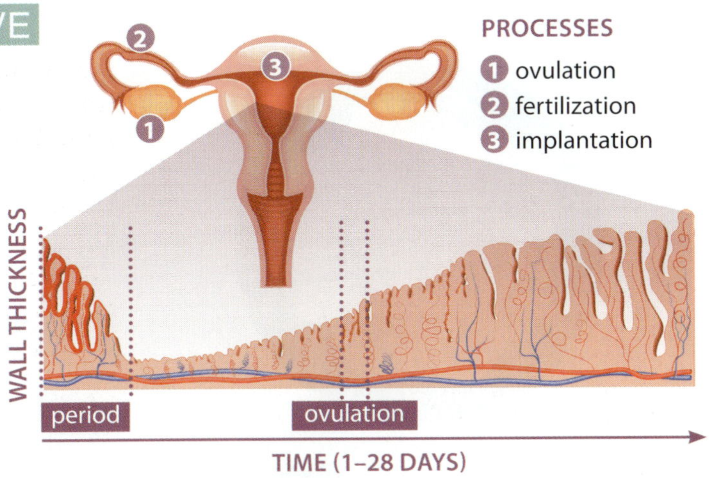

Fertilization

To occur, fertilization needs to happen within a few days after a female has ovulated. It also requires the presence of sperm. When an egg from the female and sperm from the male come together, the egg is fertilized (see image at right). This happens in the tube connecting the ovary to the uterus. Now the ovum is a *zygote*. Scientists have actually observed a brief flash of light at fertilization. A new life has now begun! The fertilized cell begins to divide and moves down to the uterus. And now a woman's pregnancy begins.

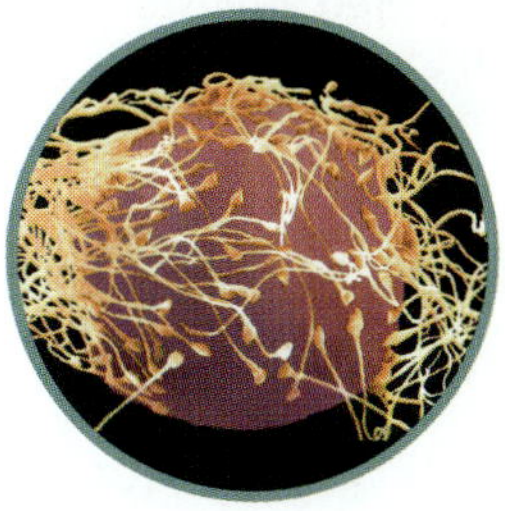

2 *Ovaries.* Females have two ovaries that alternate to produce one ovum each month. Since females carry the XX chromosome pair, they can pass only one chromosome to each of their children, an X chromosome.

3 *Uterus.* This female organ nourishes and protects a developing baby. When an ovum is fertilized, it moves to the uterus, where it attaches and grows until it is big enough to live on its own. The uterus is also called the *womb*. If a fertilized egg gets an X chromosome from the father, the child will be female. If the egg receives a Y chromosome, the child will be male.

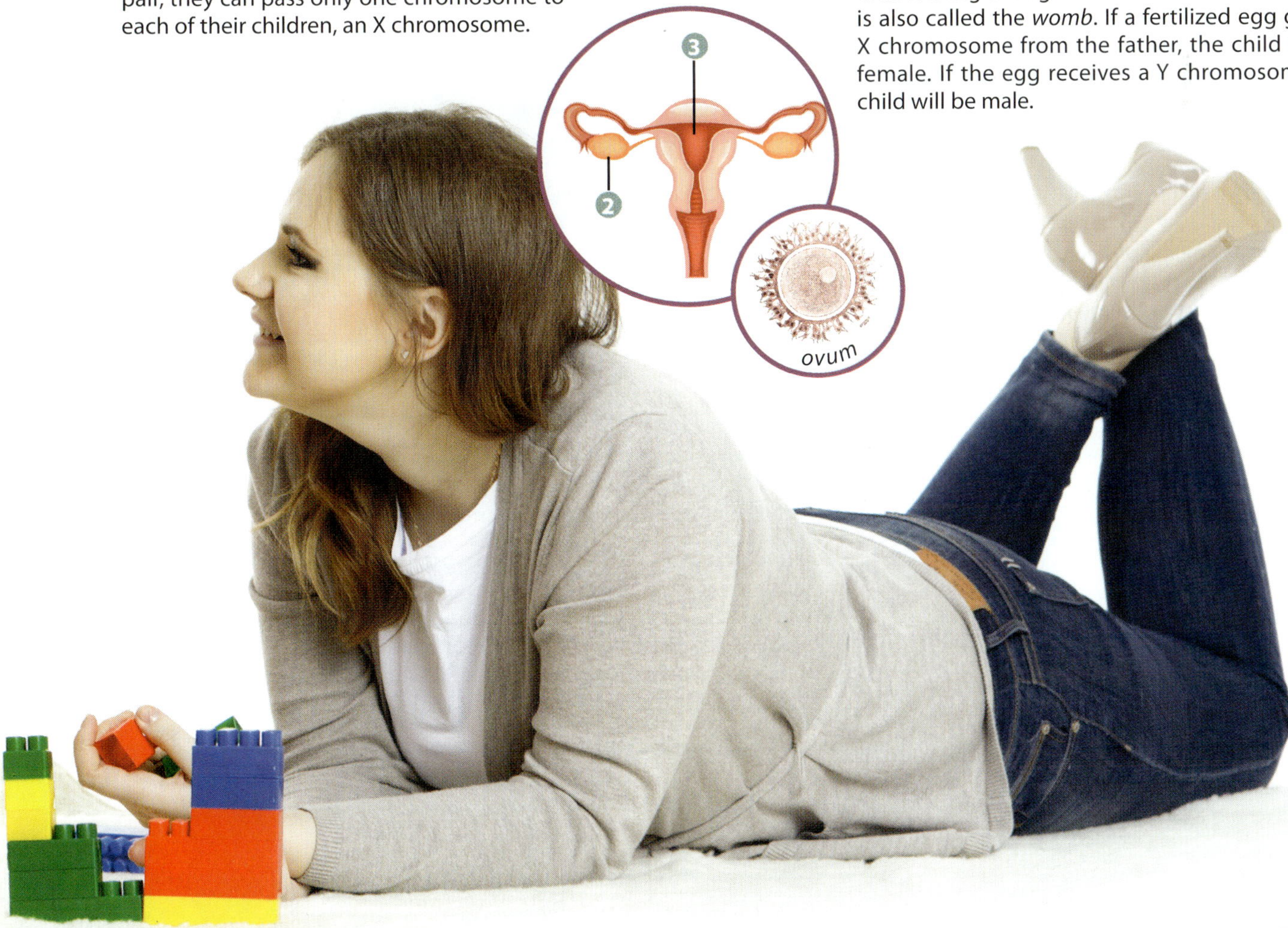

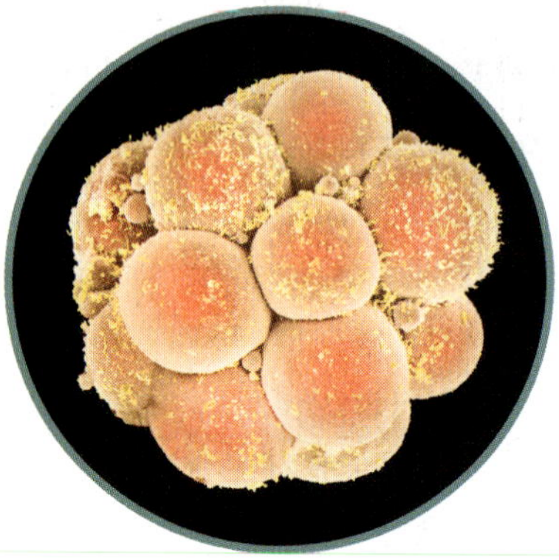

17-1

A dividing fertilized cell is called a *blastocyst*. This baby is ready to attach to Mom!

Implantation

After the fertilized cell has used up its reserves of energy, it needs a new supply. After about a week, it gets nourishment when it attaches to the thickened walls of the uterus. From now on, baby will get nourishment from Mom. The baby is considered an **embryo** until it is eight weeks old. From this point, the baby's development takes off.

17B REVIEW QUESTIONS

1. Is human reproduction good or bad? Explain.
2. The main organs of the reproductive system are the ______ (female) and the ______ (male).
3. Where does a growing child develop?
4. What is the main goal of the reproductive system?
5. Order the following events chronologically: fertilization, implantation, and ovulation.

ETHICS Prenatal Screening

In Iceland, there are almost no children born with the genetic condition known as Down syndrome. You might ask, "Why not? Have they discovered some miracle cure?" No. Sadly, expectant Icelandic mothers of unborn children with Down syndrome almost always abort their babies. Even in the United States, 67% of women who find out that their unborn child has Down syndrome choose to have an abortion. How do they know that their child is affected? The answer is *prenatal screening*.

Prenatal screening allows doctors to examine an unborn baby. The tests can be used to reveal whether a baby is a little boy or a little girl. They can also reveal whether a baby has any major health conditions such as Down syndrome or heart defects. Some prenatal screening involves taking pictures of the baby with sound waves while others involve obtaining the baby's genetic information.

People use prenatal screening for many purposes.

Use the Ethics box above to answer Questions 6–7.

6. Research, plan, and outline a three-paragraph essay explaining the scriptural principles and outcomes to argue whether prenatal screening is ethical. Examine the different purposes that people use prenatal screening for. Are these purposes good or bad? Do the good purposes justify the use of prenatal screening? Or do the bad purposes outweigh the good ones?
7. Now, using your outline as a guide, write your response.

17C

HUMAN GROWTH AND DEVELOPMENT

How will I change as I get older?

17C Questions

- What are the stages of growth?
- How is each stage of growth different?

17.6 PRENATAL DEVELOPMENT TO CHILDHOOD

It takes about forty weeks of pregnancy for a human life to develop from a single cell into an adorable newborn. Doctors split that period into three stages called **trimesters**, each of which lasts about thirteen weeks.

PRENATAL DEVELOPMENT AND BIRTH

THE FIRST TRIMESTER

Weeks 1–13. The first trimester brings many changes to the rapidly dividing zygote. After the baby attaches to its mother, an organ called the **placenta** develops. This organ allows the baby to get nourishment from the mother without their blood actually mixing. The **umbilical cord** is a tube that connects the baby to the placenta. It delivers nutrients and takes away waste products from the embryo. From when the baby reaches nine weeks old until he is born, he is considered a **fetus**. The baby in this picture is six weeks old and is the size of a large pea.

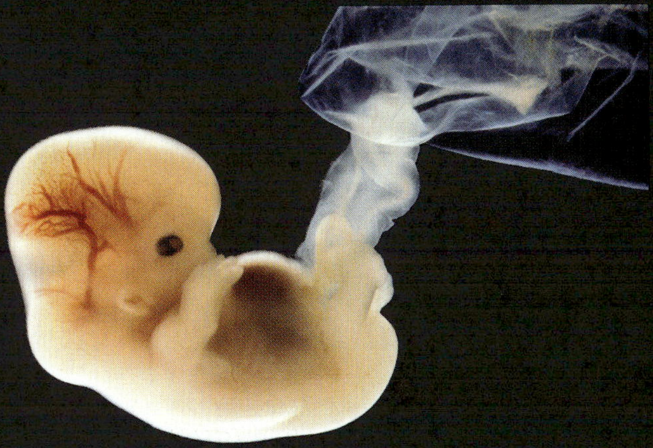

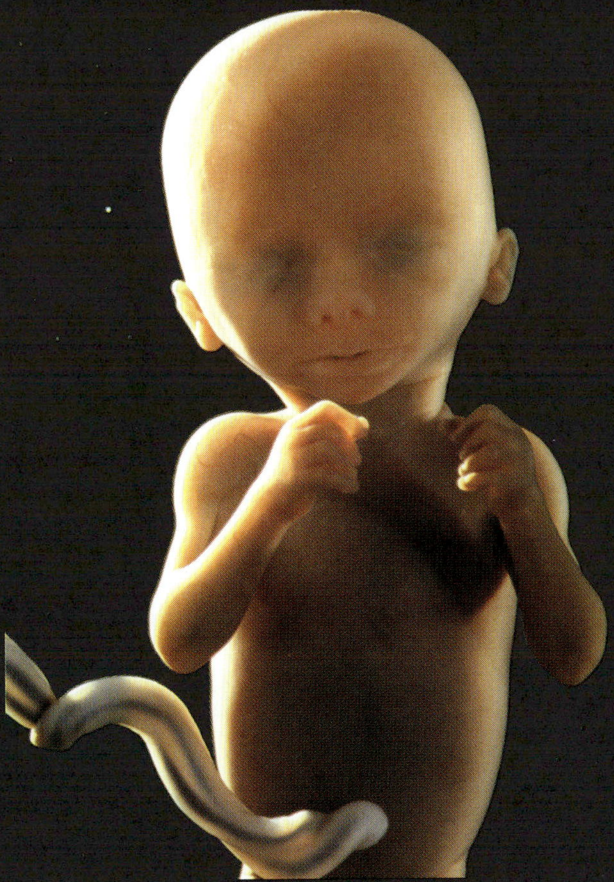

THE SECOND TRIMESTER

Weeks 14–27. The fetus is definitely starting to look like a baby now! Bones, brain, and personality are developing. The baby can suck his thumb, make faces, and leave fingerprints. This baby is sixteen weeks old and is about the size of a pear.

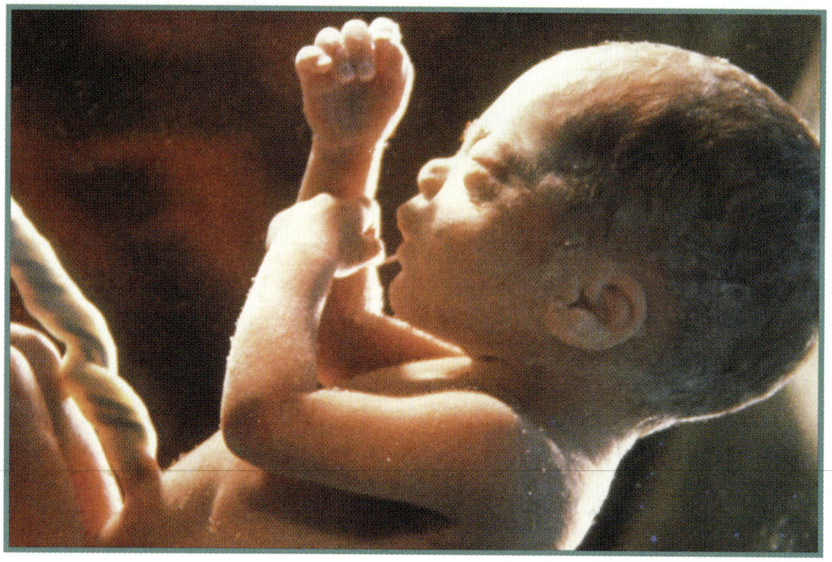

THE THIRD TRIMESTER

Weeks 28–40. Baby is getting ready to enter the world! He's growing hair and opening and closing his eyes. He's getting big enough that Mom is getting rather uncomfortable. This baby is thirty-four weeks old and about the size of a cantaloupe.

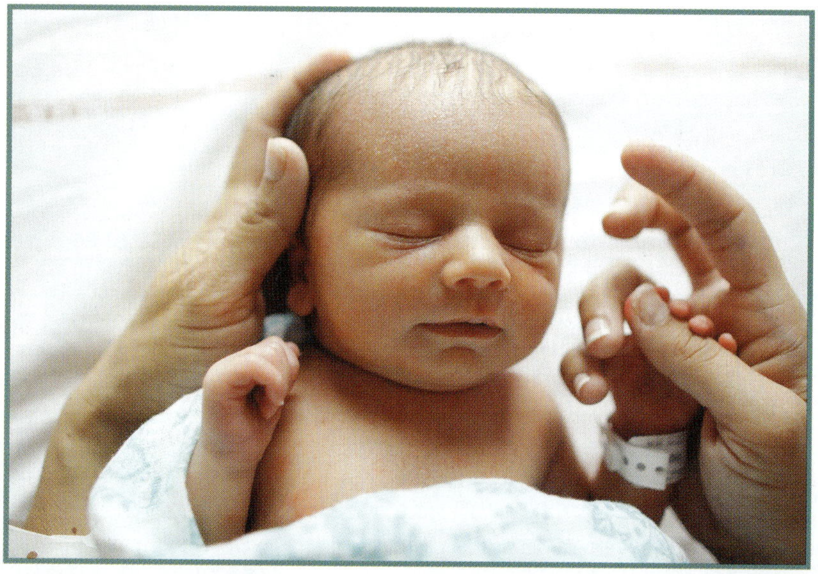

BIRTH

It's time! Mom's endocrine system fires up, producing hormones that start labor. Muscles in the uterus repeatedly squeeze the baby until he emerges into the world. The placenta is expelled afterward.

During your first year of life, you changed so much. You learned to eat rather than get nutrition from your mother. You learned to move on your own. You learned to communicate with others. You became and are still becoming more independent.

When you grew from a baby into a child, your body proportions began to change. Your arms and legs got longer as you became more coordinated. You developed mentally too. All these changes affected your social interactions. And in a person's teen years—a period known as *adolescence*—physical, mental, and social changes peak.

17.7 ADOLESCENCE AND ADULTHOOD

We've already mentioned the changes that your body undergoes with puberty because of the action of the endocrine system. You also change mentally and emotionally as you transition to adulthood. Hopefully you also change spiritually as you mature in your walk of faith with Jesus.

By the time you are in your late teens, you will have reached physical maturity. When you reach your twenties, you will be strongest physically. People usually begin to have children when they are in this phase of life.

After your twenties, your body doesn't stop changing. Your metabolism will begin to slow. You will begin to experience changes in muscles, bones, hair, and skin as you age. This may seem to you like a long time away! Consider that the average American lives to about eighty years of age. But though our bodies get older and slow down, we can still have rich experiences. Think of how older people have invested in you. You can invest in people the same way as you get older.

God has given you a certain number of days to live. You don't know how much time God has given you, so it's important to be wise with the time that you have (Ps. 90:12). How will you choose to live your life? You can be sure that real blessedness comes from following God's plan rather than listening to the ungodly (Ps. 1).

Although we will all face death, it is not the end. The Bible teaches that those who have trusted in Jesus' death and resurrection for redemption will be raised. They will have glorified bodies that do not suffer the effects of the Fall. Death can be the beginning of a new life with God. When God's redemption of this world is complete, we can enjoy the work of managing His creation without the frustrating effects of the Fall. And who knows? Maybe you can use what you've learned about life science even then!

LAB EXERCISE

EXPLORING GROWTH CURVES

Equipment

computer with Internet access

a diverse group of students

a tape measure

How does the growth of different people compare?

You've probably noticed that not everyone in your class is the same height even though you are all basically the same age. Why is that? Some of it has to do with the genetic information that you got from your parents. Doctors have graphs called *growth curves* that they use to determine whether a person is on track for growth. But these graphs take into account that people are different and they grow differently.

PROCEDURE

❶ Do a keyword search on the Internet for "growth curves." Make sure that you find one for both guys and girls and print them out.

1. Why are there different growth curves for guys and girls?

❷ Now take a good look at your graphs.

2. What are the three kinds of data displayed on these graphs?
3. What do each of the lines have beside them?

The title of a growth chart tells us that the lines represent *percentiles*. The idea here is the same as the results you get from your standardized tests. A person with an average height for a twelve-year-old would be in the 50th percentile, which means that about 50% of twelve-year-olds are taller than this value and 50% are shorter. When he plots his age and weight on the graph, the dot will fall exactly on the line that has a 50 by it. About 70% of twelve-year-olds will have heights that range between 7.6 cm less and 7.6 cm more than this height.

❸ Measure the height of the people in your group using the tape measure or work with the data that your teacher gives you.

4. Record the height data in centimeters along with the age data in years.
5. Use this data to find the percentiles for each of your data points.

ANALYSIS

6. Using the data in the growth curves, do you expect the people in your group to keep growing? Explain.
7. How do the heights of an average twelve-year-old guy and girl compare?
8. How do the growth curves of an average twelve-year-old guy and girl compare?

Now look at the steepness of the lines. This is called the *slope*. Maybe you've seen this topic in math class! The slope of these lines shows *how fast* people are growing. The steeper the line, the faster the growth.

9. Do people grow at the same rate their whole life? Describe what you observe.
10. At what age will most guys and girls stop getting taller?

CONCLUSION

People are different! God shows His creative design by the unity and diversity in His creation. When we can value people that are diverse, we can return glory to Him. Your height is something that God knows and governs, so don't worry about it! Instead, honor Him by being content with how He has created you (Ps. 139:14–16).

GOING FURTHER

Want to learn more? Look at your graphs to see how weight changes with age. You can also find growth curves for head circumference. Both weight and head circumference change differently with age than height does.

17C REVIEW QUESTIONS

1. List in order the stages of prenatal development, including birth.
2. Associate the following features with prenatal development, childhood, adolescence, or adulthood.
   a. placenta
   b. puberty
   c. slowing metabolism
   d. changing body proportions
   e. umbilical cord
   f. developing verbal communication
   g. having children
3. What is the difference between an embryo and a fetus?
4. What causes a child's body to begin to develop into an adult?

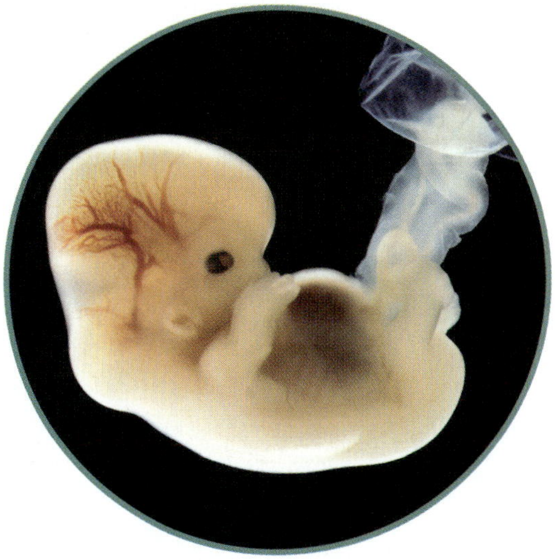

CHAPTER 17

CHAPTER SUMMARY

17A THE ENDOCRINE SYSTEM

- God has created every person as either male or female down to each single cell.
- The endocrine system is made of glands that manage and produce hormones. These include the hypothalamus, pituitary gland, thyroid, pancreas, adrenal glands, and gonads.
- The glands in the endocrine system send hormones through the bloodstream.
- Feedback systems manage the amount of hormones that endocrine glands produce.
- Both guys and girls change during puberty: both experience voice deepening, guys start sperm production, and girls experience a menstrual cycle.

17A TERMS

hormone	353
hypothalamus	353
pituitary gland	353
thyroid	353
pancreas	353
adrenal gland	353
puberty	355
menstrual cycle	355
ovum	355
ovulation	355

17B THE REPRODUCTIVE SYSTEM

- God created mankind with the ability to reproduce within the biblical model of the family to fulfill the Creation Mandate for His glory and mankind's own pleasure.
- The function of the reproductive system is different from all the other systems because it doesn't work to maintain homeostasis; instead, its goal is to produce and protect new life.
- The human body produces new life when a sperm fertilizes an ovum and the fertilized ovum implants in the female's uterus.

17B TERMS

gonad	358
embryo	360

CHAPTER SUMMARY

17C HUMAN GROWTH AND DEVELOPMENT

- The stages of human growth and development are prenatal development, childhood, adolescence, and adulthood.
- A person's rate of growth, physical features, mental ability, and personality change in different ways in the different stages of growth and development.

17C TERMS

trimester	361
placenta	361
umbilical cord	361
fetus	361

CHAPTER REVIEW QUESTIONS

Recalling Facts

1. List five of the glands in the endocrine system.
2. The function of the endocrine system is to produce and release ______ to maintain homeostasis.
3. Name one change to a person's body that comes with puberty.

In Questions 4–12, match the descriptions to the labeled structures in the diagrams at right.

4. the male gonads
5. the female gonads
6. produces estrogen
7. produces testosterone
8. produces ova
9. produces sperm
10. location of ovulation
11. location of fertilization
12. location of implantation

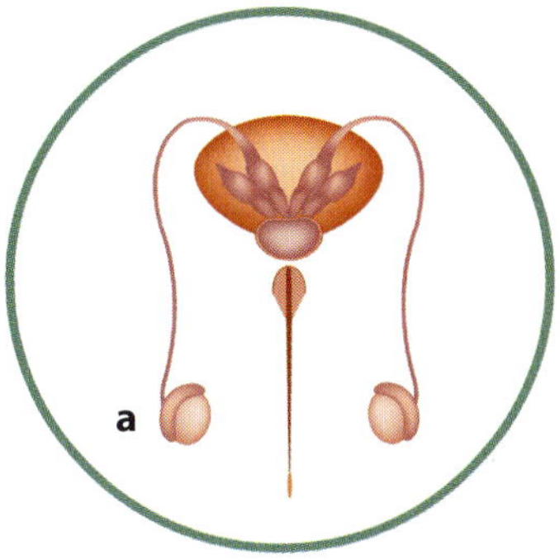

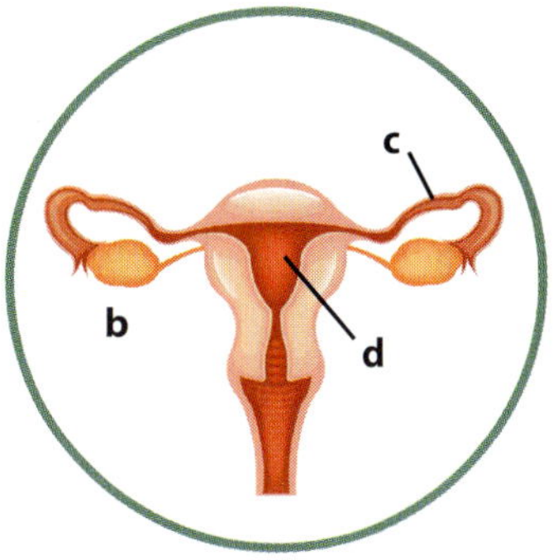

13. List the stages of human development including those before birth.
14. In which stage of growth and development does a person develop the most mentally?

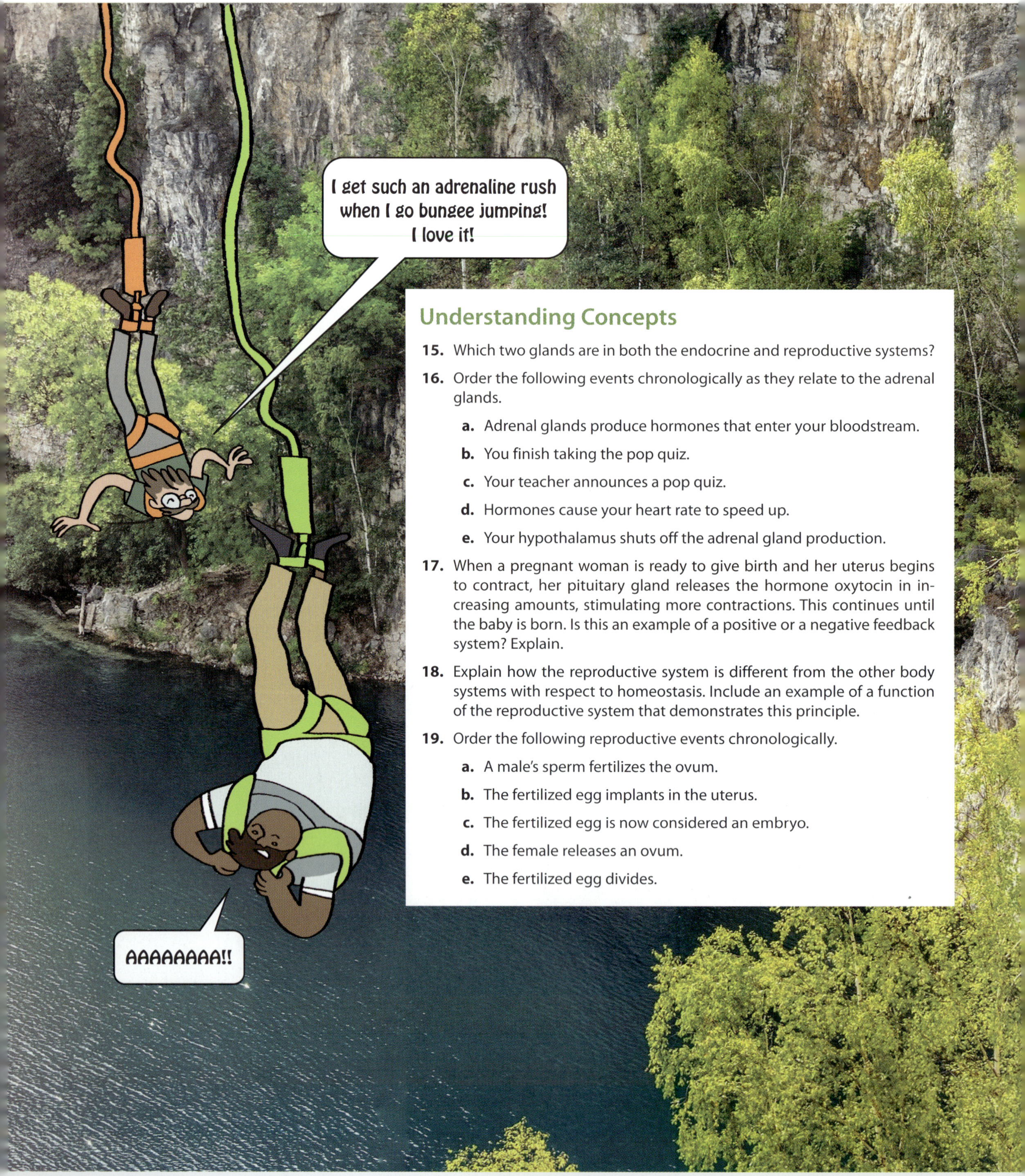

Understanding Concepts

15. Which two glands are in both the endocrine and reproductive systems?

16. Order the following events chronologically as they relate to the adrenal glands.

- **a.** Adrenal glands produce hormones that enter your bloodstream.
- **b.** You finish taking the pop quiz.
- **c.** Your teacher announces a pop quiz.
- **d.** Hormones cause your heart rate to speed up.
- **e.** Your hypothalamus shuts off the adrenal gland production.

17. When a pregnant woman is ready to give birth and her uterus begins to contract, her pituitary gland releases the hormone oxytocin in increasing amounts, stimulating more contractions. This continues until the baby is born. Is this an example of a positive or a negative feedback system? Explain.

18. Explain how the reproductive system is different from the other body systems with respect to homeostasis. Include an example of a function of the reproductive system that demonstrates this principle.

19. Order the following reproductive events chronologically.

- **a.** A male's sperm fertilizes the ovum.
- **b.** The fertilized egg implants in the uterus.
- **c.** The fertilized egg is now considered an embryo.
- **d.** The female releases an ovum.
- **e.** The fertilized egg divides.

Critical Thinking

The parathyroid glands are four small glands usually located on each side of the neck behind the thyroid. They produce parathyroid hormones that move calcium from the bones to the blood if blood calcium levels get too low.

Use the graph below (showing the levels of calcium and parathyroid hormone in the blood) to answer Questions 20–22.

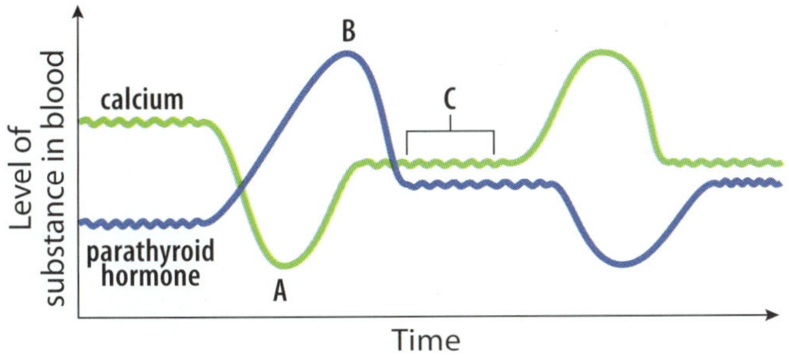

20. Does this hormone exhibit positive or negative feedback? Explain.
21. Which letter indicates what happens to parathyroid hormone levels after blood calcium levels drop too low?
22. Which letter indicates homeostasis of calcium and parathyroid hormone levels?
23. How should a Christian think about human reproduction?
24. How will your body change as you get older?

ETHICS Artificial Wombs

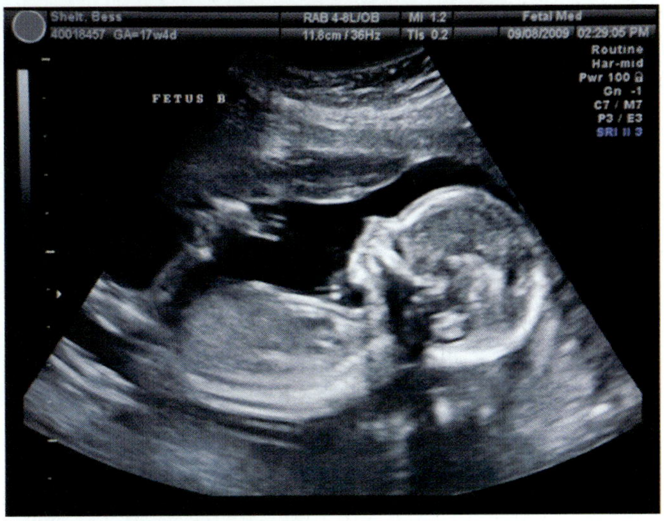

Babies born before twenty weeks old can't survive outside the womb with current medical technology. Babies born about twenty-three or twenty four weeks old have a fifty-fifty chance of surviving. Most of them will have medical troubles such as cerebral palsy, blindness, and paralysis.

Pediatricians at Thomas Jefferson University are working on this problem. They have been creating and experimenting with artificial wombs, bags that could enclose a fetus and supply nutrients and carry away wastes. They have been experimenting with fetal lambs, which have proved to be good models for human fetuses. But would it be ethical to even test this on human babies, much less use them on a wide scale? And how would this affect legislation on abortions?

Use the Ethics box above to answer Question 25.

25. Write a three-paragraph response on the ethical use of artificial wombs. In the first paragraph, discuss principles from Scripture that may guide Christians in this area. In the second paragraph, discuss the possible outcomes of heeding or ignoring the guidelines of Scripture. In the third paragraph, give your opinion on how scientists should use artificial wombs.

Unit 5 INTERACTING WITH THE BIOSPHERE

FROM A MARINE BIOLOGIST TO YOU

I love science! But when I was younger there was a conflict between what I was being taught in school about evolution and what I read about science in the Bible. Everyone I knew, as far as I was aware, believed in evolution. When I got to college, my professors had a lot of hard questions for me! When I finally met some people who did not accept evolution, I could hardly believe it. But with their encouragement I began to rethink what I believed.

Studying the natural world has become a lifelong passion. I have had the pleasure of studying coral reefs for my doctoral work. After hundreds of scuba dives and thousands of hours learning about the weird and wonderful creatures that live there, I can testify with the psalmist, "For thou, Lord, hast made me glad through thy work: I will triumph in the works of thy hands" (Ps. 92:4).

Dr. Rob Carter

18 EXPLORING THE BIOSPHERE

PARKS IN PERIL

America's national parks are places where people can get back to nature. In places like Yosemite, Yellowstone, and the Grand Canyon, folks can find fresh air, scenic beauty, and plenty of open space. These days, there's something else people can find in our national parks—lots of other people! How can we keep wild places wild and still have them be accessible to millions of visitors? All those visitors can have a lasting and harmful impact on the ecology of the parks they visit. Exercising good and wise dominion over God's creation includes exploring ways to protect natural places for people to enjoy.

18A

FACTORS IN THE ENVIRONMENT

What makes one environment different from another?

18A Questions

- What is an ecosystem?
- What things make up an ecosystem?

18.1 ECOSYSTEMS AND ECOLOGY

Let's think for a moment about the things you can expect to find in a national park like Yosemite. There are natural features like mountains, rivers, and glaciers. There are also living things like bears, trout, and pine trees. These particular organisms and the place where they live together make up an **ecosystem**—a limited area in which living and nonliving things interact. The study of the interactions between those living and nonliving things is called **ecology**.

18-1
What things affect the life of this trout?

At first, the study of ecology might not seem very complicated. Think of a trout living in one of Yosemite's streams, for example. What factors affect its life? Simple, right? There's the trout, and there's the stream that it lives in. But it's not actually as simple as that. For instance, where did the water in the stream come from? Originally it fell as either rain or snow. But where did *that* come from? The water vapor that fell as rain or snow had risen into the air through evaporation far out over the Pacific Ocean. So the water in the stream is affected by events that happen many hundreds of miles away. Complex, isn't it? The never-ending tangle of actions and interactions within an ecosystem is what makes ecology fascinating!

18.2 FACTORS IN THE ENVIRONMENT

Ecosystems consist of many parts that interact with each other. Some of those parts are living things and others are not. We call the living parts of the ecosystem the **biotic factors.** The nonliving parts are the **abiotic factors**. Let's take a look at some of these.

ABIOTIC FACTORS

Sunlight. The amount of sunlight that an ecosystem receives determines how much energy is available. It controls how much photosynthesis can take place and also affects the daily pattern of temperatures and local climate.

Air. Of course, the oxygen that living things need is found in Earth's atmosphere. But air also contains the carbon dioxide that plants need for photosynthesis and the water vapor that forms rain or snow.

Water. All living things need water. The kinds and numbers of organisms that can live in an ecosystem are governed by *how much* water is available and *when* it is available.

Altitude. This mountain peak is bare rock because of its altitude. High-altitude ecosystems are colder than those at lower elevations. Fewer plants and animals are adapted to living in such harsh conditions.
Soil. Whether rich loam or bare rock, the soil or substrate in an ecosystem affects what kinds of plants can grow there.
Climate. The **climate** of an ecosystem is its long-term pattern of weather. Climate is determined mainly by how much sunlight and water an ecosystem receives.

BIOTIC FACTORS

Plants. Plants play a crucial role in most ecosystems by making the sun's energy available to other organisms in the form of sugars that are produced during photosynthesis.

Animals. Only a small number of animals in an ecosystem are as large as this mule deer. Most of the animals in an ecosystem, such as insects, are far smaller.

People. Of course, humans can have a huge impact on an ecosystem too. Visitors to national parks, such as Yosemite, need transportation, food, and lodging. In 2016 alone, Yosemite roads, restaurants, hotels, and campgrounds had to accommodate over five million park guests.

PROTECTING PARKS AND PEOPLE: SERVING AS A PARK RANGER

"Protect the park from the people, the people from the park, and the people from the people" is an unofficial park ranger motto. Yosemite is a wild and sometimes dangerous place, and people have to be protected. Besides natural hazards such as wild animals, rock falls, icy trails, and cold, swift rivers, some visitors face the extra dangers of high-risk adventures like rock climbing. On September 11, 1993, a rock climber fell to his death from Yosemite's Lost Arrow Spire. Ranger Mary Litell Hinson was on the job. She was part of the search-and-rescue team that found the body. Later she broke down in tears as she informed the climber's friend of the tragedy.

But protecting people from the park isn't a ranger's only job. Sometimes they have to protect the park from the people or the visitors from each other. The day's challenge might be picking up litter, monitoring campfires, or closing trails to protect mother bears and their cubs. Park rangers have to be ready for anything!

18A REVIEW QUESTIONS

1. What is an ecosystem?
2. What is ecology?
3. List three abiotic factors in an ecosystem.
4. Would bacteria living in a stream be an abiotic or biotic factor in an ecosystem? Explain.
5. Choose one factor in an ecosystem and describe how it might affect one other factor in the same ecosystem.

LAB EXERCISE

CARP QUESTION

Can one abiotic factor affect another?

E **Equipment**

bowls or beakers (2)

probeware with temperature and dissolved oxygen probes

Both trout and carp can be found in lakes that ice over during winter. But carp are also perfectly happy in warm ponds and lakes where no trout can live. Is there another abiotic factor besides temperature that affects each fish's ability to survive in either warm or cold water?

PROCEDURE

Day 1:

❶ Fill two bowls or beakers with water. Allow one to chill in a refrigerator overnight; keep the other at room temperature.

Day 2:

❷ Copy Table 18-1 on a separate sheet of paper.

❸ Use the probeware to measure the temperature and the dissolved oxygen concentration of the water in each container. Record your results in the *Day 2* row of Table 18-1.

❹ Place the container of room-temperature water in the refrigerator and allow it to chill overnight. Leave the container of chilled water out to achieve room temperature.

1. Why is oxygen important for living things like trout and carp?
2. Do both containers have the same dissolved oxygen concentration? If not, which container of water has a lower dissolved oxygen concentration?

Day 3:

❺ Measure the temperature and dissolved oxygen concentration of each container. Record your results in the *Day 3* row of Table 18-1.

3. Are your results the same as those obtained on Day 2?
4. What appears to be the relationship between the temperature of the water and its dissolved oxygen concentration?
5. On the basis of your results, give two possible reasons why carp can survive in warm water while trout cannot.
6. Imagine you work for a company that raises trout to sell for stocking ponds and lakes. A farmer comes to you in the spring and asks about the possibility of stocking trout in his farm pond. Using the results in your experiment, what do you think you would need to know before making a recommendation to the farmer? Explain.
7. Why do you think it was important for this experiment to switch the two samples and retest them on Day 3?

Table 18-1

Dissolved Oxygen Concentration Versus Temperature in Water

	Chilled		Room Temperature	
	Temperature (°C)	Dissolved Oxygen (mg/L)	Temperature (°C)	Dissolved Oxygen (mg/L)
Day 2				
Day 3				

18B

BIOMES AND ECOSYSTEMS

How do scientists study the environment?

18.3 MODELING THE ENVIRONMENT

God designed Earth to be a home for living things. And there is life almost everywhere on Earth. Protists live 11,000 m deep in the Marianas Trench, and bar-headed geese fly 6500 m high through the Himalayas. Saharan silver ants forage in one of the world's hottest deserts, while polar bears patrol the pack ice of the Arctic Ocean. These places, along with all the other habitable portions of the earth, are called the earth's *biosphere*.

But the biosphere is extremely large and diverse. At present, it's impossible to model every interaction that takes place between all the biotic and abiotic elements within the entire biosphere. There's just too much information out there! It makes sense then for ecologists to break down the earth into smaller chunks for study. One way to do this is to group together parts of the biosphere that have similar biotic and abiotic factors. These smaller parts of the biosphere that share similar climates and kinds of living things are known as **biomes**.

Of course, there are some factors that can affect the entire biosphere. But thinking about the biosphere as a collection of different types of biomes makes it easier for ecologists to learn about and model the interactions within each type. Even so, biomes can still be dauntingly large and complex. To really understand what happens in them, scientists can define even smaller parts of biomes for study. You'll see some of these in Subsection 18.5.

18B Questions

- How do ecologists model the environment?
- What kinds of biomes are there?
- How are habitats and niches related?

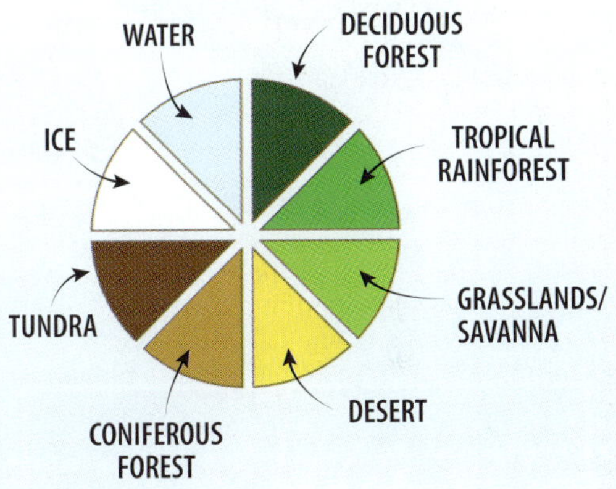

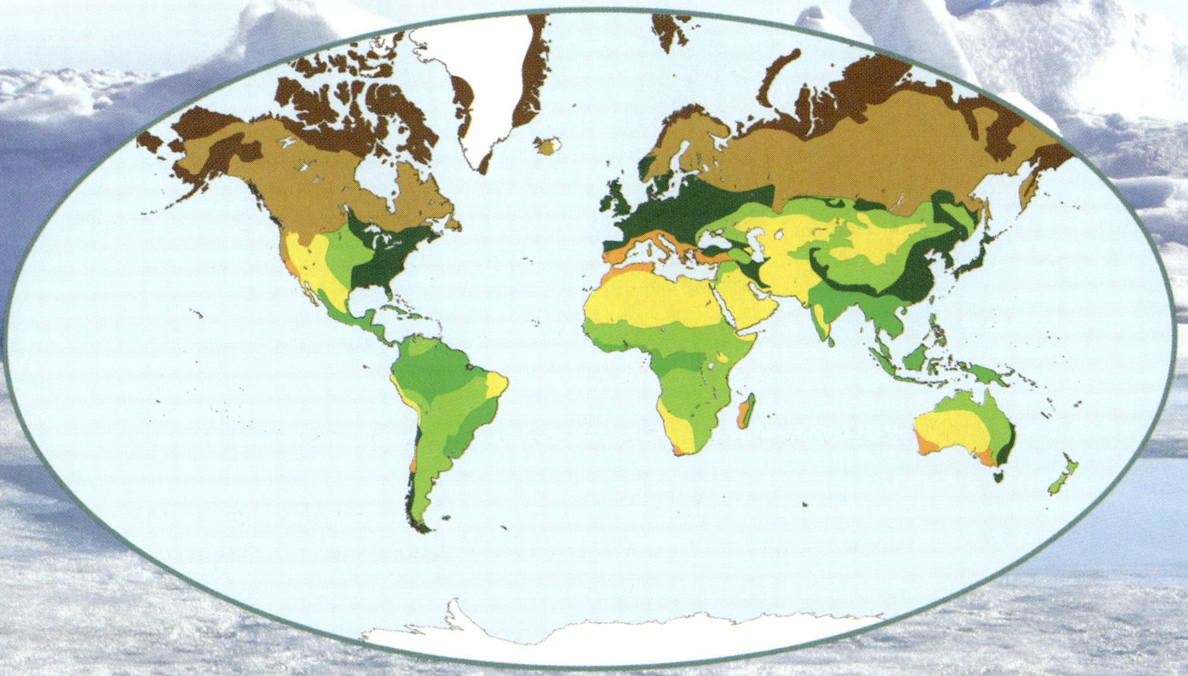

Coniferous Forest

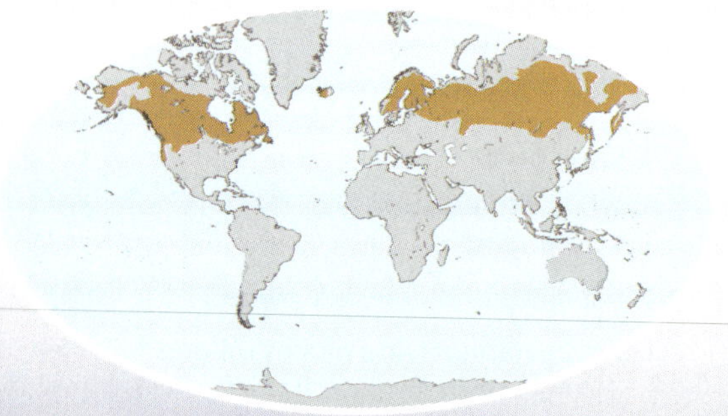

In far northern North America, Europe, and Asia, winters are long, snowy, and cold, while the short summers are warm and moist. **Coniferous forests**, also called *boreal forests* or *taiga* in that part of the world, cover vast areas of the region. In fact, coniferous forests are Earth's largest land biome. Remember that coniferous trees are those that bear their seeds in cones. The needles and scales of conifers are designed to survive the cold of winter. Downward-sloping branches effectively shed snow. Thickly wooded coniferous forests typically allow little light to reach the forest floor. And their fallen needles also create poor soil conditions. As a result, few other kinds of plants can grow on the forest floor.

greenish warbler

Only very hardy animals, such as lynxes, moose, and great gray owls, can live and remain active in the northern forests year-round. Less sturdy residents, such as mice and squirrels, may spend the winter months hibernating. Many migratory birds, such as warblers, spend only their summers here. They come to take advantage of the huge swarms of flies, gnats, and mosquitos that breed in the forest's abundant bogs and marshes.

moose

Deciduous Forest

The trees in a **deciduous forest** lose their leaves at the end of each growing season. These forests are found across large portions of North America, Europe, and Asia. Deciduous forests have four seasons, mild temperatures, and around 100 cm of rainfall each year.

The decayed leaves of deciduous trees make rich soil. Good soil and ample rainfall combine to produce a thick and often tangled *understory* (the plants of the forest that grow beneath the *canopy* formed by tall trees).

The far-flung deciduous forests of eastern North America are home to many kinds of animals. Here you might see black bears, white-tailed deer, opossums, and turkeys. The same sort of forest in India may be home to tigers, gaurs (similar to bison), Asian elephants, and peacocks.

Tundra

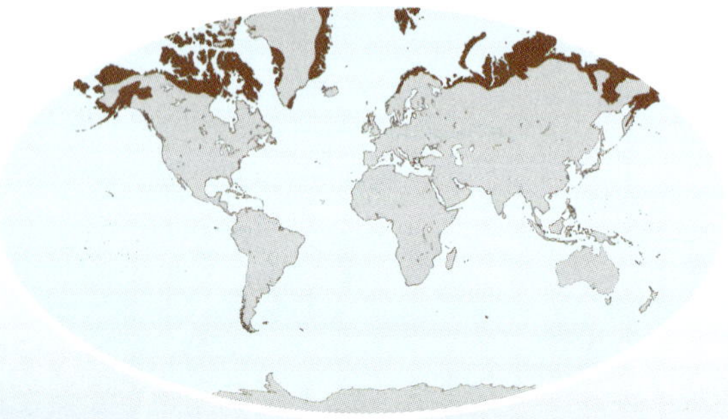

If you were to travel northward through the boreal forests of the Northern Hemisphere, in time you would reach what's called the *tree line*. North of this line is a region where only a few inches of soil thaw out during the short summer growing season. Beneath that is a layer called *permafrost*, which remains frozen year-round. Rainfall is also scarce here. The thin soil and lack of rain prevent the growth of tree roots. Instead of trees, huge areas are covered by low-growing shrubs, grasses, and lichens. This biome is called **tundra**. In the north it's known as *Arctic tundra*, but it also exists in high mountains. There it is known as *alpine tundra*.

Though it is a harsh environment, the Arctic tundra supports a surprising amount of wildlife. Gray wolves prey on caribou and musk oxen. Arctic foxes pursue smaller prey such as lemmings and tundra voles. Small ponds and streams formed by summer meltwater produce vast clouds of insects. Astonishing numbers of ducks, geese, and other migratory birds take advantage of this seasonal abundance of food and nesting habitat to raise their young. Before the winter cold returns, these birds take wing for warmer climes, sometimes as far away as South America. Their annual migrations often cover many thousands of kilometers.

Grasslands

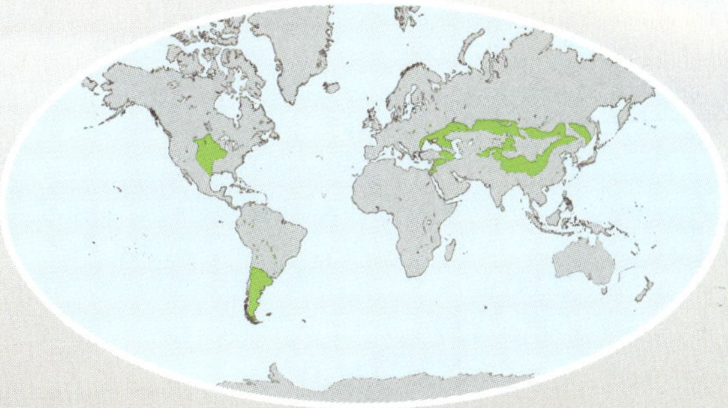

Grasslands are often found in the interiors of continents, especially in North and South America and Asia. Interior summers tend to be warm, while winters can be cold and windy. Also called *prairies* or *plains*, these areas don't receive enough rainfall to support the growth of forests. Many grasslands receive 25–75 cm of rain each year. Some grasslands get more than this, but dry-season fires prevent tree growth. Yearly rainfall determines what kinds of grasses will grow on a grassland. Less rain normally permits only shorter grass species to grow. Deep, fertile grassland soils have often served humans well as productive farmland.

Grasslands provide plenty of food for grazing animals. Bison and pronghorns once roamed the Great Plains of North America in large numbers. In the South American *pampas*, one can find guanacos, while the Eurasian *steppes* support wild horses and gazelles. Sometimes the herds aren't mammals—instead, they're large, swift-running birds such as emus or rheas. Abundant grains also feed large numbers of rodents like mice, gophers, and prairie dogs. These in turn are preyed upon by foxes, badgers, and other small carnivores. From the air, or perched on a rare tree or utility pole, hawks and falcons also scan the seas of grass for unwary rodent meals.

American badger

Swainson's hawk

bison

Desert

Would it surprise you to learn that high temperatures are *not* the defining feature of a desert? It's true! **Deserts** are biomes that receive very little water. Each year they can expect to see about 25 cm or less. The Sahara in Africa is perhaps the world's most famous desert. When we think of it, we often imagine scenes of blistering-hot sand dunes. But the frigid interior of Antarctica is also a desert. It receives only about 5 cm of water each year in the form of snow.

The extreme temperatures of the scorching Sahara dunes and the frozen heart of Antarctica limit the kinds of life that can survive there. But the more temperate deserts found elsewhere are home to a wide variety of plants and animals. All of them have special features or behaviors designed for living in an environment short on liquid water. Desert plants often have thick, fleshy stems and leaves for storing water. Many also conserve water by not blooming every year. Instead, they produce flowers only during rare wet years. Desert mammals can keep cool by being more active at night. Reptiles warm their bodies in the morning sun, then seek shade as the daytime heat increases.

Only a few large mammals are suited for desert life. These include mountain lions, oryx, and, of course, camels. Most desert mammals, like jackrabbits, jerboas, and kangaroo rats, are smaller. Small- to medium-sized desert hunters, such as jackals and bobcats, take advantage of these food sources. Desert birds include hawks, owls, and quail. But deserts are perhaps best known for their diverse kinds of reptiles. Among these are two well-known species of poisonous lizards—the Gila monster and Mexican beaded lizard.

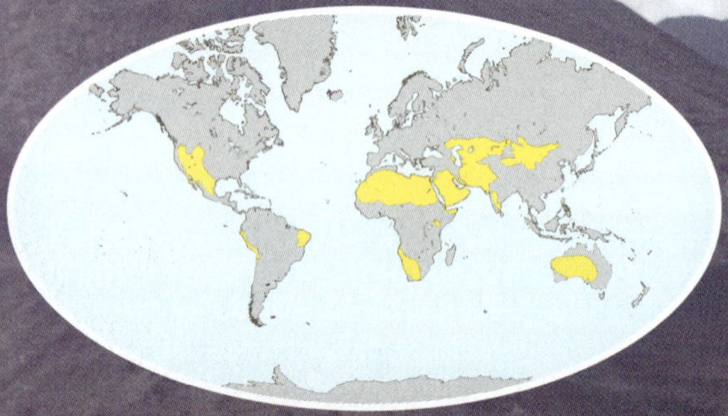

Fabiani lizard

vicunas

Tropical Rainforest

Tropical rainforests combine year-round warmth and large amounts of rainfall with a lack of a dry season. They get a lot of rain—over 170 cm per year and often much more. Some rainforests receive over 10 *meters* of rain in a year!

The many and varied plants in a rainforest make up four distinct, vertical zones. The highest zone, the *emergent layer*, consists of the tops of the tallest trees. Beneath this is a thick *canopy layer* made of the crowns of shorter trees. Little sunlight can get through this layer, so the lower layers exist in perpetual shade. Closer to the ground is the understory made of small trees, bushes, and large-leafed plants. Finally there is the *forest floor* itself, rich in fungi and bacteria that decompose the abundant plant and animal material.

Rainforests are well known among the world's biomes for having the greatest diversity of living things. A few hectares of rainforest may have hundreds of species of trees. Insects, birds, amphibians, and reptiles each show their greatest variety of form and color here. Most of these creatures live in the rainforest's trees. Knowing that, it's no surprise then to learn that some of the most common rainforest mammals are tree-dwelling primates (monkeys, lemurs, and apes) and bats. In the rainforest, even plants live on other plants! Held high above the soil by their host plants, epiphytes such as orchids and bromeliads obtain moisture and nutrients from the damp rainforest air.

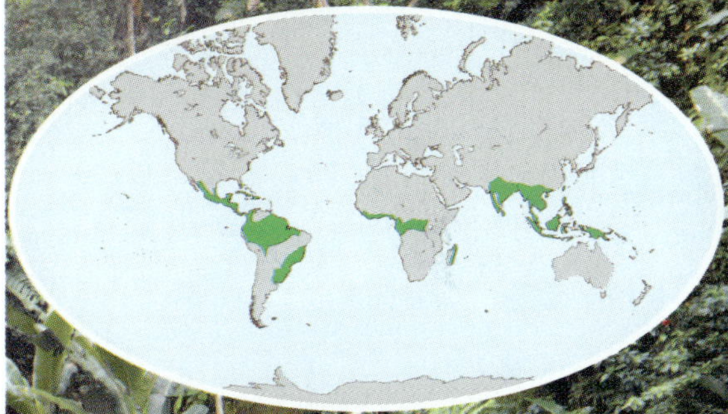

Chaparral

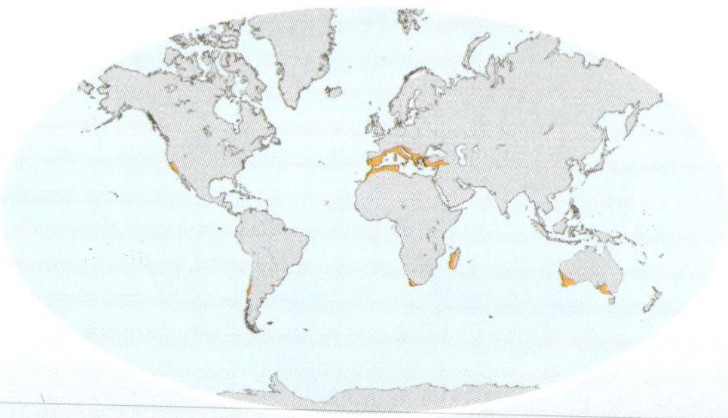

Chaparrals are found in coastal areas with mild, wet winters and hot, dry summers. Dry summers mean that chaparrals are also subject to cycles of wildfires and regrowth. One of the largest regions of chaparral is the coastlands around the Mediterranean Sea. For this reason, chaparrals are often said to have a Mediterranean climate.

Chaparral plants and animals must be able to cope with both fire and drought. The most common plants in chaparrals are shrubs. They often have tough, waxy leaves to conserve water. Many have thorns that deter browsing animals. Unlike other biomes, a chaparral's growing season is in winter and early spring when water is available. Many chaparral plants depend on fires to return nutrients to the soil and to clear the ground for new growth.

Chaparral animals are adapted to seasonal changes. Large animals like deer may escape summer heat by migrating to cooler elevations in the mountains. Smaller mammals, such as jackrabbits, become less active during the summer or restrict their activities to the coolness of night. Amphibians survive the dry weather by entering a dormant state called *torpor*. They find a damp, sheltered place to hide until the winter rains return.

Savanna

The **savannas** of the world are regions that are not quite either forests or grasslands, but rather something in between. Their widely spaced trees don't form a closed canopy like those in a forest. The understory is mostly grasses. Like chaparrals, savannas have a wet season and a dry season. Rainfall totals are modest. Savannas near the Equator have warm temperatures throughout the year. They are sometimes called *tropical grasslands*.

Tropical savannas cover large portions of South America, but the best-known examples are in Africa. Here one finds endless seas of grass dotted by acacia and baobab trees. The savanna supports a wondrous variety of grazing animals. Wildebeests, zebras, giraffes, rhinoceroses, and many kinds of antelope abound. These are preyed upon by prides of lions, stealthy leopards, and swift cheetahs.

Life on the African savannas is strongly linked to the changing availability of water. Wet-season rains refill lakes and streams and allow new grasses to grow. Grazing animals can forage over wide areas. As the dry season arrives, water supplies dry up. Animals are forced to linger near the few remaining watering holes or migrate to find other sources of water. The yearly migration of wildebeest alone involves about 1.5 million animals!

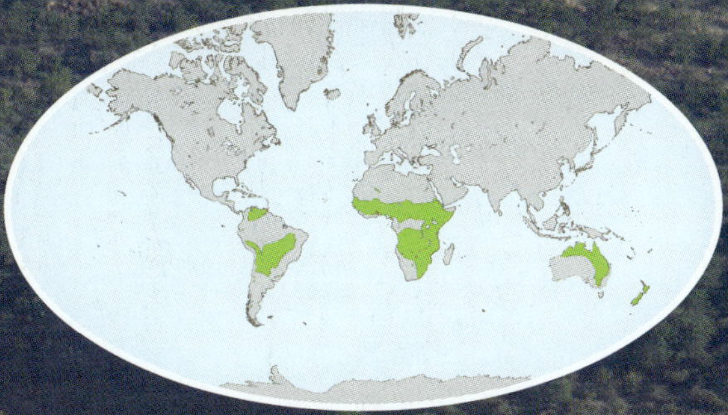

secretary bird

African wild dog

Freshwater Biomes

Freshwater biomes are diverse, but they cover less than 2% of the earth's surface. These biomes can be split into two types according to water flow. The first includes biomes composed of standing water, such as lakes, ponds, swamps, and bogs. The second includes biomes that contain running water (streams and rivers). The differences between the two types can be great. In many cases, species that live in one type cannot survive in the other.

On top of the factors that affect land biomes, aquatic biomes (those consisting of water) are greatly affected by two added factors. One is depth. The sunlight needed for photosynthesis is quickly scattered as it passes through water. The deeper the water, the less light that is available for photosynthesis. Depth also determines how much oxygen is available for living things. Deep water contains very little oxygen. In deep lakes, enough oxygen and sunlight to support much life is found only in the topmost layers.

The other major factor in aquatic biomes is the supply of nutrients needed by protists and algae. Nutrients tend to collect in low-elevation aquatic biomes, so they are often thick with algae and have murky water as a result. In contrast, high mountain lakes and streams tend to be crystal clear because few nutrients are available for algae and other organisms.

Marine Biomes

About 70% of the earth is covered by salt water. In the ocean, living things can be found from the shoreline to the bottoms of the deepest trenches (over 10 km below the surface). Like freshwater biomes, marine biomes are largely governed by depth and the supply of nutrients for photosynthesis. Because of this, most marine life is found in the upper layers of the world's oceans. Along ocean shores, rising and falling tides also regularly cover and then uncover areas of land. Organisms that live in this *intertidal zone* must be fit for life both in and out of water. Also along the coasts, rivers empty fresh water into the sea. This creates bays and marshes where the two waters mix. The water here is *brackish*—its salt level is somewhere between that of fresh water and seawater.

Marine organisms can be placed into four large groups according to their manner of living and location in the ocean. Animals such as worms and clams form part of the *benthos*. They live on the ocean's bottom. *Plankton* are animals that drift in the ocean's currents, mainly near the surface. Plankton include tiny algae and small animals such as copepods and larval squid. Animals that can move about freely, the ocean's *nekton*, are those such as whales and adult fish. *Transient animals* spend only part of their time in the ocean but depend on it in some way. These include animals like seals and marine birds.

Because of the ocean's vast size and the great variety of interactions between its abiotic factors, ocean biomes occur in many forms. At one extreme are brightly colored coral reefs, rich in number and variety of life forms. At the other extreme are the deep abyssal plains, nearly devoid of oxygen and cloaked in lasting darkness. Here, one may search endless hectares of sea floor and happen upon only a few urchins or crinoids.

18.5 ZONES, HABITATS, AND NICHES

Breaking down the biosphere into biomes helps to make modeling it easier. Still, different regions of a biome can be very different from each other. Australian savannas, for example, have very different plants and animals than savannas in Africa. This is where the definition of ecosystems (p. 373) comes in handy. Australian and African savannas are the same kind of biome, but their unique combinations of biotic and abiotic factors make them distinct ecosystems. One has eucalyptus trees and kangaroos, while the other has baobabs and rhinoceroses—very different!

Zonation

You don't have to travel from one continent to another to see a change from one kind of ecosystem to another. Take the mountains of California's Sierra Nevada for instance. Their elevations range from just above sea level in the foothills to over 4000 m at the tallest peaks. As one travels up into the mountains, the scenery gradually changes from grassland, to forest, and lastly to mountain peaks of bare rock. This gradual change caused by increasing elevation is an example of *zonation*. The creation of distinct zones within an ecosystem can be caused by other abiotic factors besides elevation. You saw earlier that tropical rainforests have distinct zones governed by the heights of the forest trees and plants. Rivers can be divided into pools and riffles that are based on the speed of flowing water. Ocean zones are determined in part by the sea's depth.

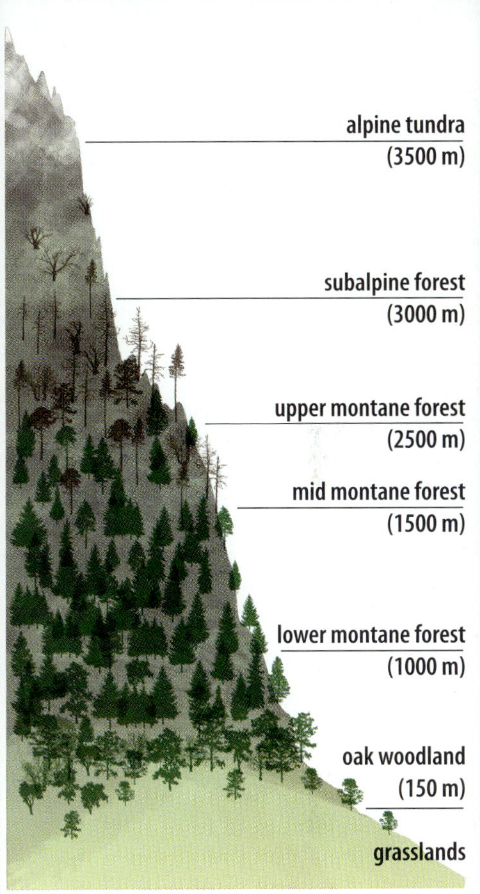

SIERRA NEVADA BIOTIC ZONES

Habitat

A single ecosystem can be *huge*! Just think of the Amazon jungle or the East African savanna. Most animals don't use all the space within an ecosystem to find food, shelter, and mates. They use only some part of it. The portion that they use is their **habitat**. In the Sierra Nevada, an example of this is the American pika, a small, rabbit-like animal. The pika makes its home only in the boulder fields found high in the subalpine forest. Pikas make their nests in the nooks and crannies between the boulders. Nearby grasses serve as food. Everything the pika needs to survive can be found in this rocky portion of the subalpine zone.

Niches

Of course, pikas aren't the only things that live in their habitat. The plants that the pikas feed on grow there. Predators such as golden eagles and long-tailed weasels hunt there—especially for pikas! Each of these organisms plays a different role in the daily life of the mountains. The parts played by pikas and plants and predators within their habitat are their **niche**—their role within the ecosystem. By eating grasses and being eaten in turn by predators, pikas serve as an important link in the transfer of energy through an ecosystem. Golden eagles have a different niche. They don't eat plants, but they do help keep the pika population in check. Every organism in an ecosystem has a slightly different niche.

18B REVIEW QUESTIONS

1. What is the biosphere?
2. What do different parts of the world that are classified as a particular type of biome have in common?
3. The floor of Yosemite Valley within Yosemite National Park is dominated by coniferous forest. Would it be more correct to define Yosemite Valley as a biome or an ecosystem? Explain.
4. Copy Table 18-2. Fill in your copy with the missing descriptors for each biome.

Table 18-2

	Coniferous Forest	Deciduous Forest	Tundra	Grassland	Desert	Tropical Rainforest	Chaparral	Savanna
Temperatures	cold winters, warm summers	mild year-round		warm summers, cold winters			mild winters, hot summers	warm year-round near Equator
Precipitation	moist year-round		scarce		very little			wet and dry seasons
Plants		deciduous trees			plants with fleshy stems and leaves	many and varied		

5. Sea turtles spend their entire adult lives at sea except for when females briefly return to land to lay their eggs. Are the beaches where they lay their eggs part of their habitat? Explain.
6. Mountain lions are the top predators in the mountains of the American West. Jaguars are the top predators in the rainforests of Central America. Which of the following statements is true for both mountain lions and jaguars?
   - **a.** They live in the same kind of biome.
   - **b.** They live in the same kind of habitat.
   - **c.** They have the same niche within their ecosystems.

CHAPTER SUMMARY

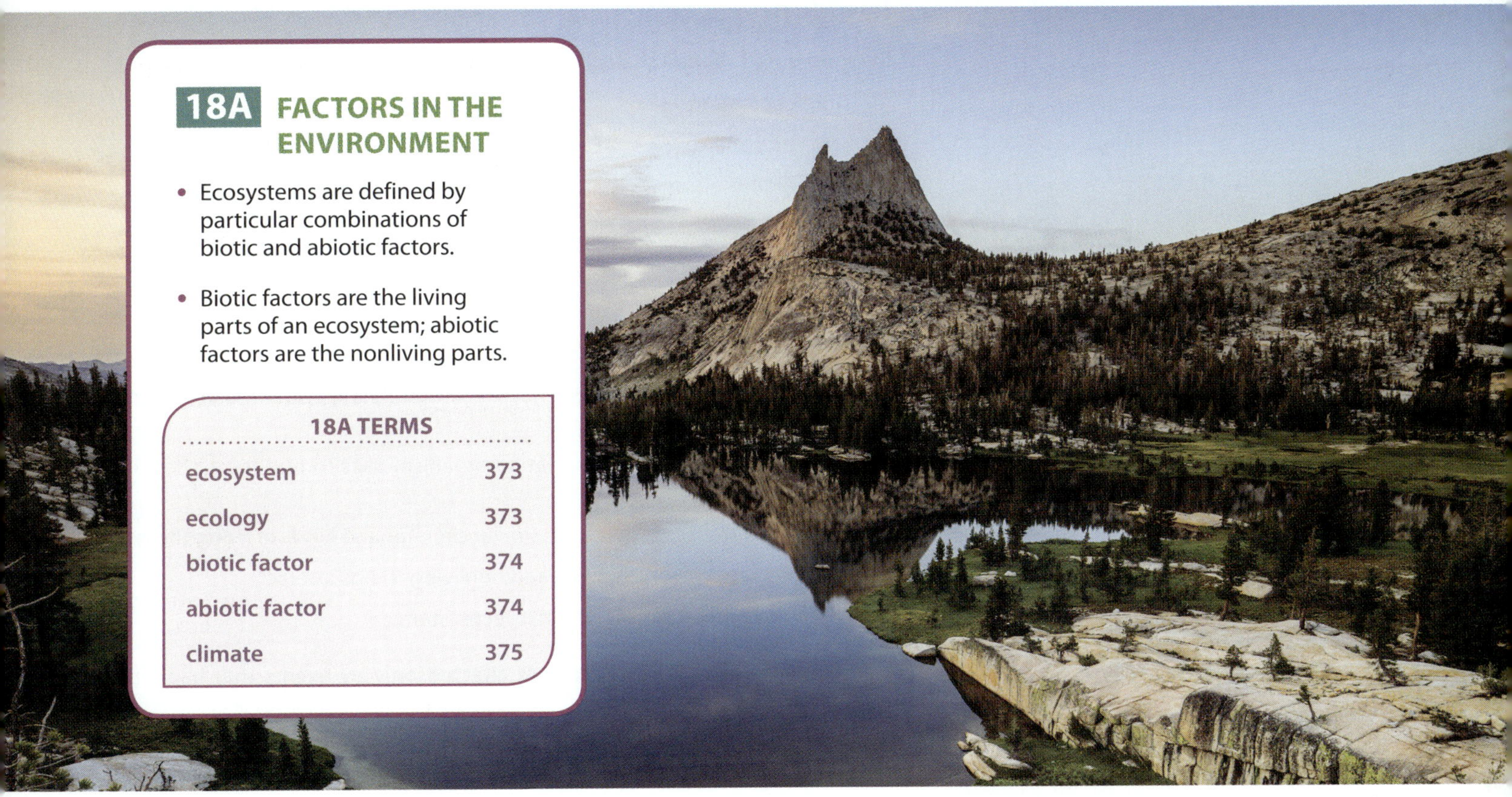

18A FACTORS IN THE ENVIRONMENT

- Ecosystems are defined by particular combinations of biotic and abiotic factors.
- Biotic factors are the living parts of an ecosystem; abiotic factors are the nonliving parts.

18A TERMS	
ecosystem	373
ecology	373
biotic factor	374
abiotic factor	374
climate	375

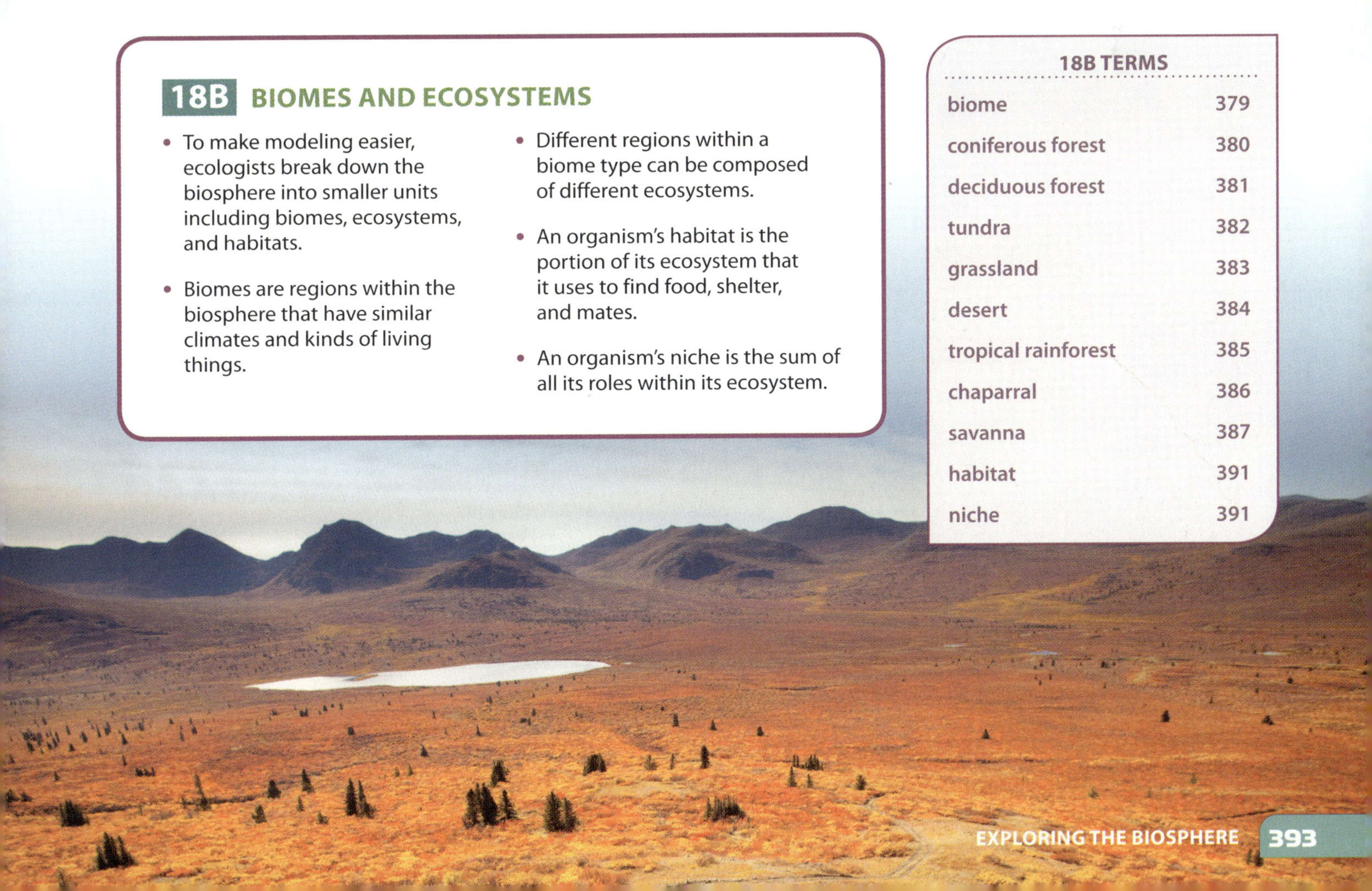

18B BIOMES AND ECOSYSTEMS

- To make modeling easier, ecologists break down the biosphere into smaller units including biomes, ecosystems, and habitats.
- Biomes are regions within the biosphere that have similar climates and kinds of living things.
- Different regions within a biome type can be composed of different ecosystems.
- An organism's habitat is the portion of its ecosystem that it uses to find food, shelter, and mates.
- An organism's niche is the sum of all its roles within its ecosystem.

18B TERMS	
biome	379
coniferous forest	380
deciduous forest	381
tundra	382
grassland	383
desert	384
tropical rainforest	385
chaparral	386
savanna	387
habitat	391
niche	391

CHAPTER REVIEW QUESTIONS

Recalling Facts

In Questions 1–6, match each term with its definition.

- **a.** ecosystem
- **b.** biosphere
- **c.** biome
- **d.** habitat
- **e.** niche
- **f.** zone

1. the habitable portion of the earth
2. a distinct area within an ecosystem caused by a gradual change in an abiotic factor
3. the parts of an ecosystem that an organism uses to find food, shelter, and mates
4. regions of the earth that have similar climates and kinds of living things
5. a limited area in which living and nonliving things interact
6. the role of an organism within its ecosystem

7. (True or False) An ecosystem is affected only by events that happen within that ecosystem.
8. In what ways does the amount of sunlight received affect life in an ecosystem?
9. Why are fewer organisms adapted to living in mountain ecosystems?

Understanding Concepts

10. Which of the following things would an ecologist study? Choose all that apply.
    - **a.** the effect of cosmic rays on gases in Earth's upper atmosphere
    - **b.** the competition for food between pikas and long-tailed voles
    - **c.** the patterns of seed distribution by migratory birds
    - **d.** the long-term changes in annual snowfall in the Sierra Nevada
11. Briefly describe three ecosystems near where you live.
12. Is an aquarium filled with fish an ecosystem? Explain.
13. Classify the following factors in a tropical rainforest as biotic or abiotic.
    - **a.** humidity
    - **b.** breezes from a waterfall
    - **c.** vegetation of varying heights
    - **d.** nutrient-poor soil
    - **e.** predators like the jaguar
14. Identify three abiotic factors and three biotic factors in the image at left.
15. Why is it wrong to describe climate as the high and low temperatures recorded on a certain day?
16. (True or False) Because it is nonliving, soil has no effect on the living things in an ecosystem.

17. Why is Earth's atmosphere considered part of the biosphere?
18. Why do ecologists divide the biosphere into smaller units such as biomes for study?

Use the Case Study below to answer Questions 19–22.

CASE STUDY RIPARIAN ZONES

Many small creeks and streams flow through coniferous forests. There is typically more water available for plant roots along a stream's banks than farther away from the stream. The banks themselves are also often subject to flooding. These constantly wet conditions allow plants to grow near streams that are different from those plants found deeper in the forest. This narrow strip of land along the stream corridor is called a *riparian zone*. Streamside plants include sedges, horsetails, butterbur, and willow and alder trees. The plants within the riparian zone are critical to the life of the stream. Their roots hold bank soil in place. During floods, their stems slow down the floodwaters, creating safe refuges for small fish. In the spring, mayfly larvae leave the stream and climb up the plants before molting into adults. Riparian trees shade the stream from the hot summer sun. In autumn, falling riparian leaves return nutrients to the water.

19. What is the abiotic factor that defines the limits of a riparian zone?
20. Is a riparian zone an example of an ecosystem or a biome? Explain.
21. Are the leaves that fall into the stream a biotic factor or an abiotic factor?
22. Describe an alder tree's niche.

Critical Thinking

23. Why is it important for Christians to study ecology? (*Hint*: Think of the Creation Mandate.)

24. You are planning to study a forest biome near your home for a project during your class's ecology unit. You prepare a list of the different species of plants and animals that live in the forest. Your project will consist of a description of each species. What is wrong with your project? Suggest two ways that you can improve it.

ETHICS Fish Fight

You've probably learned something in school about the Revolutionary War, the Civil War, and World War II. But do you know about the Cod Wars? Cod is a very popular food fish found in the North Atlantic Ocean. Fishing for cod has been an important industry throughout Northern Europe for centuries. In 1958, the small island nation of Iceland created a 12 nautical-mile-wide *exclusion zone* around the country (1 nautical mile, or *nmi*, is equal to 1.85 km). This zone was intended to protect fishing grounds that Icelanders considered to be their own. Great Britain recognized only a 4 nmi exclusion zone. Tempers flared when British fishing boats entered inside Iceland's 12 nmi limit escorted by warships of Britain's Royal Navy.

If the cod had been raised in a pond in Great Britain or in Iceland, there would be no argument over whose fish they were. The problem with the North Atlantic cod is that they don't live in either country; they live out in the open ocean. So whose fish are they?

Use the Ethics box above to answer Question 25.

25. Write a three-paragraph response on the ethical use of resources on the high seas, such as cod or other fish stocks. In the first paragraph, discuss principles from Scripture that may guide Christians in this area. In the second paragraph, discuss the possible outcomes of heeding or ignoring the guidelines of Scripture. In the third paragraph, give your proposal on how such resources should be shared.

RHYTHMS IN ECOSYSTEMS 19

MASTERS OF THE SKIES

High in the late summer skies above Greenland, a peregrine falcon *stoops*, or dives, toward a ptarmigan (TAR meh guhn) far below. Quickly gaining speed, the peregrine exceeds 320 kph before striking the other bird. The falcon needs the nourishment because in a few weeks it will begin a different type of flight, one much longer and slower. It will travel 7000 km from Greenland to the warm coasts of Central America while its summer home experiences the icy darkness of an Arctic winter. Then when the sunlight returns to Greenland, the peregrine will once again take to the skies for the long return journey to its home in the far north.

The peregrine's yearly journey is just one example of natural rhythms. Animal migrations, reproduction times, and periods of hibernation are all cycles that God has established for the earth.

19A

ENVIRONMENTAL CYCLES OF MATTER

Why don't we ever run out of water?

19.1 WATER CYCLE

19A Questions
- What is the path of the water cycle?
- How are the oxygen and carbon cycles connected?
- What is the nitrogen cycle?

Natural rhythms don't occur just in living organisms. They also occur in nonliving processes. Matter moves through the environment in large cycles that operate on a global scale. These cycles often involve both the physical environment and living populations. Let's take a look at four important cycles.

REPLENISHING EARTH'S WATER

The **water cycle** is the continuous movement of water from the earth to the atmosphere and back to the earth.

2 *Condensation.* In the atmosphere, water vapor cools and becomes liquid again in a process called condensation. The tiny drops of water form clouds that become part of the moving air masses in the atmosphere.

3 *Precipitation.* The water in the clouds eventually returns to the earth by falling as precipitation. Rain, snow, sleet, and hail are all forms of precipitation.

4 Water that falls as precipitation makes its way into ponds, lakes, rivers, and eventually the ocean, completing the cycle.

The water cycle is responsible for moving water from the oceans to the rest of the earth. Because of the water cycle, no new water needs to be created. Water molecules that you recently drank might have flowed through Rome in the days of the apostle Paul.

19.2 CARBON AND OXYGEN CYCLES

Carbon and oxygen also cycle through the ecosystem. The **carbon cycle** and the **oxygen cycle** are so closely connected that they are often considered together. They meet in the processes of photosynthesis and cellular respiration.

JOINED CYCLES

1 ***Photosynthesis.*** When plants and algae perform photosynthesis, they remove carbon dioxide from the atmosphere to make glucose and release oxygen back into the atmosphere. The plants then use the carbon-containing glucose to make other molecules, which are then used to make plant structures.

2 ***Cellular Respiration.*** All organisms use carbon-containing glucose to power cellular respiration, which takes oxygen from the atmosphere and releases carbon dioxide. They also use carbon-based molecules like carbohydrates, fats, and proteins to build structures.

3 ***Fossil Fuels.*** Fuels such as gasoline, oil, and coal contain large amounts of carbon. When they are burned, the carbon reacts with oxygen from the atmosphere to form carbon dioxide, which is then released back into the atmosphere.

4 ***Decay.*** Dead organisms are decomposed by organisms such as bacteria and fungi. Using oxygen from the atmosphere, they convert many of the dead organisms' carbon-based structures into carbon dioxide, which is cycled back into the atmosphere.

19.3 NITROGEN CYCLE

Nitrogen is one of the elements that make up amino acids. Amino acids join together to form proteins. Since proteins are involved in nearly everything a cell does, all living things need nitrogen to survive and reproduce. Nitrogen cycles through the environment in the **nitrogen cycle**.

NITROGEN CYCLE

Atmospheric Nitrogen. Nitrogen gas makes up 78% of the atmosphere. But living things cannot use nitrogen gas.

Nitrogen Conversion. To be useful to organisms, nitrogen gas must be converted to nitrogen compounds. Bacteria in the soil and on the roots of some plants perform most of this task; lightning does some as well.

Integration. Plants absorb the nitrogen compounds from the soil and use them to make proteins. Animals then eat these plants and digest the proteins to use in their own bodies.

The earth has only a limited supply of water, carbon, oxygen, and nitrogen. To survive, organisms depend on these substances being recycled. God has created the cycles of matter to ensure that these molecules are always available.

19A REVIEW QUESTIONS

1. What are the three primary parts of the water cycle?
2. Draw a diagram of the water cycle. Include plants, clouds, oceans, rain, and the three primary parts from Question 1 as labels.
3. Which two biological processes connect the carbon and oxygen cycles?
4. Indicate whether each of the following parts of the carbon and oxygen cycles produces carbon dioxide or oxygen: cellular respiration, decay, and photosynthesis.
5. Why do plants need bacteria to convert nitrogen gas to nitrogen compounds?
6. Legumes are plants whose roots have bacteria that convert nitrogen gas to nitrogen compounds. Why might a farmer plant a crop of inedible legumes in his field?

Reconversion. Other soil bacteria convert the nitrogen compounds back into nitrogen gas. The gas escapes back into the atmosphere, and the cycle continues.

Decay. Animal waste and the dead bodies of organisms are broken down by soil bacteria and become useful nitrogen compounds once again.

19B Questions

- How does energy flow through an ecosystem?
- Why isn't there an "energy cycle"?
- How does the amount of available energy change as it flows through an ecosystem?
- Why is there death and suffering in God's world?

19B ENERGY IN THE ENVIRONMENT

Does energy also cycle through the environment?

19.4 FOOD CHAINS

Matter, such as water, carbon, oxygen, and nitrogen, cycles through ecosystems. But what about energy? For most ecosystems, the sun provides a constant supply of energy. Producer organisms use photosynthesis to store the sun's energy in sugars. They then use the sugars to power cellular respiration and to build other structures.

Consumer organisms cannot store energy directly from the sun. Instead, they obtain the energy they need by eating other organisms. Many consumers eat mostly producers, while some consumers mainly eat other consumers. The passing of materials and energy from producers through consumers is called a **food chain**.

Every living organism is part of at least one food chain.

WHAT'S ON THE MENU?

In this food chain from the American grassland, plants are the main producers.

Rabbits eat plants, making the rabbit a type of consumer called a *herbivore*. For the rabbits to survive, there must be many more plants than there are rabbits.

The hawk is a *carnivore*, so it will try to eat the rabbit. There must be more rabbits than hawks so that some of the rabbits will survive to reproduce.

Organisms such as bacteria and fungi that obtain nutrients and energy by breaking down the bodies of dead organisms are called *decomposers*.

19.5 ENERGY IN A FOOD CHAIN

Scientists often use **energy pyramids** to illustrate how energy flows through a food chain. An energy pyramid is divided into layers called **trophic levels**. Each level represents a different type of organism in the food chain, and each is dependent on the level below it for its energy.

ENERGY PYRAMIDS

Notice that the producer area 1 of the pyramid is much larger than the consumer area. That is because each time that energy is transferred from one level to the next, some usable energy is lost.

The plants obtain their energy from the sun. When a rabbit 2 eats a plant, the plant's energy passes to the rabbit. But the rabbit cannot use all the energy in the plant since it cannot digest all its molecules. And much of the energy that the rabbit obtains from the plant is converted to heat and lost.

If a hawk 3 eats the rabbit, the hawk can use some of the energy stored in the rabbit. But again, the hawk cannot digest all the rabbit's molecules. And much of the energy that the hawk does obtain from the rabbit is again lost as heat.

Finally, when the hawk dies, decomposers break down the molecules in the hawk's body to obtain energy. When each decomposer dies, its cell membrane bursts. The energy that passes from the decomposer to the soil is not usable by other organisms, so it is lost. Once energy is lost, organisms cannot use it again.

In fact, the organisms in each trophic level use or lose about 90% of the energy that they receive. That means that only 10% of the energy is available for the organisms in the trophic level above. And eventually the remaining energy is released by decomposers and lost. Energy doesn't cycle through the ecosystem. It flows through it. As a result, the earth needs a constant supply of energy from the sun for life to continue.

LAB EXERCISE

TRACKING ENERGY IN A FOOD CHAIN

Equipment
calculator

In Chapter 13, you learned that the energy in food is often measured in Calories.

❶ Draw an energy pyramid. Using words or drawings, place *American kestrel*, *grass*, and *mouse* in the appropriate trophic level of the energy pyramid.

1. Over the course of its lifetime, a deer mouse eats 1.6 kg of grass. Each kilogram of grass contains about 1750 Calories. How many Calories of energy were in the grass that the mouse consumed?

❷ Place your answer to Question 1 in the grass trophic level of the energy pyramid.

2. But the mouse is able to use only about 10% of the energy in the grass. How many Calories does the mouse actually use?

❸ Place your answer to Question 2 in the mouse trophic level of the energy pyramid.

3. An American kestrel captures and eats a deer mouse. The kestrel is able to use only 10% of the energy available from the deer mouse. How many Calories is the kestrel able to use?

❹ Place your answer to Question 3 in the kestrel trophic level of the energy pyramid.

4. If a kestrel eats a mouse each day, how many kilograms of grass are needed to support one kestrel for a year?
5. On the basis of your calculations above, why do you think there must be more producers than herbivores?
6. In the food chain described in the energy pyramid above, what would happen if all the grass suddenly died?
7. On the basis of how you answered Question 6, why is it important for us to monitor the health of different populations of living things?
8. How is the work of monitoring plant and animal populations related to the Creation Mandate in Genesis 1:28?

The American kestrel is a small bird of prey that commonly feeds on small rodents.

19.6 FOOD WEBS

Food chains and energy pyramids are useful tools, but they have a major problem—they are too simplistic. Each member of a food chain is actually part of multiple food chains, not just one. Animals often eat whatever they can without being too picky. Many animals move around and can be part of very different food chains. A bald eagle may eat a squirrel one day, making it part of a land food chain. The next day, it may eat a fish, making it part of an aquatic food chain.

Many animals are omnivores, which eat both plants and animals, so they could fit on several levels of an energy pyramid. A grasshopper might eat a blade of grass. And a mouse then eats the grasshopper and is in turn eaten by a kestrel. Or the mouse might eat the grass and be eaten by the kestrel. Depending on the circumstances, the mouse can act as either a herbivore or a carnivore.

To make matters even more confusing, carnivores may eat herbivores and other carnivores. So the kestrel might eat the grasshopper and skip the mouse. Because the kestrel eats both herbivores and carnivores, it could be placed on two different levels of the energy pyramid.

All the interrelated food chains that are possible in an ecosystem form a **food web**. So a food web of the American prairie looks like the diagram below.

In many ways a food web gives a more accurate picture of what happens in an ecosystem than a food chain does. But often a food web shows only the primary organisms within an ecosystem. Showing all of them would make the food web too large and confusing to be useful.

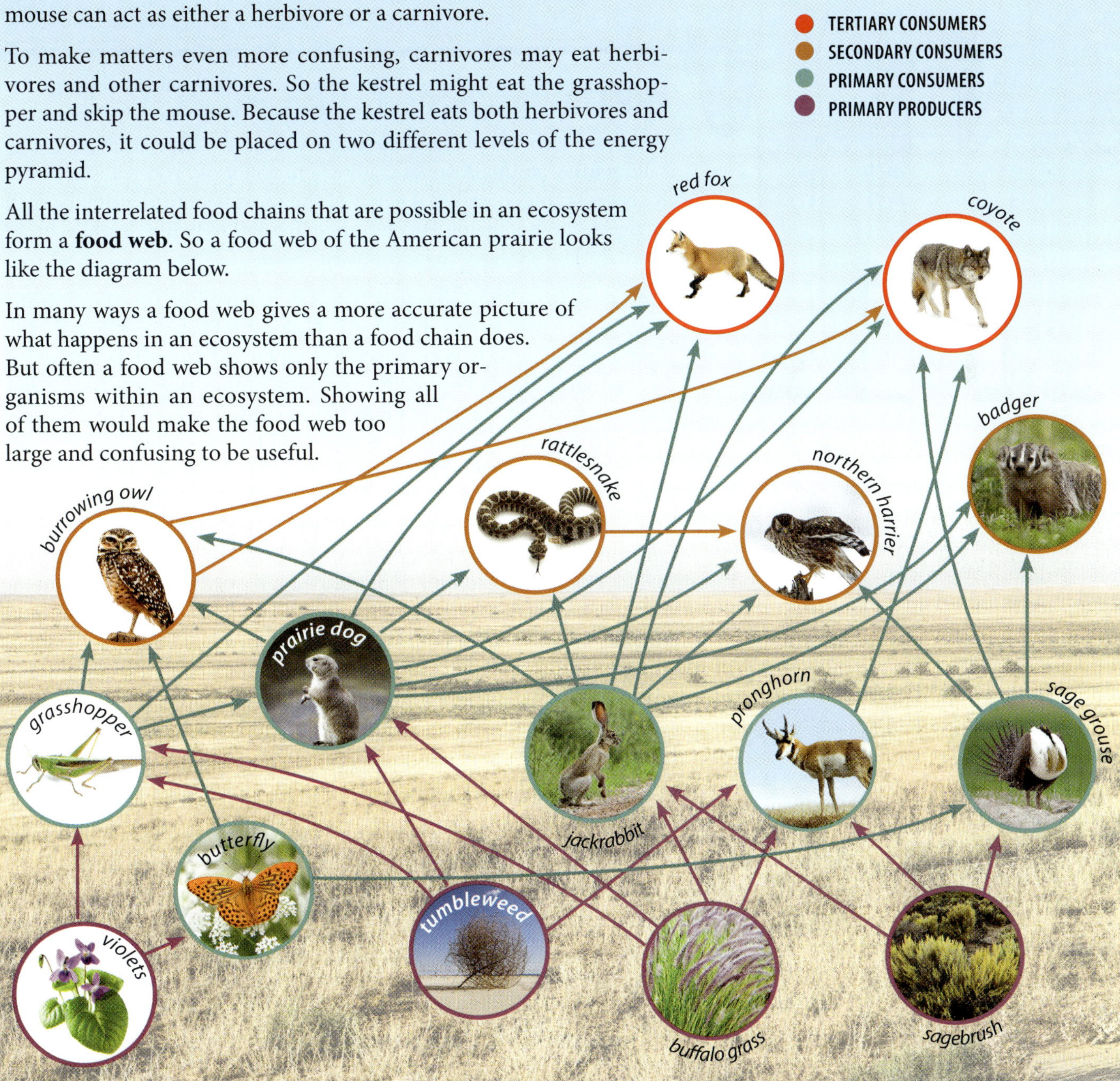

19.7 CREATION, DEATH, AND FOOD WEBS

Any food web has at least one carnivore in it. These animals live by killing other animals. In fact, anywhere you look, you will see death. It seems strangely sad that some of the most beautiful animals, such as tigers and peregrine falcons, are equipped to bring death to other animals. Does it have to be this way?

We know from the Bible that it wasn't always like this. God didn't design the world to be full of death. Genesis 1:31 says that everything that God made was very good. But when Adam and Eve rebelled against God, death entered the world. Animals began killing other animals for food. Creation longs to be released from its present state of death and decay (Rom. 8:18–25). God has promised that one day He will restore the goodness and peace of His original creation. Then "the wolf also shall dwell with the lamb … and the calf and the young lion and the fatling together" (Isa. 11:6) in the eternal kingdom with all who trust Jesus for salvation.

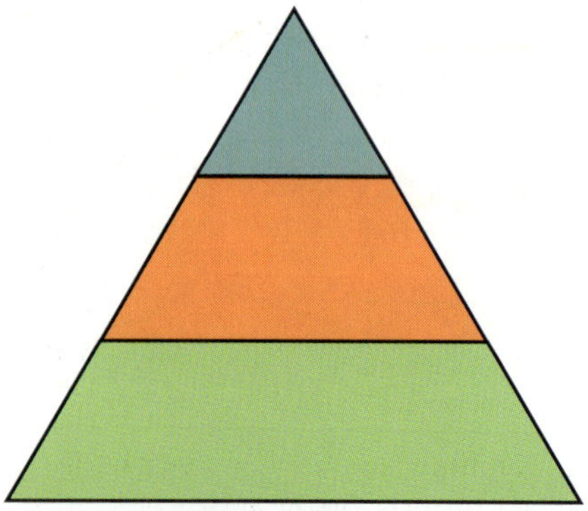

19B REVIEW QUESTIONS

1. What type of organisms stores energy from the sun in sugar molecules?
2. Why can't energy cycle through an ecosystem?
3. In the energy pyramid at left, what is the color of the trophic level that would represent a herbivore?
4. If an organism in the green trophic level of the energy pyramid contains 10,000 Calories, how many Calories of that 10,000 are available to an organism in the orange trophic level?
5. Why is a food web a more useful tool than a food chain?
6. For lunch you have a grilled cheese sandwich with a bowl of tomato soup. Draw a food web including wheat, tomato, grass, a cow, and you! Use arrows to show the flow of energy in your food web.
7. How do you think food pyramids looked before Adam sinned?

19C

BIOTIC RHYTHMS

Do populations have rhythms?

19.8 RELATIONSHIPS BETWEEN ORGANISMS

Individual organisms interact with one another. A deer browses the leaves of a small shrub. Ants carry off the seeds of a plant. Mosquitos draw blood from vertebrates and sometimes transmit tiny disease-causing organisms. These are all examples of relationships between organisms.

Most organisms do not interact with one another in a conscious way. Only animals and people can do that. But even microscopic organisms can have an immense effect on other organisms.

19C Questions

- How do organisms interact in an ecosystem?
- What is the difference between primary and secondary succession?
- What issues make managing organisms difficult?

RELATIONSHIPS WITHIN A SPECIES

Some animals—the jaguar is an example—are relatively solitary. They spend most of their lives alone except when they mate.

Some animals, such as bald eagles, mate for life, spending most of their lives with their mates.

Other animals, such as honey bees, live in larger groups. Many of these groups cooperate, or work together, in finding food, protecting the young, and defending the group from predators.

Individuals and populations of different species also interact with one another in different ways. Let's take a look at some of them.

RELATIONSHIPS BETWEEN SPECIES

In **predation**, one organism eats another.

Two individuals or populations, such as this horse and this duck, that do not impact each other have a relationship known as **neutralism**.

When two populations, like elk and bison, have similar lifestyles, they are in **competition** for resources. Individual organisms also compete with other members of their population. Animal groups can similarly compete with other groups of the same species for food, mates, and territory.

19.9 SYMBIOSIS

Most interactions between animals of different species occur over a relatively short amount of time. A lion chases a zebra, and the zebra either is eaten or gets away. If the zebra escapes, that particular lion may never interact with that one zebra again. In symbiosis, though, individuals of different species interact frequently over a long period of time, often living together. Let's examine the most common types of symbiosis.

TYPES OF SYMBIOSIS

Two organisms that live together with both benefiting from the relationship exist in a form of symbiosis called **mutualism**.

Sometimes two organisms live together and one harms the other. This is called **parasitism**. The organism that benefits is called the *parasite*, and the organism harmed is called the *host*.

Commensalism occurs when two organisms live together, with one benefiting from the relationship while the other is not affected.

But ecology is much more complex than it might first appear. A clownfish hides from predators among the stinging tentacles of a sea anemone. The tentacles do not harm the fish but will kill its potential predators. For many years, scientists thought that this was an example of commensalism, but now many think that the fish may help the anemone in some way. This would make their relationship one of mutualism.

The uncertainty and complexity of the example above demonstrate one of the problems in managing natural resources. It's hard for scientists to know exactly how organisms affect each other. Many times they do not know how a decision regarding one species will affect other species in the same ecosystem.

19.10 SUCCESSION

Matter cycles through an ecosystem, and energy flows through it. Do populations of organisms experience any rhythms? Unless an ecosystem has gone through some major change in the recent past, the community of organisms that live in it will not change much. This kind of community is called a *climax community*.

If something occurs to disturb an ecosystem, the community that lives there will often change. But over time, a climax community will return. The process of a disturbed ecosystem returning to its climax community is called **ecological succession**.

primary succession

If an ecosystem begins without any soil, the succession is called **primary succession**. Soil and organic matter are gradually deposited on the rock by wind and water or by living creatures. Small, hardy organisms called *pioneer species*, like lichen that can live on rock, help to break up the rock into smaller bits. Mosses and fungi begin to grow and further break up the rock. This helps to increase the organic matter in the layer of soil that is forming. Over time, more and more organisms begin to live on what once was nothing but rock.

Primary succession is fairly rare. It occurs mainly in the aftermath of volcanic eruptions, such as the 1980 Mount St. Helens eruption.

Most succession doesn't start with bare rock. When succession starts in an area with soil, it is called **secondary succession**. This type of succession occurs when a fire burns through an area or when an area cleared for agriculture or construction is abandoned. The pioneer species are often grasses and small shrubs. Gradually, these pioneer species are replaced by a climax community.

The time it takes to complete the succession process varies. Primary succession takes much longer than secondary succession does. And some ecosystems are in a state of constant change. Many grassland ecosystems survive best when they are swept by periodic fires. These fires are followed by secondary succession followed by another fire.

For many years, scientists tried to put out fires that started in grasslands. But then they noticed that trees were starting to take over the ecosystems. So now they allow fires to burn as long as people and houses are not in danger. This illustrates another issue that makes managing ecosystems difficult. Aspects that we assume are negative might actually have a positive impact on the ecosystem. To properly manage any environment, we must carefully study them to see the rhythms that God has created in them.

This field is undergoing secondary succession. After a century has passed, it will be almost impossible to tell that the land was ever anything but a forest.

19C REVIEW QUESTIONS

1. Indicate whether the relationships between the following organisms illustrate predation, competition, or neutralism.
   - **a.** a camel and a vulture
   - **b.** a lion and a gazelle
   - **c.** a bobcat and a coyote
   - **d.** an anteater and termites
2. A tapeworm spends its entire life inside the intestine of another animal, absorbing nutrients from the animal's food. What type of symbiosis is this?
3. Epiphytes are plants that grow on another larger plant, such as a tree. The epiphyte benefits from having better access to water, minerals, and sunlight. The tree gets nothing in return but is not harmed either. What type of symbiosis is this?
4. Classify each example of succession below as primary or secondary succession.
   - **a.** An old photo shows a small island composed of rock without any vegetation. Explorers visiting the island recently found the island covered with grasses.
   - **b.** A person visits the location of his great-grandfather's garden and finds trees growing in it.
   - **c.** An orchard was planted on a prairie. A fire destroyed the trees, and prairie grasses grew in their place.

ETHICS — Engineering For Migrations

Many fish migrate during their lifetimes. Salmon are born in freshwater streams, spend most of their lives in the ocean, and return to the same streams where they hatched to lay their own eggs. American and European eels hatch in the ocean, live much of their lives in freshwater streams, and then return to the ocean to lay eggs. Many other fish also migrate, though most migrations are not as spectacular as those of salmon and eels. Some fish live in lakes but lay eggs in streams, while others spend the day deep in the water but then move nearer to the surface at night.

When fish migrate, they can face many obstacles. One such obstacle is a culvert that carries a stream under a road. Fish migrating up such a stream must swim through the culvert. But culverts can cause problems for migrating fish.

Many culverts are now made so that fish can pass through them more easily. But sometimes a culvert that is good for allowing fish to swim *upstream* is not good at passing debris *downstream*. Is there a way to enable fish migration while keeping our roads intact?

Use the Ethics box above to answer Question 5.

5. Research and write a three-paragraph essay explaining how a culvert affects migrating fish. Discuss some of the ways that a culvert can be designed to help fish swim through it. Also explain how these designs may make a culvert less effective for a road that passes over it. Finally, explain how the biblical worldview balances people's need for good roads and the need of fish to migrate.

CHAPTER SUMMARY

19A ENVIRONMENTAL CYCLES OF MATTER

- Water evaporates, condenses as clouds, and precipitates as part of the water cycle.
- The carbon and oxygen cycles are connected through cellular respiration and photosynthesis.
- Nitrogen gas is converted to nitrogen compounds that plants use to make proteins. These same compounds are eventually converted back to nitrogen gas.

19A TERMS	
water cycle	398
carbon cycle	400
oxygen cycle	400
nitrogen cycle	402

19B ENERGY IN THE ENVIRONMENT

- Energy cannot cycle through an ecosystem.
- Each time energy is transferred between trophic levels, less of it becomes available to do work.
- Food webs more accurately depict energy flow through an ecosystem than food chains do.
- The curse of death resulted from the Fall and will be removed after Christ's return.

19B TERMS	
food chain	404
energy pyramid	405
trophic level	405
food web	407

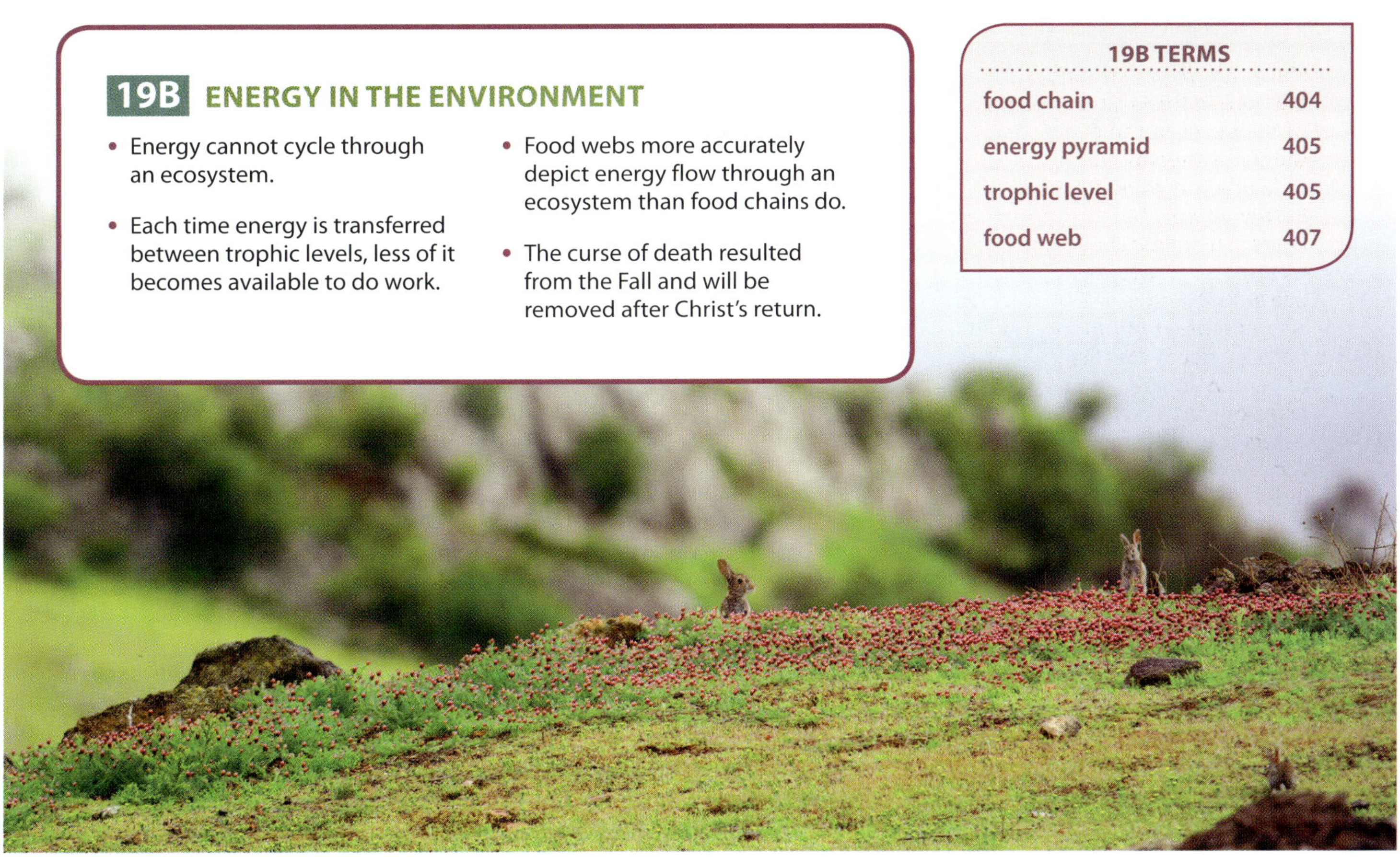

CHAPTER 19

CHAPTER SUMMARY

19C TERMS	
predation	410
neutralism	410
competition	410
mutualism	411
parasitism	411
commensalism	411
ecological succession	412
primary succession	412
secondary succession	413

19C BIOTIC RHYTHMS

- Succession that starts without any soil is called primary succession; succession that begins with soil is referred to as secondary succession.
- Predation, competition, and neutralism are the primary relationships between organisms of different species.
- Symbiosis occurs when two organisms of different species live together. One benefits while the other may be benefited, harmed, or unaffected.
- Biotic rhythms such as migration can make managing organisms difficult.

CHAPTER REVIEW QUESTIONS

Recalling Facts

1. Which environmental cycle is composed of evaporation, condensation, and precipitation?
2. Clouds are made of tiny droplets of liquid water. Which part of the water cycle are clouds part of?
3. What two cycles of matter are very closely tied together?
4. Which two cycles involve decay releasing matter back into the ecosystem?
5. Which organisms are responsible for cycling energy through an ecosystem?
6. How much of the energy that an organism in a food chain obtains is available for the next organism in the food chain?
7. Does the diagram at left show a food web or a food chain?
8. Why did the eating of animal flesh not exist in the animal world before the Fall?
9. What is the name of the relationship between members of different species that do not affect each other in any way?

10. Lichens are composed of algae and fungi, neither of which can survive without the other in the harsh conditions that lichens often inhabit. What type of symbiosis is this?

Understanding Concepts

11. Draw a diagram of the nitrogen cycle. Include labels for *atmospheric nitrogen*, *nitrogen conversion*, *integration*, *decay*, and *reconversion*.
12. When grass absorbs energy from sunlight, only a portion of that energy is available for herbivores, which eat the grass. What happens to the energy that is *not* available to herbivores?
13. A crow is an omnivore. Which level(s) of the energy pyramid on the right could the crow be placed in?
14. How do we know that death was not part of the original creation?
15. Why is it difficult to manage animal populations?
16. If succession occurred in the area shown below, would it be an example of primary or secondary succession?

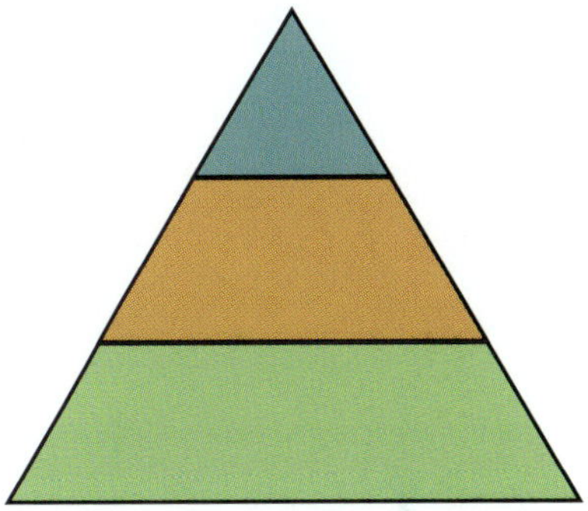

Critical Thinking

17. What would happen to the oxygen cycle if the carbon cycle didn't exist?
18. How would you respond to someone who says that death and suffering indicate that the world was not created by a loving God?
19. Write a one-paragraph argument for a mosquito's relationship with a human as either predation or parasitism.
20. Examine the food web below. Identify each relationship indicated.
    a. limpets and mussels
    b. octopuses and seagulls
    c. mussels and octopuses
    d. fish and starfish

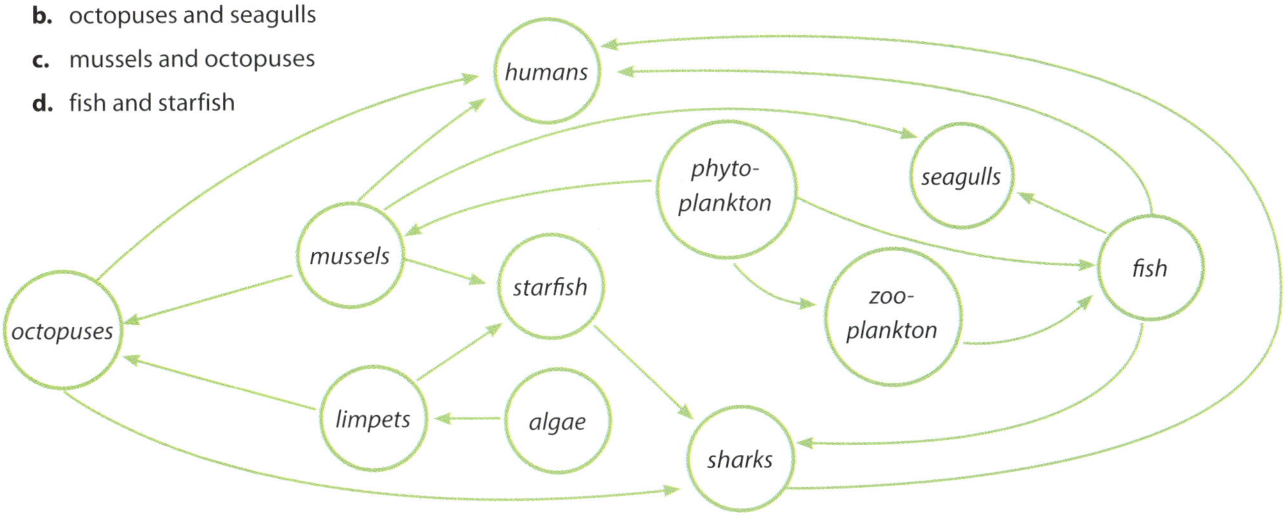

Use the Case Study below to answer Questions 21–22.

CASE STUDY CHERNOBYL'S GHOST TOWN

On April 26, 1986, northern Ukraine experienced the worst nuclear disaster in history. One of the four reactors at the Chernobyl Nuclear Power Plant exploded. The escaped radiation spread throughout Eastern Europe. The 53,000 inhabitants of the nearby city of Pripyat were evacuated a few days after the disaster. Eventually, everyone living within 30 km of the nuclear power plant was evacuated. This area continues to be uninhabited due to the high levels of radiation still present in the area.

21. Forest is now taking over the streets and concrete buildings of Pripyat. Write a paragraph arguing for this being a case of either primary or secondary succession.

22. Many feral dogs live in the evacuated area. Some of these dogs have been equipped with GPS locators and Geiger counters that detect and transmit radiation levels. This allows researchers to keep track of the amount of radiation in the evacuated area around Chernobyl. Evaluate this use of feral dogs.

Use the Case Study below to answer Questions 23–25.

CASE STUDY URBAN PEREGRINES

If you wanted to see a peregrine falcon, where would you go? Traditionally, peregrines nest on cliffs and other remote places. But these days, one of the best places to see peregrine falcons might be a city. Urban peregrines nest on the sides of skyscrapers, which are similar to the cliffs that form the falcon's natural nesting sites. Generally, peregrines have been welcomed by city residents both because of their beauty and because they eat nuisance species like pigeons. In fact, in some areas of the world, peregrines in cities may be faring better than peregrines in rural areas.

23. Peregrine falcons have proved useful in controlling pigeons. In the past, some cities have tried using poisons to control the pigeons. Why would peregrines be a better option?

24. Name a domestic animal that a peregrine falcon might be competing with by eating pigeons.

25. Write a short paragraph imagining the peregrine falcon as part of the new heaven and new earth in eternity.

20 MANAGING GOD'S CREATION

BREAKFAST WITH GIRAFFES

At Giraffe Manor outside Nairobi, Kenya, endangered Rothschild's giraffes often come to the door to be fed by hotel guests. They even stick their heads in the large windows to be fed from the breakfast table. The giraffes belong to the nearby Giraffe Center, whose breeding program raises giraffes for release into the wild. The two-story-tall animals are free to roam over to the Manor and meet guests.

Giraffe Manor is an example of ecotourism. This relatively new industry uses tourism to raise awareness for conservation. In some cases, the profits from ecotourism directly support conservation efforts. But this raises an interesting question: what is the best way to manage the environment?

20A
PROTECTING THE ENVIRONMENT

How can we wisely reduce pollution?

20A Questions

- What are the different forms of pollution?
- Why should we work to curb pollution?

20.1 THE POLLUTION PROBLEM

The Creation Mandate instructs us to rule and care for the earth. But because we are fallen, we often fail to obey these commands. One way that mankind has often failed to care for the biosphere is by refusing to deal with the waste that we produce. Any time we use a resource, we produce waste. These wastes can be harmful if improperly released into the environment. In such cases, the wastes are called *pollutants*. The improper release of pollutants into ecosystems is **pollution**.

Pollutants can be harmful to living things. In extreme cases, pollutants can damage an ecosystem so much that it can no longer support life. Most cases of pollution are not this severe, but even small amounts can change an ecosystem. This can make it difficult for native organisms to thrive. For instance, some factories use water to cool their machinery. If the heated water is released back into a cool stream, it may change the overall temperature of the stream. Even if

20-1
The air of this city is heavily polluted.

the heated water is clean, cool-water species may die or be forced to leave while warm-water species move to the now warm parts of the stream.

Sometimes the pollution directly affects organisms at one trophic level only. But the changes may also affect all the other organisms in the ecosystem. Excess fertilizer that runs off a field may collect in a pond and cause the algae population in the pond to grow rapidly. Later when the algae die, the population of decomposer organisms will increase as they begin to decompose the algae. When this happens, the decomposers often use all the available oxygen in the water. Many of the fish and other organisms in the water die as a result. Although the fertilizer did not harm the fish directly, it produced a chain reaction that led to their death.

TYPES OF POLLUTION

Substance pollutants are solids, liquids, or gases that pollute an ecosystem. They are divided into biodegradable and nonbiodegradable pollutants.

Nonbiodegradable pollutants, such as plastic and glass, cannot be broken down in the environment. They quickly accumulate until they cause serious problems.

Biodegradable pollutants, such as fertilizer and paper, can be broken down by organisms in the environment. Decomposers may be able to break down a small amount of a biodegradable pollutant that enters an ecosystem. But if a large amount enters, the decomposers can't keep up.

Energy pollutants are forms of energy that can negatively affect the environment. Heat, light, and nuclear radiation are all examples of energy pollutants. New York City's annual Tribute in Light must be turned off for short periods of time to allow birds attracted to the light to escape.

20.2 SOLVING THE POLLUTION PROBLEM

Since pollution is such a problem, is there anything we can do about it? There are several different ways of dealing responsibly with pollutants. Let's look at the main ones.

PREVENTING POLLUTION

Reduce. Some wastes can be greatly reduced or even eliminated. These cooling towers cool hot water before it is released into the river, reducing energy pollution. More careful use of fertilizer can reduce the amount that enters nearby streams.

Recycle. Other wastes can be recycled. Nonbiodegradable glass and plastic can be recycled and turned into other items. This shoe was made from recycled plastic. A few countries even recycle some of the radioactive waste produced by nuclear power plants.

Treatment. Other wastes can be treated to make them less harmful. Some sewage treatment plants treat sewage so that pure water is released at the end of the process.

Dispose. Some wastes cannot be prevented, recycled, or treated. In those cases, they should be disposed of in a way that limits their harm to the environment. This is the role of modern sanitary landfills.

20.3 POLLUTION, THE ENVIRONMENT, AND THE CHRISTIAN

If pollution is such a problem, why does it still occur? The answer lies in people's priorities. Properly dealing with waste can be expensive. Some individuals avoid the cost by simply dumping pollutants rather than by recycling, treating, or properly disposing of them.

Other people may pollute because they don't think that they produce enough waste to make any difference. After all, the earth is very large, and one person's contribution to the total waste produced is small. Or they think that since biodegradable pollutants can be broken down in the environment, then those pollutants don't really do any harm.

The Creation Mandate forbids all those attitudes. Christians must remember that the Creation Mandate commands humans to care for the world and its creatures. We should take responsibility for the wastes we produce. Taking care of God's world means that we must consider how our wastes affect other people and the environment. We must let our actions be governed by the Bible, not by the actions or expectations of other people.

Each one of us can make a difference. Turning off lights when they are not in use saves energy and money. Recycling used bottles, cans, and newspapers instead of throwing them away can help keep landfills from filling up too quickly. And properly disposing of trash instead of littering keeps substance pollution from affecting the wildlife that lives around us. Doing your part to manage wastes glorifies God while helping humans to flourish and the creation to thrive.

20A REVIEW QUESTIONS

1. What are the two forms of pollution?
2. What are the two forms of substance pollutants?
3. Name two methods for responsibly dealing with wastes.
4. Why should Christians be concerned with properly dealing with pollution?
5. Give two reasons why people don't take responsibility for the wastes they produce.

LAB EXERCISE

AIR FILTRATION: BALANCING FILTRATION WITH AIR FLOW

E Equipment

- prepared shoe boxes (2)
- hair dryer
- quilt batting
- scouring pad
- coffee filters (2)
- tape, clear plastic
- aluminum foil
- thread
- baking soda
- protractor
- balance
- graduated cylinder, 10 mL

PROCEDURE

Dealing with pollution is always a balancing act. While it is important that we reduce, recycle, treat, and properly dispose of wastes, it is always vital that we provide for the needs of people. Sometimes the short-term needs of people result in producing some pollutants.

Ventilation in a building is a good example of this type of balancing act. Dirt, dust, and other particles are often carried by the air in a ventilation system. These particles often cause respiratory problems for people who live or work in the building. The perfect filter would remove every particle carried by the air. But such a filter would severely restrict the amount of air flowing through the ventilation system. A reduced airflow would allow carbon dioxide to build up to toxic levels. Ventilation designers and engineers try to design filters that stop most airborne particles without severely restricting airflow. Today you will have a chance to experiment with different types of filtering materials in a model of an air duct.

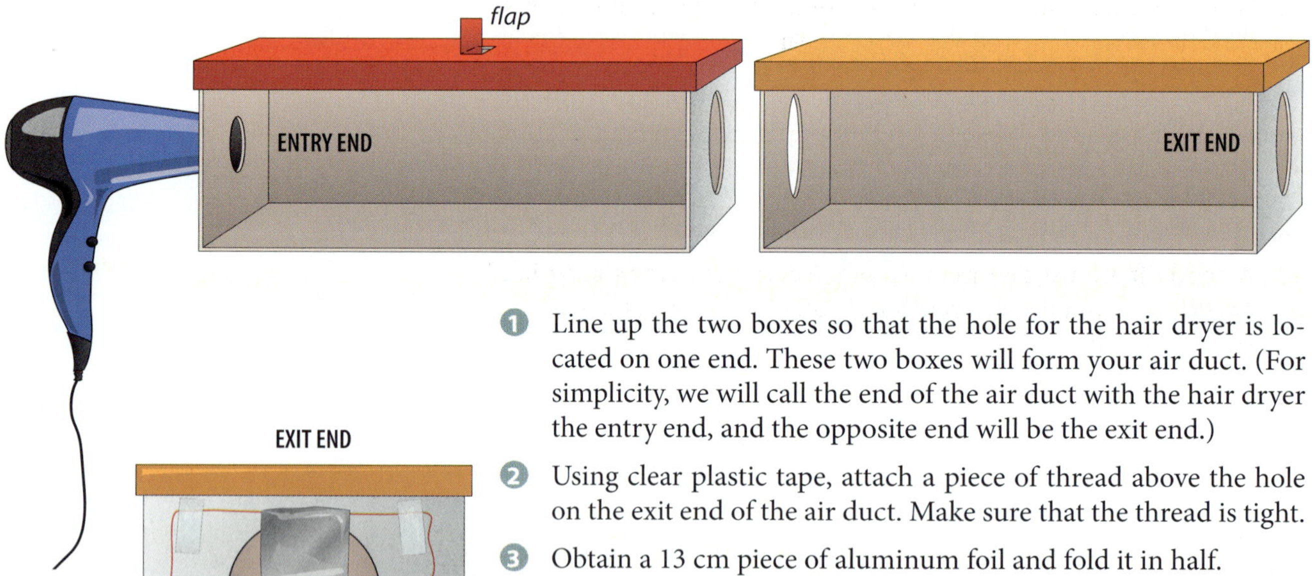

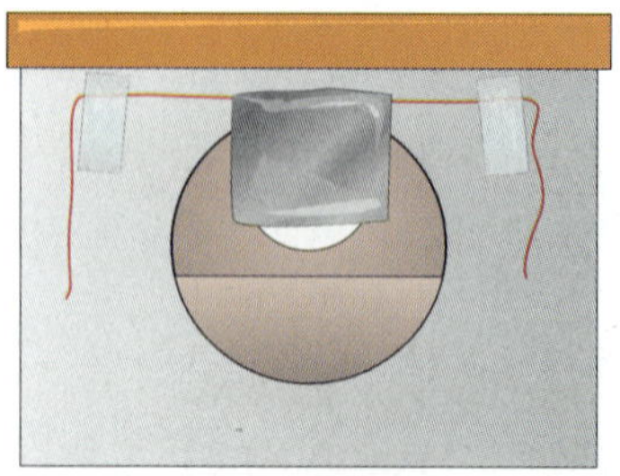

1. Line up the two boxes so that the hole for the hair dryer is located on one end. These two boxes will form your air duct. (For simplicity, we will call the end of the air duct with the hair dryer the entry end, and the opposite end will be the exit end.)
2. Using clear plastic tape, attach a piece of thread above the hole on the exit end of the air duct. Make sure that the thread is tight.
3. Obtain a 13 cm piece of aluminum foil and fold it in half.
4. Hang the folded piece of aluminum foil over the thread. Crimp the ends of the foil together. The foil strip should be able to pivot freely on the string.
5. Insert the end of the hair dryer in the hole in the entry end and turn it on high.
6. Use a protractor to measure the angle of the foil from vertical while the hair dryer is on. Record the angle in the *Foil Angle* column of the table on the next page.
7. Obtain a piece of quilt batting (about 25 cm long) and determine its mass. Record the mass in the *Mass 1* column of the table.
8. Fold the batting in half and place it between the two boxes of the air duct so that it completely covers the holes.

9. Turn on the hair dryer and measure the angle of the foil. Record the measurement in the table.
10. Using your scale, measure and pour 1 g of baking soda into the graduated cylinder.
11. With the hair dryer running, pour the baking soda through the flap in the lid of the first box.
12. Determine the mass of the quilt batting. Record the mass in the *Mass 2* column of the table.
13. Repeat Steps 9–12 with the coffee filter and the scouring pad.

Filter	Mass 1	Mass 2	Foil Angle
No filter			
Quilt batting			
Scouring pad			
Coffee filter			

A good filter must allow most of the air to pass through it. For our purposes, we will assume that the angle of the foil should not be less than half the angle of the foil when the air duct does not have a filter.

1. Divide the angle of the foil with no filter (measured Step 6) by 2. What is the minimum angle of the foil that is acceptable for a filter?
2. Which filter(s) allowed at least half the air to pass through them?
3. A good filter must capture most of the particles in the air. Which of the three filters captured the greatest mass of baking soda? How do you know?
4. Which material that you tested made the best filter?
5. Why is the science behind air filtration important for fulfilling God's commands?

20B
USING RESOURCES

Can managing resources actually help the environment?

20B Questions
- What are the differences between renewable and nonrenewable resources?
- What are the benefits of natural resource management?

20.4 TYPES OF NATURAL RESOURCES

God gave humans a world with everything that we need to flourish. Materials that we can obtain from the physical world are called **natural resources**. Minerals such as gems and iron ore are natural resources. So are land, water, and even living things.

Some natural resources are the result of processes that replenish the resources even as they are used. These are **renewable resources**. For instance, when trees are cut down, new trees will grow in their place. It's important to realize, though, that renewable resources can be misused if not managed properly. If all the trees in a large area are cut down, it may take years for them to regrow on their own. But if seedlings are planted to replace the cut trees, the forest may grow back in just a few years.

Other resources such as oil and coal are not being formed today. These are **nonrenewable resources**. Once the coal is removed from one location, more must be found in another location.

It is possible for people to exhaust the world's supply of a nonrenewable resource. When that happens, we must find a different resource to take its place.

20.5 MANAGING NATURAL RESOURCES

Managing natural resources can be difficult. God's world is very complex. We must be careful that in managing one natural resource we do not harm another. To complicate matters, the Fall has marred both the world and us. We often make wrong or unwise decisions.

In spite of the difficulties, managing natural resources is a necessary and rewarding task. It allows humans to use natural resources without destroying the world that God has given us. It also ensures that natural resources will be available for future generations.

There are different ways of thinking about how to manage natural resources. On one extreme are those who believe that people should use natural resources with little thought for whether it harms the environment. After all, the world is here for our use, right? These people fail to recognize that though the world is available to us, it doesn't belong to us. On the other extreme are those who think we should use as few resources as possible with little regard for human need. In their view, humans pervert the natural order. Happily, very few people accept these two extreme positions. Most people are somewhere in the middle.

The thinking of most people is a blend of two philosophies of natural resource management. **Conservation** is the idea that natural resources should be used in ways that avoid long-term harm to the environment. The other philosophy is one of **preservation**. This is the idea that humans should leave natural resources in as near to their natural state as possible. Let's examine these ideas about resource management.

CONSERVATION

A Mind for the Future. Conservation is not about not using a resource. It's just about using it *carefully*. As a result, the resource should be available for use by future generations and by more people.

Sustainability. Conservation of renewable resources includes using them at rates that allow the resources to replace themselves. Logging companies must plan how many trees they will cut down each year so that they give the forest time to grow back. Many logging companies also plant trees to replace the ones they harvest. This allows the forest to regrow more quickly.

Efficiency. The goal of conservation is to ensure that natural resources are available over the long term. So the conservationist view includes the idea of using the resource as efficiently as possible.

Restoration. A conservationist mindset seeks to undo some of the damage of the past. American bison were once hunted nearly to extinction. Strict protections have helped the bison population grow from around 1000 animals to 500,000 in 100 years.

PRESERVATION

Protection. The goal of preservation is to leave natural resources alone. This is often done to give people a place to go to enjoy the outdoors. Sometimes it is done to save endangered species such as the whooping crane.

Parks. State and national parks may preserve areas that have important historical or natural features. The Gettysburg National Military Park was established to preserve the site of one of the bloodiest and most significant battles of the American Civil War.

Wilderness Areas. Wilderness areas, such as Colorado's Eagles Nest Wilderness Area, are a form of preservation. These parts of public lands are left in their wild state. People can visit them, but the areas generally do not have any buildings or roads.

People debate over the best way to manage natural resources. These arguments display a range of opinion on the topic of **environmentalism**, a concern for the proper stewardship of natural resources. Not everyone agrees on what proper stewardship means. As a result, people often disagree on whether an area and its resources should be used or set aside, that is, conserved or preserved.

Both those who seek short-term profits and those who focus only on long-term preservation are showing misplaced priorities. The Bible offers a different perspective. It reveals God's glory as the highest priority. God is glorified when His image bearers obey the Creation Mandate and govern the world in ways that use natural resources to ensure both human flourishing and a thriving creation.

WORLDVIEW SLEUTHING: SPRUCING UP THE FOREST

Many Americans in the late nineteenth and early twentieth centuries were not concerned about conservation. They thought that natural resources were limitless. Often this view had devastating effects on both people and the environment.

A sad example of this is the red spruce tree. People valued red spruce for its strength and light weight. Large quantities were harvested in West Virginia and neighboring states during the first decades of the twentieth century. The trees seemed so numerous that timber companies didn't think it necessary to leave some behind to regrow the forest. Nor did they think to plant new ones. Eventually, most of the red spruce were cut down, and the timber companies were forced to close as well. The people who worked for these companies had to find other work and often had to move away from the area.

The removal of most of the red spruce caused unforeseen environmental problems as well. The red spruce ecosystem was very different from the deciduous forests that replaced it. Private groups and government agencies are now beginning to replant red spruce trees in West Virginia's mountains.

INTRODUCTION

You are a news reporter writing a short article on the Central Appalachian Spruce Restoration Initiative (CASRI). Your assignment is to explain to your readers the goals and methods of this organization.

TASK

Research the CASRI and write a two-paragraph report on it. In the first paragraph, explain their purpose and the methods they use to accomplish that purpose. In the second paragraph, evaluate the plan. If possible, include where the plan has been tried before and whether it has succeeded.

PROCEDURE

1. Do a keyword search using the words "Central Appalachian Spruce Restoration Initiative."
2. Note how the worldviews of the website authors affect their view of the plan.
3. Plan and write your report. Read it and make appropriate revisions to confirm that you have completed the task as outlined above. Have one person proofread your report.
4. Share your report with your class or family.

20B REVIEW QUESTIONS

1. What is the difference between renewable and nonrenewable resources?
2. Classify each of the following natural resources as renewable or nonrenewable.
   a. sunlight
   b. coal
   c. wood
   d. beef
   e. stone

3. What are the two main benefits of managing natural resources?
4. What are the two philosophies of managing natural resources?
5. According to the Bible, what should be our highest priority?

Use the Case Study below to answer Questions 6–9.

CASE STUDY WHOSE TREES ARE THESE?

Even for California, which has a long history of bad fire seasons, the 2017 fire season was really bad. Over 1.2 *million* acres of forest and chaparral burned. Thousands of homes and other structures were destroyed, and over forty people lost their lives. One 2017 fire, the Thomas Fire, grew into the largest ever recorded in California.

In a sense, many of the trees that burned during the 2017 fire season were *your* trees. They had grown on National Forest land, which is land owned by the United States Federal Government. National Forests cover about 8.5% of the total land area of the United States. They are part of the country's *public lands*—lands that are owned by the government for use by the people. That includes you!

By law, public lands are supposed to be used. There are two principles written into federal law that are meant to guide the management of National Forests. The first, called *multiple use*, states that National Forests are to be managed with five major products and services in mind: timber, rangeland, water, wildlife, and recreation. The second principle, *sustained yield*, means that National Forests should be managed to produce as much of these products and services as is possible without harming the land.

A number of factors combined to produce the conditions needed for California's tragic fires. First, several groups challenged some of the federal government's timber harvesting policies in court. New rules were enacted that made it difficult to cut trees on public lands, so fewer trees were harvested. Then, starting in 2012, five years of severe drought rocked the state. As the lack of water weakened the trees, they became increasingly vulnerable to infestation by tiny but deadly bark beetles. By 2016, an estimated 102 million trees had perished. These dead, standing trees were the perfect fuel source for the firestorms of 2017.

Managing Earth's natural resources is a very complex task. As events in California showed, even good intentions can have unintended and lethal consequences. We still have much to learn about how best to manage our world.

6. Were the groups that sued the federal government to change the timber harvesting policies trying to conserve or preserve natural resources?
7. Fires occur naturally in California's forests, burning dead trees. But for years people have tried to put out as many fires as possible. How did this make the fires of 2017 worse?
8. When timber harvest is allowed, how does it replace the work of fire?
9. Write a paragraph suggesting a plan to avoid such a terrible fire year.

20C Questions

- Why do humans need to manage the environment?
- What are practical principles for managing God's living world?

20C

MANAGING CREATION

How should we use God's living world?

20.6 WHY US?

Why are humans the only creatures that manage the environment? Even the animals that use tools—like chimpanzees—do not cut down trees for houses or extract iron ore for nails. Dolphins don't establish parks to ensure that they don't eat all the fish.

The truth is that we are different from all other creatures. God created us in His image and gave us the Creation Mandate. That alone justifies and requires humans to manage the ecosystems of the world. At times people have ignored the command to rule and care for the earth, and it has always ended up hurting both people and the environment.

20.7 OUR JOB AS MANAGERS

We are all involved in management to one degree or another. We make decisions about how long to run the heat or air conditioning. Some people belong to organizations that support conservation. Landowners make decisions about how best to use their land. We even have some say on how management occurs on public lands. We vote for leaders to make some of the decisions regarding natural resource management. Some government agencies allow the public to comment on a proposed decision before it goes into effect. So whatever vocation God calls you into, you do share in the responsibility of managing natural resources. Let's get an overview of a few major principles that should guide our decisions regarding natural resource management.

MANAGEMENT PRINCIPLES

Maximize usefulness. Management policies should maximize a natural resource's benefit to humans and wildlife. The needs of people should be met while producing as little waste as possible. Management practices should also allow creation to thrive while ensuring that resources are available to meet the future needs of people.

Minimize hidden costs. Any management decision has costs. Some are obvious, but others, called *hidden costs*, are not. These hidden costs can often be very high. Damming a lake can provide many benefits, but the valley behind the dam will become covered with water and can no longer be used for any other purpose. Careful research before making a decision can help reduce hidden costs.

Maximize long-term health of the resource. God created natural resources for us to use *and* care for. So good management should ensure that living resources thrive. Artificial reefs are an example of this. Large objects such as old ships or oil rigs have been sunk into the ocean to provide habitat for sea life.

20C REVIEW QUESTIONS

1. What passage of Scripture commands people to manage the earth?
2. How would you respond to someone who argues that humans should change the environment no more than other mammals?
3. Name three principles to follow for managing God's world.

ETHICS Population Growth

Earth's human population is growing. When Jesus walked the earth, an estimated 250 million people lived on the planet. By 1650, the population had doubled to 500 million. Two hundred years later, in 1850, the population had doubled again, reaching 1 billion people. By 1930, only eighty years later, the population had reached 2 billion. In less than 45 years, the population doubled again, reaching 4 billion in 1974. Most estimates predict that the global population will exceed 8 billion around the year 2025.

This growth has caused some people to predict horrible doomsday scenarios where humans become so numerous that they destroy the earth. Such claims are not new. An English economist, Thomas Malthus, made similar claims in 1798 when the human population was around 800 million. The human population is now more than nine times larger than in Malthus's day, but the predicted disaster has not occurred.

Use the Ethics box above to answer Question 4.

4. Research and write a three-paragraph essay explaining the biblical worldview on human population growth. Include how the Bible views human reproduction. Also include ways that humans have avoided the ecological doomsday that Malthus predicted. Make suggestions on how the earth might support a higher number of people.

CHAPTER SUMMARY

20A PROTECTING THE ENVIRONMENT

- Pollutants are wastes that harm the environment.
- Pollutants are classified as either energy or substance pollutants. Substance pollutants may be biodegradable or nonbiodegradable.
- The production of pollutants can be reduced. Those that must be produced can be recycled, treated, or safely disposed of.

20A TERMS	
pollution	420
substance pollutant	421
nonbiodegradable pollutant	421
biodegradable pollutant	421
energy pollutant	421

20B USING RESOURCES

- Natural resources are the materials that we obtain from the physical world.
- Renewable resources are being replenished by natural processes; nonrenewable resources are not replenished.
- The two philosophies of managing natural resources are conservation, the careful use of natural resources, and preservation, the very limited or even nonuse of natural resources.

20B TERMS	
natural resource	426
renewable resource	426
nonrenewable resource	426
conservation	427
preservation	427
environmentalism	429

20C MANAGING CREATION

- God has given humans the right and the responsibility to manage the environment.
- Good management maximizes the usefulness of the resource, minimizes hidden costs, and maximizes the long-term health of the resource.

CHAPTER REVIEW QUESTIONS

Recalling Facts

1. Name the four methods of responsibly dealing with pollutants.
2. Is milk a renewable or a nonrenewable resource? Explain.
3. What is environmentalism?
4. How does the Creation Mandate justify humans managing the environment?
5. Choose one of the three management principles and explain it.

Understanding Concepts

6. Classify each of the following as either an energy pollutant (E), a biodegradable substance pollutant (B), or a nonbiodegradable substance pollutant (NB).
   a. loud sounds from a factory
   b. a plastic soda bottle floating down a stream
   c. a cardboard box falling from a truck to the side of the road
   d. a cigarette butt thrown from a car window
7. Draw a hierarchy chart showing the relationship between the following terms: energy pollutants, biodegradable pollutants, pollutants, nonbiodegradable pollutants, and substance pollutants.
8. Explain how dumping radioactive wastes in a forest is an example of both substance and energy pollution.
9. How would you respond to someone who defends throwing a paper bag out the window of a car by saying, "It's just one paper bag and it's even biodegradable."
10. A company produces a waste in its operation. It can recycle it or dispose of it in a landfill. Which should the corporation do?
11. A resource is becoming scarce in an area. Explain why people should consider whether the resource is renewable or nonrenewable before determining a proper course of action.
12. What happens to renewable resources when humans do not practice wise natural resource management?
13. Explain how people adopting a conservation view of natural resource management can result in an increased number of bison in an area where they have become sparse due to excessive hunting.
14. Environmentalism is a very broad topic for debate. On one side of the discussion are those who argue that humans do not need to worry about the environment as we extract natural resources. On the other side are those who argue that people should not disturb the environment. How does the Creation Mandate respond to both of these extremes?

15. Suppose state officials choose to preserve all of a state's public forests. A large timber company can no longer cut trees, and the company closes down. Its three hundred employees are now out of work. Which management principle from page 433 should the state officials have considered before making their decision? Explain.

Critical Thinking

16. Defend the existence of sanitary landfills.
17. Use the Creation Mandate to argue that improperly dumping biodegradable pollutants is wrong.

Use the information below to answer Questions 18–20.

John Oliver and his family were the first white settlers to move to Cades Cove, Tennessee, to homestead after the War of 1812. He built a log cabin in 1822 in the middle of the forest. He cut down some of the trees and used the logs to build the cabin. He also harvested rocks around the cabin for a chimney and left trees on the border of his property as a wind break. He made use of the soil to grow food for his family.

18. How did John Oliver apply the principle of maximizing usefulness in his project?
19. How could he have applied the principle of minimizing hidden costs to his project?
20. How could he have applied the principle of maximizing the long-term health of the resource?

CASE STUDY BLACK BEARS ARE BACK

Of all wildlife, large carnivores are often the animals that are hardest to manage. They eat game that people like to hunt. Sometimes they eat domesticated animals as well. Large carnivores are also potentially dangerous.

But large carnivore conservation can be accomplished. Black bears have greatly increased in numbers in the last fifty years. Although bear attacks do happen, they have been very rare. Overall, black bears and humans seem to live together safely.

Use the Case Study above to answer Questions 21–23.

21. Many states that had forbidden bear hunting now allow a limited bear hunting season. What does this indicate about the bear population in those states?
22. How is a hunting season a part of managing black bears?
23. Read Genesis 1:28 and Genesis 9:1–3. Would hunting black bears originally have been part of applying the Creation Mandate? Is it a legitimate part of applying the Creation Mandate now? Explain.

APPENDIX A

READING TABLES, GRAPHS, AND SCIENTIFIC DIAGRAMS

Scientists often like to compare two or more groups of observations to see whether they are related in some way. Doing so is a type of *analysis*, the process of looking for patterns and relationships within sets of data. These patterns and relationships can often be better understood by presenting them in a visual form, such as a table, graph, or diagram. Not surprisingly then, science textbooks normally contain many tables, graphs, and diagrams too. Knowing how to read them will greatly enhance your learning experience.

Tables

One of the simplest means of organizing data is to use a *table*. A table arranges data into rows and columns. Each row or column corresponds to a defined set of data or variable. For example, Table 1 shows the results of a stream survey conducted by fisheries biologists. The table has been set up to show at a glance *what kind* of fish were observed during the survey (row headings), as well as *how many* of each sex were observed (column headings).

Table 1

Stream Survey

Species	Male	Female	Unknown
Chinook	3	5	1
Coho	11	9	2
Steelhead	1	0	0

Data tables are usually read from left to right across the table but may also be designed to include useful information if read in another direction as well. Reading from left to right, we can see that the biologists observed three male Chinook salmon, five female Chinook salmon, and one Chinook salmon whose sex could not be determined. Reading from top to bottom reveals that the biologists observed a total of fifteen male fish of all species. If you wanted to know how many female coho salmon were observed, you would first find the row with the coho heading, then scan over to where that row intersects the column for female fish. Doing so shows that nine female coho salmon were observed.

Table 1 displayed the results of a survey, but tables are often used to record the results of experiments as well. A group of students used Table 2 to record the results of an experiment to test the relationship between the time needed for a car to travel a certain distance on a slant track and the angle at which the track was set.

Table 2

Car Track Slope and Time of Travel

Track Angle (°)	Trial 1 (s)	Trial 2 (s)	Trial 3 (s)	Average (s)
10	3.25	3.10	3.30	3.22
20	2.75	2.60	2.55	2.63
30	2.25	2.30	2.20	2.25

The two changing quantities, track angle and time, are called *variables*. Track angle, recorded in the first column, is the *independent variable* of the experiment since the angle of the track is set by the experimenters and does not depend on the values for time. The time to travel down the track, measured in the next three columns, is the *dependent variable* because its value changes depending on the angle of the track. Notice that in each column heading the units used to make each measurement are given (degrees for track angle and seconds for time). When the units are given in the heading, it means that all the measurements in that column are in those units; for example, every trial time is measured in seconds.

Table 2 displays the results of three trials done at each of three track settings, for a total of nine trials. The students also calculated the average time at each track setting and recorded that value in a fourth column. Can you find the result for the second trial at a 20° track setting?

Graphs

Another way to graphically display the relationship between two variables is to plot their values on a *graph*. The values of the independent and dependent variables are called the *coordinates* of the data. Each point plotted on a graph represents an *ordered pair* of coordinates, where the first number in the pair is an independent variable coordinate and the second is a dependent variable coordinate. You may have graphed ordered pairs in the form (x, y) in a math class.

Useful graphs include a title describing the graph's purpose and labels identifying the quantities and units used on each axis. Numbered scales and a grid are usually included to help you estimate values between the variables plotted on the graph.

Scientists usually plot the independent variable on the horizontal axis of a graph, also known as the x-axis, with increasing values to the right. The dependent variables are plotted on the y-axis, the vertical axis, with the values increasing upward. These are not hard and fast rules, and many graphs are arranged differently to improve the clarity of the plot. One of the first rules of reading graphs is to notice what the two variables are.

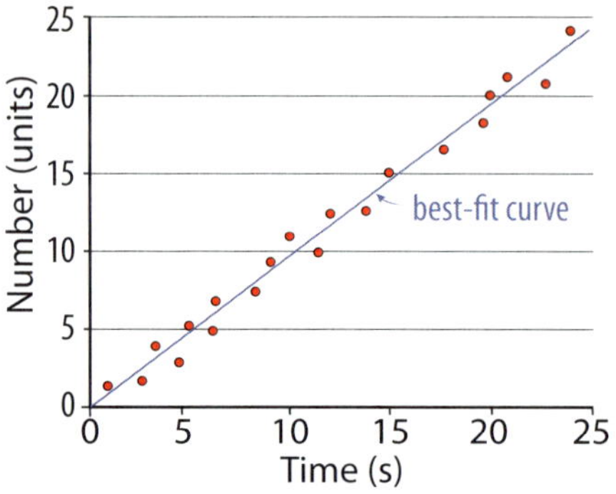

A scatterplot with a linear best-fit curve

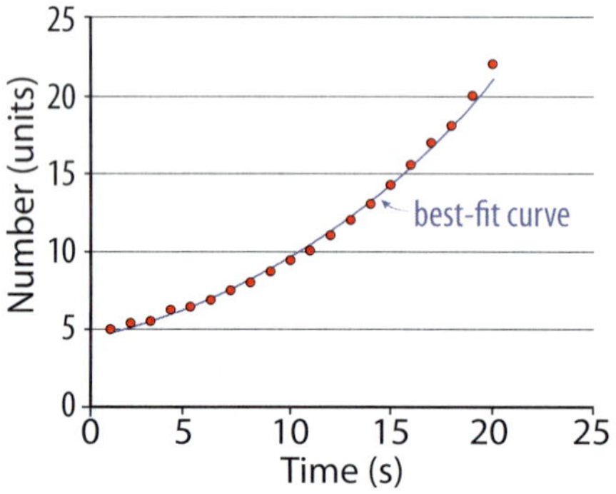

A scatterplot with a nonlinear best-fit curve

Graphs come in different forms. A simple plot of points on a graph is called a *scatterplot*. This is the starting point for many graphs. Scientists like to detect trends in the data and to create an equation that describes the trend. They draw a straight or curved line through the pattern of dots. The kind of line depends on how the pattern changes. We call the trend line a *curve of best fit*, even if it looks like a straight line. Graphs whose curve of best fit is a straight line are said to show *linear relationships*. These are fairly rare in nature. Most trend lines in nature are slightly curved to really wavy! These graphs, logically enough, show *nonlinear relationships*.

One of the most important things to learn from a graph is the rate at which the dependent variable changes in comparison to the independent variable. This rate is shown by the *slope* of the graph. A rising, or *positive*, slope shows that things are changing in a connected way—each increase in the independent variable produces an increase in the dependent variable. A horizontal slope shows that there is no relationship between the variables—a change in one does not produce a change in the other. A dropping curve has a *negative* slope, meaning that each increase in the independent variable produces a decrease in the dependent variable.

Scatterplots and trend lines can be read to obtain information not directly measured in the data set. The trend line connecting two data points represents values that are estimated, not measured. Obtaining unmeasured data this way is called *interpolation*. In addition, scientists often try to predict the values of the dependent variable if the independent variable continues to change beyond the range of the measured data. This method of analysis is called *extrapolation*. Extrapolating data depends heavily on assuming that the trend will continue as it has in the actual measured data.

Another common kind of graph is a bar graph. A *bar graph* depicts the value of the dependent variable of a data set in the form of a bar whose height (or length, depending on how the graph is set up) is determined by the variable's value. In the following graph, for example, the total rainfall for each month of the year in a rainforest can easily be seen. The shorter the bar, the less the amount of rainfall

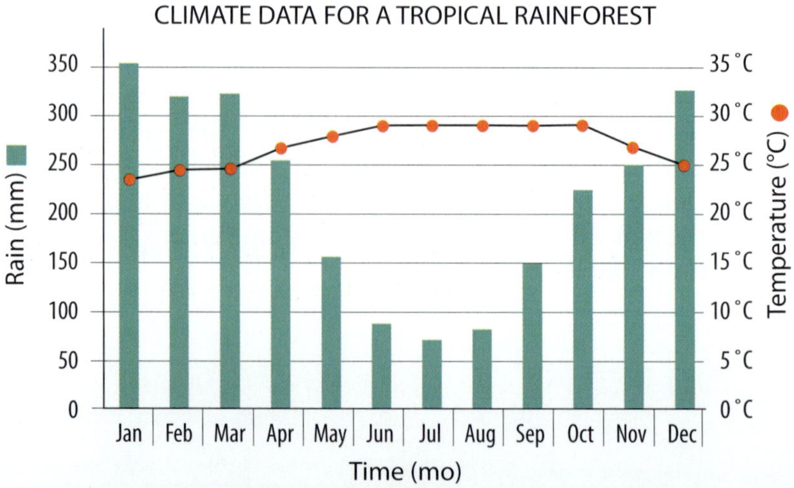

for a particular month. If you were asked to pick which months are the wettest in a rainforest, your eyes would quickly be drawn to the tallest bars on the graph. As this example shows, bar graphs are a good way to present data in which the independent variable has a nonnumerical value.

You will have a few opportunities to make graphs in exercises in the Student Activities.

Scientific Diagrams

Although biologists do make measurements during their investigations and present their findings in graphs, much of the content of this textbook is *conceptual*, meaning that ideas, facts, principles, and models are presented without using a lot of mathematics (perhaps this is a relief to you!). Biological processes and models are frequently illustrated by *diagrams*. These are visual aids whose purpose is to help make a concept more clearly understood. Diagrams are usually intended to help clarify something presented in the text, so the text and diagram should be studied together in order to best understand the idea being presented. Diagrams are used in this textbook mainly to illustrate two things: structures and processes. Let's take a look at what to expect in each type of diagram.

Structural Diagrams

A structural diagram is intended to help you identify the parts of an organism. Keep in mind that the primary purpose of a structural diagram is to *inform* the reader, not to be an *exact representation* of the structure being discussed. Structural diagrams often make use of certain enhancements to make something clearer.

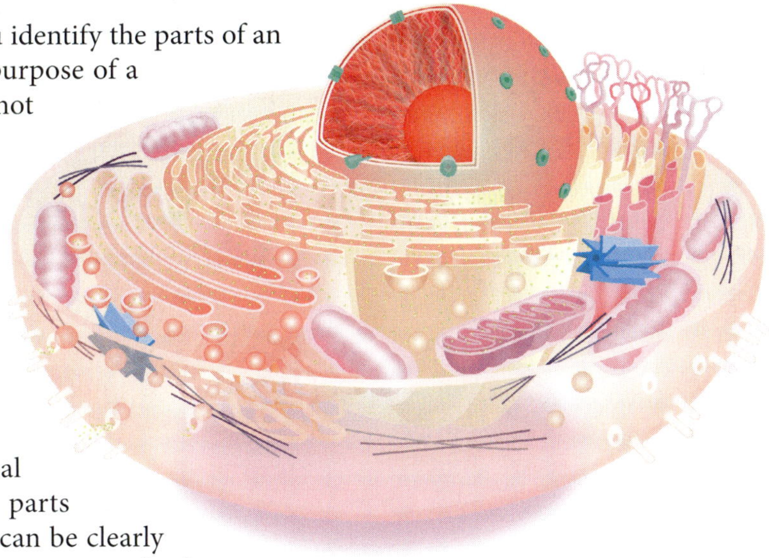

Structural diagrams can enhance color, zoom in on a structure, or cut away the exterior so that we can see inside. *Color coding* is often used to make the parts of a diagram more distinct and, in many cases, makes use of colors not found in the original structure. *Exploded diagrams* separate the parts of a structure so that each individual piece can be clearly defined. A *zoom-in* is a close-up view of one portion of a diagram, giving more focus and detail to that portion. A *cutaway* shows how a structure might look if we could see into its interior. Structural diagrams will use call-outs or numbered bullets to label parts of the diagram. Each of the features just described is a technique designed to make a diagram more information rich. In the diagram at right, notice the use of the cutaway feature to help you better understand the structure of a cell.

Process Diagrams

A process diagram is intended to show not how something looks but how it *works*. Because some biological processes are very complex, diagrams of those processes can seem overwhelming at first glance. By definition, a process has a beginning and an end—it starts with something and ends up with something else. So process diagrams, unlike structural diagrams, normally have a part that should be looked at first, followed by other parts that are viewed in an orderly progression.

Process diagrams usually give visual clues about where to begin viewing the diagram. There may be arrows leading from the start of the process to its end, or the steps of the process may be indicated by numbered bullets. Many diagrams make use of both arrows and bulleted items. Color coding is frequently used to help distinguish the parts of a process. Use the arrows or numbered bullets or both to guide you to the first step in the process, then read each step of the process along with the relevant text. Don't be concerned if you have to look at a process diagram several times to get the main ideas of what it teaches.

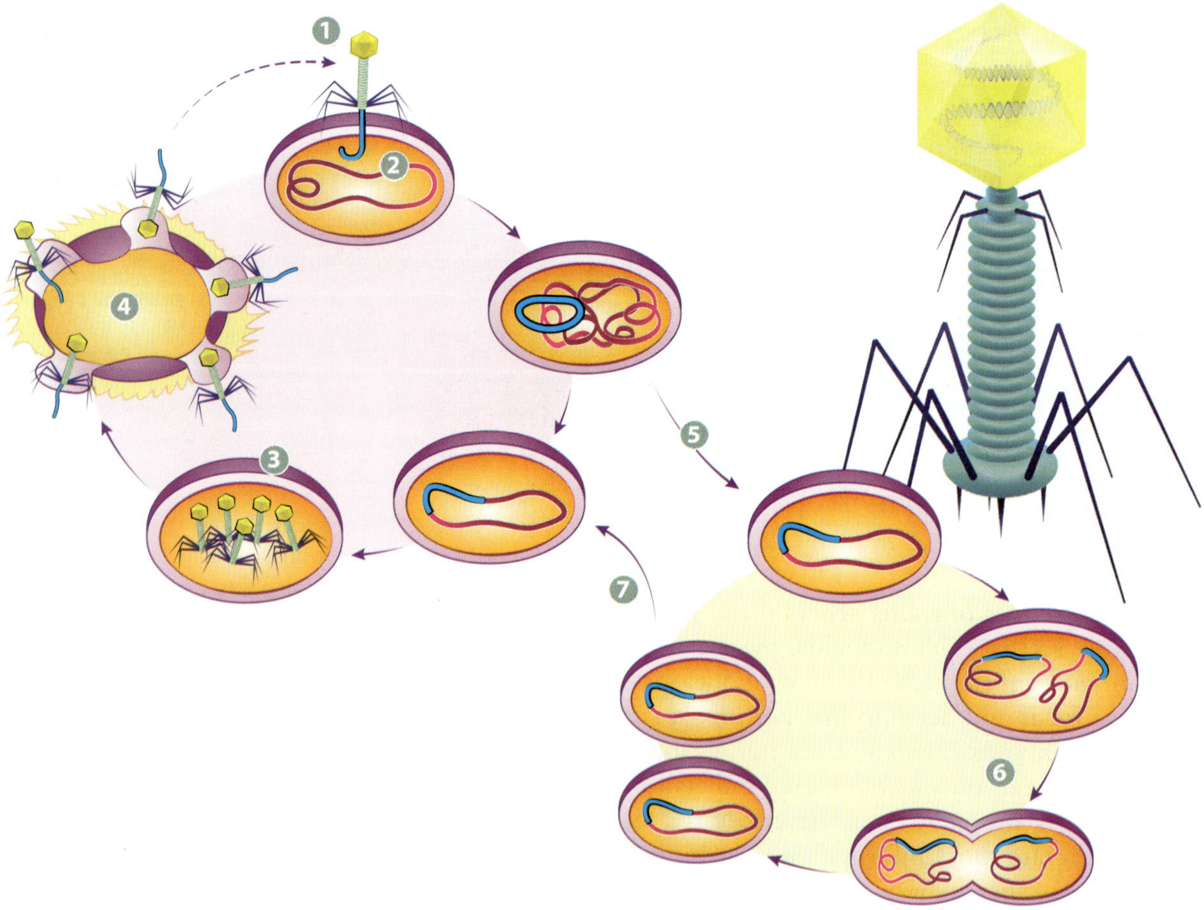

APPENDIX B

CREATING GRAPHIC ORGANIZERS

There will be several times in this textbook when you will be asked to create a *graphic organizer*—a visual way to represent data. Use this appendix as a guide to help you create different kinds of graphic organizers to show how concepts in life science compare, how they are structured, or how they move through a process.

Table/T-Chart

A table is often used in science to organize data, but it can also be used to organize descriptions and other information. A table can be a helpful way to compare two or more concepts in life science. For example, the four main types of plants are compared in Table 1 below. If you are comparing only two concepts, this kind of table with only two columns can be called a *T-chart* because of its shape.

Table 1

Comparing Types of Plants

	Bryophytes	Seedless Vascular Plants	Gymnosperms	Angiosperms
Reproductive structures	spores	spores	seeds	seeds
Primary structure	gametophyte	sporophyte	sporophyte	sporophyte
Specialized reproductive structures	none	none	cones	fruit
Common example	moss	fern	pine tree	apple tree

When you create a table, position the concepts that you are comparing in the top row. Place the characteristics under consideration in the leftmost column. Fill in the cells, considering how each characteristic or category differs for the concepts that you are comparing.

Hierarchy Chart

A *hierarchy chart* is one that shows the relationships between several concepts. Hierarchy charts are especially useful in life science when studying classification.

When you create a hierarchy chart, think of the different categories that a larger category includes. The example at right shows how all the objects in the solar system can be classified as the sun, dwarf planets, planets, small solar system bodies, or moons.

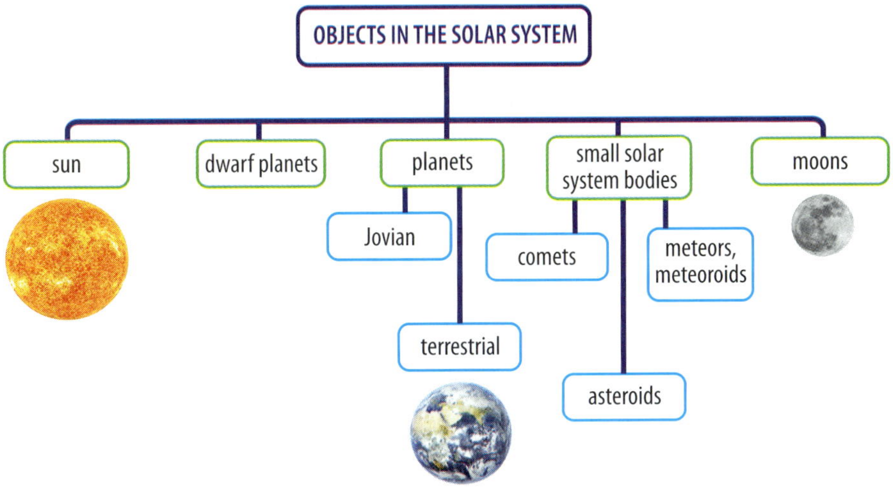

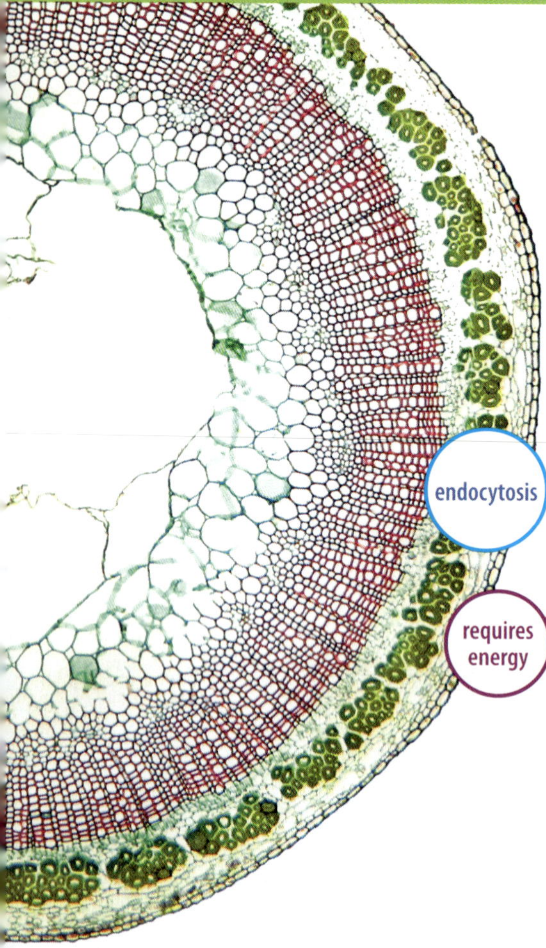

Concept Map

A *concept map* is a freeform graphic organizer that relates concepts, sub-concepts, and their related characteristics. It's similar to an outline in visual form. Because concept maps are so flexible, they can be more difficult to create.

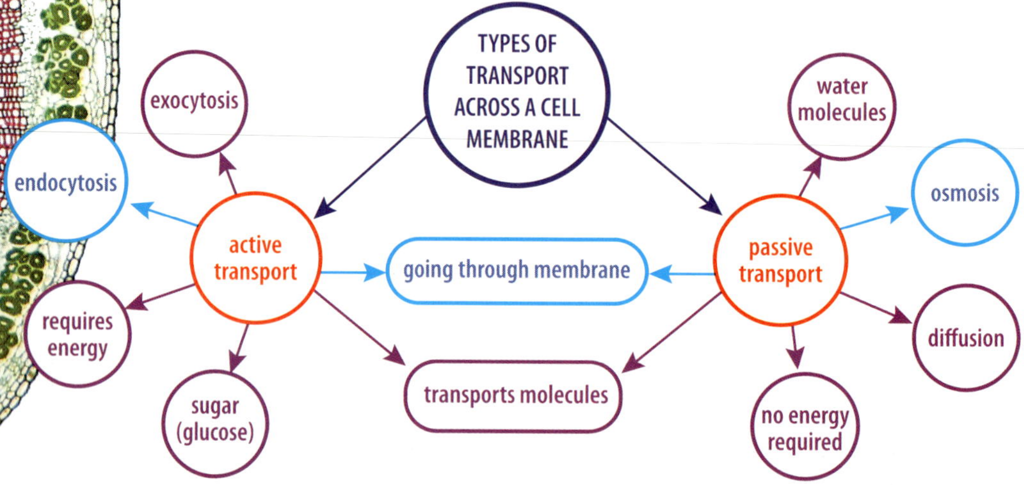

To make a concept map, start with the biggest concept and put it in a circle. Use lines or arrows to connect this main concept with smaller concepts. You can use linking words to help a viewer see the connection between two concepts and read the concept map like a sentence. The further you get from the main idea, the more specific the information should become.

Process Map/Flowchart

A *process map* illustrates a specific chain of events in a life science process. Because of its purpose, it almost always includes arrows to indicate the direction of the process.

When you create a process map, think of the process that you are illustrating and the different steps that are part of this process. Make sure that you place events in the correct order! You can show a process in a line of events such as a timeline. If the process repeats itself, you can show it in a circle as a cyclical process that doesn't have a beginning or an end.

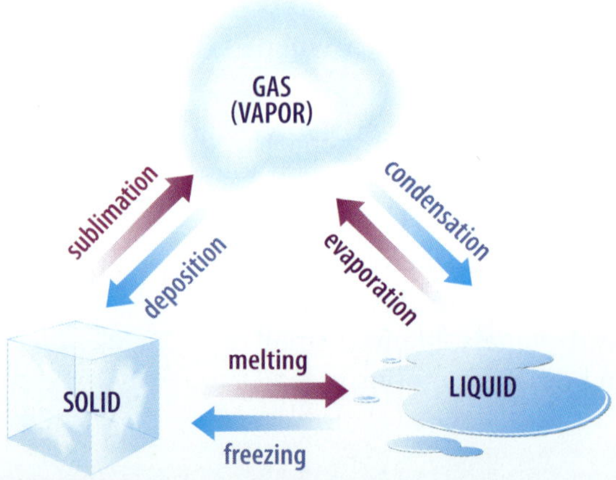

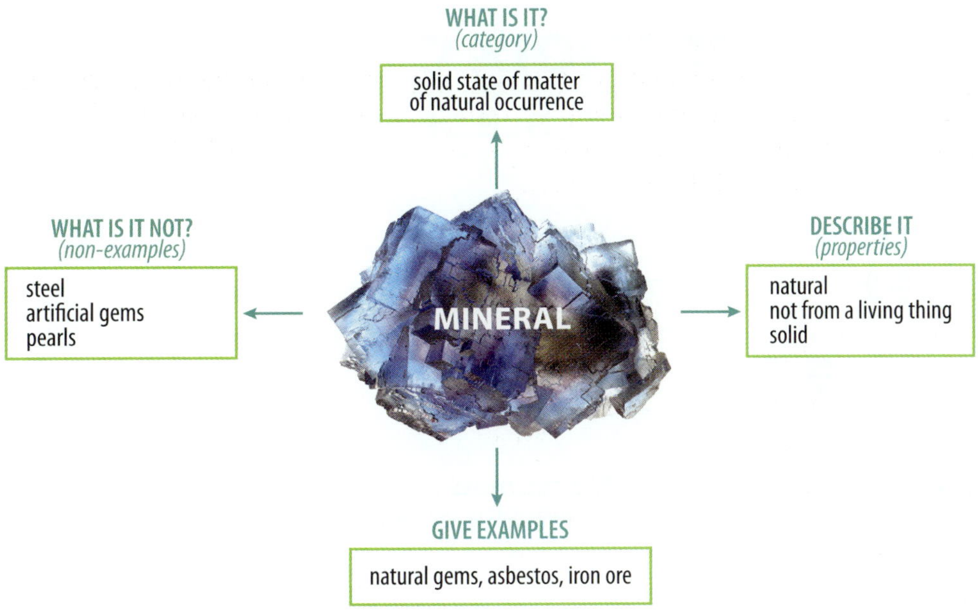

Concept Definition Map

A *concept definition map* takes a concept and considers four questions: What is it? What is it *not*? What are its properties? What are some examples?

Think through these four questions for a concept when you create a concept definition map. Then put the concept in the middle of the map and organize the four questions in boxes labeled with the questions around the concept.

Venn Diagram

The last type of graphic organizer is a *Venn diagram*. This graphic organizer begins with two partially overlapping circles, one for each of two concepts. Descriptions or parts of the two concepts are placed in the circles. If a description or part goes with both concepts, it is placed in the overlapping part of the graphic organizer. If it goes with only one concept, it is placed in the nonoverlapping part of the diagram with the concept.

When you create a Venn diagram, ask yourself what the main parts or descriptions of the two concepts are. Determine whether a part or description goes with both concepts or just one. Then place it in the proper part of the Venn diagram.

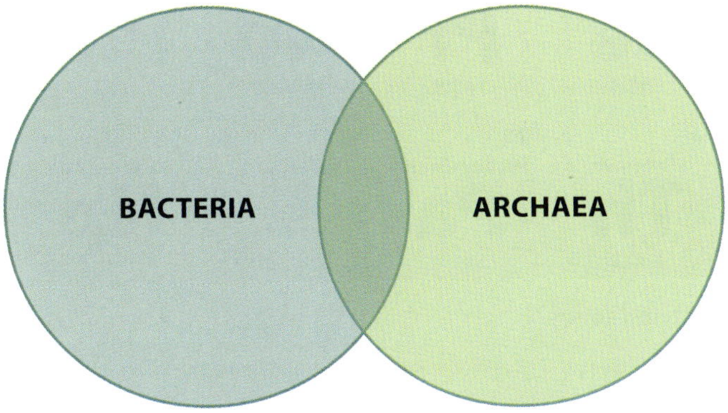

APPENDIX C

PERIODIC TABLE OF THE ELEMENTS

Atomic number: 92
Radioactive: ☢
Name: Uranium
Symbol: U
Atomic mass: 238.0
• rounded to four significant digits
• mass number of isotope with longest known half-life indicated by ( )

	1	2	3	4	5	6	7	8	9
1	1 Hydrogen H 1.008								
2	3 Lithium Li 6.94	4 Beryllium Be 9.012							
3	11 Sodium Na 22.99	12 Magnesium Mg 24.31							
4	19 Potassium K 39.10	20 Calcium Ca 40.08	21 Scandium Sc 44.96	22 Titanium Ti 47.87	23 Vanadium V 50.94	24 Chromium Cr 52.00	25 Manganese Mn 54.94	26 Iron Fe 55.85	27 Cobalt Co 58.93
5	37 Rubidium Rb 85.47	38 Strontium Sr 87.62	39 Yttrium Y 88.91	40 Zirconium Zr 91.22	41 Niobium Nb 92.91	42 Molybdenum Mo 95.95	43 ☢ Technetium Tc (98)	44 Ruthenium Ru 101.1	45 Rhodium Rh 102.9
6	55 Cesium Cs 132.9	56 Barium Ba 137.3	57 Lanthanum La 138.9	72 Hafnium Hf 178.5	73 Tantalum Ta 180.9	74 Tungsten W 183.8	75 Rhenium Re 186.2	76 Osmium Os 190.2	77 Iridium Ir 192.2
7	87 ☢ Francium Fr (223)	88 ☢ Radium Ra (226)	89 ☢ Actinium Ac (227)	104 ☢ Rutherfordium Rf (267)	105 ☢ Dubnium Db (268)	106 ☢ Seaborgium Sg (269)	107 ☢ Bohrium Bh (270)	108 ☢ Hassium Hs (277)	109 ☢ Meitnerium Mt (278)

Alkali metals
Alkaline-earth metals
Transition metals
Inner transition metals
Post-transition metals
Metalloids
Nonmetals
Halogens (also nonmetals)
Noble gases
☢ Radioactive isotopes

Lanthanide series

58 Cerium Ce 140.1	59 Praseodymium Pr 140.9	60 Neodymium Nd 144.2	61 ☢ Promethium Pm (145)	62 Samarium Sm 150.4
90 ☢ Thorium Th 232.0	91 ☢ Protactinium Pa 231.0	92 ☢ Uranium U 238.0	93 ☢ Neptunium Np (237)	94 ☢ Plutonium Pu (244)

Actinide series

	10	11	12	13	14	15	16	17	18
									2 Helium **He** 4.003
				5 Boron **B** 10.81	6 Carbon **C** 12.01	7 Nitrogen **N** 14.01	8 Oxygen **O** 16.00	9 Fluorine **F** 19.00	10 Neon **Ne** 20.18
				13 Aluminum **Al** 26.98	14 Silicon **Si** 28.09	15 Phosphorus **P** 30.97	16 Sulfur **S** 32.06	17 Chlorine **Cl** 35.45	18 Argon **Ar** 39.95
	28 Nickel **Ni** 58.69	29 Copper **Cu** 63.55	30 Zinc **Zn** 65.38	31 Gallium **Ga** 69.72	32 Germanium **Ge** 72.63	33 Arsenic **As** 74.92	34 Selenium **Se** 78.97	35 Bromine **Br** 79.90	36 Krypton **Kr** 83.80
	46 Palladium **Pd** 106.4	47 Silver **Ag** 107.9	48 Cadmium **Cd** 112.4	49 Indium **In** 114.8	50 Tin **Sn** 118.7	51 Antimony **Sb** 121.8	52 Tellurium **Te** 127.6	53 Iodine **I** 126.9	54 Xenon **Xe** 131.3
	78 Platinum **Pt** 195.1	79 Gold **Au** 197.0	80 Mercury **Hg** 200.6	81 Thallium **Tl** 204.4	82 Lead **Pb** 207.2	83 Bismuth **Bi** 209.0	84 Polonium **Po** (209)	85 Astatine **At** (210)	86 Radon **Rn** (222)
	110 Darmstadtium **Ds** (281)	111 Roentgenium **Rg** (282)	112 Copernicium **Cn** (285)	113 Nihonium **Nh** (286)	114 Flerovium **Fl** (289)	115 Moscovium **Mc** (290)	116 Livermorium **Lv** (293)	117 Tennessine **Ts** (294)	118 Oganesson **Og** (294)

63 Europium **Eu** 152.0	64 Gadolinium **Gd** 157.3	65 Terbium **Tb** 158.9	66 Dysprosium **Dy** 162.5	67 Holmium **Ho** 164.9	68 Erbium **Er** 167.3	69 Thulium **Tm** 168.9	70 Ytterbium **Yb** 173.1	71 Lutetium **Lu** 175.0
95 Americium **Am** (243)	96 Curium **Cm** (247)	97 Berkelium **Bk** (247)	98 Californium **Cf** (251)	99 Einsteinium **Es** (252)	100 Fermium **Fm** (257)	101 Mendelevium **Md** (258)	102 Nobelium **No** (259)	103 Lawrencium **Lr** (266)

APPENDIX D

COMBINING FORMS

You may find science a little intimidating because of all the long, unfamiliar words that scientists use. These larger words are usually made of smaller chunks called *combining forms*. Use the list in this appendix of combining forms and their meanings to help you decode the meanings of larger words.

For example, if you ran into the word *aerophyte*, you could separate this word into two parts, *aero-* and *-phyte*. *Aero-* means "air," and *-phyte* means "plant". So an aerophyte is an air plant, or a plant that is able to pull water out of the air and doesn't have roots embedded in the soil.

Some word roots can be used only as prefixes, others only as suffixes. Still others can be used as either. For example, *cardiologist* is the name of a heart doctor, while *tachycardia* refers to a rapid heart rate. In the list below, a hyphen (-) indicates roots that are normally used as suffixes.

A

a (Gk.)—not, without

ac, **ad**, **ag** (L.)—to, toward

acanth, **acantha**, **acantho** (Gk.)—spine, thorn

aer, **aero** (Gk.)—air

alb, **albi**, **albid** (L.)—white

allelo (Gk.)—one another, pair

amph, **amphi**, **ampho** (Gk.)—on both sides

an (Gk.)—without

ana (Gk.)—up, throughout, again

ant, **anti** (Gk.)—opposite, against

ante (L.)—before

ap, **apic** (L.)—tip, extremity

aqua, **aquari** (L.)—water

arthr, **arthro** (Gk.)—joint

-**ase**—suffix forming enzyme names

aster, **astero**, **astr** (Gk.)—star

aur, **aure** (L.)—gold, golden

aut, **auto** (Gk.)—self

B

bact, **bacter**, **bactr** (Gk.)—a rod

bacteri, **bacterio** (Gk.)—bacteria

bi (L.)—two, twice, double

bio, **bios**, **biot** (Gk.)—life

bol, **bola**, **bole**, **bolo**, **bolus** (Gk.)—a throw; clod, lump

C

card, **cardi**, **cardia**, **cardio** (Gk.)—heart

carn, **carneo**, **carni** (L.)—flesh

cat, **cata**, **cato** (Gk.)—down, downward

caud, **cauda** (L.)—tail

cell, **cella**, **celli** (L.)—chamber, cell

cent, **centi** (L.)—a hundred

centr, **centri**, **centro** (Gk.)—center

cephal, **cephala**, **cephalo** (Gk.)—head, brain

cereb, **cerebell**, **cerebr**, **cerebro** (L.)—brain

chem, **chemo** (Gk.)—juice; pour; chemistry

chlor, **chloro** (Gk.)—green

chondr, **chondro** (Gk.)—grain, corn; cartilage

chord, **chorda** (L.)—string, cord

chrom, **chroma**, **chromato**, **chromo** (Gk.)—color

chrys, **chryso** (Gk.)—gold

cide, **cis**, **cision** (L.)—to kill; cut

cili, **cilia**, **cilio**, **cilium** (L.)—eyelash, eyelid, small hair

co, **com**, **con** (L.)—with, together

coagul (L.)—drive together, curdle

crani, **crania**, **cranio**, **cranium** (Gk.)—skull

cret, **creta** (L.)—to separate

crist, **crista** (L.)—crest, tuft

cut, **cutane**, **cuti**, **cutic** (L.)—skin

cyan, **cyane**, **cyani**, **cyano** (Gk.)—dark blue

cycl, **cycle**, **cyclo** (Gk.)—circle, wheel

cyst, **cystis**, **cysto** (Gk.)—bladder, pouch

cyt, **cyte**, **cytic**, **cyto**, **cytus** (Gk.)—hollow place, cell

D

dactyl, **dactylo**, **dactylus** (Gk.)—finger, toe

derm, **derma**, **dermi**, **dermis**, **dermo** (Gk.)—skin

detrit (L.)—to wear off

di, **dis** (Gk.)—separate, apart; two, double

dia (Gk.)—across, through

dipl, **diplo** (Gk.)—double, two

dors, **dorsi**, **dorso**, **dorsum** (L.)—back

dyn, **dynam**, **dynamo**, **dynast** (Gk.)—be able; power, energy

E

e (L.)—out, without, from

echin, **echino**, **echinus** (Gk.)—spiny, sea urchin

eco (Gk.)—house

ecto (Gk.)—outside, outer, external

-ell, **-elle** (L.)—small

em (L.)—in, into

-emia (Gk.)—blood

en, **end**, **endo** (Gk.)—within, inner

epi (Gk.)—upon, over, beside

equ, **equa**, **equi** (L.)—equal

erythr, **erythro** (Gk.)—red

eu (Gk.)—good, true, well

ex (L.)—out, off, from, beyond

exo (Gk.)—out, outside, without

extra (L.)—outside, more, beyond, besides

F

fasci, **fascia** (L.)—a bundle

fer, **fera** (L.)—to bear. *See also* phor

fibr, **fibril**, **fibrin**, **fibro** (L.)—fiber

flagell, **flagellum** (L.)—whip

foli, **folia**, **folium** (L.)—leaf

G

gamet, **gamete**, **gameto** (Gk.)—marriage, spouse

gast, **gaster**, **gastero**, **gastr**, **gastri**, **gastro** (Gk.)—stomach, belly

gene, **genea**, **geneo**, **genesis**, **geno** (Gk.)—birth, origin, family, beginning

-genic, **-genous** (Gk.)—producing

genus (L.)—race, sort, kind

geo (Gk.)—earth

germ (L.)—bud, sprout, seed

gest (L.)—to carry

gli, **glia**, **glio** (Gk.)—glue

glob, **globo**, **globus** (L.)—ball, globe

gluc, **gluco** (Gk.)—sweet

glutin (L.)—glue

glyc, **glycer**, **glyco** (Gk.)—sweet

gon, **gone**, **gonia**, **gonidi**, **gonium**, **gono**, **-gony** (Gk.)—seed, generation, offspring

gran, **grani**, **grano**, **granum** (L.)—grain

graph, **grapho**, **graphy** (Gk.)—to write

gymn, **gymno** (Gk.)—bare

H

halo (Gk.)—salt, sea

hapl, **haplo** (Gk.)—single

hem, **hema**, **hemato**, **hemia**, **hemo** (Gk.)—blood. *See also* -emia

herb, **herba**, **herbi**, **herbo** (L.)—grass

hered (L.)—heir

heter, **hetero** (Gk.)—other, different

hist, **histo** (Gk.)—web, tissue

homo (L.)—man

homo, **homeo**, **homio** (Gk.)—like, same, resemble

hybrid (L.)—offspring of different parents

hydr, **hydra**, **hydri**, **hydro** (Gk.)—water

hyper (Gk.)—over, beyond

hypo (Gk.)—under, beneath

I

ichthy, **ichthyo**, **ichthys** (Gk.)—fish

il, **im**, **in**, **ir** (L.)—into, in; not

intr, **intra** (L.)—within

is, **iso** (Gk.)—equal

J

jug, **jugo** (L.)—*See* zyg

K

kary, **karyo** (Gk.)—nut, nucleus

kine, **kinema**, **kinemato**, **kines**, **kinesi**, **kinet**, **kineto** (Gk.)—move, moving, movement

L

lampr, **lampro** (Gk.)—shining, clear

later, **lateral**, **latero** (L.)—side

leuco, **leuko** (Gk.)—clear, white

libra (L.)—a balance, scales

liga, **ligam**, **ligat** (L.)—bound; a band

lip, **lipo** (Gk.)—fat

log, **logo**, **logus**, **logy** (Gk.)—word, study of

lun, **luna** (L.)—moon

lys, **lysi**, **lysio**, **lysis**, **lytic** (Gk.)—loosening, break apart

M

macr, **macro** (Gk.)—large

matr, **matri**, **matro** (L.)—mother

medi, **media**, **medio** (L.)—middle

mei, **meio** (Gk.)—less

melan, **melano** (Gk.)—black

mens, **mensa** (L.)—table

mer, **mere**, **meri**, **mero** (Gk.)—a part

mes, **meso** (Gk.)—middle

met, **meta** (Gk.)—between, with, after, change

meter, **metr**, **metra**, **metri**, **metro**, **-metry** (Gk.)—measure

micro (Gk.)—small

mill, **mille**, **milli**, **millo** (L.)—one thousand

mito (Gk.)—thread

mono (Gk.)—one

morph, **morpha**, **morpho** (Gk.)—form, shape

mult, **multi** (L.)—many

muta (L.)—change

myc, **myce**, **mycet**, **myceto**, **myco** (Gk.)—fungus

my, **myo**, **mys** (Gk.)—muscle

N

nan, **nani**, **nano**, **nanus** (Gk.)—dwarf

nast, **nasto** (Gk.)—pressed closed, solid

nem, **nema**, **nemato**, **nemo** (Gk.)—thread

nephr, **nephri**, **nephron**, **nephrus** (Gk.)—kidney

neur, **neura**, **neuro** (Gk.)—nerve

-nomy (Gk.)—the science of

nuc, **nucle**, **nucleo** (L.)—central part

O

ocul, **oculi**, **oculo**, **oculus** (Gk.)—eye

-oid (Gk.)—like, form

omni (L.)—all

oo (Gk.); **ov**, **ovi**, **ovo**, **ovum** (L.)—egg

opt, **opti**, **opto** (Gk.)—eye, vision

ora (L.)—mouth

orth, **ortho** (Gk.)—straight, correct, upright

os, **oss**, **osse**, **ossi** (L.); **ost**, **oste**, **osteo**, **osteum** (Gk.)—bone

-ose—suffix forming sugar names

osmo (Gk.)—pushing

P

par, **para** (Gk.)—beside

-parous (L.)—to give birth

ped, **peda**, **pede**, **pedi**, **pedo** (L.)—foot

per (L.)—through, by means of

peri (Gk.)—around

phag, **phage**, **phago**, **-phagy** (Gk.)—to eat

pharm, **pharmac** (Gk.)—drug, poison

pheno (Gk.)—show, seem, appear

phil, **phila**, **phile**, **phili**, **philo** (Gk.)—beloved, loving

phob, **phobia**, **phobo** (Gk.)—fear

phor, **phora**, **phore**, **phori**, **phoro** (Gk.)—to bear

phos, **phot**, **phota**, **photi**, **photo** (Gk.)—light

phyll, **phyllo**, **phylum** (Gk.)—leaf

physis (Gk.)—growth

phyt, **phyto**, **phytum** (Gk.)—plant

plan, **plani** (L.); **plat**, **plate**, **plati**, **platy** (Gk.)—broad, flat

plasm, **plasma**, **plasmato**, **plasmo** (Gk.)—something molded

plast, **plasto** (Gk.)—formed, molded

plur, **pluri** (L.)—more, several

pneumo, **pneumon**, **pneumona** (Gk.)—lungs

pod, **podo**, **pody** (Gk.)—foot

poly (Gk.)—many

poten, **potent** (L.)—powerful

pro (Gk.)—before, in front of

prot, **prote**, **proto** (Gk.)—first, original

pter, **ptero**, **pterum**, **ptery** (Gk.)—wing, feather, fin

pulmo, **pulmon**, **pulmono** (L.)—lung

purpur, **purpure** (L.)—purple

-pus (Gk.)—foot

R

radi, **radia**, **radiat**, **radio** (L.)—spoke, ray

re (L.)—back, again

ren, **rena**, **reni**, **reno** (L.)—kidney

reticul (L.)—network

rhiz, **rhiza**, **rhizo** (Gk.)—root

S

sacchar, **saccharo** (Gk.)—sugar

sarc, **sarci**, **sarco** (Gk.)—flesh

scop, **scope**, **scopo**, **-scopy** (Gk.)—to see, watch

scri, **scrib**, **script** (L.)—to write

secret (L.)—set apart

semi (L.)—half

-sis (Gk.)—the act of

soma, **somat**, **somato**, **-some** (Gk.)—a body

sperm, **sperma**, **spermato**, **spermi** (Gk.)—seed

spir, **spira**, **spiro** (L.)—coil; twist

-spire (L.)—to breathe

spor, **spora**, **-spore**, **spori**, **sporo** (Gk.)—seed

stas, **stasi**, **stasis** (Gk.)—standing, posture

stat, **stati**, **stato** (Gk.)—standing, placed

stoichi, **stoicho** (Gk.)—element

stom, **stoma**, **stomato**, **stomo** (Gk.)—mouth

stria, **striat** (L.)—furrow, streak

stroma, **stromato** (Gk.)—anything spread out, coverlet

sy, **syg**, **syl**, **sym**, **syn**, **sys** (Gk.)—with, together

T

tel, **teleo**, **telo** (Gk.)—end, complete, far

tetr, **tetra** (Gk.)—four

therm, **thermos** (Gk.)—heat

thesis (Gk.)—an arranging

thigm, **thigma**, **thigmato**, **thigmo** (Gk.)—touch

thora, **thoraco**, **thorax** (Gk.)—breastplate

thylac, **thylaco** (Gk.)—sack, pouch

tom, **tome**, **tomi**, **tomo**, **tomy** (Gk.)—to cut

ton, **tono** (Gk.)—tone; tension

top, **topo**, **topus**, **topy** (Gk.)—place

tot, **tota**, **toti** (L.)—all

toxic, **toxicum** (L.)—poison

tran, **trans** (L.)—across, through

tri (L.)—three

trop, **tropae**, **trope**, **tropo** (Gk.)—turn, change

troph, **trophi**, **tropho** (Gk.)—food, nourishment

turg (L.)—swell

typ, **-type**, **typi**, **typo** (Gk.)—blow, strike; impression

U

uni (L.)—one

ur, **ura**, **uro** (Gk.)—tail

V

vacu (L.)—empty

valen, **valent** (L.)—strength; be worth

vas, **vasa**, **vaso** (L.)—vessel, duct

vent, **venter**, **ventr**, **ventro** (L.)—belly, underside

vert, **verta**, **verte** (L.)—to turn

vor, **vora**, **-vore** (L.)—to devour, eat

X

xanth, **xantho** (Gk.)—yellow

Z

zo, **zoa**, **zoi**, **zoo**, **zoon** (Gk.)—animal

zyg, **zygo**, **zygus** (Gk.)—yoke

zym, **zyma**, **zymo** (Gk.)—yeast, leaven, ferment

GLOSSARY

A

abiotic factor A nonliving part of an ecosystem

abscisic acid A plant hormone that inhibits several of the other plant hormones

active immunity A fast-acting defense against a pathogen caused by a previous infection by that same pathogen

active transport The movement of molecules across a membrane that requires the cell to do work

adaptation An inherited trait that helps an animal survive to reproduce

adrenal gland Either of the two endocrine glands that produce hormones that help a person in stressful situations

aggression A willingness to attack another individual

alimentary canal A tube composed of several different digestive organs where digestion takes place

alternation of generations The repeating cycle of gametophyte and sporophyte structures in plants

alveolus A tiny sac where gas is exchanged between the air in the lungs and the blood

amino acid One of about twenty molecules used to build proteins

amphibian An ectothermic vertebrate that usually lives in water when young and on land as an adult

anaerobic Describes a process that does not use oxygen

anatomy The study of the organs and structures of the body

angiosperm The group of plants that produces flowers and fruit

annual ring The rings in a woody stem that result from different growth rates of xylem at different times of the year

antibody A protein that reacts against specific antigens

anticodon A set of three nucleotides in tRNA that match a codon on mRNA

antigen A chemical on the cell membrane of an organism that causes an immune system to produce antibodies to attack it

artery A blood vessel leading away from the heart

arthropod One of a large group of animals with jointed appendages and an exoskeleton

asexual reproduction Reproduction that does not involve gametes

ATP The molecule that a cell uses to power many processes in the cell that need chemical energy

atrium An upper chamber of the heart

autotroph An organism that makes its food using energy from its environment

auxin A plant hormone that causes shoots and stems to grow by causing their cells to enlarge while slowing down cell division

B

baraminology The effort to define the boundaries between the various created kinds of animals

B cell A white blood cell that produces antibodies

behavior The way that an animal interacts with other animals or with its environment

binary fission A form of bacterial reproduction in which the DNA is replicated and the cell membrane pinches together to form two identical bacteria

biodegradable pollutant A substance pollutant that can be broken down by organisms in the environment

bioluminescence Light produced by living organisms

biome The parts of the biosphere that share similar climates and kinds of living things

biotic factor A living organism, considered as part of its ecosystem

bladder A muscular storage organ for urine

blood pressure The pressure of circulating blood against the walls of arteries

bone marrow The material at the center of some bones that produces blood cells

brain stem The portion of the brain that helps control breathing, heartbeat, blood pressure, coughing, sneezing, and swallowing

bronchus Either of the two branches of the trachea that lead into a lung

C

calorie A unit of measurement often used to measure the energy that the body takes in

cap The umbrella-shaped structure at the top of a mushroom

capillary A tiny blood vessel connecting an artery and a vein

capsid The protein coat that surrounds the nucleic acid of a virus

capsule A slimy layer surrounding some bacteria that protects them from harmful substances and prevents their drying out

carbohydrate A type of nutrient used to fuel cellular respiration

carbon cycle The movement of carbon from the atmosphere into living organisms, then into the earth, and finally back to the atmosphere

cardiac muscle A striated, involuntary muscle found in the heart

carnivore An animal that gets most or all of its energy and nutrients from eating other animals

carpel The female part of the flower consisting of the sticky stigma, the tubular style, and the ovary that holds the eggs

cartilage A type of connective tissue that is softer and more flexible than bone

cell The smallest working unit of an organism

cell cycle The process of cellular growth and reproduction

cell membrane The structure that surrounds all cells and regulates what comes in and goes out

cell theory The model of the cell that states that cells make up all living things, come from preexisting cells, and perform the functions of living things

cell wall A rigid structure that surrounds and protects some cells

cellular respiration The process in organisms of breaking down molecules to produce ATP

cellulose A molecule made of long chains of sugars that makes up the cell walls of plants

central nervous system The part of the nervous system composed of the brain and spinal cord

cerebellum The portion of the brain that helps with motor control

cerebrum The largest portion of the brain, consisting of two dome-shaped hemispheres that control conscious thinking

chaparral A biome with mild, wet winters, hot, dry summers, and cycles of wildfires and regrowth

character A characteristic of a species such as size or shape

chemical digestion The form of digestion that uses chemicals to break down food molecules into nutrients

chlorophyll The green pigment that absorbs sunlight for photosynthesis

chromosome A structure that contains a cell's DNA

chyme A pasty substance that results from food being ground and treated with enzymes and acid in the stomach

cilium A small, hairlike structure that some cells use to move

circulatory system The multiple organs that work together to transport nutrients and oxygen to cells and remove wastes and carbon dioxide

climate The long-term pattern of weather in an ecosystem

cnidarian One of a group of animals with stinging cells, called *cnidocytes*, including jellyfish

cochlea The coiled tubular structure where sound vibrations are changed into nerve impulses

codominance A pattern of inheritance in a hybrid in which both traits are present but not blended

codon A set of three nucleotides in a strand of mRNA that codes for one of the amino acids

commensalism A type of symbiosis that benefits one organism while not affecting the other

common ancestor A species of organism whose descendants are believed to form two distinct species

compact bone Very hard bone made of circular layers

competition A relationship between two organisms in which they both use the same resources

concentration gradient The difference in the number of a certain type of molecule on each side of a membrane

coniferous forest A biome with long, cold winters, short, warm summers, and coniferous trees as the climax vegetation

conjugation (1) A process in which one bacterium shares a plasmid with another; (2) a method of sexual reproduction used by protists that involves exchanging DNA through their oral grooves

connective tissue Tissue that supports, connects, and protects other body structures

conservation The idea that natural resources should be used in ways that avoid long-term harm to the environment

consumer An organism that eats other organisms or particles of food

control group The group in a controlled experiment that is *not* exposed to the independent variable

controlled experiment A test of the way that something works by examining the effect of one factor at a time

cork cambium A layer of living cells that produces new cork cells just under the dead cork of a plant's bark

courtship A behavior that an animal performs while seeking a mate

Creation Mandate God's command to Adam and Eve to fill the earth and have dominion over it

cuticle The waxy coating of plants that helps prevent water loss

cytokinesis The final stage of the cell cycle during which the parent cell pinches to give each new daughter cell some cytoplasm

cytokinin A plant hormone that causes plant cells to divide

cytoplasm The material inside of a cell, excluding the nucleus

cytoskeleton Several types of long, thin fibers anchored to various parts of the cell that help the cell maintain and change its shape

D

data Information that people collect

dating The process used to estimate the age of rocks

day-neutral plant A plant that blooms whenever the conditions are right regardless of the length of night

deciduous forest A biome with four seasons, mild temperatures, around 100 cm of yearly rainfall, and deciduous trees as the climax vegetation

decomposer An organism that digests dead organisms and recycles their nutrients back into the environment

dermis The second layer of skin

desert A biome that receives very little water

dicot The group of angiosperms that have vascular bundles in a ring around the outer portion of the stem

diffusion The movement of molecules through a membrane as a result of a concentration gradient

digestion The process of breaking down food into molecules for use

digestive system The body system that breaks down food and absorbs nutrients

DNA The molecule that contains important information for making proteins and helps regulate many cellular processes

dominant trait A trait of a specific character that appears in a hybrid organism

E

eardrum The membrane that stretches over the opening to the middle ear and conducts vibrations to the hammer

echinoderm One of a group of animals with spiny skin, including sea stars, sea urchins, and sea cucumbers

ecological succession The process of a disturbed ecosystem returning to its climax community

ecology The study of the interactions in an ecosystem

ecosystem A limited area in which living and nonliving things interact

ectotherm An animal with a body temperature that is determined by its environment

embryo (1) The young plant that grows into a new sporophyte; (2) the term used to identify a baby developing in the womb until it is eight weeks old

endoplasmic reticulum A network of tubes in eukaryotic cells that help produce and transport proteins and other molecules throughout the cell

endoskeleton A living internal support structure

endospore A bacterium's DNA and ribosomes enclosed in a tough case that can survive unfavorable conditions

endotherm An animal that can regulate its internal body temperature

energy pollutant A form of energy that can negatively affect the environment

energy pyramid A model of how energy flows through a food chain

engineered system A group of things that are planned, designed, and built to work together to do a certain job

environmentalism A concern for the proper stewardship of natural resources

enzyme A protein that speeds up or slows down chemical reactions

epidermis The outer layer of the skin

epithelial tissue Tissue that lines the external and internal surfaces of the body

ethics The study of applying moral principles to life

ethylene A plant hormone gas that causes fruit to ripen and causes trees to close off the xylem and phloem to their leaves before the leaves drop off

eukaryotic cell A cell with a nucleus

evolution The model that states that simpler organisms have slowly changed into new and more complex organisms over millions of years

excretory system A system of special organs that rid an animal's body of cell wastes

exoskeleton A support structure on the outside of the animal's body

experimental group The group in a controlled experiment that is exposed to the independent variable

external fertilization A type of sexual reproduction in which the sperm and egg unite outside the parents' bodies

F

fat A type of nutrient used to protect internal structures and nerves, to form cellular membranes, and to store extra energy

fertilization The joining of gametes to form a diploid cell

fetus The term used to identify a developing baby between nine weeks and birth

fiber A thick-walled cell that helps support a plant

filter feeder An animal that strains tiny algae, protists, or other food items out of the water

flagellum A whip-like structure that some cells use to propel themselves through their environment

food chain The passing of matter and energy from producers through consumers

food web All the interrelated food chains that are possible in an ecosystem

fossil The preserved remains of plants and animals

fruit An angiosperm ovary that swells to surround and protect seeds

fruiting body The part of a fungus that produces spores

fungus Any member of kingdom Fungi having cells with cell walls made of chitin

G

gamete A sperm cell or an egg cell

gametophyte The plant structure that produces gametes and forms the primary structure of bryophytes; is microscopic in gymnosperms and angiosperms

gene A section of a DNA molecule that codes for a unique protein

genetic drift A natural process in which the ratio of alleles in a population changes randomly over time

genetic load The accumulated random mutations within the DNA of a population of organisms

genotype The letters that symbolize an organism's alleles for a specific gene

gill (1) a structure on the underside of a mushroom's cap containing cells that produces spores; (2) an organ that exchanges gases with aquatic surroundings

Golgi apparatus The organelle in eukaryotic cells that modifies many proteins and packages them into small, bubble-like structures called *vesicles*

gonad The main organ of the reproductive system

grassland A biome often found in the interior of continents with grass as the climax vegetation since it receives too little rainfall to support trees

growth plate An area of bone made of cartilage that divides in order for bone to grow

gymnosperm The group of plants that produces cones

H

habitat The portion of an ecosystem that an animal uses

hair follicle A structure that holds cells that divide to produce hair growth

heart A muscular organ for pumping blood through the body

herbivore An animal that feeds on plants

heredity The study of how traits are passed from one generation to the next

heterotroph An organism that gets energy by eating other organisms

homeostasis An organism's ability to keep its internal processes stable

homologous chromosomes The matching pairs of chromosomes in a diploid cell

homologous structures Features that are shared by distinct types of organisms

hormone (1) A small chemical made by some plant tissues that affects how other plant structures grow; (2) a chemical that acts as a messenger in the body of a human or animal

host A living organism that a parasite obtains food from or that a virus infects in order to reproduce

hybrid An organism with a mix of alleles for a specific character

hypha Filaments composed of long cells bound together that make up fungi

hypothalamus The part of the brain that links the nervous system to the endocrine system

hypothesis An explanation that is based on limited observations

I

imperfect fungus A fungus that reproduces only asexually

imprinting The process in which a young animal follows and imitates the nearest moving object

incomplete dominance A pattern of inheritance in a hybrid in which traits are blended together

incubate To warm eggs or offspring

inflammation A local response to pathogens consisting of increased blood flow and macrophages attacking the pathogens

innate behavior An unlearned type of behavior that an animal is capable of from birth

instinct A complex innate behavior that occurs over a longer period of time than a reflex does

integumentary system The body system that includes skin, hair, and nails

internal fertilization A type of sexual reproduction in which the sperm and egg unite inside the female's body

interphase The portion of the cell cycle during which the cell grows and carries out its normal functions

invertebrate An animal that lacks a backbone

involuntary muscle A muscle that is not controlled at will

irreducible complexity A feature of engineered systems in which removing one part of the system causes the whole system to cease functioning

J

joint The connection between two or more bones

K

kidney An organ in vertebrates that filters wastes from the blood

kind A distinct group of organisms that God created to reproduce

L

larynx The organ that helps prevent food from entering the trachea and contains the vocal cords

law An idea that scientists use to describe what they observe

learned behavior A type of behavior that is developed by observation and training

lichen A fungus and an alga intertwined together so that they appear as a single organism

life science The study of life that includes what is known about life and the processes used to learn about living things

ligament A tough, flexible band of connective tissue that holds bones together at a joint

long-day plant A plant that does not bloom until the nights are shorter than the critical night length

lung An organ in vertebrates that exchanges gases with the air

lymph Leaked plasma that flows through the lymphatic system and back into the circulatory system

lymph node A structure along the lymphatic system that filters dead cells, bacteria, viruses, and other particles out of the lymph

lysosome A small organelle in eukaryotic cells that act as the demolition crew within a cell

M

marsupial A mammal with a pouch in which to raise its young

mechanical digestion The form of digestion that grinds and crushes food into smaller particles

meiosis The process that reduces by half the amount of DNA present in each sperm or egg cell

melanin The brown pigment that determines human skin color

menstrual cycle The monthly cycle in women caused by the production of estrogen by the ovaries that results in thickening uterus walls until pregnancy starts or the shedding of the inside uterus layer during menstruation

metamorphosis A type of growth that involves a change to the body

migration A seasonal, long-distance trip

mineral An inorganic nutrient that the body uses to control processes and build certain structures like bones

mitochondrion The organelle in eukaryotic cells that makes most of the cell's ATP

mitosis The process of the cell cycle that ensures that each new daughter cell produced has the same number and kinds of genes as the original parent cell

model A simplified explanation, description, or representation of the world

modern synthesis The modern model of evolution that blends mutation and natural selection

mollusk One of a group of animals with soft bodies and usually a shell, including snails, octopuses, and clams

monocot The group of angiosperms that have vascular bundles evenly distributed throughout their stems

multicellular Having more than one cell

muscle tissue Tissue that contracts either to move the body or to move substances through the body

muscular system The organ system that an animal or human uses to move around

mutation Random changes in DNA

mutualism A type of symbiosis that benefits both organisms

N

nasal cavity The area inside the nose that warms and moistens incoming air

natural resource A material that humans can obtain from the physical world

natural selection The process through which organisms with certain adaptations survive to pass on their traits to a greater number of offspring than other organisms do

nephron A tiny structure in the kidney that filters blood

nerve A bundle of neurons

nerve tissue Tissue that receives and transmits electrochemical impulses to coordinate the functions of the body

nervous system The organ system that ensures that all the body systems of an animal work together

neuron Special cells that collect information and transmit electrical impulses

neutralism A relationship between two organisms in which neither affects the other

niche The roles played by an organism within an ecosystem

nitrogen cycle The movement of nitrogen from the atmosphere into living organisms, then into the earth, and finally back to the atmosphere

nonbiodegradable pollutant A substance pollutant that cannot be broken down by the environment

nonrenewable resource A natural resource that is not being formed today

notochord A flexible rod found in chordate animals

nucleotide The subunit of DNA consisting of a sugar, a phosphate, and a base

nucleus The structure that holds and protects the DNA in a eukaryotic cell

O

omnivore An animal that will eat both plant and animal matter

organ A group of tissues that work together

organelle The usually membrane-surrounded structures in a eukaryotic cell that perform special functions

organism A living thing

osmosis The movement of water across a semipermeable membrane

ovary The organ in plants and female animals and humans that produces eggs

ovulation The process of the ovaries producing an egg

ovum A female gamete; also called an egg

oxygen cycle The movement of oxygen from the atmosphere into living organisms and back to the atmosphere

P

palisade layer The tightly packed cells in a plant leaf that make most of the plant's food

pancreas The organ that helps the digestive system by producing and secreting insulin and glucagon, both of which regulate glucose in the bloodstream

parasite An organism that obtains food from a living organism

parasitism A type of symbiosis that benefits one organism and harms the other

passive immunity An introduction of antibodies from outside the body to help defend against a type of pathogen

passive transport The movement of molecules across a membrane without the cell doing any work, as in osmosis and diffusion

perfect fungus A fungus that reproduces both sexually and asexually

periosteum The tough, white tissue that covers bones

peripheral nervous system The portion of the nervous system that does *not* include the brain and spinal cord

peristalsis The contraction and relaxing of smooth muscles that moves food along the alimentary canal

petal The often colorful flower part that helps attract pollinators to many flowers

pharynx The organ that connects the passages of the nose and mouth to the esophagus

phenotype The trait that an organism expresses as determined by its genotype

pheromone A chemical that causes those near an individual to behave in specific ways

phloem The vascular tissue that moves sugars throughout a plant

photosynthesis The process by which a plant uses energy from the sun to turn carbon dioxide and water into the sugar glucose

physiology The study of how organs and structures function

pituitary gland The endocrine gland that produces a variety of hormones that help control the body's processes

placenta The organ that allows a developing baby to get nourishment from the mother without their blood mixing

placental mammal A mammal that provides its embryo with oxygen and food through a structure called the *placenta*

plankton Tiny organisms that float near the ocean's surface

plasma The fluid, composed mainly of water, that makes up slightly more than half of the volume of blood

plasmid A small, often circular strand of DNA that is separate from the main chromosome in some bacteria

platelet A structure in the blood that repairs breaks in the circulatory system

pollution The improper release of wastes into ecosystems

polygenic inheritance A pattern of inheritance that occurs when two or more genes interact to control a single character

pore A small opening in the epidermis that releases oil onto the skin surface

predation A relationship between two organisms in which one eats the other

preservation The idea that humans should leave natural resources in as nearly their natural state as possible

prey An animal that a carnivore eats

primary succession Ecological succession that begins in an area that has no soil

producer An organism that can make its own food, typically through photosynthesis

prokaryotic cell A cell with no nucleus

protein A type of nutrient used to build structures and control reactions

protein synthesis The process of building a protein by bonding amino acids together

protist Any organism that is not an animal, plant, bacterium, or fungus

pseudopod A bulge of cytoplasm that some cells use as an anchor to pull themselves forward

puberty The hormone-controlled developmental stage that produces changes in the body leading to reproductive maturity

pulse A slight change in blood pressure in an artery when the heart beats

pupil The opening in the center of the iris that allows light to enter the eye

purebred An organism that always produces descendants with the same trait for a specific character

Q

qualitative data Data communicated using words and descriptions

quantitative data Data communicated with numbers

R

recessive trait A trait of a specific character that is hidden by the dominant trait in a hybrid organism

rectum The final structure of the large intestine

red blood cell A blood cell that lacks a nucleus and carries oxygen to the cells

reflex An immediate, inborn, involuntary reaction to a stimulus

renewable resource A natural resource that is the result of processes that replenish the resource as it is used

replication The DNA-copying process

respiratory system The multiple organs that bring in oxygen and remove carbon dioxide from the body

retina The inner layer of the eye containing receptors that are sensitive to light

ribonucleic acid (RNA) A type of nucleic acid similar to DNA whose structure is based on the information in a gene

ribosome A small structure that assembles proteins found in both prokaryotic and eukaryotic cells

S

saliva A fluid produced by the salivary glands of the mouth that begins chemical digestion of carbohydrates

saprophyte An organism that obtains food from dead organic matter

savanna A biome with hot temperatures throughout the year, moderate rainfall, and a mix of grasses and trees as the climax vegetation

scavenger An animal that eats dead animals

scientific inquiry Using a variety of tools and skills to answer a question or solve a problem

scientific name A unique name that identifies an organism created by combining the names of the genus and species

secondary succession Ecological succession that begins in an area that already has soil

seed A structure produced by some plants to protect the embryo

selectively permeable Allowing some substances to pass through while keeping others out

sepal The leaflike flower part positioned outside the petals

sexual reproduction A type of reproduction in which a sperm unites with an egg to form a zygote

short-day plant A plant that does not bloom until the nights are longer than the critical night length

simple dominance A pattern of inheritance in which there are only two alleles and the dominant trait is always expressed in a hybrid organism

skeletal muscle Another name for a voluntary muscle

skeletal system The body system formed by bones

skeleton The organ system that supports the bodies of many animals

smooth muscle An involuntary muscle without striations found in internal organs

spinal cord The part of the central nervous system that extends down the back and relays impulses between the brain and the rest of the body

spiracle A tiny opening that takes in air along the side of the body of certain animals

sponge One of a group of animals with no symmetry, cephalization, or gut

spongy bone Hard and rigid bone with empty spaces throughout

spongy layer The loosely packed cells in a plant leaf that allow gases to move easily to the palisade layer

spore A structure involved in fungal reproduction that allows a fungus to survive through unfavorable growth conditions

sporophyte The plant structure that produces spores and forms the primary structure of most plants, excluding bryophytes

stamen The male part of a flower consisting of filaments and pollen-producing anthers

stem cell A special cell that divides and matures into any type of cell needed by the body

stipe The stalk of a mushroom

stoma An opening in the underside of the leaf that allows gases to move into and out of the leaf

striated muscle A muscle that has a pattern of dark and light bands, called *striations*, found in both skeletal and cardiac muscle

substance pollutant A solid, liquid, or gas that pollutes an ecosystem

symbiosis A relationship in which two organisms interact frequently over a long period of time, often living together

symmetry The property of an object that means that it can be divided by one or more imaginary lines into mirror-image halves

T

taxonomy The science of sorting organisms into groups

T cell A white blood cell that may attack a pathogen, attack an infected body cell, or activate B cells

tendon A band of connective tissue that attaches muscle to bone

territory An area that an animal claims as its own

testis The organ in male animals that produces sperm

theory An idea that scientists use to explain the data that they have collected

theory of natural selection The theory that traits that allow individuals to survive and reproduce better than other individuals without those traits will become more common in the population

thyroid The endocrine gland that produces hormones that control how fast the body burns food to produce energy

tissue A group of similar cells that work together

trachea The organ that conducts air toward the lungs

trait A form of a character such as tall or short

transcription The process that cells use to make RNA copies of a gene

transitional form A supposed organism of the past with body parts that are intermediate between early ancestral forms and later fully developed forms; also called a *missing link*

translation The process of converting the DNA code into a protein

trimester One of the three stages of pregnancy

trophic level A layer of an energy pyramid

tropical rainforest A biome that combines year-round warmth and large amounts of rainfall with a lack of a dry season

tropism A plant's growth toward or away from a stimulus

tundra A biome with little rainfall, cold temperatures, frozen soil just beneath the surface, and low-growing shrubs, grasses, and lichens as the climax vegetation

turgor pressure Pressure caused by the presence of water inside a plant cells' central vacuole that helps support a plant

U

umbilical cord The tube that connects a developing baby to the placenta

unicellular Having only one cell

uniformitarianism The concept that all of Earth's physical processes happen at the same rate in the present time as they did in the past

ureter One of a pair of very thin tubes that move urine from the kidneys to the bladder

urethra A tiny muscular tube that releases urine from the body

urine A fluid containing wastes that is filtered from the blood by nephrons

uterus The female organ where humans and most mammals protect their embryos prior to birth

V

valve A flap in a vein that allows blood to flow in only one direction

variable A factor that changes in a controlled experiment

vascular tissue Plant tissue that conducts water and other materials in some plants

vein (1) The vascular tissue of a plant's leaf; (2) a blood vessel leading toward the heart

ventricle A lower chamber of the heart

vertebrate An animal that has a backbone

vessel An organ that conducts blood through an animal's body

vestigial structure A structure that according to evolutionary theory no longer has a function

villus One of the many microscopic, finger-like structures on the internal wall of the small intestine that provide a large internal surface area for absorbing nutrients from digested food

virus A nucleic acid surrounded by a protein coat that uses other cells to reproduce

vitamin A type of nutrient that the body needs to function properly and cannot produce on its own

voluntary muscle A muscle that is controlled at will

W

water cycle The continuous movement of water from the earth to the atmosphere and back to the earth

white blood cell A blood cell that attacks foreign invaders

worldview The overarching narrative that a person uses to see and interpret the world

X

xylem The vascular tissue that moves water and nutrients throughout a plant

Z

zygote The diploid cell that results from fertilization

INDEX

Boldface page numbers denote the location of the first use of key terms.

D

E

F

S

T

PHOTO CREDITS

Key: (**t**) top; (**c**) center; (**b**) bottom; (**l**) left; (**r**) right; (**bg**) background; (**i**) inset

Cover

front Tirik Shutterstock.com; **back** IrinaK/Shutterstock .com; **spine** © iStock.com/photomaru

Front Matter

i bogdanhoda/Shutterstock.com; **ii** ThePinkPig /Shutterstock.com; **iii** Sean Locke Photography /Shutterstock.com; **iv**t Roland Magnusson /Shutterstock.com; **iv**b Stepan Kapl/Shutterstock.com; **v** Chris Mattison/Alamy Stock Photo; **vi** Adobe Stock /Ludmila Smite; **vii**t vetre/Shutterstock.com; **vii**bl khuruzero/Shutterstock.com; **vii**br Tapui/Shutterstock .com; **viii** Alexandra Mitchell/Shutterstock.com; **ix** Jinning Li/Shutterstock.com; **x**t Christian Kober/AWL Images /Getty Images; **x**b KAMONRAT/Shutterstock.com; **x**bi "Pink indian pipes" by Magellan nh/Wikimedia Commons/CC BY 3.0; **xi**t © iStock.com/imv; **xi**b Thomas Groß/EyeEm/Getty Images; **xii**tl Joel Sartore/National Geographic/Getty Images; **xii**tr Close Encounters Photo /Shutterstock.com; **xii**c Dr. Marcia Harrison-Pitaniello, Department of Biological Sciences, Marshall UniversityHuntington, WV 25755; **xii**b Hawk777 /Shutterstock.com; **xiii**t Hulton Archive/Archive Photos /Getty Images; **xiii**ti zcw/Shutterstock.com; **xiii**c © iStock .com/PicturePartners; **xiii**b (fern) © iStock.com/ka_ru; **xiii**b (moss) Peter Chadwick/Dorling Kindersley/Getty Images; **xiii**b (flower) © iStock.com/mariemilyphotos

Chapter 1

xiv–1 Jezper/Shutterstock.com; **1** Courtesy of Dr. Kevin L. Anderson; **2** Christian Kober/AWL Images/Getty Images; **3**t Martin Gabriel/NPL/Minden Pictures; **3**b © iStock.com /reptiles4all; **5** Lou Linwei/Alamy Stock Photo; **6** Adobe Stock/Sondem; **7**tl © iStock.com/anna1311; **7**tr © iStock .com/brozova; **7**cl © iStock.com/skynesher; **7**cr VW Pics /Universal Images Group/Getty Images; **7**b Robert Sijan/123RF; **8** NOAA; **9**tl © Kjorgen | Dreamstime.com; **9**tr © Marek Mis/Science Source; **9**c Science Photo Library /Media Bakery; **9**b neil harrison/Alamy Stock Photo; **10** (boy) © iStock.com/skodonnell; **10** (lungs) © Nosorogua | Dreamstime.com; **10** (blood cells) © Vivilweb | Dreamstime.com; **10** (heart) © iStock.com/angelhell; **11**bg Cultura Exclusive/Tanveer Badal/Getty Images; **11** Stephen Frink Collection/Alamy Stock Photo; **12**t DGLimages/Shutterstock.com; **12**b Hack_bsh /Shutterstock.com; **13**t MolecluarExpressions.com at Florida State University; **13**b Hack_bsh/Shutterstock.com; **14**l © iStock.com/stockar; **14**r Nick Garbutt/NPL/Minden Pictures; **15**t © iStock.com/Alex Potemkin; **15**bl Pete Oxford/NPL/Minden Pictures; **15**br Carlos Amarillo /Shutterstock.com; **16**bg JAIME CHIRINOS/SCIENCE PHOTO LIBRARY; **16**t Oli Scarff/Getty Images News /Getty Images; **16**b "Aepyornis Skeleton" by Monnier /Wikimedia/Public Domain; **17**t LightField Studios /Shutterstock.com; **17–18**b Teeraphat/Shutterstock.com; **17**bi © iStock.com/kasto80; **18**t molekuul_be/Shutterstock .com; **18**c Harry Sieplinga/Getty Images; **19** (archaea) Power and Syred/Science Source; **19** (*E. coli*) © iStock .com/Dr_Microbe; **19** (green algae) Science Stock Photography/Science Source; **19** (amoeba) Papilio/Alamy Stock Photo; **19**, **20**, **21** (mushroom) © Valentinar | Dreamstime.com; **19**, **20**, **21** (bromeliad) Sanit Fuangnakhon/Shutterstock.com; **19**, **20**, **21** (frog) Dirk Ercken/Shutterstock.com; **20** (hermit crab, left) Wow Pho /Shutterstock.com; **20** (hermit crab, center) Redmond Durrell/Alamy Stock Photo; **20** (hermit crab, right) isselee/123RF; **20** (hermit crab, bottom) Michel Gunther /Biosphoto/Getty Images; **20** (rainbow crab) Vangert /Shutterstock.com; **20** (lobster) Adobe Stock/Giuseppe Porzani; **20**, **21** (butterfly) paulrommer/123RF; **21** (fossa) "Cryptoprocta Ferox" by Ran Kirlian/Wikimedia Commons/CC BY-SA 4.0; **21** (mongoose) Iakov Filimonov/123RF; **21** (tiger) SantiPhotoSS/Shutterstock .com; **21** (rhinoceros) © iStock.com/deramarok; **22** IMAGEMORE Co, Ltd./Getty Images; **23**t robert_s /Shutterstock.com; **23**b Super Prin/Shutterstock.com; **24**t Teeraphat/Shutterstock.com; **24**b, **26**b, **27**b Rosa Jay /Shutterstock.com; **25** Michio Hoshino/Minden Pictures /SuperStock.com; **26**t Adobe Stock/sveta; **27**t Evgeniia Litovchenko/Shutterstock.com

Chapter 2

28 nobeastsofierce/Shutterstock.com; **29**t U.S. Air Force photo/Tech. Sgt. Jason Robertson; **29**b Andreas Wass /Shutterstock.com; **29**bi Garry DeLong/Oxford Scientific /Getty Images; **30**tl "Bryum capillare leaf cells" by Des_ Callaghan/Wikimedia Commons/CC BY-SA 4 International; **30**tli "Matthias Jacob Schleiden"/Wikimedia Commons/Public Domain; **30**tr Biophoto Associates /Science Source; **30**tri Mondadori Portfolio/Getty Images; **30**bl David M. Phillips/Science Source; **30**bli Mondadori Portfolio/Getty Images; **30**br D. Kucharski K. Kucharska /Shutterstock.com; **30**bri ullstein bild/Getty Images; **31**t, **33** © iStock.com/sbrogan; **31**c, **32** grebcha/Shutterstock .com; **31**b Cookie Studio/Shutterstock.com; **31**bri Jose Luis Calvo/Shutterstock.com; **31**bci Nathapol Kongseang /Shutterstock.com; **31**bli BIOPHOTO ASSOCIATES /Science Source/Getty Images; **38–39**bg Stephen B. Goodwin/Shutterstock.com; **40**l Pektoral/Shutterstock .com; **40**r picturepartners/Shutterstock.com; **41**t sabza /Shutterstock.com; **41**c Bettmann/Getty Images; **41**b kirillov alexey/Shutterstock.com; **42–43**bg CI Photos /Shutterstock.com; **42** © iStock.com/skynesher; **43** Sebastian Kaulitzki/Shutterstock.com; **45** Wiratchai wansamngam/Shutterstock.com

Chapter 3

47, **63**r Wim Van Egmond/Science Source; **48** MYN/Jerry Monkman/NPL/Minden Pictures; **48–49**bg Lidiya Oleandra/Shutterstock.com; **52**bg Denys Po/Shutterstock

.com; **52**t Mr Twister/Shutterstock.com; **52**bl Yatso /Shutterstock.com; **52**br Melinda Fawver/Shutterstock .com; **53**, **63**l Wolfgang Poelzer/WaterFrame/Getty Images; **54**bg CI Photos/Shutterstock.com; **55** all Sarah Lompe; **56**bg, **62**bg Rattiya Thongdumhyu/Shutterstock.com; **57**bg Rattiya Thongdumhyu/Shutterstock.com; **58** Choksawatdikorn/Shutterstock.com; **58–59**bg © iStock .com/panaramka; **59**l, **64**l Eric Isselee/Shutterstock.com; **59**r, **64**r © iStock.com/GlobalP; **60**t Iaremenko Sergii /Shutterstock.com; **60**b Sean Crane/Minden Pictures/Getty Images; **60**bi Reprinted from Zoologischer Anzeiger—A Journal of Comparative Zoology, Vol. 254, January 2015. Juan P. Zurano, Diego S.Ojeda, Claudio J.Bidau, Wagner F.Molina, Mario A. Ledesma, Pablo A. Martinez, "A comparison of heterochromatic regions in three species of neotropical canids," 1–7, Copyright 2015, with permission from Elsevier. Permission conveyed through Copyright Clearance Center. **61** Ryan M. Bolton/Alamy Stock Photo; **61**bg "Antonov An-2 3view" by Kaboldy/Wikimedia Commons/CC BY-SA 3.0; **62** Ed Reschke/Photolibrary /Getty Images; **66** Iain Masterton/Alamy Stock Photo

Chapter 4

67 © iStock.com/IvonneW; **68**t, **84**t Science Source/Getty Images; **68**b all, **69** (pea pods with flowers) Scisetti Alfio /Shutterstock.com; **69**bg Yasonya/Shutterstock.com; **69** (open pea pod) Labrador Photo Video/Shutterstock.com; **70–71**bg Tamara Kulikova/Shutterstock.com; **70** (axial position) yingtustocker/Shutterstock.com; **70**, **72** (terminal position) Derek Hall/Dorling Kindersley/Getty Images; **70**, **72** (tall, short plant) © iStock.com/feellife; **70** (inflated pods) Dmitrii Kazitsyn/Shutterstock.com; **70** (pinched pods) © iStock.com/Nirad; **70** (green pods) JIANG HONGYAN/Shutterstock.com; **70** (yellow pods) picturepartners/Shutterstock.com; **70** (round, wrinkled, yellow, green seeds), **71**l, c Martin Shields/Science Source /Getty Images; **70** (brown seed coat) Perutskyi Petro /Shutterstock.com; **70** (white seed coat) milka-kotka /Shutterstock.com; **71**r, **76**tr, **77**t CI Photos/Shutterstock .com; **73** © iStock.com/busypix; **74** Chursina Viktoriia /Shutterstock.com; **75**tl, tr, **84**bl, **84**bc Stephen B. Goodwin/Shutterstock.com; **75**tc, **84**br Susii/Shutterstock .com; **75**c (variegated camellia) portumen/Shutterstock .com; **75**c (red camellia) © iStock.com/meixf; **75**c (white camellia) © iStock.com/scisettialfio; **75**b © iStock.com /igaguri_1; **76**tl Phonlamai Photo/Shutterstock.com; **76–77**b © iStock.com/FatCamera; **78**t KS631/Dorling Kindersley/Getty Images; **78–79**b apiguide/Shutterstock .com; **79**t (white chicken) © iStock.com/FlamingPumpkin; **79**t (black chicken) © iStock.com/Olha_stock; **79**t, **85**t (erminette chicken) GlobalP/iStock/Getty Images; **80** Nature Picture Library/Alamy Stock Photo; **81** John E Marriott/All Canada Photos/Getty Images; **82** Dora Zett /Shutterstock.com; **83**tl Joel Sartore/National Geographic /Getty Images; **83**tr Close Encounters Photo/Shutterstock .com; **83**b Dr. Marcia Harrison-Pitaniello, Department of Biological Sciences, Marshall University Huntington, WV 25755; **85**b Sergey Ryumin/Moment/Getty Images; **86** © iStock.com/kjekol; **87** Slavko Sereda/Shutterstock.com

Chapter 5

88 Olga Melhiser Photography/Moment/Getty Images; **89** Sovfoto/Universal Images Group/Getty Images; **90**bg Jens Mayer/Shutterstock.com; **90**t © iStock.com/KiraVolkov; **90**ct © iStock.com/clu; **90**cb, **115**t © iStock.com/hsvrs; **90**bl PjrStudio/Alamy Stock Photo; **90**br Celiafoto /Shutterstock.com; **91**tl Mike Dunning/Dorling Kindersley/Getty Images; **91**tr, **96**t Eric Isselee /Shutterstock.com; **91**b Rosa Jay/Shutterstock.com; **92–93**bg Antonio Macia Martinez/Shutterstock.com; **92**l Bjoern Wylezich/Shutterstock.com; **92**c Snowshill /Shutterstock.com; **92**r © iStock.com/alice-photo; **93**t Science Source/Getty Images; **93**b GraphicaArtis/Archive Photos/Getty Images; **93**t bg, **93**b bg revers/Shutterstock .com; **94**t thoron/Shutterstock.com; **94**b "Darwin's finches by Gould" by John Gould/Wikimedia Commons/Public Domain; **94–95** (pages, photo frame, leaf, book) Lukiyanova Natalia frenta/Shutterstock.com; **94** (fern) Slavko Sereda/Shutterstock.com; **95** (feathers) Potapov Alexander/Shutterstock.com; **95** (fossils) Triff /Shutterstock.com; **96–97**b lembi/Shutterstock.com; **98** Encyclopaedia Britannica/Universal Images Group/Getty Images; **98–99**bg Westend61/Getty Images; **100**tli, tri, cli, cri Universal Images Group North America LLC/Alamy Stock Photo; **100**tl © iStock.com/Meal_meaW; **100**tr © iStock.com/Kaphoto; **100**cl Potapov Alexander /Shutterstock.com; **100**cr Joel Sartore/National Geographic/Getty Images; **100**bl Geoff Scott-Simpson /Moment Open/Getty Images; **100**bli Betty & Nathan Cohen/Visuals Unlimited/Getty Images; **100**br "Haeckel drawings" by Romanes, G. J./Wikimedia Commons/Public Domain; **101** Kues/Shutterstock.com; **102**t Wild Horizon /Universal Images Group/Getty Images; **102**b, **115**b © iStock.com/Atypeek; **102–3**bg Chris Schmid/National Geographic/Getty Images; **104**t © iStock.com/vencavolrab; **104**b © iStock.com/defun; **106–7**bg, **116**t © iStock.com /andrej67 / BJU Press; **107**t Joel Sartore/National Geographic Creative/Getty Images; **107**b © iStock.com /reptiles4all; **108**l © iStock.com/vichie81; **108**r "Darwinius masillae PMO 214.214"/Jens L. Franzen, Philip D. Gingerich, Jörg Habersetzer1, Jørn H. Hurum, Wighart von Koenigswald, B. Holly Smith/Wikimedia Commons /CC BY-SA 2.5; **108–9**bg Vlada Photo/Shutterstock.com; **109**t Lou Linwei/Alamy Stock Photo; **109**b Christian Jegou/Publiphoto/Science Source; **110–11**bg All Canada Photos/Alamy Stock Photo; **110** Jeffrey Phelps/Getty Images; **112** Krys Bailey/Alamy Stock Photo; **112–13**bg ananaline/Shutterstock.com; **113** © iStock.com/GlobalP; **114** Compassionate Eye Foundation/Jasper White/Digital Vision/Getty Images; **116**br FotoRequest/Shutterstock .com; **116**bl Michiel de Wit/Shutterstock.com; **116–17**bg (fossils) Triff/Shutterstock.com **117** Kim Taylor/NPL /Minden Pictures

Chapter 6

118–19 © iStock.com/hudiemm; **119** Courtesy John Sandford; **120** STEVE GSCHMEISSNER/Science Source; **121** Lotus_studio/Shutterstock.com; **122–23**bg Yongyut Kumsri/Shutterstock.com; **122**r BSIP/Universal Images Group/Getty Images; **122**tl STEVE GSCHMEISSNER /Science Photo Library/Getty Images; **122**bl Dennis

Kunkel Microscopy/Science Source; **123**t grebcha /Shutterstock.com; **123**b Guenter Albers/Shutterstock .com; **124–25**bg Sam Wirzba/Perspectives/Getty Images; **124** val lawless/Shutterstock.com; **125**t © iStock.com /Kondor83; **125**b Claudia Totir/Moment/Getty Images; **127** Africa Studio/Shutterstock.com; **127**i, **131**b Lee D. Simon/Science Source; **129**t James King-Holmes/Science Source; **129**b Africa Studio/Shutterstock.com; **130** © iStock.com/Chris Ryan; **131**t BSIP/Universal Images Group/Getty Images / STEVE GSCHMEISSNER/Science Photo Library/Getty Images / Dennis Kunkel Microscopy /Science Source; **132**t Fred Hirschmann/SuperStock; **132**b © iStock.com/Kondor83

Chapter 7

133 Philippe Clement/NPL/Minden Pictures; **134–35**bg Malcom Park/Science Faction/Getty Images; **134**l Trish Gant/Dorling Kindersley/Getty Images; **134**r M. I. Walker /Science Source; **135** (amoeba) Lebendkulturen.de /Shutterstock.com; **135**, **152**t (coccolithophores) Steve Gschmeissner/Science Source; **135** (Phytophthora, microsporidians) Eye of Science/Science Source; **135** (potato) vipman/Shutterstock.com; **135** (ants) BERNATSKAYA OXANA/Shutterstock.com; **135** (slime mold) "Stemonitis axifera 341275" by Dr. Lorne Stobbs /Wikimedia Commons/CC BY-SA 3.0; **136** Michael Abbey/Science Source; **136–37**bg © iStock.com/baona; **138** Ed Reschke/Photolibrary/Getty Images; **138–39**bg Merten Snijders/Lonely Planet Images/Getty Images; **139** "Karenia brevis"/FWC Fish and Wildlife Research Institute/Flickr/CC BY-NC-ND 2.0; **140** M I WALKER /Science Source/Getty Images; **140**bg www.ferpectshotz .com/Moment/Getty Images; **141**t De Agostini Picture Library/Getty Images; **141**b Lebendkulturen.de /Shutterstock.com; **142–43**bg © iStock.com /dogayusufdokdok; **142**t, **152**b bonchan/Shutterstock.com; **142**b Ake13bk/Shutterstock.com; **143**t © iStock.com /tanukiphoto; **143**ti Paul Zahl/National Geographic/Getty Images; **143**c Buiten-Beeld/Alamy Stock Photo; **143**ci "CSIRO ScienceImage 1392 Scanning Electron Micrograph of Chytrid Fungus"/Dr Alex Hyatt, CSIRO/Wikimedia Commons/CC BY 3.0; **143**b Paul Cowan/Shutterstock .com; **143**bi De Agostini Picture Library/Science Source; **144–45**bg nullplus/Shutterstock.com; **145**l, **153**l © Isselee | Dreamstime.com; **145**r Wally Eberhart/Visuals Unlimited/Getty Images; **146–47**bg TTphoto/Shutterstock .com; **146** Stefan Holm/Shutterstock.com; **147** Dennis Kunkel Microscopy/Science Source; **148**bg Fletcher & Baylis/Science Source; **148**l carballo/Shutterstock.com; **148**c Davydenko Yuliia/Shutterstock.com; **148**r © iStock .com/Givaga; **149**bg © Evgeniy Fesenko | Dreamstime; **149** © Nata7707k | Dreamstime; **150**t Valentina Razumova /Shutterstock.com; **150**b © iStock.com/DrPAS; **151** © iStock.com/tacojim; **153**r © Chien Lee/Minden Pictures; **154**t Steve Maslowski/Visuals Unlimited/Getty Images; **154**b Matt Meadows/Science Source

Chapter 8

155 smereka/Shutterstock.com; **156**i "Pink indian pipes" by Magellan nh/Wikimedia Commons/CC BY 3.0; **156–57**b, **173**t KAMONRAT/Shutterstock.com; **157**t Jouan Rius/NPL/Minden Pictures; **157**c (plant cells) © iStock.com/Nnehring; **157**c (cellulose) Michael Abbey /Science Source/Getty Images; **157**c (xylem) STEVE GSCHMEISSNER/Science Photo Library/Getty Images; **157**c (leaf) Orientaly/Shutterstock.com; **157**bi nechaevkon/Shutterstock.com; **158** Danita Delimont/Gallo Images/Getty Images; **159**bg Peter Clark/Shutterstock .com; **159**t © iStock.com/PFMphotostock; **159**c Daniel Vega/age fotostock/Getty Iamages; **159**b rdonar /Shutterstock.com; **159**bi Scott Camazine/Science Source; **160**bg Johannes Kornelius/Shutterstock.com; **160**t Pascal Goetgheluck/Science Source; **160**c Arterra/Universal Images Group/Getty Images; **160**b "Equisetopsida"/Rror /Wikimedia Commons/CC BY-SA 3.0; **161** Jupiterimages /Photodisc/Getty Images; **162**t margouillat photo /Shutterstock.com; **162**ti TheFarAwayKingdom /Shutterstock.com; **162**bl Rui Elena/Shutterstock.com; **162**br Tamara Kulikova/Shutterstock.com; **163**ti "WolffiaArrhiza2" by Christian Fischer/Wikimedia Commons/CC BY-SA 3.0; **163**t National Geographic Creative/Alamy Stock Photo; **163**c allstars/Shutterstock .com; **163**b ER_09/Shutterstock.com; **164**r Bob Gibbons /Science Source; **164**ri Ray Coleman/Science Source; **164**lt © iStock.com/MirekKijewski; **164**lb © iStock.com/Zerbor; **164**lbi © iStock.com/CQYoung; **165**t © iStock.com /PicturePartners; **165**c © iStock.com/ka_ru; **165**bl Peter Chadwick/Dorling Kindersley/Getty Images; **165**br © iStock.com/mariemilyphotos; **166** Dusan Petkovic /Shutterstock.com; **166**i, **173**b Katrina Leigh/Shutterstock .com; **167**bg © iStock.com/tbradford; **167**l © iStock.com /ttbphoto; **167**r Neil Fletcher & Matthew Ward/Dorling Kindersley/Getty Images; **168**r © iStock.com /chengyuzheng; **168**ri Rattiya Thongdumhyu/Shutterstock .com; **168**l bergamont/Shutterstock.com; **168**li Mike Rosecope/Shutterstock.com; **169**tl © iStock.com /chengyuzheng; **169**tc Coprid/Shutterstock.com; **169**tr © iStock.com/mg7; **169**bl © iStock.com/i-Zet; **169**br Fotofermer/Shutterstock.com; **169**bri hayakato /Shutterstock.com; **170** Visuals Unlimited, Inc./Jeffrey Wickett/Visuals Unlimited/Getty Images; **170–71**bg H. Mark Weidman Photography/Alamy Stock Photo; **171** David Cook/blueshiftstudios/Alamy Stock Photo; **172**ti Lars Evans/Alamy Stock Photo; **172**t Stephen Barnes/Arts and Crafts/Alamy Stock Photo; **172**c Colin Leftley/Dorling Kindersley/Getty Images; **172**b Wild Wonders of Europe /Ruiz/NPL/Minden Pictures; **175**t DIVINE IMAGES | SOT/Media Bakery; **175**b © iStock.com/ckhudoliy

Chapter 9

176 Dave Pattison/Alamy Stock Photo; **177**, **188**t Westend61/Dieter Heinemann/Media Bakery; **178**bg Mark Boulton/Alamy Stock Photo; **178**t anitasstudio /Shutterstock.com; **178**b Nigel Wallace/Shutterstock.com; **179**t Cathyrose Melloan/Alamy Stock Photo; **179**c Madlen/Shutterstock.com; **179**b Stephen B. Goodwin /Shutterstock.com / Susii/Shutterstock.com / Diana Taliun /Shutterstock.com / Vilor/Shutterstock.com; **180**t Yoshio Otsuka/Sebun Photo/amana images/Getty Images; **180**c © iStock.com/ChrisAt; **180**b Martin Shields/Science Source /Getty Images; **181** Hawk777/Shutterstock.com; **182** Keith Ladzinski/National Geographic/Getty Images; **182**i "Moss Gametophytes Sporophytes" by Bob Blaylock/Wikimedia

Commons/CC BY-SA 4.0; **183**, **188**c © iStock.com/joloei; **184**l, **189**r © iStock.com/DNY59; **184**rt © iStock.com /wwing; **184**rc © iStock.com/ivstiv; **184**rb © iStock.com /Chushkin; **185** vilax/Shutterstock.com; **185**ti © iStock .com/SumikoPhoto; **185**bi © iStock.com/eyewave; **186**tl FLPA/Alamy Stock Photo; **186**tr arka38/Shutterstock.com; **186**b Kate Aedon/Shutterstock.com; **187**t Hulton Archive /Archive Photos/Getty Images; **187**b zcw/Shutterstock .com; **188**b © iStock.com/helga_sm; **189**l © iStock.com /flavijus

Chapter 10

190–91 Jane Burton/NPL/Minden Pictures; **191** Answers in Genesis (www.AnswersinGenesis.org); **192** Khoroshunova Olga/Shutterstock.com; **193** Reinhard Dirscherl/Alamy Stock Photo; **194**t © iStock.com /DPFishCo; **194**bi, **214**b Thomas Groß/EyeEm/Getty Images; **194**b Michael Stubblefield/Alamy Stock Photo; **195**t Kent Wood/Science Source; **195**b Joel Sartore /National Geographic/Getty Images; **196**tl Gerald Robert Fischer/Shutterstock.com; **196**tr "Dingo (Canis lupus dingo) skeleton at the Royal Veterinary College anatomy museum white background"/Mrjohncummings /Wikimedia Commons/CC BY-SA 3.0; **196**bg © iStock .com/Onfokus; **196**b, **205**cl © iStock.com/GlobalP; **197**l, **213**t © iStock.com/abzerit; **197**r © iStock.com/aaron007; **198–99**t Dejan Patic/The Image Bank/Getty Images; **198**c DEA/M. SEEMULLER/De Agostini Picture Library/Getty Images; **198**b © iStock.com/Mirko_Rosenau; **199**bl Martin Valigursky/Shutterstock.com; **199**bcl MattiaATH /Shutterstock.com; **199**bcr © iStock.com/KevinPanizza; **199**br Valentina Proskurina/Shutterstock.com; **200** Mr. SUTTIPON YAKHAM/Shutterstock.com / Demskoy Studio/Shutterstock.com / motorolka/Shutterstock.com; **201**bg LiskaM/Shutterstock.com; **201**t Norbert Wu /Minden Pictures; **201**cl Colin Marshall/FLPA/Minden Pictures; **201**cr NatureDiver/Shutterstock.com; **201**b (flatworm) Pong Wira/Shutterstock.com; **201**b (segmented worm) Alexander Semenov/Science Source; **201**b (roundworm) Frank Fox/Science Source; **202**bg © iStock .com/Michael Zeigler; **202**t NATURAL PLANET | NATIONAL GEOGRAPHIC/Media Bakery; **202**c Mauritius/Superstock; **202**bl Martin Strmiska/Alamy Stock Photo; **202**br Errorcz/Shutterstock.com; **203** (lacewing), **203** (moth), **205**c Eric Isselee/Shutterstock.com; **203** (grasshopper) Matt Jeppson/Shutterstock.com; **203** (crab) Tadas_Jucys/Shutterstock.com; **203** (shrimp) Borut Furlan/WaterFrame/Getty Images; **203** (hornet) irin-k /Shutterstock.com; **203** (lobster), **213**b © iStock.com /trumzz; **204**tl © iStock.com/EcoPic; **204**tc © iStock.com /pong6400; **204**tr © iStock.com/Antagain; **204**c (sponge) © iStock.com/dsabo; **204**c (jellyfish) © iStock.com /Bridgendboy; **204**c (worms) Nataly Studio/Shutterstock .com; **204**c (urchin) NatalieJean/Shutterstock.com; **204**c (squid) ullstein bild/Getty Images; **204**c (beetle) Demskoy Studio/Shutterstock.com; **204**b © iStock.com/epantha; **205**t © iStock.com/JohnPitcher; **205**cr Johannes Kornelius /Shutterstock.com; **205**bl MaxkateUSA/Shutterstock.com; **205**br © iStock.com/garymilner; **206**t © iStock.com/Arsty; **206**b Tony Wu/NPL/Minden Pictures; **206**bg Rich Carey /Shutterstock.com; **207**tl © iStock.com/Farinosa; **207**tr "Paratype of Paedophryne amauensis (LSUMZ 95004)"/E. N. Rittmeyer et al. (2012). "Ecological Guild Evolution and the Discovery of the World's Smallest Vertebrate" /Wikimedia Commons/CC BY 2.5; **207**b Dan Suzio /SCIENCE SOURCE; **207**bg yhelfman/Shutterstock.com; **208–9**bg National Geographic Creative/Alamy Stock Photo; **208**t © iStock.com/luamduan; **208**bl DAN SUZIO /Science Source/Getty Images; **208**br, **214**t © iStock.com /imv; **209**t defpicture/Shutterstock.com; **209**c chloe7992 /Shutterstock.com; **209**b Nigel Pavitt/AWL Images/Getty Images; **210**t Drakuliren/Shutterstock.com; **210**b Keren Su/Corbis Documentary/Getty Images; **211** (dolphin) Brian J. Skerry/National Geographic/Getty Images; **211** (kitten) Tony Campbell/Shutterstock.com; **211** (capybara) Anan Kaewkhammul/Shutterstock.com; **211** (impala) © iStock.com/Friedemeier; **211**b © iStock.com /JohnCarnemolla; **211**bi © iStock.com/Shmenny50; **212**l © iStock.com/CraigRJD; **212**r SOMMAI/Shutterstock.com; **215**t Rosa Jay/Shutterstock.com; **215**c © iStock.com /johnaudrey; **215**b "Lipogramma trilineatum, Adult (Threeline Basslet)"/Smithsonian Institution/Wikimedia Commons; **216**t Joel Sartore/National Geographic/Getty Images; **216**b Holli/Shutterstock.com; **217**t Randimal /Shutterstock.com; **217**b Mark Conlin/Oxford Scientific /Getty Images

Chapter 11

218 AfriPics.com/Alamy Stock Photo; **219**tr Rich Carey /Shutterstock.com; **219**br hsagencia/Shutterstock.com; **219**l, **241**t Rich Carey/Shutterstock.com; **220–21**bg Stu Porter/Alamy Stock Photo; **220**l Merlin D. Tuttle/Science Source; **220**r © iStock.com/MoMorad; **221**l Nigel Cattlin /FLPA/Minden Pictures; **221**c © iStock.com/smuay; **221**r Michael Thomas Clark/Shutterstock.com; **222**bg Iakov Filimonov/Shutterstock.com; **222**tl WaterFrame/Alamy Stock Photo; **222**tr Andrew J Martinez/Science Source /Getty Images; **222**b © iStock.com/MattTheHoople; **223**t Stephen Dalton/Minden Pictures; **223**c Scott Hurd/Alamy Stock Photo; **223**b Nature Picture Library/Alamy Stock Photo; **224**bg Bridget Calip/Shutterstock.com; **224**l © iStock.com/KevinPanizza; **224**r L Newman & A Flowers /Science Source/Getty Images; **225**bg © iStock.com /BRJ1975; **225** Val Duncan/Kenebec Images/Alamy Stock Photo; **226–27**bg paula french/Shutterstock.com; **226** © iStock.com/Velvetfish; **227**t Hoberman Collection /Universal Images Group/Getty Images; **227**b Images of Africa Photobank/Alamy Stock Photo; **228** Paul Glendell /Alamy Stock Photo; **229**t © iStock.com/Goruppa; **229**c, **231**br © iStock.com/GlobalP; **229**b © iStock.com /arlindo71; **230–31**bg Panoramic Images/Getty Images; **230** Norbert Wu / Minden Pictures; **231**l, **241**b J.A. Gonzalez de Rueda/Biosphoto/Minden Pictures; **231**tr bluehand/Shutterstock.com; **232–33**bg Ariadne Van Zandbergen/Lonely Planet Images/Getty Images; **232**tl Seal salamander (Desmognathus monticola) by Leif Van Laar/Wikimedia Commons/CC by-SA 3.0; **232**tr guy42 /Shutterstock.com; **232**bl © iStock.com/troutnut; **232**br © iStock.com/Mirko_Rosenau; **233** © iStock.com /adogslifephoto; **234–35**bg Arterra Picture Library/Alamy Stock Photo; **235**t Mint Images - Frans Lanting/Getty Images; **235**b Colin Keates/Dorling Kindersley/Getty

Images; **236**tl kzww/Shutterstock.com; **236**tli Cultura Images/Gregory S. Paulson/Media Bakery; **236**tr Portuguese Man-O-War (Physalia physalis) by NOAA /Wikimedia Commons/Public Domain; **236**cl © iStock .com/OceanBodhi; **236**cli Colin Marshall/FLPA/Minden Pictures; **236**b (armadillo), **237**b (monkey) Eric Isselee /Shutterstock.com; **236**b (dog) Iakov Filimonov /Shutterstock.com; **236**b (rat) photolinc/Shutterstock.com; **236**b (deer) taviphoto/Shutterstock.com; **236**b (fish) © iStock.com/VSanandhakrishna; **236–37**b (elephants) Lukiyanova Natalia frenta/Shutterstock.com; **237**t guy42 /Shutterstock.com; **237**c Stephen Dalton/Minden Pictures; **237**b (turtle) Mariyana M/Shutterstock.com; **237**b (chicken) TTstudio/Shutterstock.com; **237**b (cheetah) Svetlana Foote/Shutterstock.com; **237**b (gecko) Avesun /Shutterstock.com; **237**b (iguana) Luis Louro/Shutterstock .com; **237**b, **242**tl (dog) alexei_tm/Shutterstock.com; **237**b (fox) Jim Cumming/Shutterstock.com; **237**b (birds) TCreativeMedia/Shutterstock.com; **238** © iStock.com /adamkaz; **239**t Richard Du Toit / Minden Pictures; **239**bl Dimj/Shutterstock.com; **239**bc Mark Brandon /Shutterstock.com; **239**br Valentina Razumova /Shutterstock.com; **240**t JTB MEDIA CREATION, Inc. / Alamy Stock Photo; **240**b Butterfly Hunter/Shutterstock .com; **242**cl Maciej Czekajewski/Shutterstock.com; **242**r photowind/Shutterstock.com; **242**bl © iStock.com/JacobH; **243**tl O.Bellini/Shutterstock.com; **243**tr © iStock.com /Okea; **243**c fivespots/Shutterstock.com; **243**b Charoenkrung.Studio99/Shutterstock.com

Chapter 12

244 Ivmonica/Shutterstock.com; **245**t Anan Kaewkhammul/Shutterstock.com; **245**c Bildagentur Zoonar GmbH/Shutterstock.com; **245**b Annetje /Shutterstock.com; **246**bg Ramon Navarro/NiS/Minden Pictures; **246** Design Pics Inc/Alamy Stock Photo; **246**i Minden Pictures/SuperStock.com; **247**t fivespots /Shutterstock.com; **247**b EMJAY SMITH/Shutterstock .com; **249**bg M. Watson/Science Source; **249**l Danita Delimont/Gallo Images/Getty Images; **249**r © iStock.com /hanhanpeggy; **250**t © iStock.com/franckreporter; **250**li Michael Evan Sewell Visual Pursuit/Photolibrary/Getty Images; **250**ri © iStock.com/KenCanning; **250**b PetlinDmitry/Shutterstock.com; **251**t RUM-photo /Shutterstock.com; **251**b © Jiri Vaclavek | Dreamstime .com; **252**l Marie Read/Science Source; **252**r Heidi and Hans-Juergen Koch/Minden Pictures; **253**t © iStock.com /azndc; **253**b © iStock.com/flownaksala; **253**bi zeber /Shutterstock.com; **254–55**t, **257**b Jan Havlicek /Shutterstock.com; **254**ct YUSRAN ABDUL RAHMAN /Shutterstock.com; **254**cb Cathy Keifer/Shutterstock.com; **254**b Aprilphoto/Shutterstock.com; **255**ct Judith Lienert /Shutterstock.com; **255**cb Jim and Jamie Dutcher/National Geographic Creative/Getty Images; **255**b Tim Laman /National Geographic Magazines/Getty Images; **256** © iStock.com/RelaxFoto.de; **257**t bluehand/Shutterstock .com; **258**t Olga Tagaeva/Shutterstock.com; **258**b Maarten Wouters/Stone Sub/Getty Images; **259** Kurashova /Shutterstock.com

Chapter 13

260 © iStock.com/attl; **261**t © iStock.com/OcusFocus; **261**b BJU Photo Services; **262** gradyreese/E+/Getty Images; **263**, **283**t RUBBERBALL IMAGES | JESSICA PETERSON/ Media Bakery; **264–65**bg © iStock.com /isitsharp; **264**ti ANDRZEJ WOJCICKI/Science Photo Library/Getty Images; **264**bi, **278**li, ci, ri, **279**ti, bli Ramona Kaulitzki/Shutterstock.com; **265**t, **274**l, **277**i (bone), **280**i, **284**i, **284**b (muscles) Sebastian Kaulitzki /Shutterstock.com; **265**bi Nathapol Kongseang /Shutterstock.com; **266**t all, c, **267**tl, tr, bc Vecton /Shutterstock.com; **266**bl, br, **267**tc © iStock.com /ikuvshinov; **266**bc LightField Studios/Shutterstock.com; **267**bl Gwen Shockey/Science Source; **267**br Designua /Shutterstock.com; **268–69**bg Val Thoermer/Shutterstock .com; **268**, **272**t, **283**c Alexilusmedical/Shutterstock.com; **270–71** © iStock.com/FatCamera; **272**b, bi Courtesy of GDMMissions; **273**l Sebastian Kaulitzki/Shutterstock.com / Tefi/Shutterstock.com; **273**lti SCIENCE PHOTO LIBRARY | STEVE GSCHMEISSNER /Media Bakery; **273**lbi POWER AND SYRED/SCIENCE PHOTO LIBRARY/Getty Images; **273**r Billion Photos/Shutterstock .com; **273**ri, **274**tr, **275**i, **283**b © iStock.com/Eraxion; **274**br Medical Art Inc/Shutterstock.com; **275** maxpro /Shutterstock.com; **276**t © iStock.com/4x6; **276**bl Ljupco Smokovski/Shutterstock.com; **276**br studioloco /Shutterstock.com; **277**bg Greg Epperson/Shutterstock .com; **277** (striated muscle) Jose Luis Calvo/Shutterstock .com; **277**i (muscle) © iStock.com/Vac1 / Ramona Kaulitzki/Shutterstock.com; **278–79**bg © Martinmark | Dreamstime.com; **279**t Monkey Business Images /Shutterstock.com; **279**bri SCIEPRO/Science Photo Library/Getty Images; **280**t Choksawatdikorn/Shutterstock .com; **280**b, **284**t ArtFamily/Shutterstock.com; **280**i, **284**i (digestive system) Henning Dalhoff/Science Source; **281** ESB Professional/Shutterstock.com; **281** (lungs) Liya Graphics/Shutterstock.com; **281** (heart) © iStock.com /angelhell; **282** AYakovlev/Shutterstock.com; **282**i tsuneomp/Shutterstock.com; **285** Valua Vitaly /Shutterstock.com

Chapter 14

286 Image Source | David Cleveland/Media Bakery; **287**, **301**t George Dolgikh/Shutterstock.com; **288–89**bg Foxys Forest Manufacture/Shutterstock.com; **288** (glucose) Linnas/Shutterstock.com; **288** (starch) Miroslav Cvetinov /Shutterstock.com; **288** (saturated fat) Victor Josan /Shutterstock.com; **288** (unsaturated fat) MOLEKUUL /SCIENCE PHOTO LIBRARY/Getty Images; **288** (vitamin C) petarg/Shutterstock.com; **288** (niacin) ibreakstock /Shutterstock.com; **288** (phosphate), **289** (amino acid, protein) molekuul_be/Shutterstock.com; **289** (water glass) AlenKadr/Shutterstock.com; **289**i (water) Mircea Maties /Shutterstock.com; **290–91**bg Monkey Business Images /Shutterstock.com; **290** Coprid/Shutterstock.com; **291**l baibaz/Shutterstock.com; **291**r © iStock.com/Agfoto; **292** Dan Coha Photography/StockFood/Getty Images; **294** Peiling Lee/Shutterstock.com; **294–95**bg © iStock.com /gorodenkoff; **296–97**, **301**b Antonio Guillem /Shutterstock.com; **296**li Henning Dalhoff/Science Source; **296**ri, **297**li, **298**t, **301**bi, **303**t Liya Graphics/Shutterstock

.com; **297**ri © Kateryna Kon/123RF; **298**bl Alexilusmedical/Shutterstock.com; **298**br Di Studio /Shutterstock.com; **298**bri Magic mine/Shutterstock.com; **299** © iStock.com/-Robbie-; **299**ti Creations/Shutterstock .com; **299**bi Sebastian Kaulitzk/Shutterstock.com; **300** Alison Young/Shutterstock.com; **302**t Africa Studio /Shutterstock.com; **302**b Bochkarev Photography /Shutterstock.com; **303**b © iStock.com/kirin_photo

Chapter 15

304 RAJ CREATIONZS/Shutterstock.com; **305** Ken Chernus/Taxi/Getty Images; **306–7** © Myrchella | Dreamstime.com; **306**i Universal Images Group North America LLC / Alamy Stock Photo; **307**ti crystal light /Shutterstock.com; **307**bi 3drenderings/Shutterstock.com; **308** alexkich/Shutterstock.com; **309**t, **320**t Alila Medical Media/Shutterstock.com; **309**b Glow Wellness/Getty Images; **310** © iStock.com/izusek; **311**t © iStock.com /pepifoto; **311**ti, **312**ri Liya Graphics/Shutterstock.com; **311**b Lily Kielmeyer; **312–13** Adobe Stock/Darren Baker; **312**li GraphicsRF/Shutterstock.com RAJ / CREATIONZS /Shutterstock.com; **312**ci, **315**l, **322** Sebastian Kaulitzki /Shutterstock.com; **313**i Monica Schroeder/Science Source; **314**l Wisiel/Shutterstock.com; **314**li domnitsky /Shutterstock.com; **314**tr SCIENCE PHOTO LIBRARY | DENNIS KUNKEL MICROSCOPY/Media Bakery; **314**cr, br Steve Gschmeissner/Science Source; **315**r, **320**b sheff/Shutterstock.com; **315**ri, **320**bi (circulatory system) NelaR/Shutterstock.com; **315**ri, **320**bi (lungs) Lucia Fox /Shutterstock.com; **316**t IMANE/Corbis/Getty Images; **316**b The LIFE Picture Collection/Getty Images; **316**bi B Christopher / Alamy Stock Photo; **317**t "Blood letting" /Maggie Black's The Medieval Cookbook/Wikimedia Commons/Public Domain; **317**b wavebreakmedia /Shutterstock.com; **317**bi, **318**ti Stocktrek Images/Media Bakery; **318–19** JP WALLET/Shutterstock.com; **318**ci (lymphatic system) 3drenderings/Shutterstock.com; **318**ci (lungs) Universal Images Group North America LLC / Alamy Stock Photo; **318**bi Alila Medical Media /Shutterstock.com / Blamb/Shutterstock.com; **319**i, **321**t sciencepics/Shutterstock.com; **321**b Dudarev Mikhail /Shutterstock.com; **323**tl Peter Scholey / Alamy Stock Photo; **323**tr Courtesy of TransMedics Inc; **323**b Adobe Stock/Arochau

Chapter 16

324 © Aspenphoto | Dreamstime.com; **325** Monkey Business Images/Shutterstock.com; **325**ti martan /Shutterstock.com; **325**bi Kateryna Kon/Shutterstock.com; **326–27**bg © iStock.com/yanik88; **326**t Alina R /Shutterstock.com; **326**c Kateryna Kon/Shutterstock.com; **326**b Spencer Sutton/Science Source; **327**tl Designua /Shutterstock.com; **327**bl Biophoto Associates/Science Source; **327**tr © iStock.com/art-4-art; **327**br Ermolaev Alexander/Shutterstock.com; **328**, **347**t Stokkete /Shutterstock.com; **329**t Africa Studio/Shutterstock.com; **329**b CDC/Science Source; **330**t SCIENCE PHOTO LIBRARY | ROGER HARRIS/Media Bakery; **330**b ChameleonsEye/Shutterstock.com; **331**, **332**bg © iStock .com/Rost-9D; **331**l, **332**c, **338**t, **347**b SCIENCE PHOTO LIBRARY | PIXOLOGICSTUDIO/Media Bakery; **331**r Adobe Stock/Africa Studio; **332**l © iStock .com/m-imagephotography; **332**li 3drenderings /Shutterstock.com; **332**r Henning Dalhoff/Science Source; **333**t Sebastian Kaulitzki/Shutterstock.com; **333**cr antoniodiaz/Shutterstock.com; **333**bl Adobe Stock/Al Troin; **333**bc maxpro/Shutterstock.com; **333**br rozbeh /Shutterstock.com; **334**tl Karramba Production /Shutterstock.com; **334**tr dny3d/Shutterstock.com; **334**lct Adobe Stock/Monkey Business; **334**lcb Standret /Shutterstock.com; **334**b Peter Cade/Iconica/Getty Images; **335** shipfactory/Shutterstock.com; **335**i ilusmedical /Shutterstock.com; **336** MARKA/Alamy Stock Photo; **336**i (hand) Tatiana Popova/Shutterstock.com; **336**i (iron) Neveshkin Nikolay/Shutterstock.com; **336**i (spinal cord) Alila Medical Media/Shutterstock.com; **337** DIVINE IMAGES | SCHEDIVY PICTURES INC./Media Bakery; **338**c Scott Cunningham/Getty Images Sport/Getty Images; **338**bl © iStock.com/AdShooter; **338**bc dean bertoncelj /Shutterstock.com; **338**br CE Photography/Shutterstock .com; **339** Lev Dolgachov/Alamy Stock Photo; **340–41**bg, **348**t Adobe Stock/Max Topchii; **340**t Subbotina Anna /Shutterstock.com; **340**b, **341**l sciencepics/Shutterstock .com; **341**r Alexilusmedical/Shutterstock.com; **342–43**bg Blend Images | Hill Street Studios/Media Bakery; **342**, **349**t Medical Art Inc/Shutterstock.com; **343** Peter Treanor /Alamy Stock Photo; **344–45**bg nd3000/Shutterstock.com; **344**t HENNING DALHOFF/Science Photo Library/Getty Images; **344**cl Science History Images/Alamy Stock Photo; **344**cr, b Gwen Shockey/Science Source; **345** Designua /Shutterstock.com; **346** oliveromg/Shutterstock.com; **348**b Chromakey/Shutterstock.com; **349**b © iStock.com /stocknroll; **350** © iStock.com/olaser

Chapter 17

351 © iStock.com/omgimages; **352**bg © iStock.com /Philartphace; **352**t, b © iStock.com/Dr_Microbe; **353**t Sebastian Kaulitzki/Shutterstock.com; **353**b, **366**t © iStock .com/nilimage; **353**i (thyroid) © iStock.com /AlonzoDesign; **353**i (adrenal glands), **353**i, **354**t, **356**b (pancreas) © iStock.com/EgudinKa; **353**i, **358**ti, **359**t, bi, **367**bl, br (ovaries, testes) © iStock.com/PrettyVectors; **354–55**b © iStock.com/Petar Chernaev; **355**t Michal Sanca/Shutterstock.com; **356**t © iStock.com/mixetto / © iStock.com/studiocasper; **357**t Courtesy of Jeremy Byrd; **357**b Altrendo/Getty Images; **358–59**b, **366**b Dmitry Melnikov/Shutterstock.com; **358**bi CLIPAREA l Custom media/Shutterstock.com; **359**ti © iStock.com/ttsz; **359**c Dr. David M. Phillips/Visuals Unlimited/Getty Images; **359**bi (ovum) "Gray3" by Henry Gray (1918) Anatomy of the Human Body/Wikimedia Commons/Public Domain; **360**t Dr. Yorgos Nikas/Science Source; **360**b Blend Images | JGI /Tom Grill/Media Bakery; **361**t, **365** Neil Harding/The Image Bank/Getty Images; **361**b Tissuepix/Science Source; **362**t Petit Format/Science Source; **362**c Catherine Delahaye/Taxi/Getty Images; **362**b, **367**t © iStock.com /RuslanDashinsky; **363**bg Adobe Stock/ARochau; **363** Monkey Business Images/Shutterstock.com; **364** Glow Images, Inc/Glow/Getty Images; **368** natureman30 /Shutterstock.com; **369** Tim Hale/Stone/Getty Images

Chapter 18

370–71 Pablo Scapinachis/Shutterstock.com; **371** Dr. Robert Carter, Creation Ministries International, Creation .com; **372** © iStock.com/B&M Noskowski; **373**bg Gary C. Tognoni/Shutterstock.com; **373** © iStock.com/gmcoop; **374–75, 393**t Jerry Sanchez/Shutterstock.com; **376–77**bg Csondy/E+/Getty Images; **377** Jeff Schultz/Perspectives /Getty Images; **378**t Olga Nikonova/Shutterstock.com; **378**b Julia_Lelija/Shutterstock.com; **379** Christopher Wood/Shutterstock.com; **380**bg Heike Odermatt/Minden Pictures; **380**t Stoked | Dethan Punalur/Media Bakery; **380**b Paul Tessier/Shutterstock.com; **381**bg Amrut Atre /Shutterstock.com; **381**t RealityImages/Shutterstock.com; **381**b © iStock.com/King_Louie; **382**bg, **393**b Pi-Lens /Shutterstock.com; **382**l BoBi Photographer/Shutterstock .com; **382**r NaturesMomentsuk/Shutterstock.com; **383**bg © iStock.com/KenCanning; **383**l MelToppi/Shutterstock .com; **383**r Robbie George/National Geographic Magazines/Getty Images; **384**bg Krys Bailey/Alamy Stock Photo; **384** Prisma by Dukas Presseagentur GmbH/Alamy Stock Photo; **385**bg Leonid Serebrennikov/Alamy Stock Photo; **385**t Ondrej Prosicky/Shutterstock.com; **385**b © iStock.com/bibi57; **386**bg Amit Basu Photography /Moment Open/Getty Images; **386**t © iStock.com/Calinat; **386**b Tim Zurowski/Shutterstock.com; **387**bg © iStock .com/FrankvandenBergh; **387**t © iStock.com/Angelika; **387**b Francois van Heerden/Shutterstock.com; **388**bg © Tony Campbell | Dreamstime.com; **388**t DaLiu /Shutterstock.com; **388**b © iStock.com/Kathryn8; **389**bg © iStock.com/RainervonBrandis; **389**t Secret Sea Visions /Photodisc/Getty Images; **389**c Rich Carey/Shutterstock .com; **389**b Volvox volvox/Oxford Scientific/Getty Images; **390–91**bg Ed Callaert/Alamy Stock Photo; **390**t Maciej Czekajewski/Shutterstock.com; **390**b totajla/Shutterstock .com; **391**t Ian Maton/Shutterstock.com; **391**b Zhecho Planinski/Shutterstock.com; **392** Warren Metcalf /Shutterstock.com; **394** gary yim/Shutterstock.com; **395** Justin Lewis/Taxi/Getty Images; **396** Vlada Photo /Shutterstock.com

Chapter 19

397 TPCImagery - Mike Jackson/Shutterstock.com; **398–99, 415**t Anna Om/Shutterstock.com; **400–401** Ms Deborah Waters/Shutterstock.com; **402–3** © Panaramka | Dreamstime.com; **402**i Worachat Tokaew/Shutterstock .com; **403**ti Science Photo Library | SCIENCE PHOTO LIBRARY/Media Bakery; **403**bi Science Photo Library | DENNIS KUNKEL MICROSCOPY/Media Bakery; **404–5, 405**bi, **415**b Adobe Stock/Julián Maldonado; **404**i, **405**ti tea maeklong/Shutterstock.com; **405**ci William Booth/Shutterstock.com; **406** Richard Fitzer/Shutterstock .com; **407**bg © iStock.com/ET1972; **407** (buffalo grass) © Wernerl | Dreamstime.com; **407** (tumbleweed) Sarnia /Shutterstock.com; **407** (violet) bonchan/Shutterstock .com; **407** (grasshopper) happymay/Shutterstock.com; **407** (butterfly) © iStock.com/MarkMirror; **407** (prairie dog) Papa Annur/Shutterstock.com; **407** (jackrabbit) Steve Byland/Shutterstock.com; **407** (pronghorn) Breck P. Kent /Shutterstock.com; **407, 416**b (coyote), **407** (red fox) Jim Cumming/Shutterstock.com; **407, 416**b (sage grouse) Tom Reichner/Shutterstock.com; **407, 416**b (sagebrush) Brendan Bucy/Shutterstock.com; **407** (rattlesnake) fivespots/Shutterstock.com; **407** (northern harrier) Jausa /Shutterstock.com; **407** (burrowing owl) © iStock.com /Waldemar Seehagen; **407** (badger) James Hager /robertharding/Getty Images; **408** ary Dublanko / Alamy Stock Photo; **409** © iStock.com/Pedro Hélder da Costa Pinheiro; **409**ti FLPA / Alamy Stock Photo; **409**bi © iStock .com/DiyanaDimitrova; **410** Jupiterimages/PHOTOS .com>> / Getty Images Plus/Getty Images; **410**li Jassmine Anderson/EyeEm/Getty Images; **410**ri Karel Bartik /Shutterstock.com; **411**tl © iStock.com/mgfoto; **411**tr © iStock.com/myshkovsky; **411**c © iStock.com/mandy127; **411**b Cate Finney / EyeEm/Getty Images; **412–13**bg, **416**t Jordana Meilleur / Alamy Stock Photo; **412** © iStock.com /Lightstone-Media; **414**t Damsea/Shutterstock.com; **414**b Echunder/Shutterstock.com; **417** J. Lekavicius /Shutterstock.com; **418**t Anton Petrus/Moment/Getty Images; **418**b © iStock.com/FRANKHILDEBRAND

Chapter 20

419 Design Pics / Keith Levit/Perspectives/Getty Images; **420–21**bg Roland Nagy / Alamy Stock Photo; **421**t, **435**t © iStock.com/Thiago Santos; **421**bl © iStock.com/Jitalia17; **421**bc © iStock.com/aofxta; **421**br © iStock.com/rabbit75 _ist; **422–23**bg RelaxFoto.de/E+/Getty Images; **422**t Rothy's; **422**c Shay Levy / Alamy Stock Photo; **422**b © iStock.com/gece33; **423** Naypong/Shutterstock.com; **425** urbans/Shutterstock.com; **426–27**bg Gartforu /Shutterstock.com; **426, 435**c Monty Rakusen/Cultura /Getty Images; **428–29**bg triggerflicks/Moment/Getty Images; **428**l © iStock.com/zhongguo; **428**r © Darren Baker | Dreamstime.com; **429**t Klaus Nigge/National Geographic/Getty Images; **429**c © iStock.com/catnap72; **429**b "Vail, Colorado - Eagles Nest Wilderness Area - 2" by Hawthorne Ave/Flickr/CC BY 2.0; **430** Kitchin and Hurst /All Canada Photos/Getty Images; **431** AFP/Getty Images; **432–33**bg Brent Waltermire / Alamy Stock Photo; **433**l, **435**b © iStock.com/nstanev; **433**r Mike Veitch / Alamy Stock Photo; **434** Arthimedes/Shutterstock.com; **436–37**b dvoevnore/Shutterstock.com; **437**t AlanHaynes.com / Alamy Stock Photo; **437**c © iStock.com/dssimages

Appendixes

438 Sekar B/Shutterstock.com; **439** AHMAD FAIZAL YAHYA/Shutterstock.com; **443**t Valentyn Volkov /Shutterstock.com; **443**bl Igor Filonenko/Shutterstock .com; **443**bc robert_s/Shutterstock.com; **443**br Daniel Fung/Shutterstock.com; **444** Jubal Harshaw/Shutterstock .com; **445** Albert Russ/Shutterstock.com; **448** Wanida_Sri /Shutterstock.com

Maps from Map Resources